W9-DCU-666

Optoelectronics/ Fiber-Optics Applications Manual Second Edition

Prepared by The Applications Engineering Staff of the
HEWLETT-PACKARD OPTOELECTRONICS DIVISION

Stan Gage, *Applications Engineering Manager*
Dave Evans, *Applications Engineer*
Mark Hodapp, *Applications Engineer*

Hans Sorensen, *Applications Engineer*
Dick Jamison, *Applications Engineer*
Bob Krause, *Applications Engineer*

McGraw-Hill Book Company

New York   St. Louis   San Francisco   Auckland   Bogotá
Hamburg   Johannesburg   London   Madrid   Mexico
Montreal   New Delhi   Panama   Paris   São Paulo
Singapore   Sydney   Tokyo   Toronto

Library of Congress Cataloging in Publication Data
Hewlett-Packard Company. Optoelectronics Division.
    Applications Engineering Staff.
    Optoelectronics/fiber-optics applications manual.

    Published in 1977 under title: Optoelectronics
applications manual.
    Includes index.
    1. Light emitting diodes.   2. Optoelectronic
devices.   3. Fiber optics.   I. Gage, Stan.   II. Title.
TK7871.89.L53H48 1981   621.36′7   80-19814
    ISBN   0-07-028606-X

567890   HDHD   8987654321

Hewlett-Packard assumes no responsibility for the use of any circuits described herein and makes no representations or warranties, express or implied, that such circuits are free from patent infringement.

TABLE OF CONTENTS

TABLE OF CONTENTS (Continued)

TABLE OF CONTENTS (Continued)

TABLE OF CONTENTS (Continued)

TABLE OF CONTENTS (Continued)

TABLE OF CONTENTS (Continued)

TABLE OF CONTENTS (Continued)

PREFACE

Over the last decade, the commercial availability of the Light Emitting Diode has provided electronic system designers with a revolutionary component for application in the areas of information display and   optoisolators.

As the use of the LED and associated derivative products has become more common, many electronic engineers have encountered the need for a resource of information about the application of and designing with LED products. This book is intended to serve as an engineering guide to the use of a wide range of solid state optoelectronic products. Wherever appropriate, throughout the book, use has been made of approximations to simplify the formulas and design equations. As such, some formulas will differ substantially in form from those found in a physics book on the subject; however, the end result will be an excellent approximation of the behavior of real devices.

To treat the various major aspects of LED products, the book is divided into sections covering each of the generalized LED product types. Additional sections treat such peripheral information as contrast enhancement techniques, photometry and radiometry, LED reliability, mechanical considerations of LED devices, photodiodes and LED theory. In the second edition, sections covering the latest technological advances in the field of optoelectronics have been integrated into the original material. Such areas as data transmission systems using fiber optics and optoisolators, backlighting using LED devices, sunlight viewable displays, optical sensing systems, optoisolators for use in industrial control systems, and interfacing displays to microprocessors are reviewed in detail in separate sections. In addition, a portion of Section 3 covering CTR degradation has been updated to present an in depth perspective on this aspect of optoisolators and how to account for it in systems design.

The authors wish to thank Denise Dow for her patience and understanding in the preparation of the typed manuscript and Clair Lee and Gail Freedman for their excellent work on the illustrations for the 2nd edition. The outstanding efforts of Tom Lee, George Liu, and Al Petrucello, who breadboarded and performance checked all of the circuits are also gratefully acknowledged.

About the Authors

Stan Gage has more than 10 years experience in the design and application of optoelectronic devices and the design and processing of linear integrated circuits. Dave Evans has been an influential contributor to the development of sunlight viewable displays, LED light bars, and automotive LED display devices. Mark Hodapp has originated many new applications of optoisolators and interfaces between LED displays and microprocessors as well as two alphanumeric display system products. Hans Sorensen participated in the development of optoelectronic technology and was instrumental in establishing industry acceptance of the millicandela as the unit of measurement for LEDs. Richard Jamison has been involved with product characterization, measurements, and applications of analog, digital, and optical, subsystems as well as emitter/detector and optocoupler products. Bob Krause has been involved extensively in the development of the optocoupler and emitter/detector products.

Optoelectronics/ Fiber-Optics Applications Manual

Section 1
LED Theory

1.0   LED THEORY

Electroluminescence in solids is a phenomenon which has been well known and intensively studied for many years. Perhaps the most commonly utilized application of electroluminescence is in the screen of a television set. Better known as cathodluminescence, this form of electroluminescence is caused by the collision of high energy electrons with the phosphor coating which lines the inside surface of a cathode ray tube. Though numerous other less common types of electroluminescence have been observed, one particular type has given rise to an entirely new field of technology. This is the phenomenon of p-n junction injection electroluminescence. The emission of light (photons) from a p-n junction was first noted in naturally occurring junctions by Lossew in 1923. More recently, (circa 1962) studies of GaAs revealed that it was quite feasible to achieve relatively high levels of electroluminescent emission from p-n junctions. Starting in 1962, an intensive development program was undertaken at Hewlett-Packard and elsewhere to produce a useful, manufacturable, visible-light emitting p-n junction device. An electroluminescent device of this type was thought to be attractive because of:

1)   The low currents and voltages required to produce useful light output.

2)   The precision with which the light emitting area could be defined through the use of semiconductor photo lithographic processes.

3)   The high speed at which the device could be switched.

1.1   The Theory of P-N Junction Electroluminescence

As a background to the theory of electroluminescent diodes, it is useful to review the basic concepts of current flow in p-n junctions.

1.1.1   Semiconductor Energy Gap

In semiconductor materials, as in all crystalline solids, electrons can assume only certain levels of energy. The Valence band and the Conduction band are the terms assigned to the two bands having the highest energy levels for electrons in normal semiconductor materials. The separation between the top of the valence band and the bottom of the conduction band is called the energy gap. In a pure semiconductor material, electrons cannot exist in (i.e., assume energy levels within the range of) this forbidden gap.

1.1.2   Semiconductor Doping

It is possible, however, to introduce impurities into the semiconductor which will produce electronic states in the forbidden gap. Impurities which add electrons to the conduction band are called donors and give raise to n-type conductivity. Impurities which produce electron vacancies, holes, in the valence band are called acceptors and give raise to p-type conducitivity. The energy bands and the relative donor and acceptor levels are shown in Figure 1.1.3-1a.

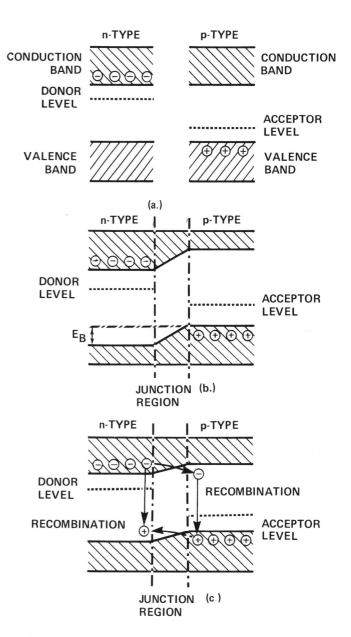

Figure 1.1.3-1   Schematic Representation of the P-N Junction

1.1.3 The P-N Junction

A p-n junction can be formed in a semiconductor material by doping one region with donor atoms and an adjacent region with acceptors. In this situation, electrons and holes will flow in opposite directions across the junction (without any applied bias) until an equilibrium is reached. This will give rise to a built-in potentional barrier, E_B, which is slightly less than the energy gap. Figure 1.1.3-1a and -1b represent schematically the material before and after this equilibrium is reached. Note that the natural potential difference across the junction makes it difficult for an electron, for instance, to move from the n to the p region. If, however, an external electric bias is applied across the junction in a manner such as to counteract the built-in potential, additional electrons and holes will flow (be injected) across the junction boundary. See Figure 1.1.3-1c. These carriers will then recombine via one of the mechanisms described in Section 1.1.4. This recombination, may give rise to photon emission as well as current flow. Within a few diffusion lengths of the junction, virtually all of the injected carriers will have recombined and the bulk current flow will occur by means of majority carrier diffusion.

1.1.4 Recombination

The injected electrons or holes are annihilated by a carrier of opposite type through one of several possible recombination processes. These recombination processes are either "radiative" or "non-radiative". The energy release in a radiative process is in the form of photons (light); whereas, the energy release in a non-radiative process is in the form of phonons (heat). The two important radiative processes in LEDs are shown schematically in Figure 1.1.4-1. Band-to-band recombination, 1, is the direct recombination of an electron near the bottom of the conduction band with a hole near the top of the valence band. The photon energy is then approximately equal to the band-gap energy of the crystal. This mechanism of radiative recombination is predominant in direct band-gap materials such as GaAs.

The second type, 2, of radiative recombination occurs by the formation and annihilation of a bound exciton at an "iso-electronic" center. Iso-electronic centers (associated with specific impurities in the crystal) are normally neutral, but introduce a local potential which is attractive to electrons. In p-type material, an injected electron is first trapped at the center. The negatively charged center then captures a hole from the valence band to form the bound exciton. The subsequent annihilation of this hole-electron pair yields a photon with an energy equal to the band gap minus an energy approximately equal to the binding energy of the center. This mechanism of radiative recombination is predominant in indirect band-gap materials such as GaP. The photon energy can be converted to wavelength by equation 1.1.4-1:

$$\lambda = \frac{1240}{\Delta E} \text{ (nm)} \qquad (1.1.4\text{-}1)$$

where ΔE is the energy transition in electron volts.

1.1.5 Materials Available for LED Devices

Equation 1.1.4-1 relates the expected wavelength of photon emission to energy differential experienced by an electron undergoing recombination. The maximum possible energy of the emitted photons is determined by the band-gap energy of the solid in which the p-n junction is formed. There are numerous elements and elemental compounds which have band-gap energies that lie in the region which could produce emission ranging from ultraviolet to infrared. However, very few of these materials are viable candidates for practical LED devices. Table 1.1.5-1 lists a few of the available materials and the associated band gap and emission wavelength. Some of these materials are not useful because they cannot be doped to form a p-n junction, some do not have emission at a useful wavelength, some have too low a conversion efficiency to be useful. At the present time, the only commercially available LED devices are manufactured using GaAs, GaP or the ternary compound Ga(As,P).

MATERIAL	BAND GAP ENERGY	EMISSION λ nm	TRANSITION TYPE
	eV		
Ge	0.66	1880	INDIRECT
Si	1.09	1140	INDIRECT
GaAs	1.43	910	DIRECT
GaP	2.24	560	INDIRECT
$GaAs_{60}P_{40}$	1.91	650	DIRECT
Al Sb	1.60	775	INDIRECT
In Sb	0.18	6900	DIRECT
Si C	2.2-3.0	563-413	INDIRECT

Table 1.1.5-1 Some of the Materials Available for LED Devices

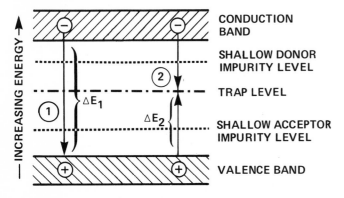

Figure 1.1.4-1 Recombination Processes in the P-N Junction

1.1.6 Direct and Indirect Band-Gap Materials

In Section 1.1.4, the concepts of recombination were presented. In this section, the concepts of direct and indirect band-gap transitions are considered. The conduction band edge and the valence band edge as a function of momentum for allowed electronic states in the $GaAs_{1-x}P_x$ compound system (where x = mole fraction) are plotted in Figure 1.1.6-1 for different values of x. As indicated, there are two wells or minima; one designated as direct and the other as indirect. Electrons in the conduction band will generally occupy states in the lowest energy minimum, while holes will occupy states near the valence band maximum. Electrons in the direct minimum and holes at the top of the valence band have equal momentum;

whereas, electrons in the indirect minimum have different momementum. Since momentum is conserved, band-to-band transitions may occur with high probability for electrons in the direct minimum. The probability of a band-to-band transition for an electron in an indirect minimum is nearly zero, since a third component (phonon) must participate in the process in order to conserve momentum. Note that GaAs and $GaAs_{1-x}P_x$ up to x ≅ .4 are primarily direct gap materials in contrast to $GaAs_{1-x}P_x$ with x > .4 and GaP which are primarily indirect gap materials. (Also note that the band gap energy increases with increasing x.) Thus, without the incorporation of special recombination centers, indirect gap materials, like GaP, are very inefficient light emitters, since the dominant recombination processes are non-radiative.

1.1.7 Enhanced Photon Emission in Indirect Gap Materials

More recent developments in LED technology have led to significantly enhanced radiative recombination in indirect gap materials, such as GaP. These have been achieved by the incorporation of appropriate impurities to form iso-electronic trapping centers, as discussed in Section 1.1.4. Because the trapped electron is highly localized at the center, its momentum is diffused. Thus, momentum can be conserved and the probability of direct recombination is greatly enhanced.

In GaP, two types of iso-electronic centers are utilized. One is formed by replacing a Phosphorous atom with an equivalent Nitrogen atom in the lattice. Another type is formed by replacing an adjacent Gallium and Phosphorous pair of atoms with a Zinc-Oxygen pair having the same total valence electrons. The resulting trap states will lie in the band gap at energies somewhat below the edge of the conduction band. This structure is illustrated schematically in Figure 1.1.7-1.

For N doped GaP, bound exciton recombination gives rise to emission at 565 nm (green). For Zn-O doping, the corresponding emission wavelength has a peak at about 700 nm (red). Nitrogen doping can be utilized in indirect gap $GaAs_{1-x}P_x$ to similarly produce yellow and red emission with relatively high quantum efficiencies.

1.2 Quantum Efficiency of LED Devices

The efficiency with which an electroluminescent material can convert current flow into detectable photon emission is of paramount importance in determining whether usable devices can be manufactured from that material. The percentage of the current flow which results in recombinations which give rise to photons of the desired

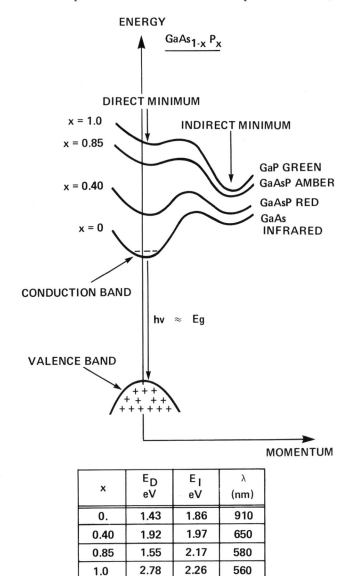

x	E_D eV	E_I eV	λ (nm)
0.	1.43	1.86	910
0.40	1.92	1.97	650
0.85	1.55	2.17	580
1.0	2.78	2.26	560

Figure 1.1.6-1   Plot of Momentum vs. Bandgap Energy for Various Compounds of the GaAs/GaP System

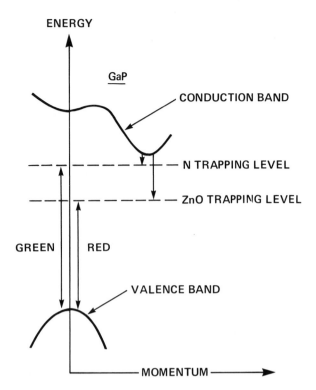

Figure 1.1.7-1 Plot of Momentum vs. Bandgap Energy for Indirect GaP Materials Showing Special Trapping Levels

wavelength is a measure of the internal conversion efficiency, η_{int} of the diode. Obviously, a material which has a very low η_{int} would be of little interest as a practical electroluminescent device. However, even a material which has an η_{int} of 100% may not be useful if the emitted photons cannot be efficiently coupled from the devices to a detector. Two major factors control the internal to external coupling coefficient. One factor is direct reabsorbtion of the emitted photon in the bulk material, basically a measure of the opacity of the material. The other factor is internal reflection at the crystal/air interface which causes the photon to be reflected back into the crystal and subsequently reabsorbed.

No matter how efficient a device is in terms of η_{ext} and η_{int}, the output power cannot be detected unless the wavelength is matched to available detectors. In the vast majority of LED applications, the detector of interest is either the human eye or a silicon photodetector. The nominal range of spectral sensitivity of the silicon detector is from about 300 nm to 1100 nm. The human eye has a much narrower range of sensitivity with useful responsivity only in the range from 400 nm to 700 nm. It is generally desirable to optimize the total coupling efficiency between the emitting device input signal and detector output response. For the case of matching a standard, non-Nitrogen doped $GaAs_{1-x}P_x$ emitter to the human eye, this is accomplished by picking the wavelength at which the product of the relative response of the eye and the relative

efficiency of the diode is largest. This maximum occurs at 655 nm and x = .4. This relationship is illustrated in Figure 1.2-1. The plots of $\overline{y}_{REL}$ and η_{extREL} are normalized to the peak values. $\eta_{LUMINOUS}$ represents the product of the two values.

For diodes having shorter peak emitting wavelength, such as N doped Ga(As,P) emitters at 565 nm (green) and 585 nm (yellow), η_{ext} is substantially lower but $\overline{y}_{REL}$ is higher. This has resulted in a spectrum of red through green emitters which all produce about the same $\eta_{LUMINOUS}$ (within an order of magnitude). The relative emission bands and responsivities of several different emitters and detectors are depicted in Figure 1.2-2. When trying to achieve matching with a silicon detector, factors other than achieving maximum responsivity may come into play. These factors are covered in Section 3.1.1.

1.3 Relative Efficiency

The mechanism of current flow in an LED as discussed in Section 1.1.3 represents the function of an ideal diode. Just as in silicon semiconductor technology, a p-n junction in Ga(As,P) material exhibits numerous discrepancies between the ideal and the actual diode. Surface recombination, tunneling phenomena, space charge recombination, current crowding and bulk recombination due to anomalous impurities tend to reduce the efficiency of photon emission of an LED. The relative effects of these non-radiative phenomena tend to be dependent on the current density (amps per square centimeter of junction area) and the perimeter to area ratio of the diode. At low forward currents (<1 A/cm^2), nearly all of the current flowing in a practical diode may result from one of these non-ideal mechanisms.

As the current density in the junction is increased to the value of 10 amps/cm^2, the non-ideal mechanisms will tend to have saturated, so that non-radiative current becomes a progressively smaller function of total current. As the junction current density becomes very large (>500 amps/cm^2) current crowding in the junction will tend to reduce emitting efficiency. The net effect of these mechanisms is to cause an LED to have a peak operating efficiency at a current which is dependent on the area and geometry of the junction and the size of the electrical contact. Figure 1.3-1 is a plot of normalized operating efficiency vs. current density for an .011" square diode. Note that at low current densities, a doubling of current may lead to a factor of 5 increase in luminous flux whereas at high current densities, doubling the current may result in slightly less than a doubling of light output.

The practical benefits of this aspect of LED devices are discussed in Sections 2 and 5.

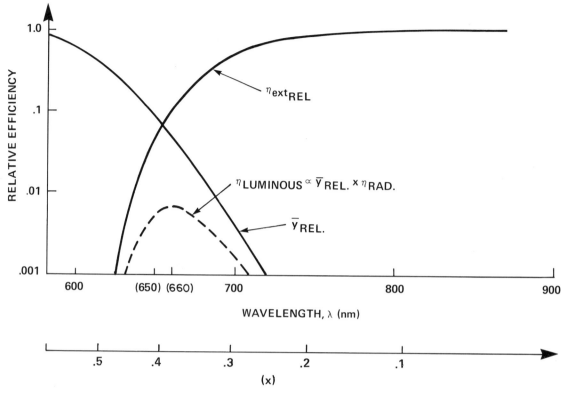

Figure 1.2-1 Relative Luminous Efficiency of the Human Eye and GaAs$_{1-x}$P$_x$ as a Function of Wavelength

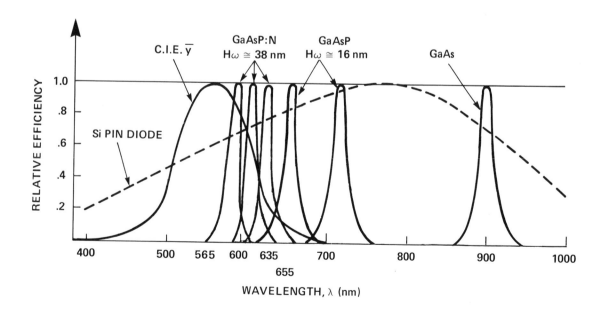

Figure 1.2-2 Normalized Responsivities of Different Emitters and Detectors

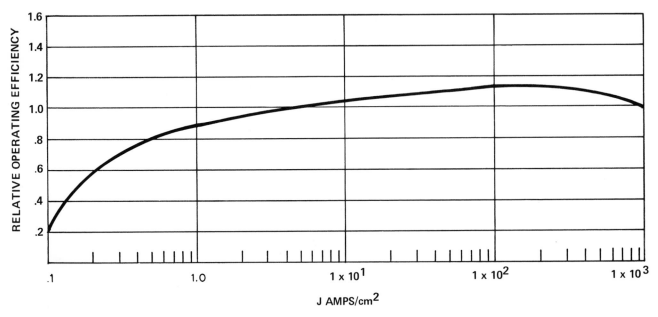

Figure 1.3-1 Normalized Operating Efficiency vs. Current Density for an LED

1.4 Material Processing

The processes used to manufacture LED devices are basically an outgrowth of the techniques perfected for the manufacture of silicon semiconductor devices. Crystal growing, epitaxial deposition, controlled impurity diffusions, photolithography, and vapor deposition of thin films all play an important role in the production of LEDs.

1.4.1 LED Structure

Since Ga(As,P) is a non-congruently melting material, a single crystal having proper GaP concentration is extremely difficult, if not impossible, to produce. The Ga(As,P) is, therefore, produced as an epitaxial layer grown on a substrate of either GaAs or GaP. The substrates are wafers sliced from ingots grown by the Czochralski method. These wafers are mechanically and chemically polished to provide a crystal substrate which is nearly free of lattice imperfections and unwanted chemical impurities. During the first stage of the epitaxial growth, the crystal layer has the same composition (GaP or GaAs) as the substrate. As the growth of the layer proceeds, the composition is changed gradually so as to maintain monocrystallinity in the epi layer. During the final mil (25 μm) of growth, the P/As ratio is held at the desired level so that later when the p-type material is diffused into the epitaxial layer, the resulting p-n junction will exist in homogeneous GaAs$_{1-x}$P$_x$ of the desired composition (typically x = .4 for direct-gap 655 nm RED LEDs). The n-type doping and any special impurities required to generate the bound exciton states are also introduced in the gas stream of the vapor-phase epitaxy system used to grow the layer.

Following growth of the epi layer, the wafer is coated with silicon nitride which acts as a barrier to the p-type diffusant and passivates the resulting p-n junction. To produce shaped p-n junctions, photolithography is used as in silicon wafer processing to shape the openings in the silicon nitride. Through the openings in the silicon nitride, p-type diffusant (generally Zinc enters the Ga(As,P) to form the p-n junction. It is not possible to use a layer of silicon dioxide because the p-type diffusant used in LED processing penetrates silicon dioxide too rapidly. Figure 1.4.1-1 depicts the crossectional and top view of a typical LED device.

1.4.2 Transparent vs. Opaque Substrate

The photons generated at the junction of a p-n electroluminescent diode are emitted in all directions. If the diode substrate is opaque, as in the case with GaAs, only those photons which are emitted upward within a critical angle defined by Snell's Law will be emitted as useful light. All other photons emitted into or reflected into the bulk crystal will be absorbed. This phenomena is illustrated in Figure 1.4.2-1a. GaP is nearly transparent by comparison with GaAs. Diodes formed in an epitaxial layer grown on a GaP substrate will exhibit improved efficiency due to the emission of photons which would be absorbed in the GaAs substrate structure. The resulting structure is depicted in Figure 1.4.2-1b. In actual manufacturing practice, indirect gap devices are generally fabricated on GaP substrates while direct-gap devices are fabricated on GaAs substrates.

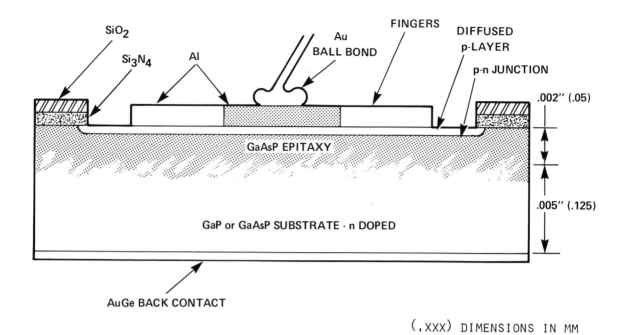

$(.XXX)$ DIMENSIONS IN MM

Figure A. CROSS SECTION OF AN LED

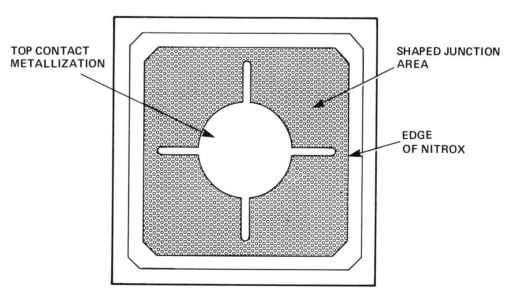

Figure B. PLAN VIEW OF AN LED

Figure 1.4.1-1   Crossectional and Top View of a Typical LED

1.7

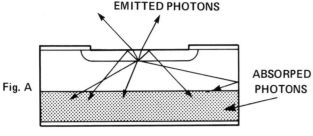

Fig. A

EMITTED PHOTONS

ABSORBED PHOTONS

GaAsP ON OPAQUE GaAs SUBSTRATE

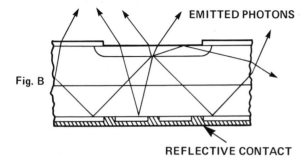

Fig. B

EMITTED PHOTONS

REFLECTIVE CONTACT

GaAsP OR GaP ON TRANSPARENT GaP SUBSTRATE

Figure 1.4.2-1 Effects of Transparent and Opaque Substrates on Photons Emitted at the Junction

1.5 The Effect of Temperature Variation on LED Parameters

The physical parameters of light emitting diodes, as in all semiconductor devices, exhibit a dependence on absolute temperature. The forward voltage/forward current relationship, quantum efficiency, and emitted wavelength are the temperature variant parameters of greatest interest to the LED user.

1.5.1 Forward Voltage as a Function of Temperature

The forward voltage/forward current relationship for an LED can be expressed by equation 1.5.1-1

$$I_F = I_o \exp\left(\frac{q\,V_F}{n\,kT}\right) \qquad (1.5.1\text{-}1)$$

where n is a function of temperature, I_F, and the nature of the recombination mechanism. Empirical results for both direct and indirect gap LEDs exhibit temperature coefficients of -1.3 mV/°C to -2.3 mV/°C depending on forward current. Figure 1.5.1-1 depicts this relation.

1.5.2 Change in Peak Wavelength as a Function of Temperature

The effective energy gap in both direct and indirect gap semiconductors tends to become slightly smaller with increasing temperature. This will result in slight increases in

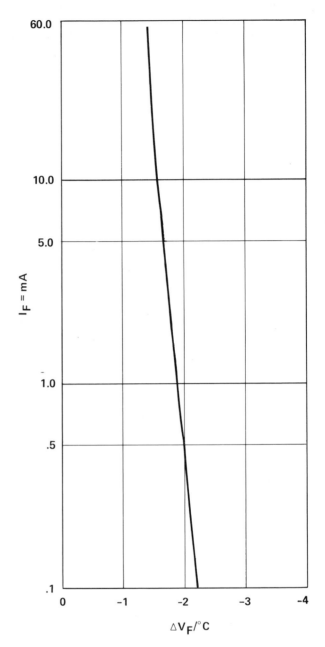

Figure 1.5.1-1 Temperature Coefficient of Forward Voltage as a Function of Forward Current

the emitted wavelength. For direct gap emitters wavelength will increase by 0.2 nm/°C. The wavelength of Nitrogen doped indirect gap emitters shows a somewhat lower dependency with typical positive variations of about 0.09 nm/°C.

1.5.3 Change in Output Power vs. Temperature

The radiant power of an LED decreases as a function of increasing temperature. Variations on the order of -1% per °C are typical for both direct and indirect gap materials. For LED devices in applications using the eye as a detector, the variation in responsivity of the eye as a function of wavelength must be added to the temperature dependent variation of the LED radiant emission. In the 650 nm Red region, the eye response is changing at about -4.3%/nm. In the 565 nm Green region, the responsivity is changing by about -.86%/nm. If the change in wavelength as a function of temperature is .2 nm/°C for a 655 nm direct gap device, then apparent optical intensity will be decreasing by about:

(1.5.3-1)

$$\frac{\Delta I_V}{\Delta T} \left(\frac{\%}{°C}\right) = \left(\frac{-4.3\%}{nm}\right)\left(\frac{.2\ nm}{°C}\right) - \frac{1\%}{°C} = -1.86\ \%/°C$$

Conversely, an indirect gap device in the 565 nm green region will exhibit:

(1.5.3-2)

$$\frac{\Delta I_V}{\Delta T} \left(\frac{\%}{°C}\right) = \left(\frac{-.86\%}{nm}\right)\left(\frac{.09\ nm}{°C}\right) - \frac{1\%}{°C} = -1.08\ \%/°C$$

The change in luminous intensity of an LED exhibits a logarithmic relationship over large changes in temperature. Calculation of expected changes should be done using the following relationship:

$$I_{V_{TEMP\ 1}} = I_{V_{TEMP\ 0}}\ exp\ k\ \Delta T \qquad (1.5.3-3)$$

where

$I_{V_{TEMP\ 1}}$ = luminous intensity at temperature 1

$I_{V_{TEMP\ 0}}$ = luminous intensity at reference temperature

$\Delta T = T_0 - T_1$

k = ln (1 − Temperature Coefficient)

Section 2
Lamps

2.0   LED LAMPS

2.1   Physical Properties of an LED Lamp Device

Incandescent, flourescent and neon lamps have been used for a long period of time in a wide variety of applications. As a result, their physical properties are well understood.

In recent years, the semiconductor LED lamp has been replacing these earlier devices in many applications. Also, new applications, designed specifically for the LED lamp, are being developed almost every day. Therefore, in order to effectively utilize an LED lamp in an application, a designer should be familiar with the physical properties of an LED device.

2.1.1   Plastic Encapsulated LED Lamp

Most commercial LED lamps are manufactured by encapsulating an LED chip inside a plastic package with a lens surface directly above the LED junction. If the plastic is undiffused, this configuration forms an immersion lens. The effect of this immersion lens construction is to enlarge the apparent size of the emitter. The magnification is a direct function of the index of refraction of the encapsulating material. The immersion construction increases the light output from the LED chip by reducing fresnel loss and increasing the critical angle.

Not every photon generated within the LED's p-n junction emerges from the surface of the junction to reach the eye of an observer. Three separate loss mechanisms contribute to reduce the quantity of emitted photons: 1) loss due to absorbtion within the LED chip material, 2) fresnel loss and 3) critical angle loss.

GaAsP/GaAs LED devices are opaque to light and absorb approximately 85% of the photons emitted at the junction, allowing a material efficiency factor of $\eta \approx .15$. A significant improvement is observed with GaAsP/GaP devices where $\eta \approx .76$.

2.1.2   Fresnel Loss

When light passes from a medium whose index of refration is n_1 to a medium whose index of refraction is n_2, a portion of the light is reflected back at the medium interface. This loss of light is called fresnel loss. The reflection coefficient is:

$$R = \left(\frac{n_2-n_1}{n_2+n_1}\right)^2 \qquad (2.1.2-1)$$

Since the numerator (n_2-n_1) is squared, the same reflection loss occurs whether the light is passing from a low-index to a high-index medium or from a high-index to a low-index

medium. The transmission coefficient across the medium interface is:

$$T = 1 + R = 1 - \left(\frac{n_2-n_1}{n_2+n_1}\right)^2 \qquad (2.1.2-2)$$

$$T = \frac{4 n_2 n_1}{n_2^2 + 2 n_2 n_1 + n_1^2}$$

The Fresnel Loss Efficiency Factor, η_{Fr}, is obtained by dividing numerator and denominator of equation 2.1.2-2 by $n_2 n_1$:

$$\eta_{Fr} = \frac{4}{n_2^2 + n_2/n_1 + n_1/n_2} \qquad (2.1.2-3)$$

For an unencapsulated GaAsP LED chip with an index of refraction of 3.4 emitting directly into air with index of refraction of 1.0, the fresnel loss efficiency factor is 0.702:

$$\eta_{Fr} = \frac{4}{2 + 1/3.4 + 3.4/1} = .702$$

Thus, only 70.2% of the light reaching the chip surface is transmitted across the chip/air interface.

If the LED chip is coated with an intermediate material having a suitable index of refraction, η_{Fr} can be improved. Ideally, this material should have an index of refraction $n_x = \sqrt{n_1 n_2} = \sqrt{(3.4)(1)} = 1.84$. Then $\eta_{Fr} = T_1 T_2$, where:

$$T_1 = \frac{4}{2 + n_1/n_x + n_x/n_1}$$

$$T_2 = \frac{4}{2 + n_2/n_x + n_x/n_2}$$

With an ideal coating having $n_x = 1.84$,

$$T_1 = T_2 = \frac{4}{2 + \sqrt{\dfrac{n_1}{n_2}} + \sqrt{\dfrac{n_2}{n_1}}} = .912$$

and the overall transmission is: $\eta_{Fr} = T_1 T_2 = (.912)^2 = .832$. This is an improvement of 18.5%.

Substantial improvement is obtained when a LED is encapsulated in plastic having an index of refraction of 1.5. The fresnel efficiency factor is calculated to be 0.816:

$$\eta_{Fr} = T_1 T_2$$
$$= \left[\frac{4}{2 + 3.4/1.5 + 1.5/3.4}\right] \left[\frac{4}{2 + 1/1.5 + 1.5/1}\right]$$
$$= [.850] \, [.960] = .816$$

This is an improvement of 16.2% over that of an unencapsulated LED and just 2.3% less than ideal.

2.1.3 Critical Angle Loss

The third efficiency loss is due to total internal reflection of photons incident to the chip surface at angles greater than the critical angle. This effect is shown diagramatically in Figure 2.1.3-1. As is depicted in Figure 2.1.3-1a, a ray of light passing from the interior of the crystal to the outer surface is refracted according to Snell's Law:

$$n_1 \sin \theta_1 = n_2 \sin \theta_2 \qquad (2.1.3\text{-}1)$$

where: θ_1 = The angle of incidence inside, at the surface of the crystal.

θ_2 = The angle of refraction outside, at the surface of the crystal.

n_1 = The index of refraction of the crystal.

n_2 = The index of refraction of the medium outside of the crystal.

The angle of incidance, θ_1, at which the angle of refraction θ_2 is equal to 90°, is called the critical angle, θ_c. The critical angle is calculated for an LED chip as follows:

$$n_1 \sin \theta_c = n_2 \sin 90° \qquad (2.1.3\text{-}2)$$

$$\sin \theta_c = n_2/n_1$$

$$\theta_c = \sin^{-1}(n_2/n_1) = \sin^{-1}(1/3.4)$$

$$\theta_c = 17.1°$$

Light rays from within the crystal reaching the surface at an angle greater than 17.1° are totally reflected back into the crystal.

The Critical Angle Efficiency Factor for an LED emitting into air is 0.0865:

$$\eta_{Cr} = \left(\frac{n_2}{n_1}\right)^2 = \left(\frac{1}{3.4}\right)^2 = .0865$$

Encapsulating an LED in a medium with a high index of refraction, n_x, increases the amount of flux that can escape from the crystal. However, if a flat surface is used, this increase in flux is lost because the refraction angle, θ_x, at the crystal-to-encapsulant interface becomes the incidence angle at the encapsulant-to-air interface, see Figure 2.1.3-1b. The value of the critical angle is not improved.

By shaping the encapsulant into a dome lens, the incidence angle at the encapsulant-to-air interface is less than the refraction angle at the crystal to encapsulant interface. As shown in Figure 2.1.3-1c, flux that would have been trapped by a flat surface is permitted to pass through the dome lens. If no flux is trapped within the encapsulant, the critical angle efficiency for a plastic dome lens become 0.195:

$$\eta_{Cr} = \left(\frac{n_x}{n_1}\right)^2 = \left(\frac{1.5}{3.4}\right)^2 = .195$$

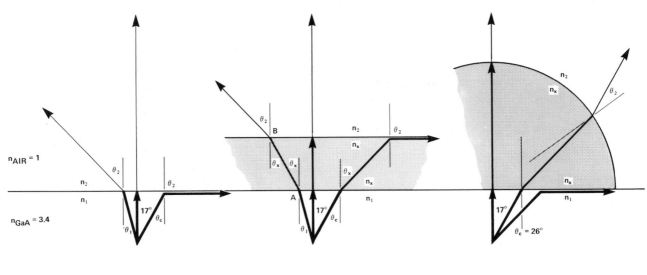

Figure 2.1.3-1   The Effects on Optical Efficiency Produced by an Optical Coating.

This improvement of 2.25:1 over an unencapsulated LED chip. The critical angle at the LED chip to dome interface is increased from $\theta_c = 17°$ to $\theta_c = 26°$.

$$\theta_c = \sin^{-1}(n_x/n_1) = \sin^{-1}(1.5/3.4) = 26.2°$$

2.1.4 Optical Efficiency

The optical efficiency of an LED lamp is the product of the absorbtion efficiency, fresnel efficiency and critical angle efficiency:

$$\eta_{optical} = \eta_A \cdot \eta_{Fr} \cdot \eta_{Cr} \qquad (2.1.4\text{-}1)$$

The optical efficiency for a T-1 3/4 clear high-efficiency red lamp is 0.121:

$$\eta_{optical} = (.76)(.816)(.195) = .121$$

2.1.5 External Quantum Efficiency

External quantum efficiency, $\eta_{q_{ext}}$, is defined by the following expression:

$$\eta_{q_{ext}} = \frac{\phi_e \text{ (W)}}{I_F \text{ (A)} \left[\dfrac{1240}{\lambda(nm)} \cdot \dfrac{eV}{photon}\right]} = \frac{photon}{electron} \qquad (2.1.5\text{-}1)$$

The value 1240 is the product of Plank's Constant and the speed of light:

$$\lambda = \frac{hc}{E_g}$$

$$= \frac{(6.626 \times 10^{-34} \text{ joules} \cdot sec)(2.998 \times 10^8 \text{ meters/sec})}{E_g (1.602 \times 10^{-19} \text{ joules/eV})}$$

$$= \frac{1.240 \times 10^{-6} \text{ m} - eV}{E_g \text{ (eV)}}$$

$$\lambda \text{ (nm)} = \frac{1240}{E_g(eV)}$$

A T-1 3/4 high-efficiency red lamp with $2\theta_{1/2} = 35°$ and produces 12 mcd at 10 mA radiates 96.6 μW. The external quantum efficiency for this lamp is 4.95×10^{-3} photons/electron.

$$\eta_{q_{ext}} = \frac{96.6 \times 10^{-6} \text{ W}}{(.010A)\left(\dfrac{1240}{635 \text{ nm}}\right)} = 4.95 \times 10^{-3} \frac{photons}{electron}$$

2.1.6 Internal Quantum Efficiency

The current flowing through an LED is composed of two components, a radiative component and a non radiative component. The internal quantum efficiency, $\theta_{q_{int}}$, is the ratio of the radiative component to the total current. The internal quantum efficiency may be calculated by dividing the external quantum efficiency by the optical efficiency:

$$\eta_{q_{int}} = \frac{\eta_{q_{ext}}}{\eta_{optical}} = \text{photons/electron} \qquad (2.1.6\text{-}1)$$

2.1.7 Calculating Radiated Flux

2.1.7.1 Luminous Efficacy and Power Per Unit Solid Angle

The ratio of luminous flux (lumens) to radiant flux (watts) is called luminous efficacy, η_v (lm/w). For a complete discussion of luminous efficacy, see the section on photometry. The value of luminous efficacy is given on each lamp data sheet. These values for luminous efficacy are:

LED Luminous Efficacy, η_v (lm/W)			
Std.Red	Hi-Eff. Red	Yellow	Green
60	135	460	630

In some applications, such as providing a light source for a lens focusing system which subtends a specified solid angle, the amount of radiated power in microwatts per steradian, I_e (μW/sr), is important. The value of radiated power per unit solid angle may quickly be calculated from the following equation:

$$I_e \text{ (μW/sr)} = \frac{I_v \text{ (mcd = m lm/sr)}}{\left(\dfrac{\eta_v \text{ (lm/W)}}{1000}\right)} \qquad (2.1.7.1\text{-}1)$$

For a T-1 3/4 clear high-efficiency red lamp with $2\theta_{1/2} = 35°$ and $I_v = 12$ mcd, the radiated power per sterdian is 89 μW/sr.

$$I_e = \frac{(12 \text{ m lm/sr})(1000)}{135 \text{ lm/W}} = 89 \ \mu\text{W/sr}$$

2.1.7.2 Calculating Total Power

In applications where the total radiated flux is to be utilized, the lamp radiation pattern must be known. The section on photometry provides a detailed discussion on calculating the total radiated flux using the method of zonal integration. In this section, an illustrative example of this method is presented which calculates the total flux for a T-1 3/4 undiffused high-efficiency red lamp with $2\theta_{1/2} = 35°$ and producing 12 mcd.

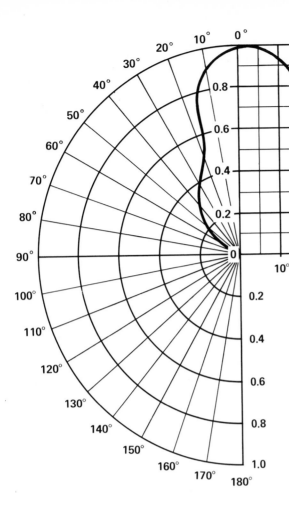

Figure 2.1.7.2-1 Radiation Pattern for an Undiffused T-1 3/4 High-Efficiency Red Lamp.

$$\phi_e = \frac{1000\ I_{vo}\ \varphi}{\eta_v} = \mu W \qquad (2.1.7.2\text{-}2)$$

where: I_{vo} = On-axis luminous intensity (mcd = m lm/sr).
φ = Total solid angle into which the flux is radiated (sr).
η_v = Luminous efficacy (lm/W).

For the above lamp with I_{vo} = 12 mcd at I_F = 10 mA:

$$\phi_e = \frac{1000\ (12)\ (1.0869)}{135} = 96.6\ \mu W$$

Figure 2.1.7.2-1 is the polar/linear plot of the radiation pattern for the above lamp. The linear graph is used to determine the relative luminous intensity points for use in the calculation. Since the grid is marked off in 5° increments, we shall take our summation over 5° intervals.

$$\Delta = 5° = \frac{180}{N}\ ,\ \text{then N = 36}$$

$$\varphi\left(\frac{mA}{I_o}\right) = \frac{1}{2}\ I_r\ (M\Delta)\ C_Z\ (M\Delta) + \sum_{m=1}^{m-1} I_r\ (m\Delta)\ C_Z\ (m\Delta)$$

$$(2.1.7.2\text{-}1)$$

where: $C_Z\ (M\Delta) = \dfrac{2\pi^2}{N}\ \sin\ (M\Delta) = .5483\ \sin\ (M\Delta)$

$I_r\ (M\Delta)$ = **Relative Luminous Intensity from linear graph at the angle mΔ.**

The calculation is set up in a tabular format. The calculation needs to include angles only up to mΔ = 70°, M = 14.

m	mΔ(°)	I_r (mA)	C_Z (mA)	I_r (mA) C_Z (mA)
1	5	.981	.0478	.0469
2	10	.938	.0952	.0893
3	15	.801	.1419	.1137
4	20	.516	.1875	.0968
5	25	.435	.2317	.1008
6	30	.379	.2742	.1039
7	35	.342	.3145	.1076
8	40	.298	.3524	.1050
9	45	.248	.3877	.0961
10	50	.199	.4200	.0836
11	55	.149	.4492	.0669
12	60	.099	.4749	.0470
M−1 ▶13	65	.056	.4969	.0278
			$\sum\limits_{m=1}^{m-1}$	1.0854
M ▶14	70	.006	.5152	.0031

The total radiated power is now calculated from the following equation:

$$\varphi\left(\frac{M\Delta}{I_o}\right) = \frac{1}{2}\ (.0031) + 1.0854 = 1.0869\ \text{sr}$$

2.4

2.1.8 Magnification and Luminous Intensity

The luminous intensity of an LED lamp is a function of the magnification provided by the immersion lens. Using the parameters of Figure 2.1.8-1 and assuming paraxial light rays, the magnification may then be defined as the ratio of the image size to the object size and approximated by the following formula:

$$\text{Magnification} = m \stackrel{\triangle}{=} \frac{y_2}{y_1} \approx \frac{1}{1 - \frac{x_1}{r}\left(1 - \frac{n_2}{n_1}\right)} \qquad (2.1.8-1)$$

For a T-1 3/4 lamp, the focal length is: X_1 = 4.70 mm, r = 2.44 mm and n_1 = 1.53. The magnification is approximately 3:

$$m \approx \frac{1}{1 - \frac{4.70 \text{ mm}}{2.44 \text{ mm}}\left(1 - \frac{1}{1.53}\right)} = 3.01$$

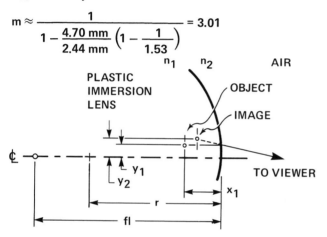

Figure 2.1.8-1 Parameters of a Spherical Immersion Lens.

The magnification may also be calculated from the focal length of the immersion lens. If the surface of the dome lens is regarded as the principal plane, a focal point may be defined as that point on the lens axis from which paraxial light rays emerge from the dome lens parallel to the lens axis. The principal plane is the locus of equivalent refracting points for a lens where extended entering and emerging rays intersect. The focal length of the lamp's spherical dome lens may be determined from the following formula:

$$fl = \frac{r}{1 - \left(\frac{n_2}{n_1}\right)} \qquad (2.1.8-2)$$

$$fl = \frac{2.44 \text{ mm}}{1 - \left(\frac{1}{1.53}\right)} = 7.044 \text{ mm}$$

The magnification may then be calculated from this focal length:

$$m = \frac{1}{1 - \left(\frac{X_1}{fl}\right)} \qquad (2.1.8-3)$$

The effect of magnification on luminous intensity and radiation pattern for an undiffused package is illustrated in Figure 2.1.8-2. If the LED is encapsulated very close to the dome surface, curve A, the dome surface is effectively flat. There is no magnification and the radiation pattern remains the lambertian pattern of the unencapsulated LED chip. In the lambertian radiation pattern, the luminous intensity varies as the cosine of the off-axis angle, θ.

$$I(\theta) = I_o \cos \theta \text{ (Lambertian radiator)} \qquad (2.1.8-4)$$

where: I_o is the on-axis luminous intensity

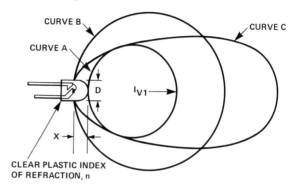

CURVE	DISTANCE X	ON-AXIS LUMINOUS INTENSITY, I_{VO}	LUMINOUS FLUX, ϕ_V	RADIATION PATTERN
A	≈0	I_V: ≈I_{VO} of unencapsulated LED	πI_{V1}	Lambertian
B	D/2=r	$n^2 I_{V1}$	$\pi n^2 I_{V1}$	Lambertian
C	>D/2	>$n^2 I_{V1}$	<$n^2 \pi I_{V1}$	Non Lamberian

Figure 2.1.8-2 Effect of Magnification on Luminous Intensity and Radiation Pattern for an Undiffused Plastic Lamp.

The luminous flux is equal to π times the luminous intensity.

At distances from the dome surface of o<x<r, the LED chip is magnified and the luminous intensity is directly proportional to the square of the magnification. When x = r, the magnification is equal to the index of refraction of the encapsulant and the luminous intensity is then equal to that of the unencapsulated chip multiplied by the index of refraction squared. For a spherically shaped dome lens, the radiation pattern is still lambertian, curve B. The total luminous flux increased by the factor of index of refraction squared.

At distances greater than the radius of the dome lens further increases the magnification above that value equal to the index of refraction. The luminous intensity increases but the radiation pattern narrows reducing the viewing angle, curve C. The maximum practical magnification is that at which the image of the LED's emitting area equals the diameter of the dome lens.

An aspheric dome lens can be used to achieve a high value of on-axis luminous intensity along with a wider radiation pattern than is obtainable with a spherical dome lens. An aspheric dome lens is used as the package of T-1 3/4 low profile lamps. The comparison of the radiation pattern of an aspheric dome lens with a spherical dome lens of equal magnification is illustrated in Figure 2.1.8-3.

At this point, the reader is reminded that when evaluating the specifications for an LED lamp, it is necessary to take into account both the on-axis luminous intensity and the lamps radiation pattern. Even though two lamps may have the same luminous intensity at a specified forward current, one may have a wider radiation pattern. The lamp with the wider radiation pattern may have a more efficient dome lens or a more efficient LED chip.

2.1.9 Diffused and Undiffused LED Lamps

The immersion lens concept applies to a lamp which has the LED encapsulated in undiffused plastic. The result is a beam of light with a high value of luminous flux which is concentrated in a narrow radiation pattern. This lamp configuration is especially useful for backlighting applications and for applications requiring a concentrated light source.

A front panel indicator lamp requires a very wide off-axis viewing angle. To achieve this wide viewing angle, diffusant is added to the lamp to disperse the light rays emitting from the LED. The result is a lamp with a wide radiation pattern and a reduced value of on-axis luminous intensity. Figure 2.1.9-1 pictorially illustrates the differences between an undiffused and diffused lamp. Dye coloring is added to tint the diffused lamp to enhance on/off contrast.

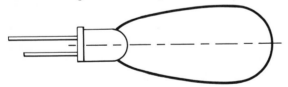

T-1 3/4 LAMP WITH SPHERICAL DOME LENS

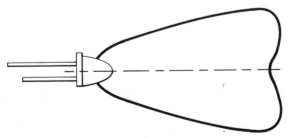

T-1 3/4 LOW PROFILE LAMP
WITH AN ASPHERIC DOME LENS

Figure 2.1.8-3   Radiation Patterns for Undiffused Lamps with Spherical and Aspheric Dome Lenses.

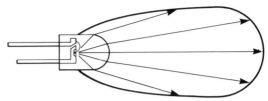

BOTH LAMPS HAVE THE SAME
LED CHIP AND THE SAME SHAPED
DOME LENS

UNDIFFUSED PLASTIC LAMP: HIGH VALUE OF ON-AXIS LUMINOUS INTENSITY
WITH A NARROW RADIATION PATTERN

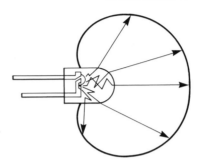

DIFFUSED PLASTIC LAMP: REDUCED VALUE OF ON-AXIS LUMINOUS
INTENSITY WITH A WIDE RADIATION PATTERN

Figure 2.1.9-1   Comparison Between Undiffused and Diffused Plastic LED Lamps.

2.2   LED Lamp Packaging

The function of a lamp package is to utilize the various physical properties described in Section 2.1 to effect the best coupling of the emitted light from an LED to an observer or electronic detector. The desired package configuration depends upon the specific requirements of each application. These requirements will determine such parameters as the size of the device, if the lamp is to be tinted or untinted and whether the encapsulating epoxy is to be diffused or undiffused. Some requirements may dictate a lamp device where epoxy encapsulating is not desired, such as in a high-reliability application.

It is obvious that there could well be as many lamp package configurations as there are applications. However, certain lamp packages have been developed by the optoelectronics industry which cover a wide range of the more common applications. Prior to discussing the commonly available package configurations, it is a benefit to the reader to first understand the basic lamp packaging process.

2.2.1   Lead Frame Packaging

Immensely popular because of its low cost, the lead frame technique is used for the packaging of LED lamp devices. Basically, in lead frame packaging, LED dice are die attached to one element of a metal frame and a wire bond is made from the top of each die to another element of the frame. During the die-attach and bonding operations, the elements are joined for mechanical support by metal straps called "dam bars". Plastic encapsulation applied around the devices surrounds also the elements of the frame while the dam bars remain outside of the plastic. When cured, the plastic provides mechanical support to the elements and the dam bars may then be sheared away. Shearing of the dam bars leaves the opposite ends of the lead frame protruding from the plastic; these then become the external leads of the finished device. Only after the dam bars are removed can the devices be directly energized for testing.

The material of the lead frame is thick enough that the dice can be attached on the edge (rather than the face) of the frame material. This places the die on the end of an element, called a "die-attach post", which is canted, allowing the die to be centered with respect to those portions of the lead frame which later become the external leads, as shown in Figure 2.2.1-1. The bonding wire is then connected from the top contact of the die to the end of the adjacent element, called a "bonding post". After the bonding wires are in place, the lead frame, bearing a number of devices is clamped into a fixture which controls precisely the distance by which the posts of the frame are lowered into cavities that have been previously filled to a

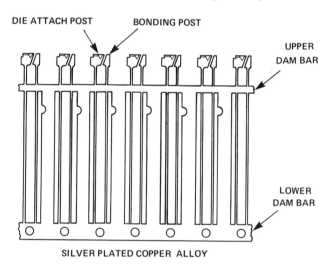

Figure 2.2.1-1   A Typical Lead Frame for LED Lamps.

precise level with the uncured liquid epoxy. The shape of these cavities, called "mold cups", determines the shape of the lenses of the lamp packages, and the distance to which the posts are inserted in the mold cups determines the magnification and hence the radiation pattern of the finished product. Dye is used to tint the epoxy to absorb ambient light and make the lamp appear darker when it is off; since the dye does not absorb appreciably the light from the LED, this raises the ON/OFF contrast ratio. In some devices, a diffusant material is added to the epoxy to cause light scattering, thereby increasing the effective viewing angle.

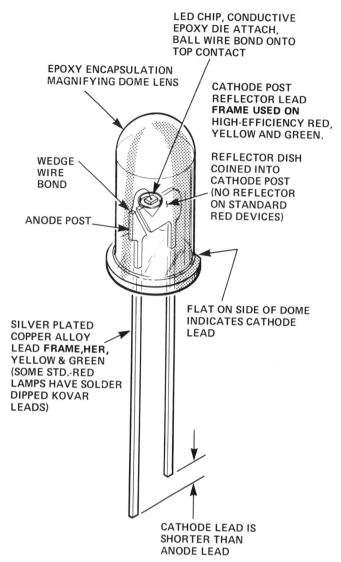

Figure 2.2.2-1   Construction Features of a T-1 3/4 Plastic LED Lamp.

Curing of the epoxy is done in two stages. A pre-cure hardens the plastic to a point which permits removal from the mold cups and shearing of the lead frame and dam bars. The post-cure is then applied, which further develops and stabilizes the mechanical properties of the epoxy, enabling the lamps to withstand the abuse they may later receive when being installed and connected into circuits. Except for testing, manufacture is now complete.

2.2.2   The Industry Standard T-1 3/4 & T-1 LED Lamps

When LED lamp devices were first introduced, two established miniature sized packages were borrowed from the incandescent lamp. These are the T-1 and T-1 3/4 sizes. In the T-X designation, the number indicates the lamp diameter in 1/8ths of an inch. These two sizes have now become standard in the optoelectronic industry.

The construction features of a T-1 3/4 lamp are illustrated in Figure 2.2.2-1. The lead frame for a GaP transparent substrate LED has a dish shaped reflector coined into the top of the cathode post. A GaP transparent substrate LED emits light from the sides of the chip as well as from the top surface. The reflector directs this side emitted light towards the dome lens in an analogous fashion to the common hand held flashlight. The result is a lamp with a front emitter equal to the larger area of the reflector dish. A standard red lamp does not require a reflector, since a GaAsP/GaAs LED emits light only from the top surface of the junction.

A variation of the standard T-1 3/4 lamp is the low profile T-1 3/4 lamp. The construction is identical to the standard lamp except for the height and shape of the dome lens. The dome is 2/3rds the height of the standard lamp and is an aspheric lens which provides high magnification coupled with a wide viewing angle. For a visual comparison, see Figure 2.1.8-3.

The construction of the T-1 lamp is essentially the same as for T-1 3/4 lamp. However, the reflector dish is too large to be successfully encapsulated inside this small lamp package. It is for this reason, that a more compact lead frame, without a reflector, is used in all of the T-1 lamp products.

Initially, the two standard lamp sizes were used only as indicators on front panels and printed circuit boards. However, as the variety of uses expanded into other areas, such as backlighting and various array configurations, the T-1 and T-1 3/4 packages could not be easily adapted to meet the shape and size requirements for all of these more sophisticated applications. As a result, new lamp packages have been developed to conform to many of these needs.

2.2.3   The Subminiature LED Lamp

Many applications require a very small, low cost LED lamp to be used in a location where space is at a premium. Such might be the case for the colon in a small desk clock, for an indicator on a small hand held instrument or for use in an array where a high packing density is required. To meet this need, the subminiature lamp package has been developed.

The subminiature device offers the customer a lamp that is low in cost, smaller in size than a T-1 lamp and has superior optical consistency device-to-device. The small size allows for a very high packing density in an array, see Section 2.4.4 on Arrays. The optical consistency of each lamp is derived   from the transfer molding process used to encapsulate the device.

Figure 2.2.3-1 illustrates the construction features of the subminiature lamp package. The subminiature package uses a radial, flexible, rectangular lead frame suitable for

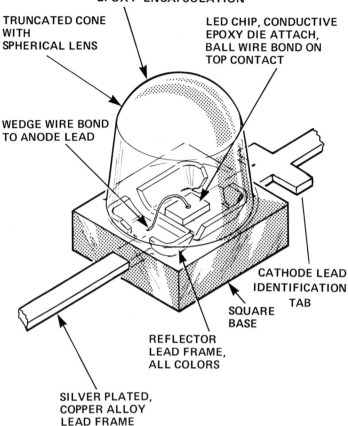

Figure 2.2.3-1   Construction Features of a Subminiature Plastic LED Lamp.

handling in the transfer molding process. From a user's point of view, this lead frame offers two important benefits. First, unlike the T-1 lead frame which does not have a reflector, a small reflector cup is formed by bending up the sides of the die attach and bonding pads. The optical effect is similar to the T-1 3/4 lamp, producing a small lamp with very high sterance. The second advantage is a choice of installation procedures; the leads may be left in the radial position for soldering to pads on the face of a printed circuit board, or the leads may be bent 90° into the axial position for insertion into plated through holes and subsequent flow soldering or for insertion into a socket.

The package dome is constructed as a truncated cone with a spherical lens to obtain high magnification for high intensity combined with a very wide viewing angle. The package base is square shaped to provide firm support for the lead frame, resulting in a high degree of mechanical reliability, and to aid in positive alignment in high density array applications.

2.2.4   The Rectangular LED Lamp

The rectangular package is designed for use in those specific applications where a cylindrically shaped device is not effective. For instance, the rectangular LED lamp is most effective for illuminating a legend by directly backlighting a transparent character so that it will stand out more vividly. As another example, a bar graph is easily implemented with rectangular LED lamps, as they may be either end stacked or side stacked to form a continuous bar of light.

The construction of the rectangular lamp is an extention of the T-1 3/4 package, as both use the same reflector lead frame. As shown in Figure 2.2.4-1, the difference is in the encapsulation. The encapsulating epoxy is formed into a rectangular light pipe with an integral diffusing layer at the top. The light pipe is the optical path through which the LED light travels to diffusing layer. The diffusing layer spreads the light to form an evenly lighted rectangular source.

2.2.5 The Hermetic LED Lamp

Although plastic devices are used in the majority of LED lamp applications, they become vulnerable when placed into adverse environments. The problem is the limit to which the encapsulating epoxy can withstand temperature extremes, moisture or other detrimental environmental conditions without losing its optical properties or exerting failure causing stresses on the LED die attach and wire bonds. Therefore, a hermetically sealed device is used for many military lamp applications and applications where the lamp may be exposed to an adverse industrial environment.

The hermetic lamp is assembled in a TO-18 package as illustrated in Figure 2.2.5-1. After die attach and wire bond, a coating of silicone jell is applied. At the top of the metal cap is an optical window with a hermetic seal at the glass to metal interface. An epoxy dome is cast on top of the optical window to increase viewing angle and to provide good on/off contrast. The metal cap is welded to the TO-18 header to complete the assembly. The exterior of the package is gold plated to resist corrosion.

2.2.6 LED Lamps that Include Other Components

The preceding sections have described the LED lamp package as a device containing only one component. Electrically, an LED lamp is a two terminal device and is capable of utilizing a limited number of other components, in addition to the LED chip, to perform certain functions. Such devices are termed "integrated LED Lamps".

A commonly used integrated lamp is the resistor LED lamp. In this device, the package contains an integral current limiting resistor, that is die attached to the anode post and wire bonded to the LED top contact. This integral resistor is a nominal 215Ω, allowing the lamp to be driven directly from 5.0 volt supply when controlled by a TTL gate.

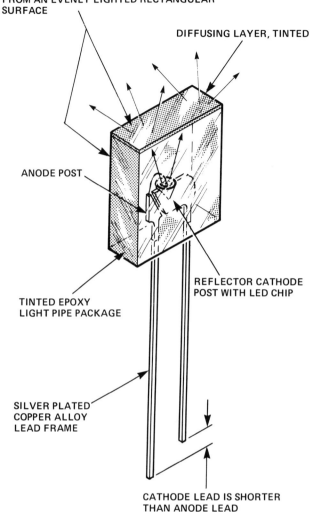

LIGHT RAYS FROM LED TRAVEL THROUGH LIGHT PIPE TO DIFFUSING LAYER. THE DIFFUSING LAYER SPREADS LIGHT TO FROM AN EVENLY LIGHTED RECTANGULAR SURFACE

DIFFUSING LAYER, TINTED

ANODE POST

REFLECTOR CATHODE POST WITH LED CHIP

TINTED EPOXY LIGHT PIPE PACKAGE

SILVER PLATED COPPER ALLOY LEAD FRAME

CATHODE LEAD IS SHORTER THAN ANODE LEAD

Figure 2.2.4-1   Construction Features of a Plastic Rectangular LED Lamp.

Another integrated lamp device is the voltage sensing LED lamp. In this device, an integrated circuit turns the lamp on or off depending upon the level of the applied voltage. A reference threshold level, V_{TH} is built into the integrated circuit. If the applied voltage is above V_{TH} the lamp is ON, and below V_{TH} the lamp is OFF. This device is commonly used as a battery test indicator.

Figure 2.2.6-1 shows schematic representations for these two integrated lamp devices.

2.3   LED Lamp Characterization Information

A lamp data sheet contains specific characterization information to aid the designer in selecting the correct lamp for his application, determining the maximum worst case operating limits and establishing nominal operating

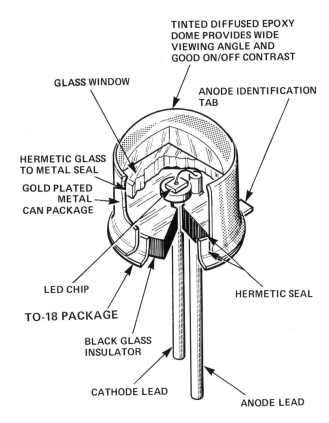

TINTED DIFFUSED EPOXY
DOME PROVIDES WIDE
VIEWING ANGLE AND
GOOD ON/OFF CONTRAST

GLASS WINDOW

ANODE IDENTIFICATION
TAB

HERMETIC GLASS
TO METAL SEAL

GOLD PLATED
METAL
CAN PACKAGE

LED CHIP

TO-18 PACKAGE

HERMETIC SEAL

BLACK GLASS
INSULATOR

CATHODE LEAD

ANODE LEAD

Figure 2.2.5-1   Construction Features of a Hermetic LED Lamp.

conditions. The data sheet contains this information in the following basic sections:

Product Description and Package Dimensions

Absolute Maximum Ratings

Electrical/Optical Characteristics

Operating Curves

Once the lamp package configuration has been selected, the designer is ready to consider those characteristics of most importance:

1.   Light output and color matching

2.   Maximum temperature derated operating limits

3.   Pulsed operating conditions

4.   Time average luminous intensity

Other considerations may also be of importance such as reverse breakdown voltage, capacitance or speed of response, depending upon the application.

For a designer to be able to effectively utilize the information contained in a data sheet, he needs an appreciation of what the numbers mean and on what basis they have been derived.

2.3.1   Light Output and Color Matching

Of initial concern to a designer is the selection of that lamp which is of the proper color and has the most light output at a specified current. The electrical/optical characteristics contain four parameters that quantitatively aid the designer in making this selection. These are peak wavelength, dominant wavelength, luminous intensity and the included angle between half intensity points.

The color of the lamp, as perceived by the eye, is dependent upon the radiated spectrum of the LED. Not only is the radiated spectrum of importance for determining color in visual applications, but it is very important in determining the coupling to a detector in non-visual applications. Peak wavelength and dominant wavelength are the two quantitative parameters which describe the radiated spectrum.

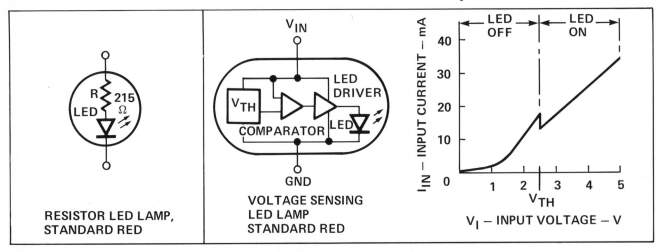

RESISTOR LED LAMP,
STANDARD RED

VOLTAGE SENSING
LED LAMP
STANDARD RED

Figure 2.2.6-1   Schematic Representation of Two Common Integrated LED Lamps.

Peak wavelength, λ_p, is that wavelength at the peak of the radiated spectrum. It is maintained within narrow limits during the growth of the epitaxial layer, with each wafer being inspected by a photo-luminescence measurement on a production basis. To the designer of a non-visual application, λ_p becomes important when determining the coupling efficiency of the LED light to a photodetector, signal loss through a fiber optic conductor or photographic film sensitivity. In visual applications, the total amount of LED emitted light passing through an optical filter, used for contrast enhancement, is approximately equal to the filter's relative transmission at λ_p. Therefore, the relative transmission at λ_p is a quantitative measure of the optical density of a contrast filter.

The color of an LED device is especially important to the designer if lamps from various manufacturers might possibly be installed in the same array. Not only does the eye detect intensity differences, it also detects color differences. The dominant wavelength, λ_d, is a quantitative measure of the color of an LED device as perceived by the eye. Two devices of somewhat different radiated spectra will appear as the same color if they both have the same λ_d. The dominant wavelength is not necessarily dependent upon peak wavelength. Conceptually, λ_d may be envisioned as that wavelength near the centroid of the radiated spectrum. Specifically, λ_d is that wavelength when mixed with an equal amount of the light from a 6500°K lamp will be perceived by the eye as the same color as is produced by the radiated spectrum of the LED.

It has been common practice of many designers to use the data sheet luminous intensity as the only parameter necessary to evaluate light output. This is not the only important consideration in the determination of the total available luminous flux. A designer needs to consider both the axial luminous intensity and the radiation pattern in making a determination of total light output. It is the radiation pattern that has the most influence.

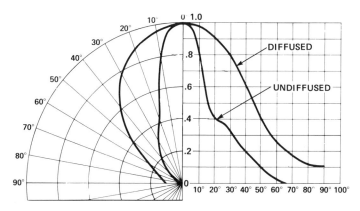

Figure 2.3.1-1   Radiation Pattern for a T-1 3/4 High-Efficiency Red LED Lamp.

Luminous Intensity, I_v, is a measure of axial light output. The unit of measure is the candela, which is luminous flux per unit solid angle. Thus, the measure of luminous intensity gives no information concerning the radiation pattern. The axial luminous intensity of each device is measured on a production basis, using a calibrated photometer, to insure that it meets a minimum value at a specified drive current.

The importance of the radiation pattern to a designer is that it defines the apparent luminous intensity of the device when viewed at some off-axis angle. The radiation pattern is quantitatively described at a single point, which is the included off-axis angle where the luminous intensity is equal to one half of the axial intensity. This included angle is referred to as $2\theta_{1/2}$. The value of $\theta_{1/2}$ is controlled by the distance of the LED from the focal point of the dome lens. Larger values of $2\theta_{1/2}$ result in a lamp which may be viewed at greater off axis angles. Also, a lamp which has a greater $2\theta_{1/2}$, but the same axial intensity will actually have a much greater total flux. A good example of this is the comparison of two undiffused T-1 3/4 LED lamps that have the same axial luminous intensity but differ in the value of $2\theta_{1/2}$. The first lamp has $2\theta_{1/2} = 35°$ and the second lamp has $2\theta_{1/2} = 25°$. Depending upon the exact shape of the two radiation patterns with respect to each other, the first lamp could produce up to three times the luminous flux that is produced by the second lamp. An examination of the data sheet curve of the radiation pattern reveals the intensity at any specific off-axis angle, as illustrated in Figure 2.3.1-1.

Relying only on a minimum or a typical intensity value for lamps used in an array does not guarantee intensity matching. It is best to install lamps that have been categorized for light output. Each light output category should have an $I_{v\ MAX}/I_{v\ MIN}$ ratio of 2:1 or less. A single array is then assembled with all lamps from the same category. To an observer, the lamps will appear evenly illuminated throughout the array.

In summary, peak wavelength, dominant wavelength, axial luminous intensity and radiation pattern are necessary to completely specify the optical characteristics of an LED device.

2.3.2   Maximum Temperature Derated Operating Limits

Once the optical characteristics of an LED device have been established, the next step is to determine the maximum temperature derated limits at which the device may be operated. Forward current, power dissipation, thermal resistance and LED junction temperature are all interrelated in establishing absolute maximum ratings.

The absolute maximum ratings of an LED device have been determined theoretically and by extensive reliability testing and are those limits beyond which reliable operation cannot be assured. These maximum ratings cannot be taken as limits in concert, as operation at one maximum limit may preclude the operation at another maximum limit. The best example of this exclusion rule is that a device cannot be reliably operated at maximum power dissipation in the maximum allowed ambient temperature. Operation at the maximum ambient temperature is only allowed with proper power derating.

Maximum ratings are based on package temperature limitations and the current density limit within the LED junction. The package temperature limitation may be based on the glass transition temperature, T_G, of the encapsulating epoxy. Above T_G, the cross linkages between the epoxy molecules change allowing the molecules to move with respect to each other. The epoxy changes from an ordered structure to an amorphous structure, analogous to that of glass, and the coefficient of thermal expansion drastically increases. The result is that thermal stresses above T_G may be sufficient to cause a catastrophic failure. The maximum current rating is based on (1) the level of current density within the LED junction that produces no more than an acceptable amount of light output degradation, and (2) the amount of power dissipation generated by the forward current that maintains the LED junction at an acceptable temperature level below T_G.

The limiting factor for operation in any ambient temperature is the LED junction temperature, T_J. For a hermetic device, T_J should be less than 125°C to realize an acceptable rate of light output degradation, and for a plastic device, T_J should be at least 10°C below the T_G of the encapsulating epoxy to prevent catatrosphic failure. The encapsulating epoxies currently being used in the manufacture of LED lamps have a nominal T_G = 120°C.

The two parameters that effect T_J, which a designer has control of, are forward current and thermal resistance. The maximum forward current is derived from the maximum allowed temperature derated power dissipation and the thermal resistance is dependent upon the method used to install the lamp in the circuit.

The maximum allowed temperature derated power dissipation is obtained by derating the absolute maximum power rating at the rate of -1.6 mW/°C above an ambient of 50°C. This is a linear derating from maximum power at an ambient of 50°C to zero power at an ambient of 125°C, as shown in Figure 2.3.2-1.

Average power dissipation is the product of average forward current and peak forward voltage. An equivalent circuit for an LED, consisting of a dc voltage source in series with a dynamic resistance, may be derived from the diode's forward characteristics. Also, a lamp data sheet lists a value

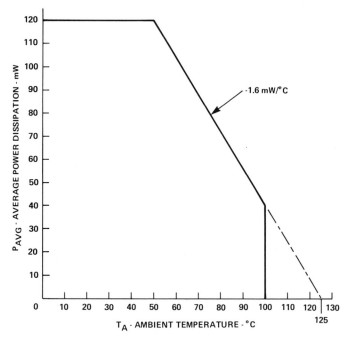

Figure 2.3.2-1   Maximum Average Power Derating for a Plastic LED Lamp.

of forward voltage at a specific current. From these two observations, the following equation may be derived to calculate the power dissipation in an LED lamp:

$$P_{AVG}(W) = I_{AVG}[V_F(DS) + R_s(I_{PEAK} - I_F(DS)] \quad (2.3.2\text{-}1)$$

where: I_{AVG} = Average forward current, amperes.

I_P = Peak forward current, amperes.

R_s = LED dynamic resistance, ohms.

$V_F(DS)$ = Data sheet forward voltage, volts.

$I_F(DS)$ = Data sheet forward current where $V_F(DS)$ is specified, amperes.

For standard red devices, $R_s = 1.6\Omega$ typical and 5Ω maximum; for GaP transparent substrate devices $R_s = 21\Omega$ typical and 35Ω maximum.

Once the maximum tolerable operating conditions have been established for pulsed operation, as described in Section 2.3.3, equation 2.3.2-1 may be used as a check to insure that the average power dissipation does not exceed the maximum derated limit. For dc operation, the dc power dissipation may be calculated by setting $I_{DC} = I_{AVG} = I_{PEAK}$.

LED junction temperature is the sum of the operating ambient temperature and the temperature rise above this ambient.

$$T_J(°C) = T_A + \Delta T_J \quad (2.3.2\text{-}2)$$

$$T_J(°C) = T_A + \theta_{JA} P_{AVG}$$

where: T_A = Ambient temperature immediately surrounding the LED lamp, °C.

P_{AVG} = Average power dissipation, Watts.

θ_{JA} = Thermal resistance LED junction-to-ambient of the lamp installed into the circuit, °C/W.

The value of θ_{JA} is the sum of the device thermal resistance, LED junction-to-lead, θ_{JC}, and the thermal resistance to ambient of the supporting structure, θ_{CA}:

$$\theta_{JA}(°C/W) = \theta_{JC} + \theta_{CA} \qquad (2.3.2\text{-}3)$$

For most printed circuit boards, θ_{CA} ranged between 35°C/W and 50°C/W, depending upon metallization pattern. Figure 2.3.2-2 illustrates this concept.

A numerical example utilzing these equations is presented in Section 2.4.3.

$$\theta_{JA}(°C/W) = \theta_{JC} + \theta_{CA}$$
$$T_J(°C) = T_A + \theta_{JA} P_{AVG}(W)$$

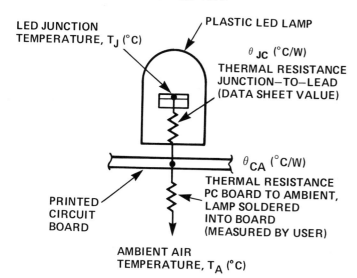

Figure 2.3.2-2   A Schematic Representation of the Thermal Resistance Paths for a Plastic LED Lamp.

2.3.3   Pulsed Operating Conditions

When a design requires an LED lamp to be operated in the pulsed mode, maximum tolerable operating limits need to be established. These maximum tolerable limits should not raise the LED junction temperature above that which would be obtained by operating the lamp at the maximum dc current. This limitation on T_J imposes a definite interrelationship between peak current, pulse duration, and refresh rate. This interrelationship is most easily obtained by establishing combinations of peak current and pulse duration for various refresh rates, maintaining the maximum T_J at that value obtained by operating at the maximum dc current. These data points are plotted on a

log-log scale to form the family of curves shown in Figure 2.3.3-1.

The curve for any specific refresh rate is the locus of maximum tolerable operating conditions which maintain the limitation of T_J. Any combination of operating

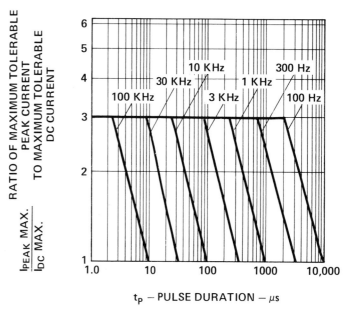

Figure 2.3.3-1   Maximum Tolerable Peak Current vs. Pulse Duration for a T-1 3/4 High-Efficiency Red LED Lamp.

conditions at, or below, a line of constant refresh rate is permissible. Operation above a line of constant refresh rate violates the limitation on junction temperature.

One procedure for determining a set of operating conditions from Figure 2.3.3-1 consists of the following five steps:

Step 1. Determine the desired duty factor, DF. Example: DF = 30%.

Step 2. Determine the desired refresh rate, f. Use DF to calculated pulse duration, tp. Example: f = 1 KHz; tp = DF/f = .30/1 KHz = 300 μsec.

Step 3. Enter Figure 2.3.3-1 at the calculated value of tp. Move vertically to refresh rate line and record the corresponding value of $I_{PEAK\ MAX}/I_{DC\ MAX}$. Example: At tp = 300 μsec and f = 1 KHz, $I_{PEAK\ MAX}/I_{DC\ MAX}$ = 2.4.

Step 4. Calculate $I_{PEAK\ MAX}$ using the data sheet maximum average forward current for $I_{DC\ MAX}$. Calculate I_{AVG} from I_{PEAK} and DF. Example: from data sheet, the maximum average current = 20 mA.

2.13

$$I_{PEAK\ MAX} = (2.4)(20\ mA) = 48\ mA$$
$$I_{AVG} = (.3)(48\ mA) = 14.4\ mA$$

Step 5. Refer to Equation 2.3.2-1 and calculate P_{AVG} as a check to insure that the above operating conditions are within the required power derating corresponding to the operating ambient temperature. Should P_{AVG} fall above the required power derating, decrease tp to reduce I_{AVG} (or reduce I_{PEAK}) in order to lower P_{AVG} to an acceptable level.

An alternate procedure is presented in the form of a numerical example in Section 2.4.3.5.

2.3.4 Time Average Luminous Intensity

The axial luminous intensity value listed on the data sheet is a dc measurement. The measurement is made by driving the lamp at the specified current for a time duration of 20 msec to 50 msec and measuring the intensity with a calibrated photometer. The luminous intensity of each lamp will be at least equal to the specified minimun, but on the average will be close to the typical value listed on the data sheet. This measurement does not give a designer any insight as to what the intensity will be at some other drive condition.

Of interest to a designer is the light output, as perceived by an observer, when the lamp is driven at some other dc current or is operated in the pulsed mode. Since the eye is a time averaging detector, it is the time average luminous intensity that is of specific interest.

For dc operation, the time average intensity is a supralinear function of forward current.

$$\frac{I_v}{I_{vo}} = \left(\frac{I_F}{I_{FO}}\right)^n \qquad (2.3.4\text{-}1)$$

where: I_{vo} = An initial luminous intensity at a reference dc current, I_{FO}.

I_v = The luminous intensity at the operating dc current, I_F.

n = A supralinear exponent that ranges between 1.1 and 1.4, depending upon the LED product and the forward current.

Equation 2.3.4 is derived in the following manner. A measurement of luminous intensity at various drive currents is obtained from a large sample of devices selected from many different lots. A curve of relative luminous intensity vs. forward current is derived from this data, normalized to 1.0 at the current where the luminous intensity is specified, and presented as a graph on the data sheet. Equation 2.3.4

is the mathematical representation of this curve, with the value of n having been derived directly from the curve.

The determination of time average intensity, observed with pulsed operation, involves the concept of relative efficiency. Once the concept of relative efficiency is understood, a designer is able to calculate the time average intensity for any given set of pulsed operating conditions.

The efficiency of an LED device may be stated as intensity per unit current, such as millicandelas/milliampere. Relative efficiency is the ratio of the efficiency at one peak current to the efficiency at another peak current for the same average current.

A graph of relative efficiency is obtained in the following manner. An LED device is measured for luminous intensity at the test current specified in the data sheet. Then the device is strobed at different peak currents, but with the same average current as specified in the data sheet, and measured for luminous intensity. The resulting curve of luminous intensity vs. peak current is normalized to 1.0 at the data sheet test current. This is the curve of relative efficiency for that device. The data sheet relative efficiency curve represents the average of a large sample of devices from many different lots.

To obtain the time average intensity of an LED lamp being operated in the pulsed mode, a designer need only multiply the ratio of the operating average current to the data sheet average current by the product of the relative efficiency at the operating peak current and the luminous intensity specified on the data sheet.

Section 2.4.2 on relative efficiency illustrates the use of the two data sheet graphs, Relative Luminous Intensity vs. Forward Current and Relative Efficiency vs. Peak Current, in determining time average intensity.

2.4 Visual Applications of LED Lamps

2.4.1 Introduction

The largest usage of LED lamps is in visual applications. LED lamps have long been used as panel mounted indicators, printed circuit status indicators, and both x-y addressable and linear arrays. With the introduction of the High Efficiency LED lamps, applications that previously could only use neon and incandescent lamps, now in many instances, can also use LED lamps. These high intensity applications include backlighting a legend or illuminating a push button. LED lamps offer many advantages to the designer. They are small, light weight, and mechanically rugged. Since they operate at low voltages and currents, they can interface directly to most digital logic families. Because LEDs are solid state devices, they have a projected operating life of over 100,000 hours. These features benefit the end user by substantially reducing field maintenance costs due to lamp replacement.

Figure 2.4.1-1 shows some of the traditional uses for LED lamps, such as panel indicators and printed circuit status indicators. The designer has the flexibility of soldering the lamps directly into a printed circuit board and positioning the board behind the front panel as shown, or by using a clip and ring to attach the lamp to the front panel and soldering or wirewrapping leads to the LED lamp. When several lamps are driven by a common LED driver, they are commonly called an array. While it is not important that the lamps be positioned together, special purpose displays can be formed by specific patterns of lighted lamps in a specific grouping. An x-y addressable array is a group of LED lamps that are connected so that one particular lamp is illuminated through the application of proper signals to an x and a y coordinate. Numeric and alphanumeric information can be displayed with 35 LED lamps arranged in a 5x7 matrix. Fewer LEDs are required if only numeric information is to be displayed (an example of a modified 4x7 matrix to display hexidecimal information is illustrated in Figure 5.1.1.1-1). Analog information can be displayed

with a linear display. Two types of linear displays are the bar graph display and the position indicator display In the bar graph display, all LED lamps that represent a value smaller than the input quantity are turned on. Only the single LED lamp that represents a value closest to the input quantity is turned on in the position indicator display. Another application of LED lamps is to highlight or backlight a printed legend. Traditionally, LED lamps have been used to attract the attention of a viewer to a message printed near the LED lamp. The introduction of the High Efficiency LED lamps, now allows LEDs to backlight a legend printed on translucent or diffused film.

Electrically, LED lamps behave similarly to silicon or germanium diodes. LED lamps emit light only when they are forward biased. Because the LED lamp has a very small dynamic resistance above the turn-on voltage of the device, LEDs are normally driven by a current source. For most applications, a battery in series with a resistor can be considered as a current source if $V_{BAT} \gg V_F$ and $R \gg R_S$. Figure 2.4.1-2 shows how the current flowing through an LED lamp can be solved graphically by superimposing a load line over the forward characteristics of the LED device. The forward current through the LED should be

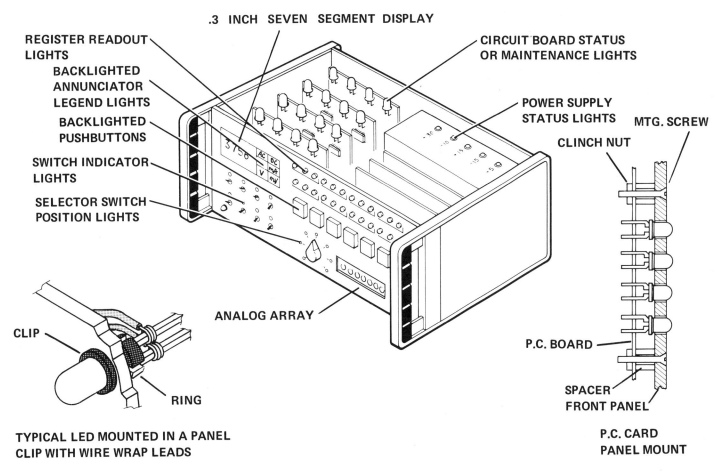

Figure 2.4.1-1   Uses of LED Indicator Lamps in a Complex Instrument.

2.15

selected to give the desired time averaged luminous intensity for worst cased values of power supply voltages and circuit tolerances. Maximum forward current is also constrained by the maximum allowable average power dissipation based on ambient temperature and the maximum tolerable peak current for a specified pulse duration and repetition rate. Section 2.4.3 outlines some of the techniques to operate an LED lamp in dc or pulsed mode and shows several worst cased circuit designs using these techniques.

Since an LED lamp is an optoelectronic device, the luminous intensity, radiation pattern, and visual appearance of the lamp should dictate the choice of LED packages and whether the lamp should be diffused, non diffused, tinted or non tinted. The luminous intensity of the lamp should be large enough to achieve a desired contrast ratio between the lamp and the background around the lamp. A lamp mounted on a reflective surface will require a higher luminous intensity than the same lamp mounted on a dull, non reflective surface. If the ambient luminous incidence is increased, the luminous intensity of the lamp will have to increase proportionately. Once the desired luminous intensity is known, the required drive current to obtain that luminous intensity can be calculated using either the figures of "Relative Luminous Intensity vs. I_F" or "Relative

Efficiency vs. I_{PEAK}" that are shown in the data sheet for that particular device. Section 2.4.2 describes these calculations in more detail.

2.4.2 Relative Efficiency

Traditionally, LED lamps have been characterized for light output (I_V) vs. dc forward current (I_F). An example of this characterization is shown in Figure 2.4.2-1. The abcissa is dc current in milliamperes and the ordinate is luminous intensity in millicandelas or luminous intensity normalized to one at a particular value of input current. The use of such a figure is relatively simple if one variable is known. For example, if an LED lamp emits 4.0 mcd at 10 mA I_F, then typically it will emit 9.9 mcd at 20 mA I_F.

Two other representations of Figure 2.4.2-1 are useful in strobed applications. An LED lamp will emit a certain number of photons per milliamp of input current. This is known as the efficiency of the LED. This efficiency varies according to the peak current through the LED. The relative efficiency curve such as Figure 2.4.2-2 is used to show this relationship. The abcissa is peak current in milliamps and the ordinate is millicandelas per milliamp normalized to one at a particular value of input current.

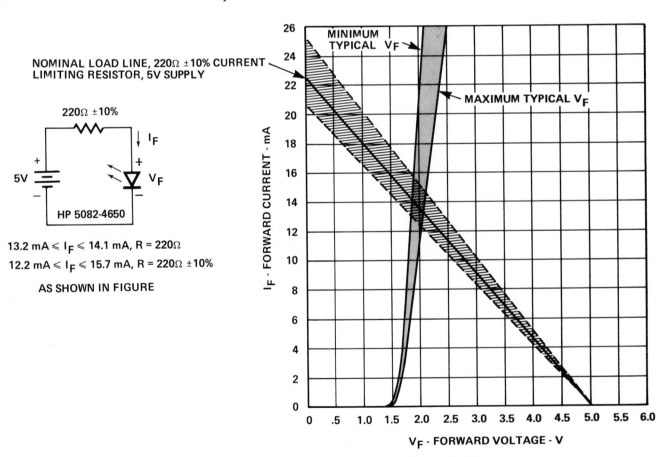

Figure 2.4.1-2   Graphical Solution of the Current Flowing through
a High Efficiency Red LED.

Figure 2.4.2-2 can be used to calculate the time averaged luminous intensity for any particular peak current and duty cycle with equation 2.4.2-1.

$$(2.4.2\text{-}1)$$

$$I_{V\,TIME\,AVG} = \frac{[I_{PEAK}]\,[DUTY\,CYCLE]\,[\eta(I_{PEAK})]\,[I_{V\,SPEC}]}{[I_{SPEC}]\,[\eta(I_{PEAK})]}$$

where I_{PEAK} is the desired peak current; DUTY CYCLE is the ratio of time the LED is "ON" to total time; η is the relative efficiency of the LED at I_{PEAK} or at I_{SPEC}; I_{SPEC} is the dc current at which $I_{V\,SPEC}$ is tested; and $I_{V\,SPEC}$ is the luminous intensity of the LED according to the test conditions.

For example, using Figure 2.5.2-2, if the desired I_{PEAK} is 60 mA, the desired duty cycle is 1/8, and the LED has a luminous intensity of 4.0 mcd at 10 mA dc, then the time averaged I_V can be calculated as follows:

$$I_{V\,TIME\,AVG}\,(mcd)$$

$$= \frac{(60\,mA)\,(.125)\,(1.54)\,(4.0\,mcd)}{(10\,mA)\,(1.00)}$$

$$= 4.6\,mcd\,at\,60\,mA\,I_{PEAK},\,1/8\,DUTY\,CYCLE$$

One final representation of Figure 2.4.2-1 is a time averaged luminous intensity curve. An example of this representation is shown in Figure 2.4.2-3. The abcissa is average current in milliamps and the ordinate is time averaged luminous intensity. Time averaged luminous intensity is then plotted for different values of peak current or duty factor. With this representation, time averaged luminous intensity can be read directly from the curve. For example, if the desired I_{PEAK} is 60 mA, the desired duty cycle is 1/8 and the LED has a luminous intensity of 4.0 mcd at 10 mA dc, then the time averaged I_V can be read directly off Figure 2.4.2-3 as 4.6 mcd (I_{AVE} = 7.5 mA). For lamps with a different luminous intensity than 4.0 mcd at 10 mA dc, the final result will need to be linearly scaled to reflect the difference. Figure 2.4.2-3 shows the advantages of strobing the LED as a means for achieving a higher luminous intensity at the same average current or by reducing the required average current and still maintaining the desired luminous intensity. For example, a typical device will emit 1.55 mcd at 5 mA as shown by Figure 2.4.2-3. The same device will emit 3.1 mcd at 60 mA peak, 5 mA average current or 1.55 mcd at 60 mA peak, 2.5 mA average current.

Ignoring the effect of junction heating due to the average power dissipation in the LED (the temperature coefficient of I_V is about -1%/°C), Figures 2.4.2-1, 2.4.2-2, and 2.4.2-3 are equivalent. The designer can derive one figure from either of the other two if he prefers. The relative efficiency

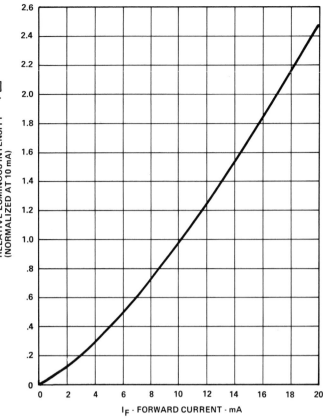

Figure 2.4.2-1   Relative Luminous Intensity vs. Current for High Efficiency Red LED.

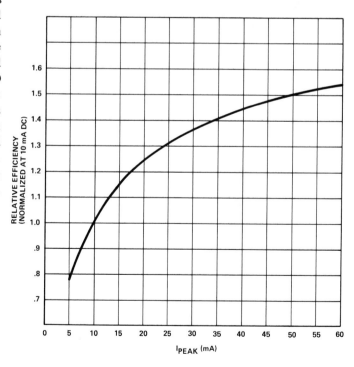

Figure 2.4.2-2   Relative Efficiency vs. Peak Current for High Efficiency Red LED.

curve is normally given on all data sheets. As a final example assume a lamp is strobed at 20 mA peak current at a 10% duty cycle and the lamp has a luminous intensity of 3.0 mcd at 10 mA dc:

Figure 2.4.2-1: $I_V = (2.48)(3.0)(.10) = .74$ mcd

Figure 2.4.2-2: $I_V = \dfrac{(20)(.10)(1.24)(3.0)}{(10)(1.00)(1.00)} = .74$ mcd

Figure 2.4.2-3: $I_V = \dfrac{(1.0)(3.0)}{4.0} = .75$ mcd

2.4.3   Driving an LED Lamp

2.4.3.1 LED Electrical Characteristics

Figure 2.4.3.1-1 shows typical electrical characteristics for standard red, high efficiency red, yellow and green lamps. Above 1.5 volts V_F, the current flowing through an LED increases very rapidly. The dynamic resistance can be considered to be the slope of the diode characteristic $(\Delta V_F/\Delta I_F)$ in the forward region. The standard red lamp has a very low dynamic resistance, while the high efficiency red, yellow and green lamps have a somewhat higher dynamic resistance. Since the dynamic resistance is so small, LED lamps should not be connected in parallel. Small variations in V_F or dynamic resistance can cause current hogging by the LED with the lowest V_F. This current hogging can cause variations in luminous intensity and excessive power dissipation in the lamp. However, LED

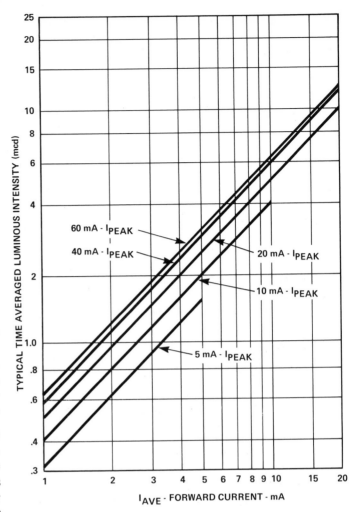

Figure 2.4.2-3   Typical Time Averaged Luminous Intensity vs. Average Current for a High Efficiency Red LED.

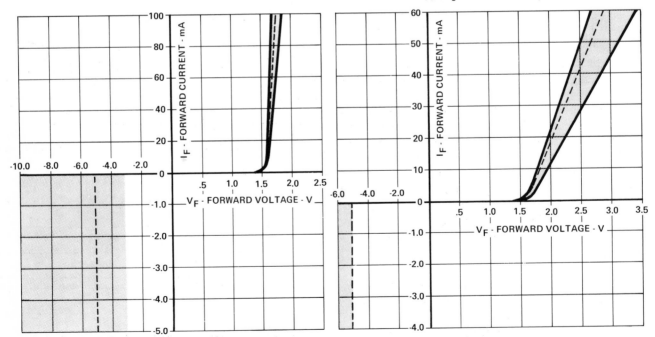

STANDARD RED LAMP (GaAsP SUBSTRATE)   HIGH EFFICIENCY RED, YELLOW, GREEN LAMP (GaP SUBSTRATE)

Figure 2.4.3.1-1 Typical Electrical Characteristics of LED Lamps.

2.18

lamps can be connected in series as long as the combined V_F doesn't exceed the power supply potential.

Negligable current flows through an LED in the reverse direction until the breakdown voltage is exceeded. Above the breakdown voltage, BV_R, the reverse current increases very rapidly, such as shown in Figure 2.4.3.1-1. Exceeding the BV_R will not harm the LED lamp as long as the reverse current is externally limited to prevent excessive power dissipation in the LED. When several LED lamps are connected in an array, reverse leakage current can cause unwanted ghosting of normally off LED's. This can be prevented by using LED drivers with a high impedance off state.

2.4.3.2 Resistive Current Limiting

When LED lamps are driven from a regulated power supply, a resistor can be used to limit the current flowing through the LED. The LED current, I_F, is determined by the following equation:

$$I_F = \frac{V_{CC} - V_F - V_{CE\ SAT}}{R} \qquad (2.4.3.2\text{-}1)$$

where V_{CC} is the power supply potential, $V_{CE\ SAT}$ is the "on" voltage of the LED driver, I_F and V_F are the forward characteristics of the LED, and R is the current limiting resistor.

If V_{CC} is considerably larger than $V_F + V_{CE\ SAT}$, then small variations of V_F or $V_{CE\ SAT}$ will have only negligable effects on I_F. However, variations of V_{CC} or R will cause a corresponding variation of I_F. For example, suppose that a standard red LED lamp is to be driven from a 5.0V supply by a transistor switch with a $V_{CE\ SAT}$ of .2V at 18 mA I_F. Then the nominal value of resistance is equal to:

$$R = \frac{5.0 - .2 - 1.65}{18} = 175\Omega$$

Figure 2.4.3.2-1 shows both numerical and graphical solutions to this example assuming tolerances of V_{CC}, R, $V_{CE\ SAT}$, and V_F.

Resistor-LED lamps are also available that are designed to operate off of a 5 volt supply.

2.4.3.3 Constant Current Limiting

In some applications, it may be desirable to drive LED lamps with a current source. The current source can be used to regulate the current through the LED regardless of power supply variations or variations in V_F between LED lamps. Figure 2.4.3.3-1 shows some examples of simple current sources constructed of npn transistors. For both circuits, the current through the LED string remains constant as long as V_{CC} is greater than $V_{CC\ (min)}$. The

NUMERICAL SOLUTION:

$$I_{F(max)} = \frac{V_{CC(max)} - V_{CE\ SAT\ (min)} - V_{F(min)}}{R(min)}$$

$$I_{F(min)} = \frac{V_{CC(min)} - V_{CE\ SAT(max)} - V_{F(max)}}{R(max)}$$

EXAMPLE:

ASSUME $V_{F(min)}$ = 1.62 @ 20 mA
$\qquad\quad V_{F(max)}$ = 1.67 @ 20 mA
$\qquad\quad V_{CC}$ = 5.0V ±10%
$\qquad\quad$ R = 180Ω ±10%
$\qquad\quad V_{CE\ SAT}$ = .2V @ 20 mA

1. V_{CC} = 4.5V, $I_{F(max)}$ = $\dfrac{4.50 - .20 - 1.62}{(180)(.9)}$ = 16.5 mA

$\qquad\quad I_{F(min)}$ = $\dfrac{4.50 - .20 - 1.67}{(180)(1.1)}$ = 13.3 mA

2. V_{CC} = 5.0V ∴ 15.8 mA ≤ I_F ≤ 19.6 mA

3. V_{CC} = 5.5V ∴ 18.3 mA ≤ I_F ≤ 22.7 mA

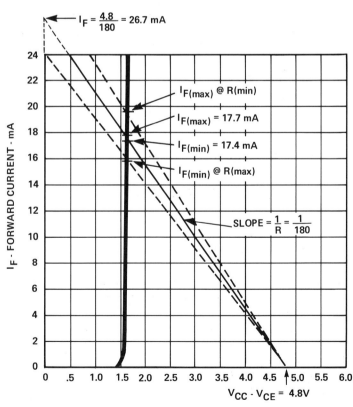

Figure 2.4.3.2-1 Numerical and Graphical Solution to Resistive Current Limiting.

first circuit uses an LED lamp as a voltage reference. Since $\Delta V_{F(T)} \cong \Delta V_{BE(T)} \cong$ -2 mV/°C, the current source will remain stable over temperature. In the second circuit, since V_{BE} varies with temperature, the current source will also vary with temperature. However, this change is typically about:

$$\frac{\Delta V_{BE}/\Delta T}{V_{BE}} \cong \frac{-2}{650} \cong -.3\%/°C$$

Commercially available current regulator IC's can also used.

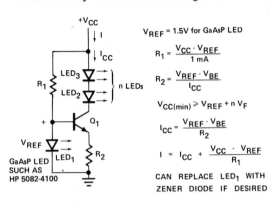

V_{REF} = 1.5V for GaAsP LED

$$R_1 = \frac{V_{CC} - V_{REF}}{1\ mA}$$

$$R_2 = \frac{V_{REF} - V_{BE}}{I_{CC}}$$

$$V_{CC(min)} \geqslant V_{REF} + n\ V_F$$

$$I_{CC} = \frac{V_{REF} - V_{BE}}{R_2}$$

$$I = I_{CC} + \frac{V_{CC} - V_{REF}}{R_1}$$

CAN REPLACE LED$_1$ WITH ZENER DIODE IF DESIRED

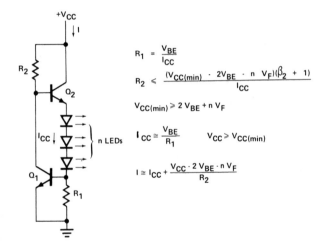

$$R_1 = \frac{V_{BE}}{I_{CC}}$$

$$R_2 \leqslant \frac{(V_{CC(min)} - 2V_{BE} - n\ V_F)(\beta_2 + 1)}{I_{CC}}$$

$$V_{CC(min)} \geqslant 2\ V_{BE} + n\ V_F$$

$$I_{CC} \cong \frac{V_{BE}}{R_1} \qquad V_{CC} \geqslant V_{CC(min)}$$

$$I \cong I_{CC} + \frac{V_{CC} - 2\ V_{BE} - n\ V_F}{R_2}$$

Figure 2.4.3.3-1 Some Examples of Constant Current LED Drivers that Regulate I_{CC} Regardless of V_{CC}.

2.4.3.4 LED-Logic Interface

Since LED lamps operate at low voltages and currents, they can interface to most digital logic families directly. Figure 2.4.3.4-1 shows some of the ways that an LED lamp can be used to interface to digital logic. TTL logic families are guaranteed to sink a minimum amount of current (I_{OL}) which can drive most LED lamps without additional buffering. Table 2.4.3.4-1 lists the guaranteed I_{OL} of most common TTL families:

74S: $I_{OL} \leqslant$ 20 mA		74LS: $I_{OL} \leqslant$ 8 mA	
74H: $I_{OL} \leqslant$ 20 mA		74L: $I_{OL} \leqslant$ 3.6 mA	
74: $I_{OL} \leqslant$ 16 mA			

TABLE 2.4.3.4-1 TTL Interface

If additional current is required, such as in a strobed application, TTL buffers are also available. CMOS buffers can also drive many LED lamps directly. Table 2.4.3.4-2 lists some of the commonly available CMOS buffers:

RCA 4049, 4050 4009, 4010	$I_{OL} \geqslant$ 3mA, V_{OL} = .4V, V_{CC} = 5V $I_{OL} \geqslant$ 8mA, V_{OL} = .5V, V_{CC} = 10V
National 74C906	$I_{OL} \sim$ 8mA, V_{OL} = .5V, V_{CC} = 4.75V $I_{OL} \sim$ 20mA, V_{OL} = .5V, V_{CC} = 10V
National 74C901 74C902	$I_{OL} \geqslant$ 3.8mA, V_{OL} = .4V,

TABLE 2.4.3.4-2 CMOS Interface

OPEN COLLECTOR GATES

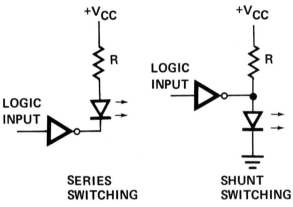

SERIES SWITCHING   SHUNT SWITCHING

ACTIVE PULLUP - TOTEM POLE GATES

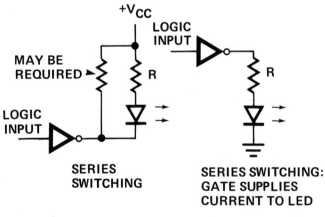

SERIES SWITCHING

SERIES SWITCHING: GATE SUPPLIES CURRENT TO LED

Figure 2.4.3.4-1 Digital Logic Can Interface Directly to LED Lamps.

Low cost transistors can be used when larger currents are required than a logic gate can supply directly. Figure 2.4.4.2-3 shows some of the commonly used LED drive schemes implemented with NPN, PNP, and FET transistors.

2.4.3.5 Worst Case Design

Regardless of whether the LED lamp design uses resistive or constant current limiting, is DC driven or strobed, the designer should consider worst case design techniques. The purpose of worst case design is two-fold. First, the design should guarantee that the LED lamps will operate within their recommended operating conditions. This will insure a long operating lifetime for the LED devices -- probably longer than the expected lifetime of the instrument. Secondly, worst case design can determine whether power supply, resistor and device tolerances will cause a noticeable variation in luminous intensity from lamp to lamp. In many applications, this requirement for luminous intensity matching is not important, but when several LED lamps are used in a closely packed array, wide variations in luminous intensity can be very objectionable to the viewer.

When a lamp is dc driven, a worst case design analysis can determine whether the lamp operates within its recommended operating conditions. The maximum operating conditions for an LED lamp under dc operation are determined by two factors. Normally each LED device is derated over temperature to keep the junction temperature below a specified maximum temperature. Typically, this maximum junction temperature is 110°C for a plastic encapsulated lamp. Secondly, the current density through the LED junction should be limited to prevent rapid degradation of light output. The maximum peak current for a standard red lamp is typically 1000 mA and 60 mA for high efficiency red, yellow, and green lamps.

Depending on the maximum expected ambient temperature for the particular application, the designer can determine the maximum average power dissipation, P_{AVG}, from the power derating curve for the device. An example of a power derating curve is shown in Figure 2.3.2-1. The average power dissipation in a lamp is the product of the average forward current times the peak forward voltage. Equation 2.3.2-1 can be used to calculate the average power dissipation in an LED lamp:

$$P_{AVG(W)} = I_{AVG} [V_F + R_S (I_{PEAK} - I_F)] \qquad (2.3.2-1)$$

where I_{AVG} is the average forward current, I_{PEAK} is the peak forward current, I_F and V_F are the LED lamp test conditions, and R_S is the LED dynamic resistance.

In a dc application, I_{AVG} is equal to I_{PEAK}. Equation 2.3.2-1 and Figure 2.3.2-1 can be used to calculate the

maximum dc current through the LED lamp as shown below:

$$(2.4.3.5\text{-}1)$$

$$P_{AVG(mW)} = I_{AVG(mA)} [V_F + R_s (I_{AVG(mA)} - I_F)/1000]$$

For a standard red lamp, $V_{F (max)} = 2.0$V @ 20 mA I_F, $R_{S (max)} = 5\Omega$

$$\therefore P_{AVG} = I_{AVG} [2.0 + 5 (I_{AVG} - 20)/1000] \qquad (2.4.3.5\text{-}2)$$

$$\therefore I_{AVG(mA)} = \frac{\sqrt{(1.9)^2 + (.02) P_{AVG(mW)}} - 1.9}{.01}$$

So at 50°C since $P_{AVG} = 120$ mW, then $I_{AVG} = 55$ mA and at 70°C where $P_{AVG} = 87$ mW, then $I_{AVG} = 41$ mA. Since the maximum average current is specified in the data sheet as 50 mA, then I_{AVG} should be restricted to less than 50 mA regardless of the results of equation 2.4.3.5-3. For additional reliability, the designer can operate at currents less than $I_{AVG (max)}$.

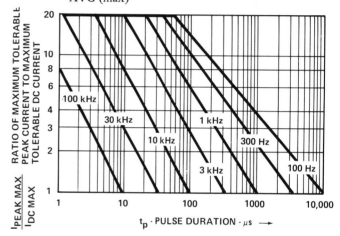

Figure 2.4.3.5-1 Maximum Tolerable Peak Current vs. Pulse Duration for a T-1 3/4 Red Lamp ($I_{DC\ MAX}$ per Figure 2.3.2-1 and Eqn. 2.4.3.5-3).

For strobed applications, a curve of maximum peak current, pulse width, and repetition rate can be used to determine the maximum recommended operating conditions for an LED lamp. The curve is determined by comparing the peak junction temperature of a lamp under strobed conditions to the average junction temperature under maximum allowable dc conditions. At any operating point, the peak junction temperature should not exceed the average junction temperature under maximum allowable dc conditions. An example of such a curve is shown in Figure 2.4.3.5-1. $I_{DC\ MAX}$ is the maximum average current calculated by equation 2.4.3.5-1 or as restricted by the data sheet. The ordinate of Figure 2.4.3.5-1 is the ratio of maximum allowable peak current to maximum allowable dc current. At any specified repetition rate, the relationship between maximum peak current and pulse width is shown. For reliable operation, the device should be operated at or

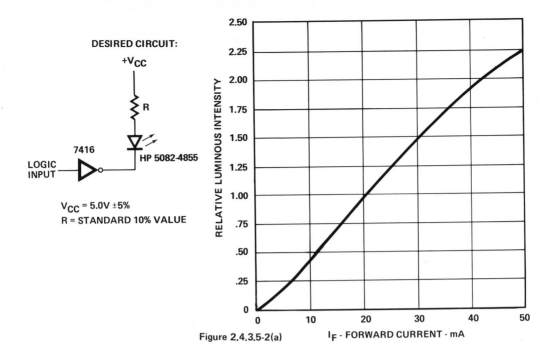

DESIRED CIRCUIT:

$+V_{CC}$

R

7416

LOGIC
INPUT

HP 5082-4855

V_{CC} = 5.0V ±5%
R = STANDARD 10% VALUE

Figure 2.4.3.5-2(a) I_F - FORWARD CURRENT - mA

1. SELECT LAMP: STANDARD RED T-1 3/4 LAMP, 1.4 mcd at 20 mA DC

2. DETERMINE NOMINAL I_F: $\dfrac{1.5 \text{ mcd}}{1.4 \text{ mcd}} = 1.07$ $\therefore I_F$ = 22 mA from Figure
    (see Figure 2.4.3.5-2(a))

3. CALCULATE MAXIMUM POWER DISSIPATION: P_{AVG} = 100 mW - 1.6 mW (70-50) = 68 mW
    (see DATA SHEET)

4. CALCULATE MAXIMUM AVERAGE CURRENT: $P_{AVG} = I_{AVG} [V_F + R_S (I_{AVG} - I_F)]$
    (see Equation 2.4.3.5-1)

    $V_F \leqslant$ 2.0V @20 mA, $R_S \leqslant 5\Omega$ $\therefore I_{AVG} = \dfrac{\left(\sqrt{(1.9)^2 + (.02) P_{AVG}} \right) 1.9}{.01} = 33$ mA

5. CALCULATE NOMINAL VALUE OF R: $R = \dfrac{V_{CC} - V_F - V_{CE\ SAT}}{I_F}$ V_{CC} = 5.0
    (see Equation 2.4.3.2-1) V_F = 1.6
                                                                           $V_{CE\ SAT}$ = .3V
    $\therefore R = \dfrac{5.0 - 1.6 - .3}{22} = 141\Omega$ $\boxed{R = 150\Omega \pm 10\%}$ I_F = 22 mA

6. CALCULATE WORST CASE I_F: $I_{F(max)} = \dfrac{5.25 - .20 - 1.61}{(150)(.9)} = 25.5$ mA $V_{F(min)}$ = 1.61V
    (see Equation 2.4.3.5-4)

    $\boxed{\therefore \text{LAMP OPERATES WELL WITHIN ABSOLUTE MAXIMUM RATINGS}}$

7. CALCULATION OF LUMINOUS INTENSITY RATIO IF DESIRED:
    (see Equation 2.4.3.5-4)
    (see Figure 2.4.3.5-2(a))

    $I_{F(max)} = \dfrac{4.75 - .20 - 1.61}{(150)(.9)} = 21.8$ mA $V_{F(min)}$ = 1.61V

    $\therefore$ RELATIVE I_V @ 21.8 mA = 1.07

    (see Equation 2.4.3.5-5)
    (see Figure 2.4.3.5-2(a))

    $I_{F(min)} = \dfrac{4.75 - .40 - 1.66}{(150)(1.1)} = 16.3$ mA $V_{F(max)}$ = 1.66V

    $\therefore$ RELATIVE I_V @ 16.3 mA = .75 $\therefore I_V$ RATIO = $\dfrac{1.07}{.75}$ = 1.43:1.00

Figure 2.4.3.5-2 Example of a Worst Case Design for a Dc Driven
LED Lamp.

2.22

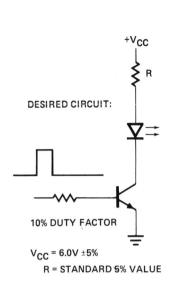

DESIRED CIRCUIT:

10% DUTY FACTOR

$V_{CC} = 6.0V \pm 5\%$

R = STANDARD 5% VALUE

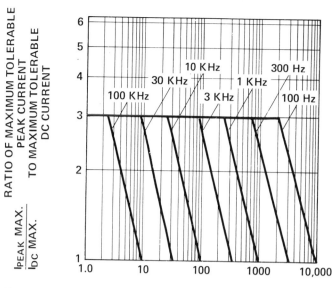

Figure 2.4.3.5-3a t_P – PULSE DURATION – μs

1. SELECT LAMP: HIGH EFFICIENCY RED LAMP - 2.0 mcd at 10 mA DC

2. DETERMINE I_{PEAK}: $\quad I_{PEAK} = \dfrac{(1.5 \text{ mcd})(10 \text{ mA})(1.00)}{(2.0 \text{ mcd})(.10)(\eta @ I_{PEAK})} \quad \therefore I_{PEAK} = 50 \text{ mA} \therefore I_{AVE} = 5 \text{ mA}$
   (see Figure 2.4.2-2)

3. CALCULATE MAXIMUM POWER DISSIPATION: $\quad P_{AVG} = 120 \text{ mW} - 1.6 \text{ mW} (70\text{-}50) = 88 \text{ mW}$
   (see DATA SHEET)

4. CALCULATE MAXIMUM AVERAGE CURRENT: $\quad P_{AVG} = I_{AVG} [V_F + R_S (I_{AVG} - I_F)]$
   (see Equation 2.4.3.5-1)

   $V_F \leqslant 3.0 @ 10 \text{ mA}, R_S \leqslant 35\Omega \qquad \therefore I_{AVG} = \dfrac{\left(\sqrt{(2.65)^2 + (.14)\, P_{AVG}}\right) - 2.65}{.07}$

   $= 25 \text{ mA, LIMITED TO 20 mA ON DATA SHEET}$

5. DETERMINE MAXIMUM ALLOWABLE PEAK CURRENT, PULSEWIDTH, REPETITION RATE

   (see Figure 2.4.3.5-3a)   >100 Hz, 10% DUTY CYCLE $I_{PEAK} \leqslant 3(20) = 60 \text{ mA}$

6. CALCULATE NOMINAL VALUE OF R: $\quad R = \dfrac{V_{CC} - V_F - V_{CE\,SAT}}{I_F}$
   (see Equation 2.4.3.2-1)

   $V_{CC} = 6.0 \qquad V_F = 2.7 \quad V_{CE\,SAT} = .3 \quad I_F = 50 \text{ mA}$

   $R = \dfrac{6.0 - 2.7 - .3}{50} = 60\Omega \text{ SELECT } \boxed{R = 62\Omega \pm 5\%}$

7. CALCULATE WORST CASE I_F: $\quad I_{F(max)} = \dfrac{6.3 - 2.5 - .2}{(62)(.95)} = 61 \text{ mA}$
   (see Equation 2.4.3.5-4)

   $\boxed{\therefore \text{ LAMP OPERATES SLIGHTLY ABOVE ABSOLUTE MAXIMUM RATINGS. DESIGNER CAN CHOOSE LARGER VALUE OF RESISTOR; TIGHTER TOLERANCES ON COMPONENTS.}}$

   DESIGNER CHOOSES 68Ω ±5% RESISTOR

8. CALCULATION OF LUMINOUS INTENSITY RATIO IF DESIRED:
   (see Equation 2.4.3.5-4)
   (see Figure 2.4.2-2) $\qquad I_{F(max)} = \dfrac{5.7 - 2.5 - .2}{(68)(.95)} = 46.4 \text{ mA}$

   $\therefore I_{V(max)} = \dfrac{(46.4)(.1)(1.48)(2.0)}{(10)(1.00)} = 1.37 \text{ mcd}$

   (see Equation 2.4.3.5-5) $\quad I_{F(min)} = \dfrac{5.7 - 3.0 - .4}{(68)(1.05)} = 32.2 \text{ mA} \quad \therefore I_{V(min)} = \dfrac{(32.2)(.1)(1.38)(2.0)}{(10)(1.0)}$
   (see Figure 2.4.2-2)

   $= .89 \text{ mcd}$

   $\therefore I_V \text{ RATIO} = \dfrac{1.37}{.89} = 154{:}1.00$

Figure 2.4.3.5-3   Example of a Worst Case Design for a Strobed LED Lamp.

below these conditions. For example, suppose the design requires a standard red lamp to be driven at 200 mA peak on a 10% duty cycle at 25°C. At 100 Hz repetition rate, the maximum allowable peak current for a 1000 μS pulse width is 3.7 x 50 = 185 mA. However at a 300 Hz repetition rate, the maximum allowable peak current is 4.6 x 50 = 230 mA on a 10% duty cycle. Thus, by operating the device at 200 mA peak current, 10% duty factor, at a 300 Hz repetition rate, the maximum junction temperature will not exceed the junction temperature obtained by operating the device at 50 mA dc. For additional reliability, the designer can derate $I_{DC\ MAX}$ below the maximum temperature derated $I_{AVG(MAX)}$ as calculated by equation 2.4.3.5-1 or as restricted by the data sheet.

When LED lamps are to be grouped together in an array, the designer should also consider the requirement for luminous intensity matching of adjacent LEDs. Variations in power supply voltages, resistor tolerances, driver tolerances, and device tolerances all contribute to variations in I_F and hence luminous intensity variations. The usual procedure to determine whether variations in I_F will cause noticeable luminous intensity variations is to calculate the worst case minimum and maximum values of I_F. Then the relative luminous intensity or relative luminous efficiency curves can be used to determine the worst case variation in luminous intensity. In general, the maximum luminous intensity ratio between LED lamps should be less than 2.0:1.0, and ratios greater than 2.3:1.0 will be objectionable to an observer. When LED lamps are driven by a common power supply, variations in the power supply voltage will cause only a small change in the maximum ratio of LED forward currents since the forward currents will all change proportionally due to power supply variations. However, the component tolerances will have the greatest effect on I_F at the minimum power supply voltage. When resistive current limiting is used, the minimum and maximum I_F can be calculated as shown below:

(2.4.3.5-4)

$$I_{F(MAX)} = \frac{V_{CC} - V_{CE\ SAT(MIN)} - V_{F(MIN)}}{R\ (MIN)}$$

(2.4.3.5-5)

$$I_{F(MIN)} = \frac{V_{CC} - V_{CE\ SAT(MAX)} - V_{F(MAX)}}{R\ (MAX)}$$

1.   Design Example

An array of standard red LED lamps is to be dc driven with a desired typical luminous intensity of 1.5 mcd at 25°C with a maximum ambient temperature of 70°C. The array will be driven by 7416 TTL hex inverters from the 5.0V TTL supply. The desired circuit and calculations are shown in Figure 2.4.3.5-2.

2.   Design Example

An array of high efficiency red LED lamps is to be strobed on a 10% duty factor with an npn transistor. The desired luminous intensity is 1.5 mcd at 25°C with a maximum ambient temperature of 70°C. The desired circuit and calculations are shown in Figure 2.4.3.5-3.

2.4.4   LED Arrays

2.4.4.1 Introduction

When several LED lamps are used in an application, the cost of the associated LED drive circuitry can often be reduced by connecting the LEDs in a multiplexed array. For example, suppose 16 LED lamps are used as status indicators on a panel. Each LED can be driven by an individual transistor or logic gate as shown in Section 2.4.3. This configuration requires 16 LED drivers, 16 current limiting resistors, and 17 address lines. The LED lamps can also be connected as a 4x4 multiplexed array. In this configuration, only 8 LED drivers, 4 current limiting resistors, and 8 address lines are required. In general, p·q LED lamps can be driven by p·q transistors or gates and p·q current limiting resistors on a DC basis. Each LED lamp can be considered to be an element, a_{ij}, where i = 1, 2, ... p and j = 1, 2, ... q, of an x-y addressable array of p·q individual LEDs, such as shown in Figure 2.4.4.1-1. If the LED lamps are connected as a multiplexed x-y array with p·q elements, then only p+q transistors or gates will be required to drive the p·q lamps and p or q current limiting resistors will be needed. In many applications, such as with a microprocessor, the information is available on a multiplexed basis and minimal logic is required for proper decoding. If dc signals are available, they can be multiplexed with inexpensive multiplexers or a bus configuration. One final advantage of the multiplexed array is that the number of wires required to connect the LED lamps to the drive circuitry is reduced. The dc driven array requires (p·q+1) wires to connect the lamps but the multiplexed x-y array requires only p+q wires.

Figures 2.4.4.1-2 and 2.4.4.1-3 show the two basic types of LED arrays. The array shown in Figure 2.4.4.1-2 is dc driven because only one LED lamp is turned on at any one time. By selecting one x axis and one y axis, the single LED specified by the coordinates (x_i, y_i) is turned on. Multiple LED lamps can be driven on a dc basis if they can all be addressed by one common x or y axis. Figure 2.4.4.1-3 shows the more general type of x-y addressable array where any combination of LED lamps can be turned on. One axis is sequentially selected on a 1/p or 1/q duty factor. While the proper axis is selected, any combination of address lines of the opposite axis can be enabled, turning on the corresponding LED lamps. If the display is refreshed at a

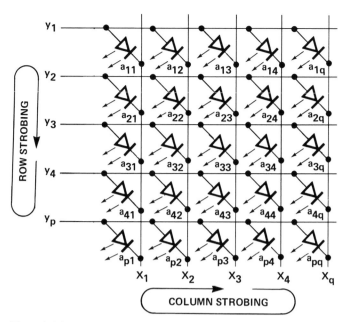

Figure 2.4.4.1-1 LED Lamps Can Be Connected as an X-Y Addressable Array of PQ Elements with P Rows and Q Columns.

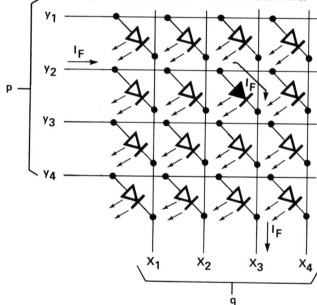

Figure 2.4.4.1-2 One LED Will Be Turned On By Applying the Proper Signal to One X Axis and One Y Axis.

fast enough refresh rate (>100Hz) then all LED lamps will appear to be dc driven. The present nomenclature used for x-y addressable arrays is that the y axis addresses (y_i; $i = 1, 2 \ldots p$) are called rows and the x axis addresses (x_j; $j = 1, 2 \ldots q$) are called columns. If the x axis addresses are sequentially selected, then the array is column strobed and if the y axis addresses are sequentially selected, then the array is row strobed.

Two basic applications exist for LED arrays. One application connects the LED lamps in an x-y addressable array as a means of simplifying the LED drive circuitry. However, each LED is individually mounted on a large surface. The second application of an x-y addressable array

is for character generation, such as an alphanumeric display, where the LEDs are mounted in close proximity to each other. In applications of this type, besides designing the drive circuitry to obtain the proper time averaged luminous intensity as described in Section 2.4.2, each LED lamp used in the array should be preselected to have less than a specified maximum luminous intensity ratio. Luminous intensity ratios of 2.3:1.00 between adjacent LEDs will be noticable by an observer. A luminous intensity ratio of 2.0:1.0 or less is recommended for this application.

2.4.4.2 Designing An X-Y Addressable LED Array

The first step in the design of an x-y addressable array is the selection of the LED lamp package that is to be used. The designer can choose between hermetic lamps, T-1 3/4 lamps, T-1 3/4 low dome lamps, T-1 lamps, subminiature lamps, or rectangular lamps. The designer can mix package styles or LED colors in the same array, although LEDs in a single row for a column strobed application or single column for a row strobed application should have similar electrical and luminous intensity characteristics. As a guide, standard red GaAsP substrate LED lamps; high-efficiency red or yellow GaP substrate LED lamps; and green GaP substrate LED lamps have substantial electrical or luminous intensity differences. To compensate for these differences, these three lamp categories should be driven in separate rows of a column strobed circuit (or separate columns of a row strobed circuit) so that a different current limiting resistor can be used for each lamp category. For character generation applications, such as an alphanumeric display, the lamp package should be chosen depending on the desired array size. Figure 2.4.4.2-1 shows the minimum LED lamp spacing for T-1 3/4, T-1, subminiature, and rectangular lamps.

Now the desired luminous intensity of the array should be specified. The desired luminous intensity sets a limit on the minimum duty factor that can be used to strobe the x-y array because of the peak current and pulse width limitations of the LED lamps. The minimum duty factor by which the x-y array can be driven is given by Equation 2.4.2-1, reproduced below:

(2.4.4.2-1)

$$\text{DUTY FACTOR} \geq \frac{[I_{V\ \text{TIME AVG}}]\,[I_{\text{SPEC}}]\,[\eta(I_{\text{SPEC}})]}{[I_{\text{PEAK}}]\,[\eta(I_{\text{PEAK}})]\,[I_{V\ \text{SPEC}}]}$$

where: all variables are defined in Section 2.4.2 and I_{PEAK} is the maximum tolerable peak current.

Table 2.4.4.2-1 shows four representative lamps and gives the maximum peak current and duty factor limitations to obtain a desired luminous intensity. The duty factors given

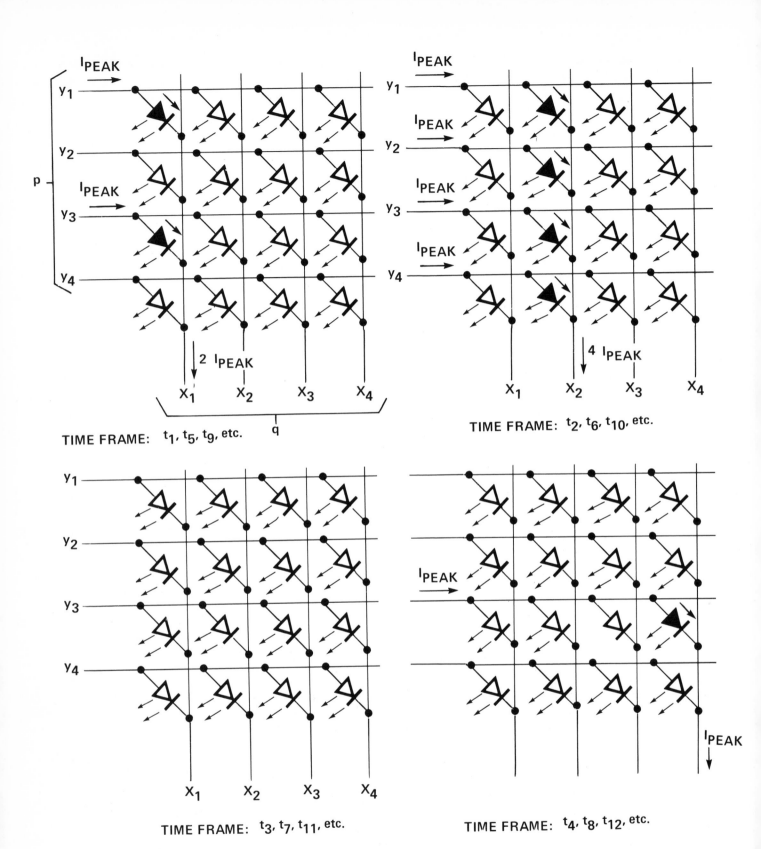

Figure 2.4.4.1-3 In a Multiplexed Array Any Combination of LEDs
Can Be Turned On by Sequentially Strobing One
Axis.

2.26

DUTY FACTOR	8-bit Latch	16-bit Latch	24-bit Latch	32-bit Latch
DC	8x 1 = 8 LEDs	16x 1 = 16 LEDs	24x 1 = 24 LEDs	32x 1 = 32 LEDs
1/2	7x 2 = 14 LEDs	15x 2 = 30 LEDs	23x 2 = 46 LEDs	31x 2 = 62 LEDs
1/4	6x 4 = 24 LEDs	14x 4 = 56 LEDs	22x 4 = 88 LEDs	30x 4 = 120 LEDs
1/8	5x 8 = 40 LEDs	13x 8 = 104 LEDs	21x 8 = 168 LEDs	29x 8 = 232 LEDs
1/16	4x16 = 64 LEDs	12x16 = 192 LEDs	20x16 = 320 LEDs	28x16 = 448 LEDs
1/32	3x32 = 96 LEDs	11x32 = 352 LEDs	19x32 = 608 LEDs	27x32 = 864 LEDs

TABLE 2.4.4.3-1 Maximum Size of X-Y Addressable Arrays that Can Be Driven by a Microprocessor.

Lamp Type	Typical Light Output	Maximum Peak Current	Minimum Duty Factor to Obtain Desired Time Averaged I_v			
			.5 mcd	1.0 mcd	1.5 mcd	2.0 mcd
GaAsP Red (655nm)	.8 mcd — 20 mA dc $\eta(1A)/\eta(20mA)=.8$	1A	1/64	1/32	------	------
GaP HER (635nm)	2.0 mcd — 10 mA dc $\eta(60mA)/\eta(10mA)=1.54$	60 mA	1/40	1/18	1/12	1/9
GaP Yellow (583nm)	1.8 mcd — 10 mA dc $\eta(60mA)/\eta(10mA)=1.54$	60 mA	1/32	1/16	1/11	1/8
GaP Green (565nm)	1.8 mcd — 20 mA dc $\eta(60mA)/\eta(20mA)=1.44$	60 mA	1/16	1/8	1/5	1/4

TABLE 2.4.4.2-1 Calculations of Minimum Duty Factor for Some Typical LED Lamps.

in the table can be decreased only by specifying brighter lamps or by exceeding the maximum tolerable peak current ratings of the LED lamps. When several different types of LED lamps are used in a single array, the lamp with the largest minimum duty factor determines the minimum duty factor for the entire array. The minimum duty factor as specified above sets an upper limit on one dimension of the x-y array. For example, suppose that an array is constructed of the standard red, high-efficiency red, yellow and green lamps specified in Table 2.4.4.2-1 and the desired time averaged luminous intensity of the display is 1.0 mcd for standard red lamp and 1.5 mcd for the other three lamps. The minimum duty factor of the array is 1/5, but a 1/4 duty factor is selected to simplify logic decoding. The other dimension of the x-y addressable array is determined by the total number of LED lamps in the array and the number of each type of LED lamp with different electrical or luminous intensity characteristics. Suppose that in the previous example, the array consists of 3 red lamps, 2 high-efficiency red lamps, 6 yellow lamps, and 1 green lamp. The x-y addressable array would be dimensioned as 4x4 even though only 12 lamps are used. Since the duty factor and desired time averaged luminous intensity for each lamp is known, the peak current can be determined for each type of LED lamp. Figure 2.4.4.2-2 shows the completely specified array described in the text.

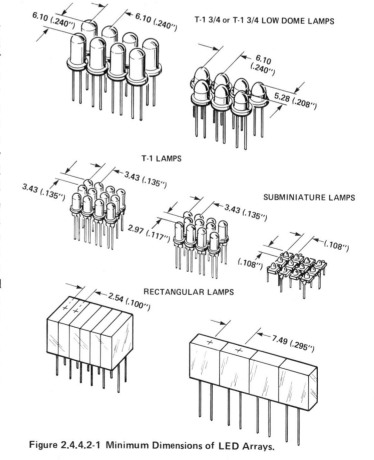

Figure 2.4.4.2-1 Minimum Dimensions of LED Arrays.

RED: I_{PEAK} = 97 mA
HIGH EFFICIENCY RED: I_{PEAK} = 24 mA
YELLOW: I_{PEAK} = 26 mA
GREEN: I_{PEAK} = 50 mA

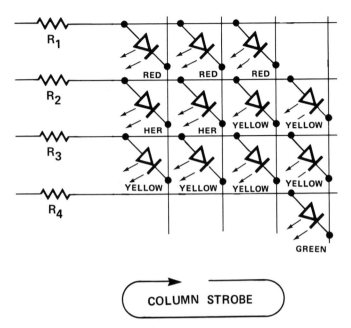

COLUMN STROBE

Figure 2.4.4.2-2 Example of X-Y Addressable Array Using 3 Red LEDs, 2 High Efficiency Red LEDs, 6 Yellow LEDs and 1 Green LED.

The final step in designing the LED x-y addressable array is the design and specification of the associated LED drive circuitry. The LED anode drivers must be able to source current to the LED array and the LED cathode drivers must be able to sink current from the array. The anode and cathode drivers can be implemented with NPN, PNP, FET transistors or commercially available LED drivers. Figure 2.4.4.2-3 shows some of the commonly used LED drive schemes implemented with NPN, PNP or FET transistors. The transistors should be selected so that the average power supply dissipation within the device or multiple transistor array is less than the manufacturers specifications. R_1 and R_2 are selected so that the transistor drivers remain in saturation under worst case conditions of V_{CC}, V_{OL}, V_{OH}, transistor H_{FE}, and resistor tolerances at the maximum peak currents specified for the LED array. Finally, the LED current limiting resistors are calculated as shown in Section 2.4.3. Figures 2.4.4.2-4 and 2.4.4.2-5 show examples of row and column strobed x-y addressable arrays using PNP anode drivers and NPN cathode drivers. The calculations of the maximum average and peak currents of each transistor assume that all LED lamps in the array are on. Active pullup drivers can also be used to drive the array provided that the B_{VR} restrictions of the LED lamps are not exceeded. If an LED lamp is connected to a low impedance driver with a potential greater than BV_R, current can flow

through the lamp in the reverse direction. This reverse leakage current will not harm the LED unless the maximum power dissipation of the lamp is exceeded, however, this reverse leakage current can flow through adjacent LED lamps in the forward direction and thus cause unwanted LED ghosting.

2.4.4.3 Design of a Microprocessor Controlled LED Array

A microprocessor can be used to control an x-y addressable array of LED lamps. The external circuitry that is required for the array is minimal and only a small portion of microprocessor time is used to refresh the LED array. The technique that is used is to periodically strobe data into a latch from the microprocessor and use the outputs of the latch to address the array by row or by column. A specified time later, new information for another row or column is strobed into the latch. If the x-y addressable array is row or coumn strobed at a refresh rate greater than 100 Hz, then the entire array will appear to be dc driven. Refresh timing for the array is handled by generating an interrupt after a specific time interval. The latch is partitioned so that n bits specify one of 2^n rows (row strobe) or 2^n columns (column strobe) and the remaining bits specify the coordinates of the LED lamps of the opposite axis which should be turned on. The total number of LED lamps that can be addressed is determined by the minimum duty factor of the array and the size of the latch. Table 2.4.4.3-1 shows the maximum number of LED lamps that can be driven with latches ranging from 8 to 32 bits and for duty factors ranging from 1/32 to dc. Figure 2.4.4.3-1 shows an example of a 5x8 LED array that is designed to interface to an Intel 8080A microprocessor. A low to high transition of output Q of the monostable multivibrator requests an interrupt from the microprocessor. The interrupt circuitry (not shown) forces a RST7 instruction into the microprocessor. Following the RST7 instruction, the microprocessor executes the program shown in Figure 2.4.4.3-2. The microprocessor then updates output port (n) which is the 8 bit latch used by the x-y addressable array. If the microprocessor uses a 2 MHz clock, then the percentage of total time required to refresh the LED array at a 100 Hz repetition rate is as follows:

(2.4.4.3-1)

$$\text{REFRESH TIME} = \frac{(143C - 1)\ R}{\text{MICROPROCESSOR CLOCK RATE}}$$
= 5.7% FOR C = 8, R = 100 Hz

where: C is the number of columns in the display and R is the refresh rate.

For the remaining 94% of the time, the microprocessor can be used to update the contents of the LED array in RAM and perform countless other tasks required by the system.

To address larger LED arrays, several 8 bit latches can be used. Each 8 bit latch is addressed by a separate output address. For N eight bit latches, each latch would be addressed as output (n+i), where i = 0, 1, 2, ... N-1. The program is modified by inserting INX HL, MOV A, M and OUT (n+i) instructions into the program as shown in Figure 2.4.4.3-2 and changing the CPI $(17)_{16}$ instruction to CPI $(OF+8N)_{16}$. The monostable should be triggered by the high to low transition of the strobe input of the final 8 bit latch. To prevent a small amount of ghosting while data is being strobed into the latches, the output of the monostable can be used to disable the 1 of N decoder and thus turn all columns off from the time that the interrupt is requested until all the data has been strobed into the latches. The extra time required to address several 8 bit

latches as compared to a single 8 bit latch is negligible. For example, suppose four 8 bit latches are used to address a 29x8 LED array. If the microprocessor uses a 2 MHz clock, the total percentage of time required to refresh the LED array at a 100 Hz repetition rate will increase to:

(2.4.4.3-2)

$$\text{REFRESH TIME} = \frac{[C(121 + 22N) - 1]\,R}{\text{MICROPROCESSOR CLOCK RATE}}$$
$$= 8.4\% \text{ FOR } C = 8, N = 4, R = 100\,\text{Hz}$$

where: C is the number of columns in the display, N is the number of eight bit latches, and R is the refresh rate.

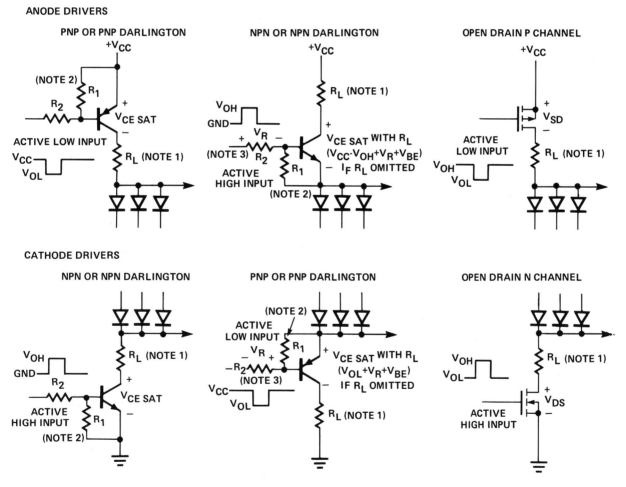

NOTES: 1. R_L OMITTED ON COLUMN DRIVERS WITH COLUMN STROBED CIRCUIT OR ON ROW DRIVERS WITH ROW STROBED CIRCUIT.
2. R_1 CAN BE OMITTED IF TRANSISTOR LEAKAGE IS SMALL.
3. R_2 CAN BE OMITTED IF R_L OMITTED AND TRANSISTOR REMAINS ACTIVE.

Figure 2.4.4.2-3 Common Transistor Drive Schemes.

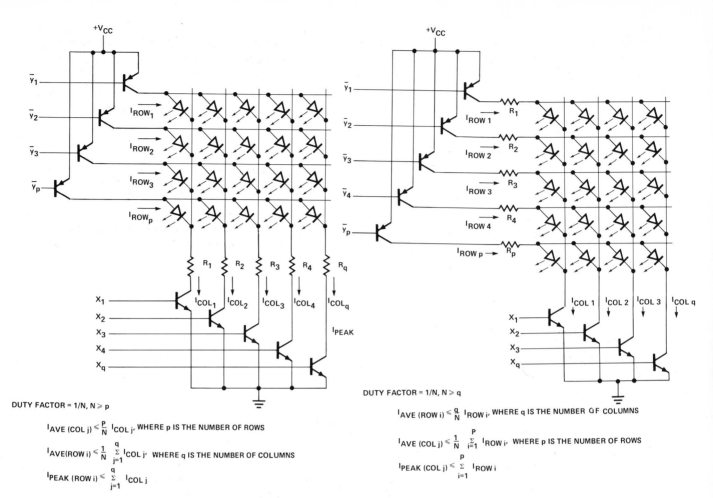

$$I_{AVE (COL\,j)} \leq \frac{p}{N}\, I_{COL\,j}, \text{ WHERE p IS THE NUMBER OF ROWS}$$

$$I_{AVE(ROW\,i)} \leq \frac{1}{N} \sum_{j=1}^{q} I_{COL\,j}, \text{ WHERE q IS THE NUMBER OF COLUMNS}$$

$$I_{PEAK\,(ROW\,i)} \leq \sum_{j=1}^{q} I_{COL\,j}$$

Figure 2.4.4.2-4 Row Strobed X-Y Addressable Array. Y_1 through Y_p are Sequentially Strobed on a 1/N Duty Cycle, N ⩾ P.

$$I_{AVE\,(ROW\,i)} \leq \frac{q}{N}\, I_{ROW\,i}, \text{ WHERE q IS THE NUMBER OF COLUMNS}$$

$$I_{AVE\,(COL\,j)} \leq \frac{1}{N} \sum_{i=1}^{p} I_{ROW\,i}, \text{ WHERE p IS THE NUMBER OF ROWS}$$

$$I_{PEAK\,(COL\,j)} \leq \sum_{i=1}^{p} I_{ROW\,i}$$

Figure 2.4.4.2-5 Column Strobed X-Y Addressable Array. X_1 through X_q are Sequentially Strobed on a 1/N Duty Cycle, N ⩾ P.

2.4.4.4 Analog Bar Graph Arrays

In many applications, analog information must be converted into a visual display. Traditionally analog panel meters have been used for low cost applications requiring only moderate accuracy and resolution. LED linear displays can often be substituted for these applications. LED linear displays consist of a linear or circular array of LED lamps that are driven by a device that decodes the varying analog or digital signal into a bar graph or position indicator display. The input signal can be decoded so that all LEDs with thresholds below the input are turned on (bar graph) or that only the LED with its threshold closest to the input is turned on (position indicator). Some examples of these type of displays are shown in Figure 2.4.4.4-1. LED linear arrays have many advantages over analog panel meters. These advantages include higher reliability, higher resistance to mechanical shock and higher visibility in low and moderate ambient lighting. Since the LED linear array is a light emitting device, it is more effective at getting the viewer's attention than a panel meter. Red, yellow and green LEDs can be used to quickly identify the proper limits of instrument operation. Distinct switching thresholds can be selected to allow the linear array to

simplify the machine operator's decision. Finally, in consumer oriented equipment, the LED linear array provides a new and distinctive selling feature for the product. Depending on the number of LEDs in the array, if the input signal is in analog form, the linear array decoder can be one or more operational amplifiers, a monolithic analog decoder, or a low cost analog to digital converter followed by a simple digital decoder. If binary or BCD information is available, such as from an analog to digital converter or a microprocessor, then only a simple digital decoder is required. Examples of several of these circuits will be shown.

When analog information is available, operational amplifiers or voltage comparators can be used to interface to the LED linear array. Since the LED lamps operate at low currents and low voltages, most operational amplifiers can drive LEDs without output buffering. Figure 2.4.4.4-2 shows examples of a bar graph display and a position indicator display. In both circuits, a five resistor voltage divider determines the switching thresholds of the LED lamps. The resistor values are not critical but the resistance ratio

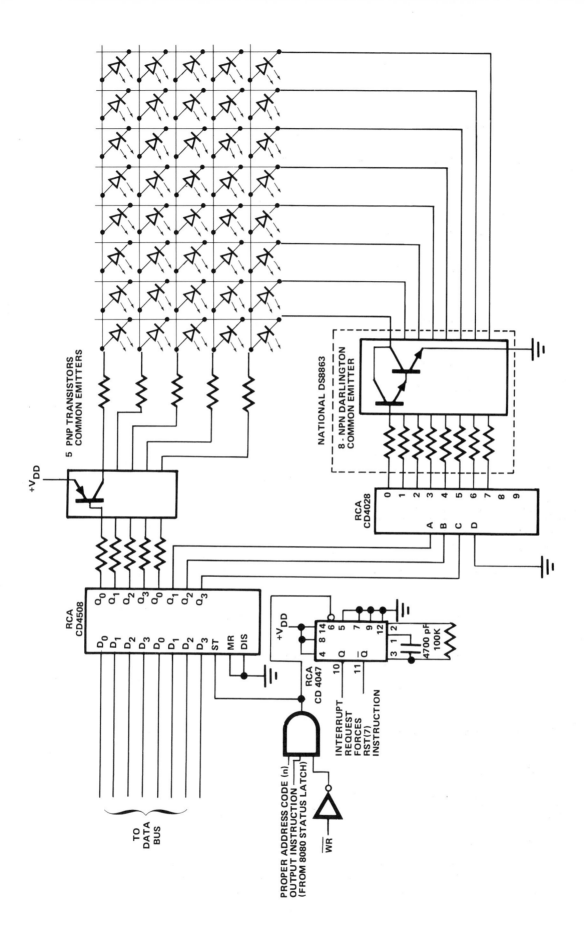

Figure 2.4.4.3-1 A Microprocessor Controlled X-Y Addressable Array.

ADDRESS	OP CODE	CLOCK CYCLES	COMMENTS
$(0038)_{16}$	PUSH PSW	11	
	PUSH HL	11	
	LHLD	16	HL = POINTER
	A_L	----	
	A_H	----	
	MOV A, M	7	A = (HL)
	OUT	10	STORES NEW ROW AND COLUMN INFORMATION
	(n)	----	
	INX HL	5	HL = HL + 1
	MOV A, M	7	A = (HL)
	OUT	10	STORES SECOND BYTE OF INFORMATION
	(n+1)	----	
ADDED FOR N EIGHT BIT LATCHES	:		
	INX HL	5	
	MOV A, M	7	
	OUT	10	STORES Nth BYTE OF INFORMATION
	(n+N-1)	----	
	MOV A, L	5	A = L
	CPI	7	COMPARES A TO $(17)_{16}$
	$(17)_{16}$	----	$(OF+8N)_{16}$ FOR N EIGHT BIT LATCHES
	JNZ	10	A $\neq$ $(17)_{16}$ JUMP TO LOOP
	$LOOP_L$	----	
	$LOOP_H$	----	
	MVI A	7	A = $(10)_{16}$
	$(10)_{16}$	----	
	STA	13	$POINTER_L$ = $(10)_{16}$
	A_L	----	
	A_H	----	
	POP HL	10	
	POP PS W	10	
	EI	4	ENABLE INTERRUPT FLAG
	RET	10	RETURN
LOOP	INX HL	5	HL = HL + 1
	SHLD	16	POINTER = POINTER + 1
	A_L	----	
	A_H	----	
	POP HL	10	
	POP PSW	10	
	EI	4	ENABLE INTERRUPT FLAG
	RET	10	RETURN

IN GENERAL,

FOR C COLUMNS, N EIGHT BIT LATCHES, THE TOTAL NUMBER OF CLOCK CYCLES REQUIRED TO REFRESH THE ENTIRE DISPLAY ONCE EQUALS:

CLOCK CYCLES = C (121 + 22N) - 1

= 1143 CLOCK CYCLES TO REFRESH THE DISPLAY IN FIGURE 2.4.4.3-1

MEMORY ADDRESS	CONTENTS (D_7-D_0)	COMMENTS
A_H A_L A_H A_L + 1	$POINTER_L$ $POINTER_H$	POINTER = NEXT COLUMN TO BE DISPLAYED
$(X X 1 0)_{16}$	0 0 0 Y_5 Y_4 Y_3 Y_2 Y_1	COLUMN 1
$(X X 1 1)_{16}$	0 0 1	COLUMN 2
$(X X 1 2)_{16}$	0 1 0	COLUMN 3
$(X X 1 3)_{16}$	0 1 1	5 BIT ROW INFORMATION — COLUMN 4
$(X X 1 4)_{16}$	1 0 0	COLUMN 5
$(X X 1 5)_{16}$	1 0 1	COLUMN 6
$(X X 1 6)_{16}$	1 1 0	COLUMN 7
$(X X 1 7)_{16}$	1 1 1	COLUMN 8

X = DON'T CARE

Figure 2.4.4.3-2 Intel 8080A Microprocessor Program used to Strobe the X-Y Addressable Array Shown in Figure 2.4.4.3-1.

between resistors will specify the threshold, V_1 to V_4, as shown in Figure 2.4.4.4-2. As shown, the linear displays can detect an analog signal greater than zero. In circuit 2.4.4.4-2a (bar graph), if $V_{IN} < V_1$, where V_1 is the switching threshold of the first lamp in the array, then all

Figure 2.4.4.4-1 Typical Applications for Position Indicator and Bar Graph Arrays.

lamps in the array will be off. As V_{IN} increases above each switching threshold, V_1 to V_4, the corresponding output of each operational amplifier conducts current to ground which turns the LED on. R is selected to limit the current through the LED to the desired value. Since each LED lamp has its own current limiting resistor, standard red, high efficiency red, yellow, and green lamps can be mixed in the same array. Voltage comparators with open collector outputs can also be used in the circuit. In circuit 2.4.4.4-2b (position indicator) only one LED lamp in the array will be on. If $V_{IN} < V_1$, the outputs of all operational amplifiers will be pulled up to V_{CC}. U_1 will source current to LED_1, turning it on but LED_2 through LED_5 will be turned off with approximately zero bias. As V_{IN} increases above V_1, the output of U_1 will conduct current to ground, turning LED_1 off but turning LED_2 on. Each LED lamp will turn on sequentially as V_{IN} increases above the specified thresholds. Either circuit can be expanded in size to accomodate any desired bar graph display. The circuits shown in Figure 2.4.4.4-2 can be operated from a single polarity power supply providing that only positive signals are to be detected. Both positive and negative signals can be detected if a dual polarity power supply is used. A dual operational amplifier with three LEDs can be connected as shown in circuit 2.4.4.4-2b to make a solid state null detector circuit, that compares V_{IN} to a specified V_{REF} determined by the resistor divider network. The LEDs would indicate $V_{IN} < (V_{REF}-\Delta)$, $(V_{REF}-\Delta) < V_{IN} < (V_{REF}+\Delta)$, and $V_{IN} > (V_{REF}+\Delta)$, or negative, zero or positive if V_{REF} is equal to zero and Δ is small. Such a device could replace a conventional analog null meter.

For longer LED linear arrays, IC decoders are available that decode the analog input signal into a bar graph or a position indicator display. Figure 2.4.4.4-3 shows an example of an LED bar graph display that uses the Siemens UAA-180 decoder and a position indicator display that uses the Siemens UAA-170 decoder. For both devices, a voltage divider network consisting of R_3, R_4, and R_5 determine

$$V_1 = \frac{R_1}{R_1+R_2+R_3+R_4+R_5} \, V_{CC}$$

$$V_2 = \frac{R_1+R_2}{R_1+R_2+R_3+R_4+R_5} \, V_{CC}$$

$$V_3 = \frac{R_1+R_2+R_3}{R_1+R_2+R_3+R_4+R_5} \, V_{CC}$$

$$V_4 = \frac{R_1+R_2+R_3+R_4}{R_1+R_2+R_3+R_4+R_5} \, V_{CC}$$

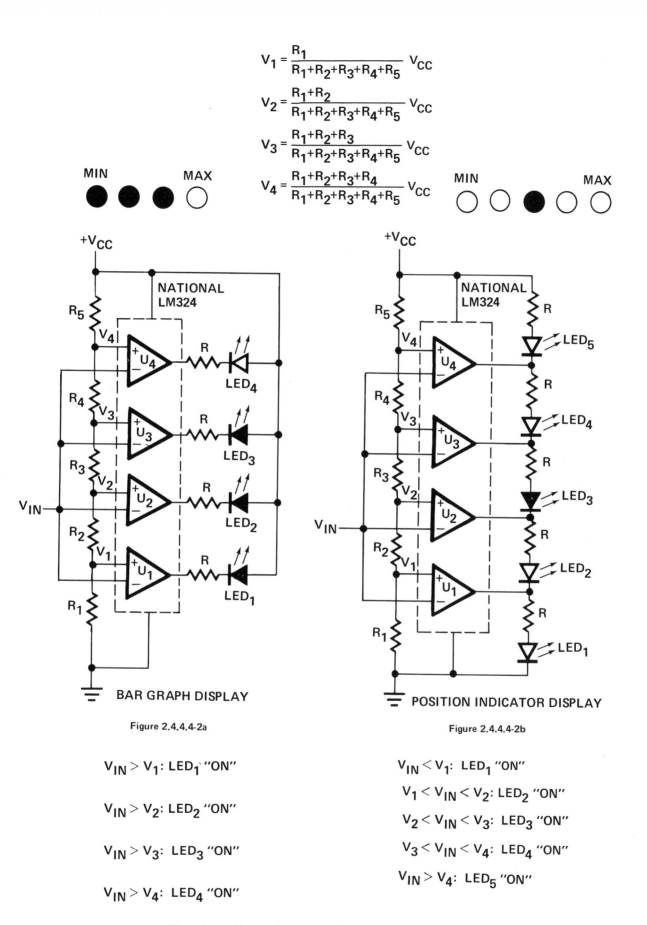

BAR GRAPH DISPLAY

Figure 2.4.4.4-2a

POSITION INDICATOR DISPLAY

Figure 2.4.4.4-2b

$V_{IN} > V_1$: LED$_1$ "ON"

$V_{IN} > V_2$: LED$_2$ "ON"

$V_{IN} > V_3$: LED$_3$ "ON"

$V_{IN} > V_4$: LED$_4$ "ON"

$V_{IN} < V_1$: LED$_1$ "ON"

$V_1 < V_{IN} < V_2$: LED$_2$ "ON"

$V_2 < V_{IN} < V_3$: LED$_3$ "ON"

$V_3 < V_{IN} < V_4$: LED$_4$ "ON"

$V_{IN} > V_4$: LED$_5$ "ON"

Figure 2.4.4.4-2 Operational Amplifiers or Voltage Comparators
Used to Decode an Analog Signal into a Bar Graph
or Position Indicator Display.

the switching range of the linear array. Using the UAA-180 decoder, if V_{IN} is less than V_{MIN}, all LED lamps will be off. If V_{IN} is greater than V_{MIN}, the input voltage will be decoded as a bar graph array with all LEDs on when V_{IN} is greater than V_{MAX}. For the UAA-170 decoder when V_{IN} is less than V_{MIN}, then LED_1 will be on. As V_{IN} increases above V_{MIN} then LED_2 through LED_{16} will turn on sequentially. When V_{IN} is greater than V_{MAX}, LED_{16} will be on. Since the UAA-170 and UAA-180 have constant current drivers for the LEDs, no current limiting resistors are required. However, the constant current drivers are programmed externally by R_1 and R_2. If longer linear arrays are desired, additional UAA-170s or UAA-180s can be cascaded in series.

Digital information can also be decoded as an LED bar graph or a position indicator display. If an analog signal is available, it may be desirable and cost effective to use a low cost analog to digital converter and then decode the digital outputs as a linear array. The position indicator display can be decoded with a one of n decoder as shown in Figure 2.4.4.4-4. One of n decoders can be cascaded to form any size of LED position indicator display desired. Since only one LED in the position indicator display will be on at a time, an x-y addressable array can be used to simplify decoding. Figure 2.4.4.4-5 shows how two one of eight decoders and 17 external components can be used to address 64 LED lamps in a position indicator display. Using commercially available 1 of 4, 1 of 8, 1 of 10, or 1 of 16

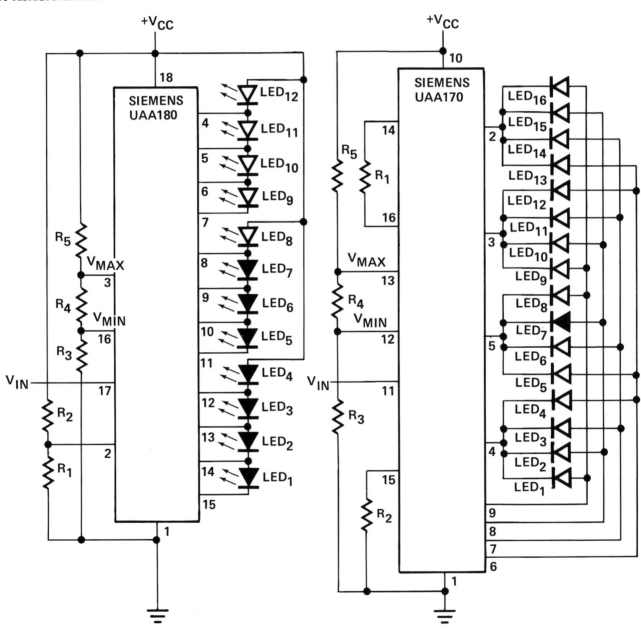

Figure 2.4.4.4-3 Use of Siemens UAA 170 and UAA 180 Position Indicator and Bar Graph Decoders.

2.34

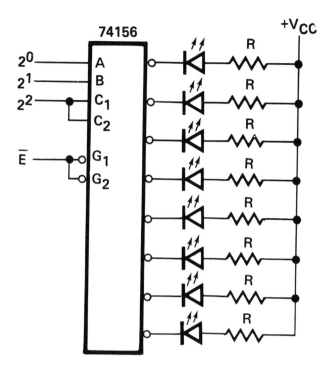

$\overline{E} = 0$ 1 OF 8 LED LAMPS WILL LIGHT

$\overline{E} = 1$ ALL LED LAMPS OFF

Figure 2.4.4.4-4 One of Eight Decoder Used as Position Indicator
Decoder/Driver.

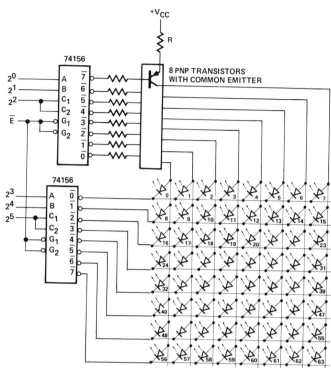

$\overline{E} = 0$ 1 OF 64 LED LAMPS WILL LIGHT

$\overline{E} = 1$ ALL LED LAMPS OFF

Figure 2.4.4.4-5 Two One of Eight Decoders Can Be Used to Decode
a 64 LED Position Indicator Display.

decoders, almost any size LED position indicator array can be decoded. Using the technique shown in Figure 2.4.4.4-4, four 1 of 16 decoders, four inverters (4/6 package) and 64 external components would be used to address the same 64 LED lamps.

Digital logic can also be used to decode an LED bar graph display. Figure 2.4.4.4-6 shows how a one of eight decoder and an And gate network can be used to decode an eight LED bar graph display. The display can be expanded to any length with the proper decoder and an extended And gate network. For large displays, a strobed x-y addressable approach is simpler. Figure 2.4.4.4-7 shows an example of such an approach. The lower three bits of the input are connected to an eight output bar graph decoder. The upper three bits of the input are continuously compared to the outputs of a divide by eight counter. The output of the counter is decoded by a one of eight decoder which row strobes the 64 LED matrix on a 1/8 duty factor. If the value of the counter is less than the upper three bits of the input, then all LEDs in that row are turned on. When the value of the counter is equal to the upper three bits of the input, the lower three bits of the input are decoded by the bar graph decoder. For values of the counter that are greater than the input, all LED lamps for that row are turned off. This technique can also be used to decode BCD inputs by substituting a decade counter, one of ten

decoder, and ten line bar graph decoder for the devices specified. The maximum size of such an array is limited only by the requirement of a certain minimum duty factor to obtain the desired time averaged luminous intensity.

Microprocessors can be interfaced directly to any of the digital decoder circuits previously shown by adding a latch to hold the input information. The microprocessor would need to update the latch only when the information is to be changed. The techniques described in section 2.4.4.3 can also be used to implement a linear array. Unless the application is microprocessor time limited, these techniques are more cost effective than the bar graph decoder circuit described in Figure 2.4.4.4-7. The microprocessor would be required to decode the input information in software and continuously refresh the linear array. The circuit described in Figure 2.4.4.3-1 can be used to implement either a bar graph or a position indicator display.

2.4.5 Backlighting

Information can be more readily assimilated from a panel by an observer if only those symbols or characters relevant to a particular condition are visible. Ideally, the panel has a "dead front" appearance for all symbols except those required to be seen, and those are made visible with light projected through them from a source in back of the panel; hence the term "backlighting", illustrated in Figure 2.4.5-1.

Red LEDs of earlier technology did not produce enough light to make backlighting practical. More recent technology not only makes red backlighting practical, but yellow and green as well.

Front lighting differs from backlighting in that it is used mainly as a substitute for or a supplement to ambient illumination. Because of scattered and stray light, it is difficult to selectively display symbols by a front lighting technique. LEDs can also be used in front lighting since the supplementary light is usually needed only when ambient light is subdued. Front lighting is, therefore, a simple design and does not require as much design attention as backlighting.

2.4.5.1 Fundamental Backlighting Requirements

There are only four parameters to consider. In the order of usual importance, they are:

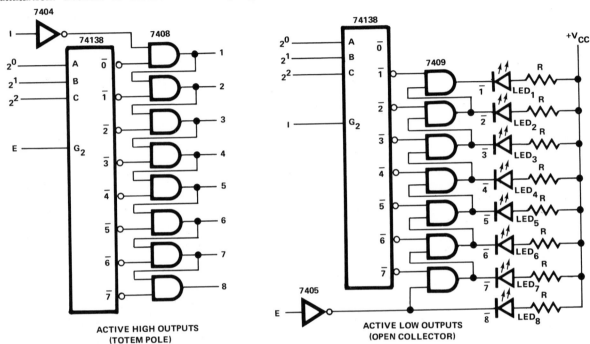

ACTIVE HIGH OUTPUTS
(TOTEM POLE)

ACTIVE LOW OUTPUTS
(OPEN COLLECTOR)

BINARY (BCD) TO BAR GRAPH DECODER

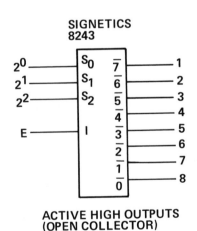

SIGNETICS
8243

ACTIVE HIGH OUTPUTS
(OPEN COLLECTOR)

I	E	C	B	A	1 2 3 4 5 6 7 8
0	0	0	0	0	0 0 0 0 0 0 0 0
0	0	0	0	1	1 0 0 0 0 0 0 0
0	0	0	1	0	1 1 0 0 0 0 0 0
0	0	0	1	1	1 1 1 0 0 0 0 0
0	0	1	0	0	1 1 1 1 0 0 0 0
0	0	1	0	1	1 1 1 1 1 0 0 0
0	0	1	1	0	1 1 1 1 1 1 0 0
0	0	1	1	1	1 1 1 1 1 1 1 0
NOTE1	1	X	X	X	1 1 1 1 1 1 1 1
1	NOTE2	X	X	X	0 0 0 0 0 0 0 0

X = DON'T CARE

NOTE1: FOR ACTIVE HIGH CONFIGURATION I = 0
FOR ACTIVE LOW CONFIGURATION I = X
NOTE2: FOR ACTIVE HIGH CONFIGURATION E = X
FOR ACTIVE LOW CONFIGURATION E = 0

Figure 2.4.4.4-6 One of Eight Decoder Used as Bar Graph Decoder/Driver.

2.36

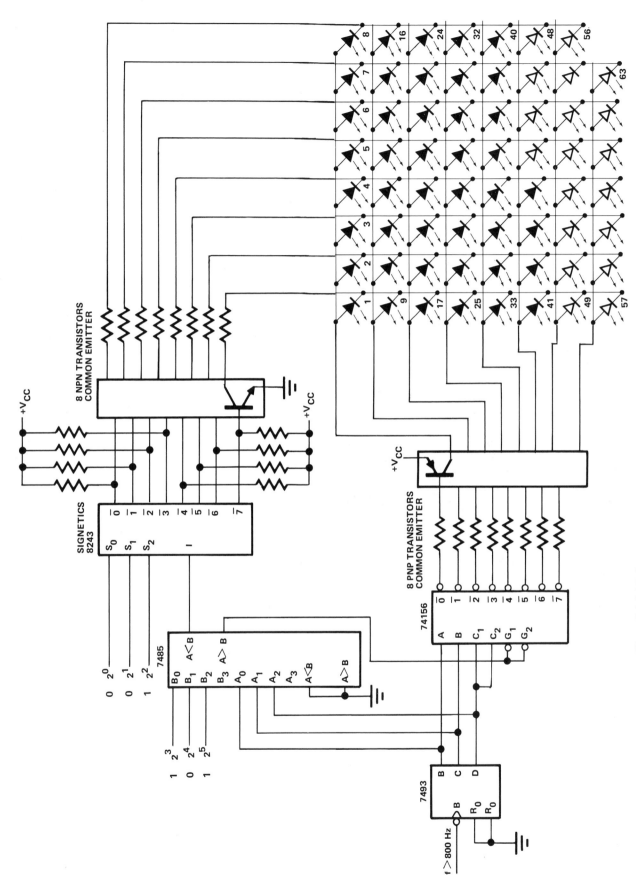

Figure 2.4.4.4-7 Strobed Bar Graph Decoder that Uses a Multiplexed X-Y Addressable Array to Simplify Decoding.

1. Contrast
2. Uniformity
3. ON Sterance*
4. Radiation Pattern

*Because this section deals only in photometric (no radiometric) quantities, the prefix luminous and subscript v may be dropped from the terms and symbols:

1. Luminous flux, ϕ_v
2. Luminous incidance, E_v
3. Luminous exitance, M_v
4. Luminous intensity, I_v
5. Luminous sterance, I_v

The radiation pattern of the backlighted symbol is ideally lambertian; but if the symbol complexity makes it unrecognizable anyway at large off-axis angles, the angular variation of intensity (or sterance) is of little importance.

Figure 2.4.5-1   Backlighting Used to Selectively Cause Special Symbols to Appear.

ON sterance describes the intensity per unit area of the symbol when backlighting is applied. For a particular background and symbol reflectance, the ON sterance determines the ambient light level where the backlighted symbol can be adequately distinguished. "Adequate" distinction of the symbol is not precisely definable because it bears a subjective relationship to the detail which must be resolved in the symbol.

Uniformity of sterance over the illuminated area is also somewhat subjective. In some applications, it has cosmetic value only, but in others, non-uniformity can cause loss of detail which may cause misrecognition of the symbol. Uniformity does, however, have an objective criteria for evaluation. Because eye response is logarithmic, a 3dB

sterance variation (2:1) is usually acceptable, and this is the basis of the developments described later.

Contrast, the most important consideration, is evaluated in terms of three contrast ratios:

a)   ON/OFF CR
b)   ON/BACKGROUND CR
c)   OFF/BACKGROUND CR

ON/OFF CR is the ratio of ON sterance to OFF sterance of the symbol. The OFF sterance is the product of surface reflectance and ambient light, so ON/OFF CR varies inversely as the ambient light level. Similarly, ON/BACKGROUND CR is the ratio of the ON sterance of the symbol to the BACKGROUND sterance, which also is the product of surface reflectance and ambient light. Neither ON/OFF CR nor ON/BACKGROUND CR should always be "as high as possible". If ON/BACKGROUND CR is too high, eye fatigue may result. This is because the ON sterance results from radiated, not reflected, flux. Both ON/OFF CR and ON/BACKGROUND CR should, of course, be adequate and ideally they are equal.

If ON/OFF CR = ON/BACKGROUND CR, it follows that OFF/BACKGROUND CR is 1.00. When tradeoffs are being made, because of high ambient conditions and marginal flux availability, the usual optimization favors ON/BACKGROUND CR, but OFF/BACKGROUND CR must not be ignored.

Adequate ON/BACKGROUND CR requirements are subjective and BACKGROUND reflectances vary widely, but an approximation to the ON sterance requirement can be drawn from Figure 2.4.5-2. The curve is drawn from:

1. Assume normal ambient, $E_A = 500\,\text{lx}$

2. Assume BACKGROUND surface reflectance, R_{BG}:

   $$M_{BG}/E_A = R_{BG} = 2\%$$

3. Assume lambertian radiation of flux reflected, so $M_{BG} = \pi\,L_{BG}$

4. Calculate BACKGROUND sterance, L_{BG}:

$$L_{BG}\left(\frac{cd}{m^2}\right) = [\frac{1}{\pi}(\frac{cd}{lm})]\ [E_A\ (lx)]\ [\frac{R_{GB}(\%)}{100\%}]$$

$$= [\frac{1}{\pi}(\frac{cd}{lm})]\ [500\ lx]\ [0.02]$$

$$= 3.18\ \frac{cd}{m^2} \qquad\qquad (2.4.5\text{-}1)$$

5. Require ON/BACKGROUND CR, (L_{ON}/L_{BG}) to be within the limits $2.5 < (L_{ON}/L_{BG}) < 10$; choose $(L_{ON}/L_{BG}) = 5$ and calculate:

$$L_{ON} = L_{BG} \, (L_{ON}/L_{BG}) \qquad\qquad (2.4.5\text{-}2)$$

$$= 3.18 \, \frac{cd}{m^2} \, (5) = 15.92 \text{ cd/m}^2$$

For the tolerable limits required in step 5, the curve in Figure 2.4.5.2 shows that an ambient ranging from 250 lx to 1000 lx can be accomodated. Less narrow requirements would, of course, broaden the range.

2.4.5.2 ON Sterance Design Considerations

There are two requirements for the ON sterance. It must be adequately high and uniform. To obtain uniform sterance requires a diffuser, as shown in Figure 2.4.5-3. The diffuser/LED combination, properly designed, gives a background of uniform sterance over which the "legend plate", bearing the symbol opening, is placed. Without proper diffusion, sterance would not be uniform over all portions of the symbol. A viewer would see the LED through only those portions of the symbol opening that lie between the viewer's eye and the LED.

A diffuser functions as diagrammed in Figure 2.4.5-3. Normal (perpendicular) incidence, E_v, on the back surface causes flux to be emitted at the front side, being scattered by the diffusant such that each increment of area on the front radiates in all directions. In an ideal (but not realizable) diffuser, each lumen impinging at the back would emerge from the front; that is, nothing would be reflected and nothing lost in the diffuser. Upon emerging, this lumen would be scattered; ideally, the scattering would be lambertian, so the ratio of total flux emitted to normal (perpendicular) intensity is π. Thus, in an ideal diffuser, an incidence of one lumen per square meter would cause emission of one lumen per square meter and normal intensity would be $1/\pi$ candelas per square meter. A "Diffusance Quotient" (similar to Intelligence Quotient) can be defined as the ratio:

$$D.Q. = \pi \, L_{VOUT}/E_{vIN} \qquad\qquad (2.4.5\text{-}3)$$

having a normal value of 1.00. D.Q. can be lowered by absorption and reflection loss, and can be raised by anisotropic (e.g., light pipe) effects in the diffuser.

For example, one type of diffusing film is described as having a transmission of 55% and a "gain" of 400%. For this film, the D.Q. is 0.55 x 4.00 = 2.2, and from equation (2.4.5-3) we can calculate:

$$\frac{L_{v \, OUT} \, (cd/m^2)}{E_{v \, IN} \, (lx)} = \frac{D.Q.}{\pi} = 0.70 \, (\frac{cd/m^2}{lx}) \qquad (2.4.5\text{-}4)$$

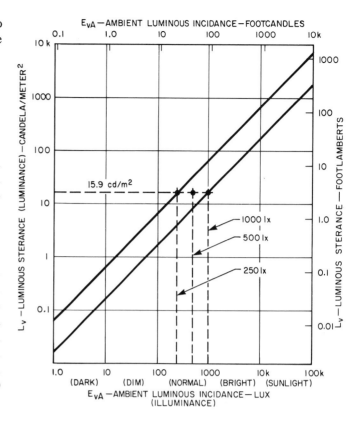

Figure 2.4.5-2   Approximate "ON" Luminous Sterance Required Under Given Ambient Light.

LAMBERTIAN LED: $\quad \dfrac{E_v(x)}{E_v(0)} = \cos^4(\theta) = \left(\dfrac{1}{1+(\frac{x}{d})^2}\right)^2$

NON-LAMBERTIAN: $\quad \dfrac{E_v(x)}{E_v(0)} = \left(\dfrac{I_v(\theta)}{I_v(0)}\right) \cos^3\theta$

Diffusance Quotient

$$D.Q. = \pi \, \frac{L_v}{E_v} \, \left(\frac{cd/m^2}{lx}\right)$$

Figure 2.4.5-3   Diffuser Effect on Flux Direction as Related to Radiation Pattern.

2.39

Industry standards for characterization of diffusers have not yet appeared, although diffusers are available such as "Light Diffusing Film" from 3M Visual Products Division, and "Chromafuse" from Panelgraphic. Tradeoffs in diffuser selection relate to D.Q. -- a high D.Q. (>1) means the radiation pattern is narrower than lambertian because transmission loss cannot be zero. It is unlikely that the radiation pattern would be broader than lambertian, so a low D.Q. (<1) means a lossy diffuser, but a narrow radiation pattern can compensate. D.Q. is therefore the one single number that comes closest to relating diffuser specifications to backlighting performance.

Adequately high ON sterance may require a tradeoff with uniformity or the addition of extra LEDs. This can be seen by analyzing, in Figure 2.4.5-3, the incidence on the diffuser at points displaced a distance, x, from the axis. The analysis uses the inverse square law*:

$$E_v = I_v/d^2 \qquad\qquad\text{(2.4.5-5)}$$

    *accuracy of the inverse square law is within 5% for distance, d, as short as two LED diameters.

At $x = 0$, $\theta = 0$, so the incidence on the diffuser is:

$$E_v(0) = I_v(0)/d^2 \qquad\qquad\text{(2.4.5-6)}$$

At any other point, x, the incidence VECTOR is:

$$|E_v| \angle\theta = I_v(\theta)/(x^2 + d^2) \qquad\qquad\text{(2.4.5-7)}$$

but the NORMAL incidence component of the incidence VECTOR is the vector magnitude multiplied by $\cos\theta$, so:

$$E_v(x) = [I_v(\theta)/(x^2 + d^2)]\ \cos\theta \qquad\qquad\text{(2.4.5-8)}$$

The non-uniformity of incidence can be seen as the ratio $E_v(x)/E_v(0)$, found by taking the ratio of equation (2.4.5-8) to (2.4.5-6):

$$\frac{E_v(x)}{E_v(0)} = \frac{[I_v(\theta)/x^2 + d^2]\ \cos\theta}{I_v(0)/d^2} \qquad\qquad\text{(2.4.5-9)}$$

$$= \frac{I_v(\theta)}{I_v(0)} = (\frac{d^2}{x^2 + d^2})\ \cos\theta$$

$$= \frac{I_v(\theta)}{I_v(0)}\ \cos^3\theta$$

In equation (2.4.5-9), the ratio $[I_v(\theta)/I_v(0)]$ is recognized as the relative intensity vs. θ, which is the radiation pattern as given in most LED data sheets. Thus, if the LED radiation pattern is lambertian, $\cos\theta$ can be substituted for $[I_v(\theta)/I_v(0)]$ to give:

(2.5.4-10)

$$\frac{E_v(x)}{E_v(0)} = \cos^4\theta \qquad \textbf{FOR LAMBERTIAN RADIATION PATTERN}$$

A curve of equation (2.4.5-10) is given in Figure 2.4.5-4 along with curves for several non-lambertian LEDs. The curves for the non-lambertian LEDs were obtained by applying equation (2.4.5-9) to the radiation patterns in their data sheets.

Here is how to use these curves and equations for a single LED backlighting a legend, assuming no reflectors:

EXAMPLE:

Legend area: 10mm x 20mm

ON Sterance, $L_v = 20$ cd/m^2

Uniformity: corner $L_v \geqslant 0.5$ x center L_v

Assume diffuser mentioned earlier: D.Q. = 2.2

SOLUTION:

1.   Refer to Figure 2.4.5-3 and determine $E_v(0)$ from equation (2.4.5-3):

$$E_v(0) = \frac{\pi \times 20\text{ cd/m}^2}{2.2} = 28.56\text{ lx}$$

2.   Find the distance from the center to the corner:

$$x = \tfrac{1}{2}\sqrt{(10\text{mm})^2 + (20\text{mm})^2} = 11.18\text{mm}$$

3.   Choose an LED and either construct a curve according to equation (2.4.5-9) or use a curve from Figure 2.4.5-4. Assume now a type 5082-4650/55. Find θ at which $E_v/E_v(0) = 0.5$:

Figure 2.4.5-4 gives $\theta = 25°$

4.   Compute LED-to-diffuser distance:

$$d = x/\tan\theta = 11.18\text{mm}/\tan 25° = 23.98\text{mm}$$

5.   Compute LED intensity required, using equation (2.4.5-6) and results of step 1, step 4:

$$I_v(0) = 28.56\text{ lx} \cdot (23.98\text{mm})^2 = 16.42\text{ mcd}$$

This is about five times the 3.0 mcd minimum specified for the 5082-4655, so a different LED is tried. Applying steps 3, 4, 5 to a 5082-4657/58:

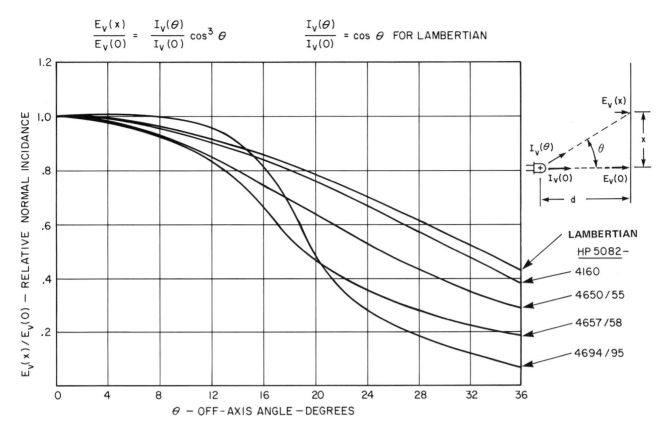

$$\frac{E_v(x)}{E_v(0)} = \frac{I_v(\theta)}{I_v(0)} \cos^3 \theta \qquad \frac{I_v(\theta)}{I_v(0)} = \cos \theta \quad \text{FOR LAMBERTIAN}$$

Figure 2.4.5-4   Normalized Incidance on Diffuser Obtained from Radiation Patterns.

3. From Figure 2.4.5-4, $\theta = 19°$

4. $d = 11.18\text{mm}/\tan 19\theta = 32.47\text{mm}$

5. $I_v(0) = 28.56\,\text{lx} \times (32.47\text{mm})^2 = 30.11\,\text{mcd}$

This is marginal for the 5082-4657, but the 5082-4658 will do, with $I_F \cong 16\,\text{mA}$.

Note that the exercise in trying the 5082-4350/55 could have been avoided by first computing the intensity of the required legend area:

$$I_{v\ OUT} = L_{v\ OUT} \times \text{LEGEND AREA} \qquad (2.4.5\text{-}11)$$

$$= 20\ \text{cd/m}^2\ (10\ \text{mm} \times 20\ \text{mm})$$

$$= 4\ \text{mcd}$$

A single LED would hardly be likely to product a legend area intensity greater than the basic LED intensity.

Note also that the solution with the 5082-4658 requires a distance behind the diffuser of 32.47 mm, which may be inconveniently large. By using more than one LED, the ON sterance and uniformity requirements over the entire legend area can be met with much less distance required behind the panel. The design procedure is a bit more complicated because it is necessary to take account of the overlapping incidance from adjacent LED(s). The 10mm x 20mm box should not be evenly subdivided (e.g. 2 LEDs 10mm apart) because the corners do not receive the overlapping

incidance. A little more "cut-and-try" on paper is much less costly, however, than "scrap-and-try-again" on the workbench.

Another POSSIBLE tradeoff would be to allow a slightly greater non-uniformity, i.e., $E_v(x)/E_v(0) < 0.5$. This would permit a larger θ, and hence a smaller d and $I_v(0)$ would be possible. Remember that $I_v(0)$ requirement varies jointly as the SQUARE of d.

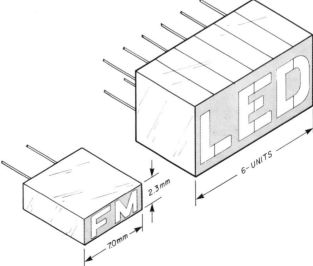

Figure 2.4.5-5   Rectangular LED Backlighting Small or Tall Legend Areas.

2.4.5.3 Backlighting Construction for Small-to-Medium Legend Areas

For illuminating small symbols, the simplest approach is the use of a rectangular LED, designed for the purpose and having built-in diffusant that meets most uniformity requirements. As shown in Figure 2.4.5-5, it can also be clustered to illuminate fairly large areas. At their specified input current levels, their ON sterance easily exceeds 40 cd/m^2. All that is needed with these is the symbol-bearing legend plate and perhaps a contrast enhancement filter. If it is desirable to have separately-illuminated areas in the cluster, thin black tape separators are adequate for crosstalk control. The bodies of these rectangular LEDs are translucent, so for good OFF/BACKGROUND CR they should be shielded from stray light behind the panel -- more black tape. Painting is not recommended.

Another simple approach uses LEDs made with undiffused encapsulant as in Figure 2.4.5-6. This "direct-illumination" method uses the basic design consideration of Section 2.4.5.2 to select the LED and spacing, d, for proper ON sterance and uniformity. This approach is typically used for fairly small legend areas (<10mm diameter); the large (10mm x 20mm) area was prescribed in the earlier examples for purposes of illustrating design tradeoffs.

The shortest distance behind the panel is achieved by using an "egg crate" reflector in which each LED is near the apex of a conically- or pyramidally-shaped reflector cavity. As seen in Figure 2.4.5-7, the reflector can be made to either combine more than one LED to backlight a large legend, or provide separation with a "web" to isolate separately energized areas. Egg-crate reflectors offer the additional benefit of re-directing side-emitted flux from the LED so that efficiency is raised and uniformity is improved. Because much of the flux reaching the diffuser comes *indirectly* from the reflector rather than directly from LED, this arrangement is described as "indirect" illumination.

Figure 2.4.5-7 also shows the recommended placement of the diffuser, legend plate, and (if used) contrast enhancement filter. Placing the legend plate behind the diffuser would "break" the symbol edges and lower the OFF/BACKGROUND CR; so unless fuzzy symbol edges are desired, the legend plate should be located as shown. Contrast enhancement filtering is always out in front.

Another method of indirect illumination uses a plastic "light pipe" as in Figure 2.4.5-8. Except for the hole to accept the LED, the plastic is solid and relies on greater-than-critical-angle reflections for its efficiency. Scratches and contact with materials other than air cause light losses at the surfaces, so these surfaces should be protected. Over a length of more than four diameters, multiple reflections cause enough diffusion to permit elimination of the separate diffuser; the front end of the "pipe" may be roughened if the length is marginal.

2.4.5.4 Backlighting Construction for Very Large Legend Areas

For very large symbols to be viewed from large distances (>10m) the uniformity requirement is sharply reduced. This permits the LEDs to be viewed directly through the symbol with NO DIFFUSER. To obtain highest possible efficiency (or ON sterance), each LED in the arrays of Figure 2.4.5-9 should be seated in a cell of an "egg crate"-type reflector. The average ON sterance is then found by dividing the single LED intensity by the area of a single cell. In Figure 3.4.5-9, notice that both arrays have the same number of LEDs, but different aspect ratios. Except for this, the "honeycomb" advantage over the "square" is rather small (a factor of only $2/\sqrt{3}$). Overlapping incidence is likely to leave smaller "cold" spots in the "honeycomb" than in the "square".

2.5   Communications and Signalling Applications

The terms "communications and signalling" are intended to categorize all those applications in which the primary function of the LED is to provide flux for detection by means other than human vision (but may include some of these as well). Such applications include:

- Card/tape reader (esp. low hole/no-hole ratio)
- Tape loop stabilizer (max and min loop sensors)
- End-of-tape sensor (reflective or transmissive)
- Optical tachometer (motor speed control)
- Assembly line monitor (parts counting/orientation)
- Bar code scanner (POS machine UPC)
- Opto-mechanical synchronizer (ignition timing)
- Safety interlock (with phase-lock loop in high ambient)
- Carriage travel sensor (beam break or reflect)
- Shaft position encoder (using arrayed devices)
- High voltage isolator (air gap or fiber optic)
- Smoke detector (both scattering- and obscuration-type)
- Densitometer (chemical analysis)
- Liquid level monitor (clear as well as opaque)

Their shorter wavelengths and higher modulation bandwidths give LEDs performance superior to IREDs in many of these applications, despite the higher quantum efficiencies of IREDs. Now new devices are available with wavelengths short enough to benefit from spectral considerations, but with quantum efficiencies so high (up to 1.5%) that they rival that of amphoteric IREDs. These new devices, emitting at 670 nm and 700 nm, are sufficiently visible that optical alignment can be done visually without the spectral viewing equipment required with IREDs.

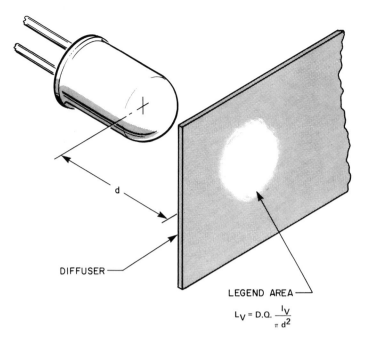

Figure 2.4.5-6   Direct Illumination; Undiffused LED Makes Sharp Edges when "d" is Small.

$$L_V = D.Q. \frac{I_V}{\pi d^2}$$

LEGEND AREA

DIFFUSER

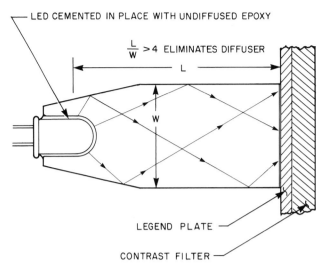

LED CEMENTED IN PLACE WITH UNDIFFUSED EPOXY

$\frac{L}{W} > 4$ ELIMINATES DIFFUSER

LEGEND PLATE

CONTRAST FILTER

Figure 2.4.5-8   Plastic Light Pipe; Reflects Side Emission and Diffuses Forward.

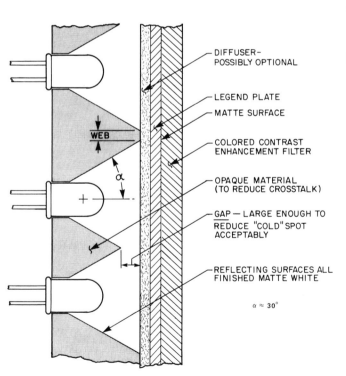

DIFFUSER– POSSIBLY OPTIONAL

LEGEND PLATE

MATTE SURFACE

COLORED CONTRAST ENHANCEMENT FILTER

OPAQUE MATERIAL (TO REDUCE CROSSTALK)

GAP – LARGE ENOUGH TO REDUCE "COLD" SPOT ACCEPTABLY

REFLECTING SURFACES ALL FINISHED MATTE WHITE

$\alpha \approx 30°$

WEB

Figure 2.4.5-7   Indirect Illumination; Egg-Crate Reflector Improves Uniformity.

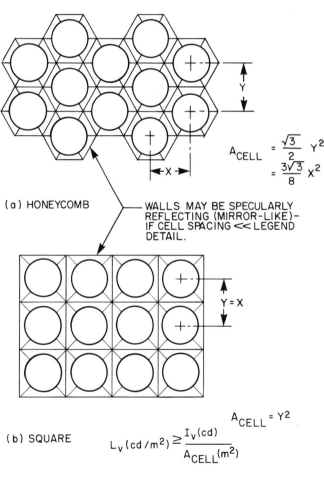

$$A_{CELL} = \frac{\sqrt{3}}{2} Y^2 = \frac{3\sqrt{3}}{8} X^2$$

(a) HONEYCOMB

WALLS MAY BE SPECULARLY REFLECTING (MIRROR-LIKE)– IF CELL SPACING $\ll$ LEGEND DETAIL.

$Y = X$

(b) SQUARE

$$A_{CELL} = Y^2$$

$$L_V (cd/m^2) \geq \frac{I_V (cd)}{A_{CELL} (m^2)}$$

Figure 2.4.5-9   LED Clusters – Honeycomb or Square Patterns for Very Large Legend Area.

2.5.1 Device Characterization for Communications and Signalling

Specifications of luminous intensity or luminous sterance are not directly applicable in most signalling applications. However, they can be converted to radiometric quantities if the luminous efficacy, η_v, is known. If η_v is not given in the data sheet it can be found by the method described in Section 7.1. Equation 2.5.1-1 gives the relationship to be used in making the conversion:

$$\eta_v\left(\frac{lm}{W}\right) = \frac{\phi_v(lm)}{\phi_e(W)} = \frac{I_v(cd)}{I_e(W/sr)} = \frac{L_v(cd/m^2)}{L_e\left(\frac{W/sr}{m^2}\right)} \qquad \text{(2.5.1-1)}$$

The most common conversion requirement is for intensity. For example, if an LED has a luminous intensity I_v = 3.5 mcd and a luminous efficacy η_v = 147 lm/W, then, since 1 cd = 1 lm/sr:

$$I_e = \frac{3.5 \times 10^{-3}\ cd}{147\ lm/W} = 23.8\ \mu W/sr \qquad \text{(2.5.1-2)}$$

Conversely, if a device used in signalling is used also for visual effects, if η_v is known, the photometric properties can be obtained from given radiometric quantities.

Another important parameter in signalling is the radiation pattern. This is usually given in the form of $I_r(\theta)$, the relative intensity normalized at θ = 0 so $I_r(0)$ = 1.0. The intensity at any angle is then the product of the axial intensity multiplied by $I_r(\theta)$.

To obtain the amount of flux that is radiated into a cone of half-angle θ, or included-angle 2θ, it is necessary to evaluate the integral:

$$\phi(\theta) = \int_0^\theta I_e I_r(\theta)\,[2\pi\sin\theta]\,d\theta \qquad \text{(2.5.1-3)}$$

Here I_e is the axial intensity. The integral can be done mathematically if $I_r(\theta)$ is a reasonably describable function. If it is not, a summation is done by the method of Section 7.3.4.

When used with an optical system for which the **numerical aperture** (N.A.) is given, the amount of flux the LED radiates into the system is found from equation 2.5.1-3, since $\theta = \sin^{-1}$(N.A.). To aid designers in quickly evaluating this flux, the data sheets for the new 670 nm and 700 nm devices give the results of equation 2.5.1-3 in normalized form:

$$\frac{\phi(\theta)}{I_e} = \int_0^\theta I_r(\theta)\,[2\pi\sin\theta]\,d\theta \qquad \text{(2.5.1-4)}$$

Thus the flux into a particular N.A. is the product of the specified axial intensity, multiplied by the normalized integral:

$$\phi(N.A.) = I_e \times \left(\frac{\phi(\theta)}{I_e}\right)\Bigg|_{\theta\,=\,\sin^{-1}(N.A.)} \qquad \text{(2.5.1-5)}$$

2.5.2 Flux Properties in Signalling

Wavelength compatibility can be of critical importance in dealing with mediums having sharply varying spectral transmittance or reflectance, or with detectors such as photoconductors, since they have relatively narrow spectral response. The effective flux coupling is found by spectral integration of the product of all spectrums involved in the system, as shown in Figure 2.5.2-1. In addition to those shown in the figure, there may also be other spectral effects, and these should be included in the product to be integrated.

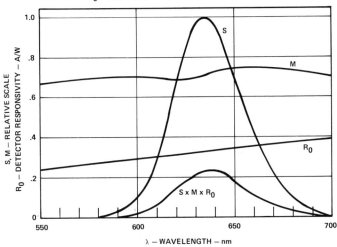

Figure 2.5.2-1   Integration to Derive Coupling of Spectral Source, Medium, and Detector.

Stability of the LED flux is sometimes also a matter of concern. It is unwise to depend on any randomly chosen LED or IRED to be a source of stable flux, even if the forward current is precisely regulated. LEDs used as optical standards would be no exception except that they are elaborately heat-sunk and are operated intermittently at a forward current that is far below rated maximum; even with these precautions, the ambient temperature must be noted and appropriate correction applied.

For applications requiring stable flux from potentially unstable LEDs (e.g. for photometer or radiometer transfer standard), the best technique is the use of a beam splitter

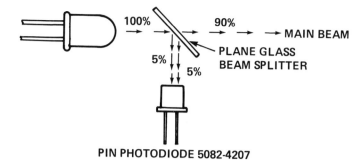

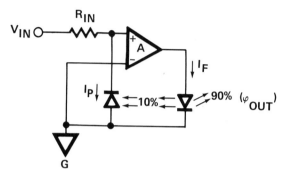

PIN PHOTODIODE 5082-4207

Figure 2.5.2-2   Servo Stabilization; Main Beam Flux Linearly Related to Input Voltage.

and a PIN photodiode in a servo system as shown in Figure 2.5.2-2. This arrangement is especially good at wavelengths below 800 nm where PIN photodiodes have a temperature coefficient of zero. If the LED used in this arrangement has a very narrow beam angle, the relative position of the photodiode should be adjusted to give the same numerical aperture (N.A.) of the optical system into which the main beam is directed. Obviously, mechanical movement of the components is not tolerable.

2.5.3   Lens System with LEDs

A fundamental principle of optics is that sterance is a constant. That is, as the image passes from one image position to another through successive lenses, the intensity per unit area of each image is reduced only by reflection loss (Fresnel loss) at the lens surface. The same is true for the image-to-object sterance ratio if the object is a lambertian emitter. It is also very nearly true when the object is an LED, as long as the cone angle for flux acceptance of the optical system is not so large that there is great deviation from a lambertian pattern. This deviation can be derived from the relative intensity, $I_r(\theta)$ of the radiation pattern, and expressed as:

$$\Delta_L = (1 - I_r(\theta)/\cos \theta) \qquad (2.5.3-1)$$

Regardless of how non-lambertian the radiation pattern is, the quantity of flux coupled into an optical system of given numerical aperture (N.A.) can be calculated from the LED radiation pattern. "Numerical aperture" is defined in Figure

2.5.3-1 and the appropriate flux integration method can be selected from Section 7.3.4.

Although N.A. is usually used in characterizing a finitely-focussed optical system, a handy relationship using the f/no relates the incidance at a target, E_T, to the sterance of a source:

$$\frac{E_T}{L_S} = \frac{\pi\tau}{4(f/no)^2(1 + \frac{d_T}{d_S})^2} \qquad (2.5.3-2)$$

In equation 2.5.3-2 the transmittance, τ, depends only on reflection losses and is usually greater than 0.7. The source-to-lens distance, d_S, and lens-to-target distance, d_T, are, respectively, the object distance, d_o and image distance, d_i relating to focal length, f, and magnification, m.

$$\frac{1}{d_o} + \frac{1}{d_i} = \frac{1}{f} \qquad (2.5.3-3)$$

$$m = \frac{d_i}{d_o} \qquad (2.5.3-4)$$

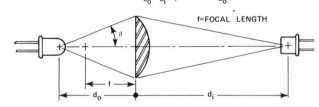

(a) BASIC DEFINITION OF NUMERICAL APERTURE, N.A. = sin θ

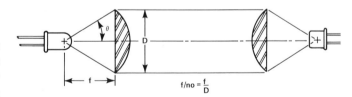

(b) FOR INFINITY-FOCUSED LENS, N.A. = $\frac{1}{2(f/no)}$

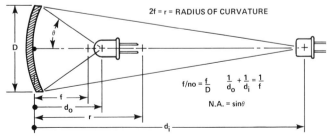

(c) THE SAME FORMULAS AS FOR THIN LENSES APPLY TO CURVED MIRRORS. LOWER f-NUMBERS ARE MORE EASILY OBTAINED WITH CURVED MIRRORS THAN WITH LENSES.

Figure 2.5.3-1   Simple Thin Lens Formulas and Definition of "Numerical Aperture".

The ratio of focal length to lens diameter is the f/no.

The relationships of equations 2.5.3-2, -3, -4 are not adequate for designing precision optics but are good enough for most signalling applications of LEDs.

One application where a lens (or lenses) enhance LED performance is in aperture or edge sensing as in Figure 2.5.3-2. The edge moves down to cut the beam between source and sensor (target). With no lens, the beam diameter is as large as the LED and the "sharpness" of the cutting depends on how small the LED or the sensor can be. Moreover, the source-to-sensor coupling is at the mercy of the inverse square law.

The ideal situation for edge or hole sensing is to have an extremely tiny beam diameter at the beam cutting edge and a large flux coupled through to the detector when the edge is not breaking the beam. Thus two lenses are ideal. The "source lens" images the LED at the plane of the beam-breaking edge, then the "target lens" images the LED image on the sensor.

Confocal lens adjustment is obtained whenever two optical systems having the same N.A. are focused on the same point from opposite directions. All the flux entering the confocal point from one lens passes through it and enters the other lens. In Figure 2.5.3-2b, if the "target" lens had a shorter focal length, it could still be confocal, but d_T would be shorter.

Applying equation 2.5.3-2 to the arrangement of Figure 2.5.3-2b, with $d_S = d_T$ yields the incidance to sterance ratio (E_T/L_S) shown in the figure. By comparing this result to that of the NO-LENS arrangement in Figure 2.5.3-2a, the coupling improvement factor of 5.26 is found. Not only is the coupling improved, but the beam diameter at the beam-cutting edge is reduced by the ratio $d_S/d_{SL} \approx 3$.

In reflective pattern sensing, it is possible to operate without lenses, but, as in edge sensing, the signal and resolution are improved by the use of lenses, as in Figure 2.5.3-3. The LEDs used to irradiate the bar code should not be imaged at the plane of the code; although this would increase the flux coupled to the photodiode, it might cause interference patterns from interaction between the code and details of the LED image. That is the reason a cluster should be used for LED 1; a single LED 1 could cause an undesirable interference pattern. That is also the reason that LED 2 (if it is used instead of LED 1) should be imaged by LENS 2 at the plane of LENS 1, or slightly to the left of it. With the image of LED2 focused at the plane of LENS1, it will be defocussed at the bar code plane and the possibility of pattern interference is reduced.

The line-resolution of the code is the diameter of the photodiode multiplied by d_{oD}/d_{iD}. In accordance with equation 2.5.3-2 the d_{iS}/d_{oS} ratio should be made as small as possible, consistent with the imaging requirements.

2.5.4 No-Lens Signalling

In some applications adequate performance is obtained without lenses to modify the size and radiation pattern of the LED optical port. Two examples are discussed here: smoke detection and tachometry.

Smoke detection by the scattering or reflective method requires only proper baffling and placement of the LED and photodetector, as shown in Figure 2.5.4-1. In both the coaxial and radial arrangement, the photodetector and baffle are so positioned that in the absence of smoke, no flux from the LED reaches the photodetector. Not shown in Figure 2.5.4-1 are the supporting walls of the smoke-sensing chamber, but these must also be designed with regard to the possibility of stray flux -- either from the LED or outside. If the stray flux reaching the photodetector is steady, sensitivity is reduced only by the square root of the stray flux amplitude. However, if the stray flux is variable, sensitivity is linearly reduced.

For room-style smoke detectors either the coaxial or the radial arrangement can be used, but for smoke detection in a flue, smoke-stack, or air-duct, the radial arrangement is preferred because there is no obstruction by the baffle to passage of air or smoke. In very large diameter ducts or stacks, projection lenses for both the LED and the photodiode should be used. The optical arrangement should maximize the volume defined by the intersection of the beam projected by the LED with the reception beam of the photodetector. This intersection volume must, of course, exclude the walls of the duct or flue.

In tachometry, a beam-breaking edge can be used, but without lensing the optical ports of the LED and photodetector might be too large to give adequate resolution. For example, if a 25mm diameter disc is to resolve a shaft rotation in one-degree increments, the line spacing at the perimeter would be:

$$x = \frac{\Delta\theta \cdot \pi \cdot D}{360^\circ} = 0.22 \text{ mm} \qquad (2.5.4\text{-}1)$$

By placing next to the rotating disc a stationary "shutter" disc having the same line spacing, when the rotating disc moves there are alternate periods of flux transmittance and no transmittance at intervals as small as whatever line spacing can be achieved. This allows the optical ports of the LED and the photodetector (on opposite sides of the disc pair) to have arbitrarily-sized optical ports. If each disc has

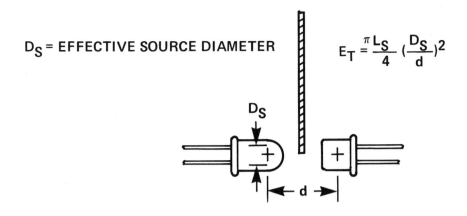

D_S = EFFECTIVE SOURCE DIAMETER

$$E_T = \frac{\pi L_S}{4}\left(\frac{D_S}{d}\right)^2$$

(a) WITHOUT LENSES

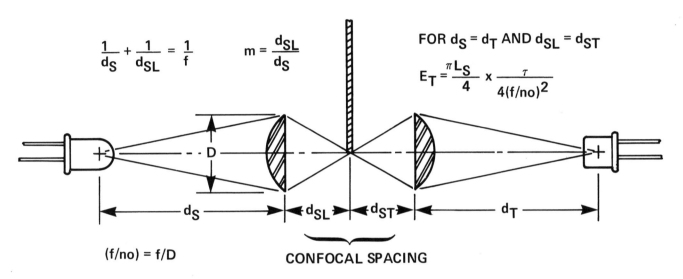

$$\frac{1}{d_S} + \frac{1}{d_{SL}} = \frac{1}{f} \qquad m = \frac{d_{SL}}{d_S}$$

FOR $d_S = d_T$ AND $d_{SL} = d_{ST}$

$$E_T = \frac{\pi L_S}{4} \times \frac{\tau}{4(f/no)^2}$$

(f/no) = f/D

CONFOCAL SPACING

(b) WITH LENSES

$$\frac{\text{LENS}}{\text{NO LENS}} \text{ INCIDANCE RATIO} = \frac{\tau}{4(f/no)^2} \times \left(\frac{d}{D_S}\right)^2 \approx 5.26$$

ASSUMING τ = .8, (f/no) = 0.65, d = 10 mm, D_S = 3 mm

Figure 2.5.3-2   Edge Sensing; Lenses Increase Coupling and Sharpen Precision.

n lines per revolution, the photodetector output will rise and fall n times per revolution. This gives shaft speed information, but no directional information.

If directional information is desired, a stationary disc with (n+1) lines next to a moving disc with n lines per revolution will produce a Moire pattern rotating n times per shaft revolution in a direction opposite that of the shaft. With equal alternating light/dark lines, the Moire pattern will be dark on one side of the shaft and have 50% transmittance diametrically opposite. If the stationary disc has (n+2)

lines, the Moire pattern has two cycles per revolution and rotates $n/2$ times per shaft revolution in opposite direction. With (n+3) lines, the Moire has three cycles per revolution and rotates $n/3$ times per revolution, again in opposite direction. To obtain Moire rotation in the same direction as the shaft, the stationary disc line spacing should be (n-1), (n-2), etc. Obviously, the stationary disc pattern need not go all the way around the shaft; it needs to cover only 180° of the Moire pattern cycle to allow LED/photodetector pairs to be spaced at 90° of the Moire cycle.

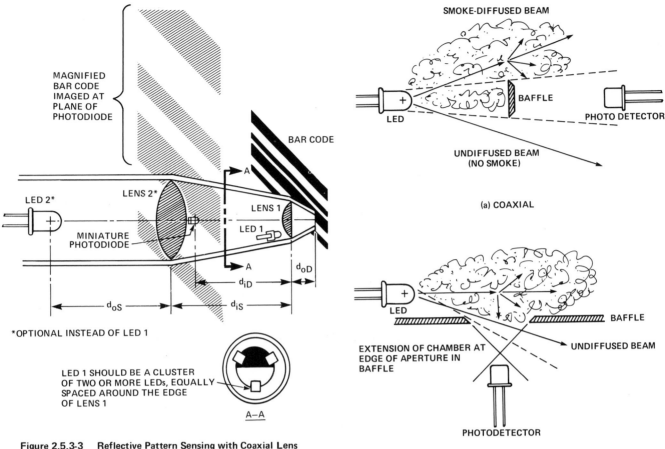

MAGNIFIED
BAR CODE
IMAGED AT
PLANE OF
PHOTODIODE

BAR CODE

LED 2*

LENS 2*

LENS 1

LED 1

MINIATURE
PHOTODIODE

d_{oD}

d_{iD}

d_{oS}

d_{iS}

*OPTIONAL INSTEAD OF LED 1

LED 1 SHOULD BE A CLUSTER
OF TWO OR MORE LEDs, EQUALLY
SPACED AROUND THE EDGE
OF LENS 1

A—A

Figure 2.5.3-3   Reflective Pattern Sensing with Coaxial Lens
Arrangement.

SMOKE-DIFFUSED BEAM

BAFFLE

PHOTO DETECTOR

LED

UNDIFFUSED BEAM
(NO SMOKE)

(a) COAXIAL

LED

BAFFLE

UNDIFFUSED BEAM

EXTENSION OF CHAMBER AT
EDGE OF APERTURE IN
BAFFLE

PHOTODETECTOR

(b) RADIAL

Figure 2.5.4-1   Reflective-Type Smoke Detector Optical
Arrangements.

The electrical circuit to be used in detecting the direction and speed of the shaft rotation is very simple, as seen in Figure 2.5.4-2. The dual J-K flip-flop in Figure 2.5.4-2(a) is the basic direction detecting element giving one pulse per Moire cycle at Q_1 for clockwise (CW) rotation and one pulse per Moire cycle at Q_2 for anticlockwise (ACW) rotation. While Q_1 pulsates, Q_2 is steady low, and vice versa, so direction and velocity information is available. If Q_1 and Q_2 are connected to the UP and DOWN inputs, respectively, of an up/down counter, the count will indicate shaft position providing there are no extraneous pulses. That is the reason for the $\overline{Q}$-to-J feedback. After a CW clocking of Q_1 (CK_1), Q_1 cannot return to zero until the Moire pattern has moved 90°, either forward (NORMAL CR_1) or backward (BACKWARD CR_1). It is therefore essential that BACKWARD CR_1 be accompanied by clock (CK_2) of Q_2. This will allow a DOWN count to cancel an UP count if the shaft is vibrating through more than 90° of the Moire cycle.

To make sure there is no race at the CK and CR inputs, hysteresis should be used in the photodetector amplifier/comparator, as in Figure 2.5.4-2(b). The feedback diodes, D_3 and D_4 are mainly for high speed operation to restrict the input and output voltage excursions. Hysteresis

can (and should) be used if they are omitted, but the values will be different. $I_{P,LH}$, the photocurrent needed to cause an L-to-H transition at the output will be the same:

$$I_{P,LH} = \frac{V_{CC}}{R_3} \qquad (2.5.4\text{-}2)$$

but the photocurrent at which an H-to-L transition occurs will be:

$$I_{P,HL} = \frac{V_{CC}}{R_3} \left(\frac{R_2}{R_1 + R_2}\right) \qquad (2.5.4\text{-}3)$$

Also, V_{OH} will rise to within a few millivolts of V_{CC}, but $V_{OL} \approx 0$.

The line patterns forming the Moire patterns need not be integrally related. Although the Moire pattern will always shift through 360° for a one-cycle shift of the moving pattern, the physical spread of the Moire pattern can be made whatever is necessary using the relationships in Figure 2.5.4-3(a) for radially aligned bars. Linear motion, of course, can be interpreted by differentially spaced parallel bars, using the same relationships.

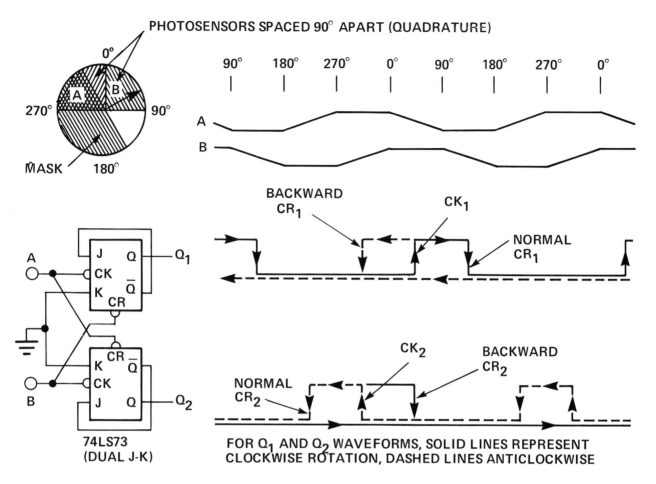

PHOTOSENSORS SPACED 90° APART (QUADRATURE)

74LS73
(DUAL J-K)

FOR Q_1 AND Q_2 WAVEFORMS, SOLID LINES REPRESENT
CLOCKWISE ROTATION, DASHED LINES ANTICLOCKWISE

(a) BASIC DIRECTIONAL SENSING. FOR CLOCKWISE ROTATION, Q_1 PULSATES;
FOR ANTICLOCKWISE ROTATION, Q_2 PULSATES.

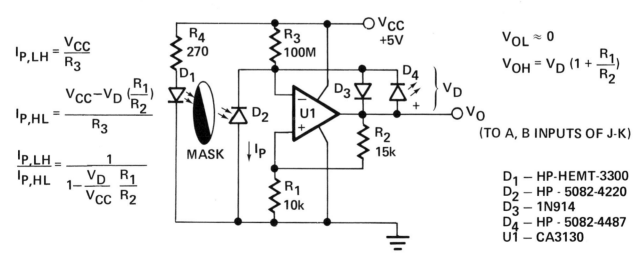

$$I_{P,LH} = \frac{V_{CC}}{R_3}$$

$$I_{P,HL} = \frac{V_{CC} - V_D \left(\frac{R_1}{R_2}\right)}{R_3}$$

$$\frac{I_{P,LH}}{I_{P,HL}} = \frac{1}{1 - \frac{V_D}{V_{CC}} \frac{R_1}{R_2}}$$

$$V_{OL} \approx 0$$

$$V_{OH} = V_D \left(1 + \frac{R_1}{R_2}\right)$$

(TO A, B INPUTS OF J-K)

D_1 – HP-HEMT-3300
D_2 – HP - 5082-4220
D_3 – 1N914
D_4 – HP - 5082-4487
U1 – CA3130

(b) COMPARATOR WITH HYSTERESIS TO PREVENT TROUBLE WITH RACES AT
J-K INPUTS WHEN BACKWARD CLEAR OCCURS.

Figure 2.5.4-2   Quadrature Phase Detector for Tachometry with
Direction Sensing.

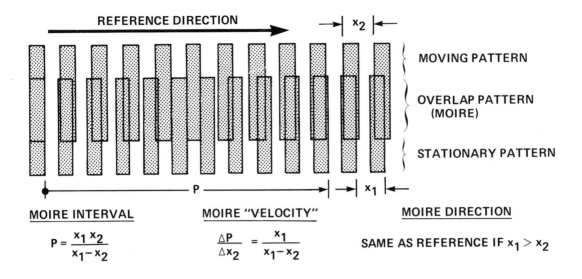

REFERENCE DIRECTION

MOVING PATTERN

OVERLAP PATTERN (MOIRE)

STATIONARY PATTERN

MOIRE INTERVAL

$$P = \frac{x_1 x_2}{x_1 - x_2}$$

MOIRE "VELOCITY"

$$\frac{\Delta P}{\Delta x_2} = \frac{x_1}{x_1 - x_2}$$

MOIRE DIRECTION

SAME AS REFERENCE IF $x_1 > x_2$

(a) PARALLEL-BAR MOIRE FROM DIFFERENTIALLY-SPACED LINES

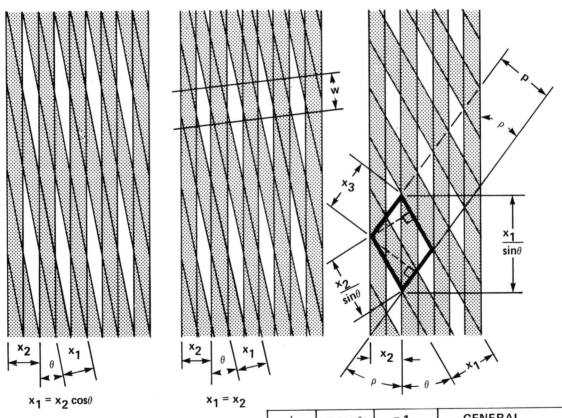

$x_1 = x_2 \cos\theta$

$x_1 = x_2$

(b) DIAMOND MOIRE FROM ANGULAR POSITIONING OF ARBITRARILY SPACED LINES

x_1/x_2	$= \cos\theta$	$= 1$	GENERAL
ρ	$\frac{\pi}{2}$	$\frac{\pi}{2} - \frac{\theta}{2}$	$TAN^{-1}\left(\frac{\sin\theta}{x_1/x_2 - \cos\theta}\right)$
P	$\frac{x_2}{\tan\theta}$	$\frac{x_2/2}{\sin\theta/2}$	$\frac{x_1 x_2}{\sqrt{x_1^2 + x_2^2 - 2x_1 x_2 \cos\theta}}$

Figure 2.5.4-3   Moire Patterns Used to Obtain Direction Information in Tachometry.

Moire patterns for quadrature direction sensing can also be produced using an angularly positioned stationary pattern of arbitrary line spacing, as in Figure 2.5.4-3(b). The geometrical derivation of the Moire cycle interval, P, and angle, ρ, from the line spacing x_1 and x_2 at angle θ is seen from the parallelogram. The parallelogram has one side $(x_2/\sin\theta)$ and a diagonal $(x_1/\sin\theta)$ with included angle θ, so the other side (x_3) of the parallelogram is found from the Law of Cosines:

$$x_3 = \sqrt{(\frac{x_1}{\sin\theta})^2 + (\frac{x_2}{\sin\theta})^2 - 2(\frac{x_1 x_2}{\sin^2\theta})\cos\theta} \qquad (2.5.4\text{-}4)$$

The two right triangles in the parallelogram are similar so:

$$\frac{P}{x_2/\sin\theta} = \frac{x_1}{x_3} \qquad (2.5.4\text{-}5)$$

Solving equations 2.5.4-4,5 for P gives the general formula in Figure 2.5.4-3(b). Notice that if $\theta = 0$, this formula becomes the same as that for the differentially spaced bars in Figure 2.5.4-3(a). The pattern angle, ρ, is derived from the triangle formed in the right half of the parallelogram by applying the Law of Sines:

$$\frac{x_2/\sin\theta}{\sin\rho} = \frac{x_1/\sin\theta}{\sin(\pi-\theta-\rho)} \qquad (2.5.4\text{-}6)$$

which, solved for ρ, gives the general formula in Figure 2.5.4-3(b).

For either parallel bar or angular Moire pattern, the recommended detector "window" width W is less than 90° of the Moire cycle, that is P/4. Making it larger does not give appreciably higher photocurrent and requires overlap of 90° spaced photodetectors. Taking the photocurrent for W = P/4 as the normal:

W/P	RELATIVE "LIGHT" CURRENT	"LIGHT / DARK" RATIO
0.25	1.000	7.01
0.10	.434	19.00
0.05	.223	39.00

A straight-line, angularly positioned stationary pattern can also be used with a radial moving pattern, yielding curved Moire patterns. However, if the radius of curvature is large enough, relative to the line spacing, quadrature spaced sensors with simple geometry can still be used.

2.5.5 Signalling Over Long Distances

Straightforward signal transmission as well as reflective pattern sensing at large distances ($>$1m) can be done with LEDs, but auxiliary optics are required.

The relationship between the sterance of the LED source and the incidance at the target was given in equation 2.5.3-2 and is repeated in Figure 2.5.5-1(a). By applying equation 2.5.3-3 the (f/no) drops out and the incidance/sterance ratio is:

$$\frac{E_T}{L_S} = \tau\,(\frac{\pi}{4}\,D_S^2)\,\frac{1}{d_T^2} \qquad (2.5.5\text{-}1)$$

Notice that the term in parentheses is just the area of the lens. It is as if an enlarged source, having an area equal to the lens and having the sterance of the LED were radiating according to the inverse square law $1/(d_T)^2$ toward the receiving detector. If another lens of diameter D_R, and focal length f_R, is placed in front of the receiver and focused on the apparent source, then again applying equation (2.5.3-2) the net result gives the incidance E_R at the receiver:

$$\frac{E_R}{L_S} = \tau_S(\frac{\pi}{4}D_S^2)\,\frac{1}{d_T^2}\cdot\times\,\tau_R(\frac{\pi}{4}\,D_R^2)\,\frac{1}{f_R^2} \qquad (2.5.5\text{-}2)$$

With a source imaged on a diffusely reflecting target, the image becomes a source which can then be imaged on a receiver. The expression for the incidance to sterance ratio in such a situation is given in Figure 2.5.5-1(b). Except for the reduction (R_T/π) due to target reflectance, the expression is very nearly the same as equation 2.5.5-2; that is, the incidance/sterance ratio varies as the product of lens areas and inversely as the square of the product of focal length times separation distance. Applying this principle yields the result in Figure 2.5.5-1(c) for coaxial mirror optics. The flat secondary mirror is not the best arrangement, but serves to illustrate the principle. In practice, the secondary mirror (the back of the source mirror) would be slightly convex to increase the focal distance to the detector so it would not require as broad a reception angle.

Both the systems of Figure 2.5.5-1(b and c) can be used for such applications as reflective scanning, remote obstruction sensing, etc. Focussed on an empty space, with low background reflectance, they can be used for smoke detection.

For straightforward signal transmission, lenses are limited by the inverse square law applied to the separation distance. Also they require a rigid mechanical structure. For these reasons, fiber optics are much more convenient and, literally, more flexible.

The principle of light transmission by a fiber optic is shown in Figure 2.5.5-2. Snell's law states:

$$n_1 \sin\theta_1 = n_2 \sin\theta_2 \qquad (2.5.5\text{-}3)$$

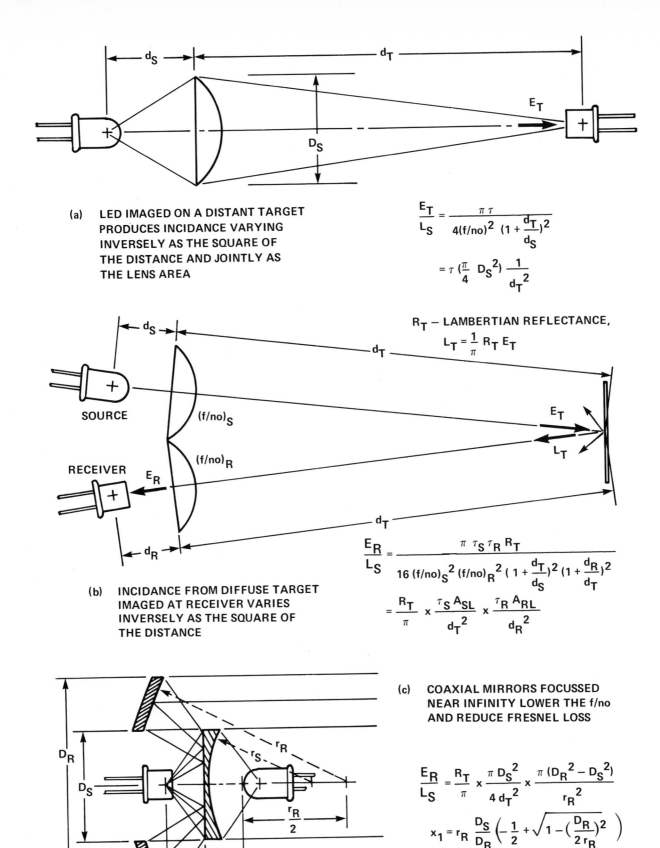

(a) LED IMAGED ON A DISTANT TARGET PRODUCES INCIDANCE VARYING INVERSELY AS THE SQUARE OF THE DISTANCE AND JOINTLY AS THE LENS AREA

$$\frac{E_T}{L_S} = \frac{\pi \tau}{4(f/no)^2 \, (1 + \frac{d_T}{d_S})^2}$$

$$= \tau \, (\frac{\pi}{4} \, D_S^2) \, \frac{1}{d_T^2}$$

R_T – LAMBERTIAN REFLECTANCE,

$$L_T = \frac{1}{\pi} \, R_T \, E_T$$

(b) INCIDANCE FROM DIFFUSE TARGET IMAGED AT RECEIVER VARIES INVERSELY AS THE SQUARE OF THE DISTANCE

$$\frac{E_R}{L_S} = \frac{\pi \, \tau_S \, \tau_R \, R_T}{16 \, (f/no)_S^2 \, (f/no)_R^2 \, (1 + \frac{d_T}{d_S})^2 \, (1 + \frac{d_R}{d_T})^2}$$

$$= \frac{R_T}{\pi} \times \frac{\tau_S \, A_{SL}}{d_T^2} \times \frac{\tau_R \, A_{RL}}{d_R^2}$$

(c) COAXIAL MIRRORS FOCUSSED NEAR INFINITY LOWER THE f/no AND REDUCE FRESNEL LOSS

$$\frac{E_R}{L_S} = \frac{R_T}{\pi} \times \frac{\pi \, D_S^2}{4 \, d_T^2} \times \frac{\pi \, (D_R^2 - D_S^2)}{r_R^2}$$

$$x_1 = r_R \, \frac{D_S}{D_R} \left(-\frac{1}{2} + \sqrt{1 - (\frac{D_R}{2 \, r_R})^2} \right)$$

Figure 2.5.5-1   Long-Range Direct and Reflective Signalling Using Lenses and Mirrors.

2.52

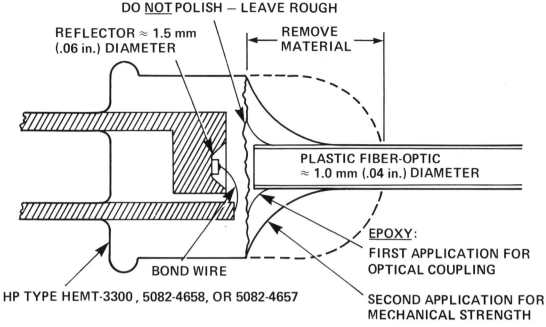

DO <u>NOT</u> POLISH — LEAVE ROUGH

REFLECTOR ≈ 1.5 mm
(.06 in.) DIAMETER

REMOVE MATERIAL

PLASTIC FIBER-OPTIC
≈ 1.0 mm (.04 in.) DIAMETER

<u>EPOXY</u>:

FIRST APPLICATION FOR
OPTICAL COUPLING

SECOND APPLICATION FOR
MECHANICAL STRENGTH

BOND WIRE

HP TYPE HEMT-3300 , 5082-4658, OR 5082-4657

(a)   EMITTER ASSEMBLY

1. REMOVE MATERIAL AS FAR AS POSSIBLE WITHOUT BREAKING BOND WIRE.

2. WET ROUGH SURFACE WITH FAST-CURING CLEAR EPOXY; ADJUST POSITION OF FIBER IN "WET" EPOXY UNTIL MAXIMUM FLUX IS OBTAINED FROM FREE END —— HOLD UNTIL CURED.

3. ADD EPOXY (2ND APPL'N) FOR MECHANICAL STRENGTH.

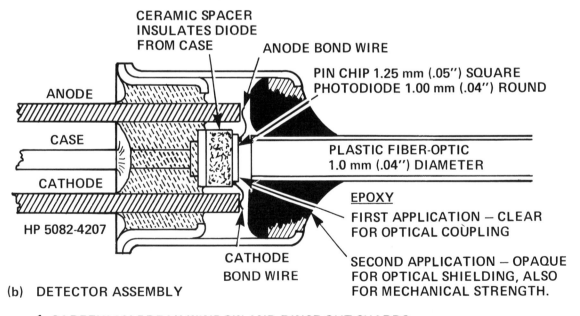

CERAMIC SPACER
INSULATES DIODE
FROM CASE

ANODE BOND WIRE

PIN CHIP 1.25 mm (.05") SQUARE
PHOTODIODE 1.00 mm (.04") ROUND

ANODE

CASE

PLASTIC FIBER-OPTIC
1.0 mm (.04") DIAMETER

CATHODE

<u>EPOXY</u>

FIRST APPLICATION — CLEAR
FOR OPTICAL COUPLING

HP 5082-4207

CATHODE
BOND WIRE

SECOND APPLICATION — OPAQUE
FOR OPTICAL SHIELDING, ALSO
FOR MECHANICAL STRENGTH.

(b)   DETECTOR ASSEMBLY

1. CAREFULLY BREAK WINDOW AND RINSE OUT SHARDS.

2. WET PHOTODIODE SURFACE WITH FAST-SETTING CLEAR EPOXY; ADJUST FIBER FOR MAXIMUM PHOTOCURRENT — HOLD UNTIL CURED.

3. ADD OPAQUE EPOXY TO MINIMIZE STRAY AMBIENT FLUX.

Figure 2.5.5-3   Fiber Optic Assembly Recommendation for
Coupling LED to PIN Photodiode.

For a ray passing from a medium of refractive index n_1 into a medium n_2, and incident at an angle θ_1 with the surface vector of the medium boundary, θ_2 is the refraction angle. For $n_2 < n_1$ as θ_1 is increased, there is some angle, θ_c, at which $\theta_2 = 90°$, and if $\theta_1 > \theta_c$ the ray will be totally reflected. Rays entering at angles less than θ_o can propagate; those entering at larger angles are lost in the cladding of the fiber. There is, therefore, a numerical aperture for fiber optics and the relationship of N.A. to the indices of refraction of the core (n_1) and cladding (n_2) is derived in Figure 2.5.5-2.

In coupling LEDs to fiber optics, there are two principles to bear in mind:

1. lenses are usually not much help

2. make the source diameter less than the fiber diameter and the fiber diameter less than the detector diameter.

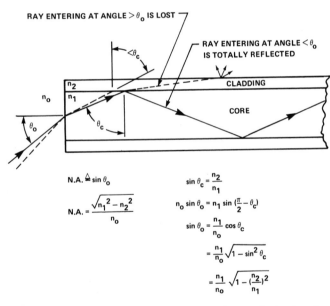

N.A. $\triangleq \sin \theta_o$

N.A. $= \dfrac{\sqrt{n_1{}^2 - n_2{}^2}}{n_o}$

$\sin \theta_c = \dfrac{n_2}{n_1}$

$n_o \sin \theta_o = n_1 \sin \left(\dfrac{\pi}{2} - \theta_c\right)$

$\sin \theta_o = \dfrac{n_1}{n_o} \cos \theta_c$

$= \dfrac{n_1}{n_o} \sqrt{1 - \sin^2 \theta_c}$

$= \dfrac{n_1}{n_o} \sqrt{1 - \left(\dfrac{n_2}{n_1}\right)^2}$

Figure 2.5.5-2   Fiber Optic Meridional Reflection Derivation of Numerical Aperture.

Lenses are not much help because, as seen in equation 2.5.3-2 the flux density cannot be made larger than it is at the surface of the source. The end of the fiber should, therefore, be as close as possible to the LED chip (die).

If the fiber diameter is less than the emitting area of the LED, flux will be lost. Even if the fiber diameter is larger than the LED emitting area, some flux is lost due to the N.A. limit. That is, fibers with large N.A. couple with lower insertion loss.

Again, at the receiving end, flux exiting from the fiber and missing the detector is lost, so a suitably large area detector should be used.

These principles are applied in the fiber optic assembly illustrated in Figure 2.5.5-3. The reason for selecting a plastic fiber optic is mostly cost, but also simplicity of assembly. Large N.A. plastic fibers of the kind shown in the figure are available from DuPont and from International Fiber Optics.

Although the plastic fiber has a large N.A., (low insertion loss) transmission loss (in dB/km) tends to be rather high. As seen in Figure 2.5.5-4, the fibers with lower transmission loss have higher insertion loss; furthermore, such fibers are usually more costly. Also shown in Figure 2.5.5-4 is the inverse-square-law coupling loss of a lens system.

When joining LEDs and photodiodes to lower-loss fiber bundles with lower transmission loss, it is usually more effective to first attach a stub ($\approx$100mm) to the device, then use a connector, such as those developed by AMP Incorporated to join the stub to the bundle. To attach a bundle directly to the LED requires first potting the end of the bundle in a binder to keep the fibers from spreading during attachment to the LED. Such spreading would raise insertion loss even further.

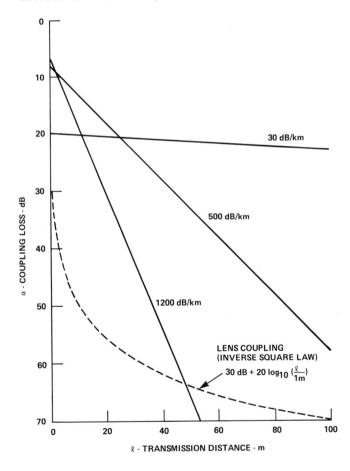

Figure 2.5.5-4   Insertion and Transmission Loss of Fiber Optics Compared with Lenses.

2.5.6 Film Exposure

In exposing photographic film with radiation from LEDs, it may or may not be necessary to take careful account of the spectral response of the emulsion. If there is doubt, the film data should be checked. The use of colored filters may be regarded as modifying the spectral response of the film.

Emulsion having a flat response over the spectrum of the LED requires only proper use of the typical exposure equation:

$$H = E t \qquad (2.5.6\text{-}1)$$

The units of E and t depend on how the film exposure requirement is given. If H is in "meter candle seconds" (mcs), then E is in lux (lx) and t is in seconds. If H is in "ergs per square centimer" (erg/cm^2) a conversion can be made to relate this to radiant incidance, E_e in μw/cm^2, and exposure time:

$$
\begin{aligned}
1 \text{ erg/cm}^2 &= 10^{-7} \text{ joule/cm}^2 \\
&= 10^{-7} \text{ w} \cdot \text{sec/cm}^2 \qquad (2.5.6\text{-}2) \\
&= 10^{-1} (\mu\text{w/cm}^2) \cdot \text{sec}
\end{aligned}
$$

Inserting appropriate units in equation 2.5.6-1 gives the exposure relationships:

$$H \text{ (mcs)} = E_v \text{ (lx)} \cdot t\text{(sec)} \qquad (2.5.6\text{-}3)$$

$$H \text{ (erg/cm}^2) = \frac{10 \cdot E_e \ (\mu\text{w/cm}^2) \cdot t \ (\text{sec})}{10} \qquad (2.5.6\text{-}4)$$

The value of either E_v or E_e depend on how the LED is coupled to the film. With no lens, and the LED a distance, d, from the film:

$$E_v \text{ (lx)} = 1000 \frac{I_v \text{ (mcd)}}{d^2 \text{ (mm}^2)} \qquad (2.5.6\text{-}5)$$

$$E_e \text{ (lx)} = 100 \frac{I_e \ (\mu\text{w/sr})}{d^2 \text{ (mm}^2)} \qquad (2.5.6\text{-}6)$$

If a lens is used, i.e., a camera, the relationship in Figure 2.5.5-1(a) applies, with d_T/d_S being negligibly small, so that:

$$E_v(\text{lx}) = \frac{\pi \tau}{4 \ (\text{f/no})^2} \cdot L_v \text{ (cd/m}^2) \qquad (2.5.6\text{-}7)$$

$$E_e \ (\mu\text{w/cm}^2) = \frac{\pi \tau}{4(\text{f/no})^2} \cdot L_e \left(\frac{\mu\text{w/sr}}{\text{cm}^2}\right) \qquad (2.5.6\text{-}8)$$

Where spectral response can be ignored, the ASA film speed is related to minimum exposure requirement, H_m by:

$$H_m = 0.8/S_{ASA}* \qquad (2.5.6\text{-}9)$$

*ref: ANSI PH2.5−1972

In equation 2.5.6-9, H_m is in lux seconds (or meter candle seconds) and can be used directly in equation 2.5.6-3 because spectral response is not considered. Combining equation 2.5.6-3, -7, -9 and assuming $\tau = 0.8$ yields the result:

$$\frac{(\text{f/no})^2}{t(\text{sec})} = \pi \ L_v \ (\text{cd/m}^2) \cdot S_{ASA} \qquad (2.5.6\text{-}10)$$

for exposure to a density 0.1 above base-plus-fog density. In most applications, however, this is a marginal exposure. For a satisfactory film record of an LED or LED display, the exposure should be 10 to 100 times the value obtained from equation 2.5.6-10, either by lowering the f/no or increasing the time, t, the sterance L_v, or the film speed used in recording. The exposure requirement in erg/cm^2 for use in equation 2.5.6-4 is obtained from the film characteristic curve (D-H curve) for wavelength ranges in which the film response is nearly constant over the LED spectrum.

To account for spectral effects, the incidance and exposure time requirements must be found by integrating:

$$\frac{E_e \ (\mu\text{w/cm}^2) \cdot t \ (\text{sec})}{10} = \frac{\int \varphi_r (\lambda) \ d\lambda}{\int S(\lambda) \cdot \varphi_r(\lambda) d\lambda} \ (\text{erg/cm}^2) \qquad (2.5.6\text{-}11)$$

The spectral sensitivity, $S(\lambda)$ is the spectral inverse of the exposure requirement for a particular density, as given by the film D-H curves and spectral response curves. $S(\lambda)$ has the units $(\text{erg/cm}^2)^{-1}$. $\varphi_r(\lambda)$ is the relative spectral output of the LED and has no units. If a spectral filter is used, its spectral transmittance should be included as a factor in the integrand of the integral in the denominator of equation 2.5.6-11.

Section 3

Optoisolators

3.0 OPTO-ISOLATORS

3.1 Optoisolator Theory

An opto isolator consists of a photon emitting device whose flux is coupled through optically transparent insulation to some sort of photodetector. The photon emitting device may be an incandescent or neon lamp, or an LED. The transparent insulation may be air, glass, plastic, or fiber-optic. The photodetector may be a photoconductor, photodiode, phototransistor, photoFET, or an integrated combination photodiode/amplifier. Various combinations of these elements result in a wide variety of input characteristics, output characteristics, and coupled characteristics. This discussion will be limited to the sort of optoisolator having an LED input, with a thin layer of transparent insulation separating it from a solid-state photodetector. The construction of such an optoisolator is shown in Figure 3.3.1-1, and the schematic representations are shown in Figure 3.3.1-2.

3.1.1 Photo Emitter

In the design of a photoemitter for an optoisolator, the main concern is optimization of the coupling to the photodetector. The parameters to be optimized are gain, bandwidth, optical port, and electrical characteristics.

A low series resistance is desirable, so a GaAs-based photoemitter is the best choice. A low forward voltage is also desirable, but not as important as optimization of the gain and bandwidth.

For GaAs-based photoemitters, with $GaAs_{1-x}P_x$ epitaxy, adjustment of x affects the wavelength, efficiency (gain), and speed of response (bandwidth). Gain, G, and bandwidth, B, normalized relative to their values at $x = 0$, are shown in Figure 3.1.1-1, along with a curve of the gain-bandwidth product, GB. It is clear that the gain-bandwidth product is optimized for $x \approx 30\%$. For this reason, all HP optoisolators use photoemitters in which the $GaAs_{1-x}P_x$ epitaxy is produced with $x \approx 30\%$, giving a wavelength of $\lambda \approx 700$ nm.

With $x \approx 30\%$, the forward voltage is very nearly that of a "standard" red LED, which is a convenience in some digital applications (Section 3.6.1).

Optical port considerations for the photoemitter of an optoisolator differ considerably from those of an LED. LEDs are made with an annular emitting region around a centered bonding pad to give a large ratio of apparent-to-actual emitting area. For an optoisolator the emitting area is as small as possible, consistent with current density consideration, and the bonding pad is offset. The offset bonding pad allows minimal obscuration (shadowing)

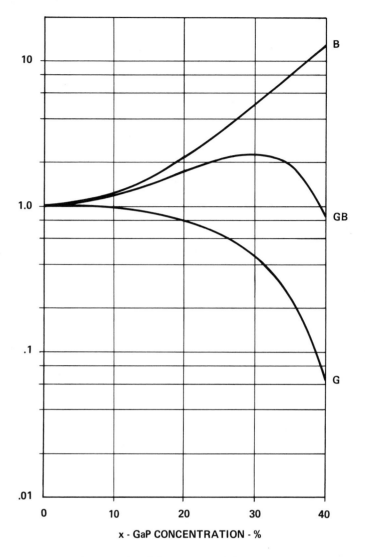

Figure 3.1.1-1   Gain (G) · Bandwidth (B) Optimization in Direct Bandgap Photoemission.

of the emitting area and allows the detector to be closely coupled. The small size reduces the flux loss at the edges and minimizes the variability of coupling due to variability in the spacing and alignment of the photodector's sensitive region.

3.1.2 Optical Medium

In selection of the optical medium, insulation is a vital consideration if the emitter-to-detector distance is very small (e.g., "sandwich" construction). If the distance is very large, such as through fiber-optics, lenses, or other medium (reflective or transmissive), the insulation is less important but spectral transmittance becomes significant, especially where plastics are involved. The effect of spectral transmittance as related to spectral properties of the emitter and the detector is discussed in Section 2.5.2.

3.1

Most optoisolators use a transparent junction coating to reduce Fresnel loss at the surfaces of the emitter and the detector. Since junction coating compounds are non-conductors, this provides insulation as well. In addition, HP optoisolators have a layer of FEP film (transparent Teflon) between emitter and detector to insure good insulation (see Section 3.3).

Fresnel losses result from reflection when flux passes from one material to another having a different index of refraction. The fraction reflected is given by:

$$R = \left(\frac{n_2 - n_1}{n_2 + n_1}\right)^2 \qquad \text{(3.1.2-1)}$$

The fraction transmitted is then derived as:

$$\tau = \frac{4}{2 + \dfrac{n_2}{n_1} + \dfrac{n_1}{n_2}} \qquad \text{(3.1.2-2)}$$

For an optoisolator having a GaAsP emitter (n = 3.6) and a silicon detector (n = 3.5), the importance of having a coupling medium other than air (n = 1) is shown in Figure 3.1.2-1, applying equation 3.1.2-2 to both the changes of refractive index.

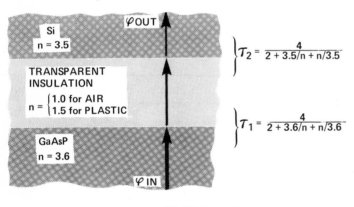

$$\frac{\varphi_{OUT}}{\varphi_{IN}} = \tau = \tau_1 \tau_2 = \begin{cases} 0.470 \text{ for } n = 1 \\ 0.698 \text{ for } n = 1.5 \\ 0.99990 \text{ for } n = \sqrt{(3.6)(3.5)} \end{cases}$$

$$\frac{\tau (n = 1.5)}{\tau (n = 1.0)} = 1.49$$

Figure 3.1.2-1   Refractive Index Effects in Optical Coupling Medium of Optoisolators.

3.1.3   Photodetector

The basics of photodetection by photodiodes are discussed in Section 4.1. Silicon photodiodes are very good detectors but usually require ancillary gain elements to produce adequate signal levels. To add gain with elements external to the optoisolator package is an inconvenience, raises the package count in the system, and degrades performance (CMR, speed). For these reasons, it is desirable to include the gain element within the optoisolator package. This can be done in either of two ways:

(a)   Hybrid - costly, but permits separate optimization of the photodiode and its amplifier;

(b)   Integrated - reduces assembly cost but compromises performance.

There are two ways to integrate the photodetector and amplifier. One of these is the use of a phototransistor, in which the photodetecting region is the collector-base junction. The other is the use of a photodiode whose photocurrent is amplified by a transistor separately integrated on the same chip. The phototransistor is less complicated to produce but has some inherent performance disadvantages relative to the photodiode/transistor: poor linearity and low speed of response.

A linearity comparison is shown in Figure 3.1.3-1. The phototransistor's non-linearity results from the flow of collector current in the collector-base junction, thus causing reduction in the collector-base depletion region which in turn reduces its responsivity. Notice that raising the collector voltage restores the depletion region and reduces the non-linearity. In the photodiode/transistor, the collector current does not flow in the photodiode -- even if the cathode of the photodiode is connected to the collector in what is called the "phototransistor connection" (because it can be operated as a two-terminal device). Best linearity, however, is obtained by keeping fixed the voltage across the photodiode.

The speed of response of a phototransistor is inherently slow because the collector-base junction must be large in order to capture photons, and therefore has a large ($\approx$20 pF) junction capacitance amplified by Miller effect (see Figure 3.1.3-2). In the photodiode/transistor, the photodiode capacitance can be made lower ($\approx$10 pF) because the junction thickness does not affect the gain of the separately integrated transistor. More importantly, however, the collector-base capacitance to which Miller effect applies is extremely small ($\approx$0.5 pF). Furthermore, as seen in Figure 3.1.3-2, much of this is contributed by capacitance between external pins and pin connections. For this reason, if no base connection is required, the rise/fall time of HP 5082-4350/51 optoisolators is improved 30% to 40% by removal of the base pin. If this is impractical, or if a base connection is required, the board layout should minimize capacitance between the base and collector leads.

Because the HP photodiode/transistor isolator is an integrated (not hybrid) device, the voltage at the emitter of the output transistor must always be at or below the voltage at any other point in the isolator IC. For example, if the emitter is driving some above-ground point, such as

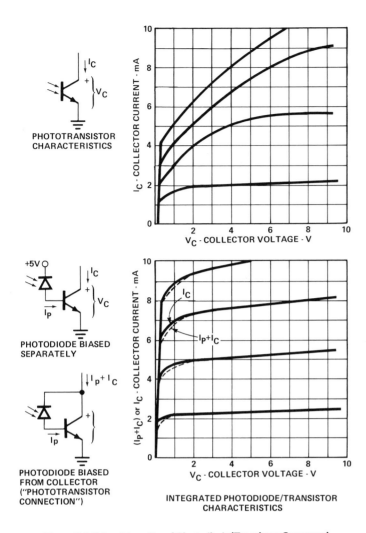

PHOTOTRANSISTOR CHARACTERISTICS

PHOTODIODE BIASED SEPARATELY

PHOTODIODE BIASED FROM COLLECTOR ("PHOTOTRANSISTOR CONNECTION")

INTEGRATED PHOTODIODE/TRANSISTOR CHARACTERISTICS

Figure 3.1.3-1 Linearity of Photodiode/Transistor Compared to that of Phototransistor.

another transistor, and if base bypass resistors are used to enhance the speed of response, the resistor to the isolator transistor base cannot be grounded. Correct connection is shown in Figure 3.1.3-3.

Nor can the emitter be permitted to float. For example, if the photodiode alone is to be employed, the emitter should be connected to the base. Failure to do so yields the results shown in Figure 3.1.3-4. The collector may float but must not be permitted to become negative with respect to the emitter, so it also should be connected to the emitter.

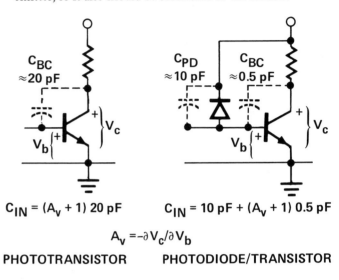

$C_{IN} = (A_v + 1)$ 20 pF

$C_{IN} = 10$ pF $+ (A_v + 1)$ 0.5 pF

$A_v = -\partial V_c / \partial V_b$

PHOTOTRANSISTOR   **PHOTODIODE/TRANSISTOR**

Figure 3.1.3-2 Phototransistor Input Capacitance Compared to Photodiode/Transistor.

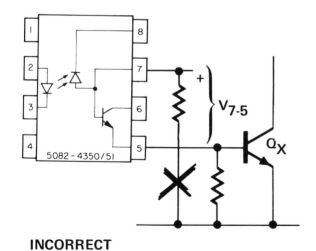

INCORRECT

RESIDUAL CHARGE ON BASE OF Q_X MAY MAKE $V_{7-5} < 0$ AND CAUSE EXTREMELY SLOW OPERATION

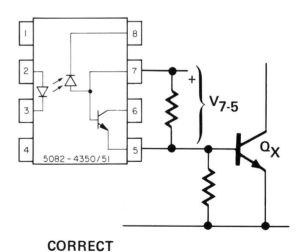

CORRECT

V_{7-5} CANNOT BECOME NEGATIVE

Figure 3.1.3-3 Correct Connection of Base Bypass Resistor with Emitter Above Ground.

3.3

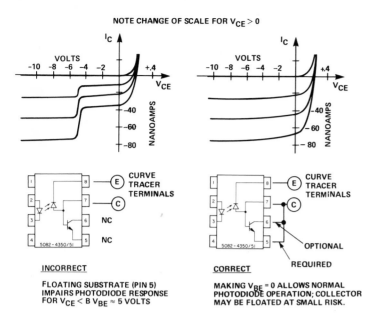

INCORRECT

FLOATING SUBSTRATE (PIN 5) IMPAIRS PHOTODIODE RESPONSE FOR $V_{CE} < B\ V_{BE} \approx 5$ VOLTS

CORRECT

MAKING $V_{BE} = 0$ ALLOWS NORMAL PHOTODIODE OPERATION; COLLECTOR MAY BE FLOATED AT SMALL RISK.

Figure 3.1.3-4   Floating Substrate Impairment of Integrated Photodiode Performance.

3.1.4   Amplifier Options

Integration of a photodiode on the same chip with amplifying transistor(s) entails some tradeoffs. The silicon resistivity cannot be as high as that for an optimal PIN photodiode because it would then be impossible to integrate reasonable gain/speed transistors. On the other hand, silicon resistivity low enough to make optimal transistors would not permit a reasonably good photodiode. Within the constraints of this tradeoff, ($C/A < 90$ pF/mm^2 for the photodiode and GB > 500 MHz for the transistor), a variety of detector-amplifier configurations can be executed.

Analog amplifiers in HP optoisolators are of two types, shown in Figure 3.1.4-1. One is the single transistor with the photodiode anode connected to the base and the cathode separated for reverse bias connection (e.g. V_{CC}). The other type is called a split-darlington; the emitter of the first transistor is connected to the base of the second transistor in the usual darlington manner, but the collector of the second is separated (split) from the collector of the first transistor to allow the output collector to drop to a lower V_{CE} in saturation. In the usual darlington circuit, the collectors are common, and the lowest possible output V_{CE} is the sum of V_{BE} of the second transistor and $V_{CE\,(SAT)}$ of the first. This results in a minimum output V_{CE} of ≈ 800 mV. With the *split*-darlington configuration, the output V_{CE} is not held up by the V_{BE} of the output transistor and may drop to less than 100 mV. This is especially important in digital applications requiring a low output V_{CE} for good noise immunity. The base of the second transistor is available for strobing and for speed enhancement with resistive bypassing -- handy features in

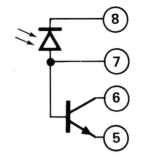

SINGLE TRANSISTOR AMPLIFIER

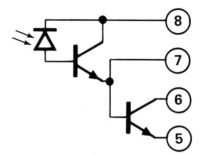

SPLIT-DARLINGTON AMPLIFIER

$$V_{CE2(SAT)} = V_{BE2} + V_{CE1(SAT)}$$

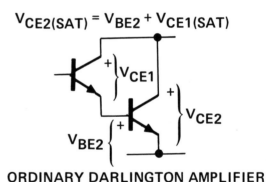

ORDINARY DARLINGTON AMPLIFIER

Figure 3.1.4-1   Analog-Type Photocurrent Amplifiers.

digital applications. The high gain makes the split-darlington amplifier useful in many low power applications and as a gain element in a closed loop where the input to the main amplifier must be isolated from the reference comparator (e.g. analog power supply regulation). However, because the base of the *first* transistor is not available for feedback connection, the split-darlington amplifier is not optimal for linear applications.

For digital applications at moderate data rates, the analog types can be used if their gain/output-current capabilities are adequate. For high data rates, the transistors must be operated at lower closed-loop gain in order to achieve the required bandwidth. HP detector/amplifiers for digital applications have a high speed linear amplifier driving a Schottky-clamped output transistor, as shown in Figure 3.1.4-2. Bias for the photodiode is decoupled from V_{CC} to reduce the possibility of "chatter" (oscillatory transition

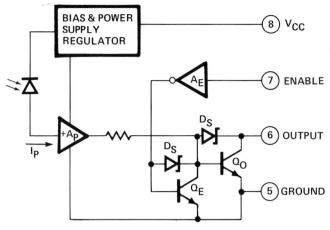

A_E — INVERTER FOR ENABLE **INPUT**
A_P — NON-INVERTING PHOTOCURRENT AMPLIFIER
D_S — SCHOTTKY DIODES CLAMPING Q_E AND Q_O
Q_E — ENABLE SWITCH (OFF, UNLESS "ENABLE" IS LOW)
Q_O — OUTPUT TRANSISTOR

Figure 3.1.4-2   High Gain, High Speed Photocurrent Amplifier for Digital Applications.

from one logic state to another due to regenerative coupling via the power supply line). The linear amplifier has a tendency to be unstable if the high-frequency impedance of the power supply is not low enough. For this reason, a low-inductance bypass capacitor (0.01 μF ceramic disc) should be installed adjacent to each isolator of this type. This, and other chatter-suppression techniques are discussed in Section 3.3.

The Schottky clamp is a metal-silicon diode in parallel with the base-collector junction of the output transistor. A metal-silicon (Schottky) diode has a lower turn-on voltage than a P-N junction, so when the transistor is driven into saturation, the Schottky diode bypasses the current which would otherwise enter the base-collector junction. With reduced current entering the base-collector junction, there is a proportionate reduction in the charge to be removed when the transistor is to be turned off, and thus the Schottky clamp reduces the turn-off delay. The attendant drawback is the 400 mV higher $V_{CE\,(SAT)}$ for a Schottky clamped transistor.

The "enable" input has threshold voltage and input current levels resembling a TTL input. However, it is not necessary to apply a pullup resistor to insure its remaining high. Unless the enable is connected to a strobe, it may simply be left open. The strobe applied to the enable input may be either open-collector or active-pullup.

With the enable high, analog operation is possible because there is no hysteresis. However, the dynamic range is limited. The lower limit is the threshold input current for operation in the active region -- this threshold may be as high as 4 mA. The upper limit is the maximum dissipation rating on the output. Because of a touchy bias situation, analog separation is not recommended for designs to be mass produced.

The reason for omitting hysteresis was not to permit analog operation, but rather to permit maximum data rate. With hysteresis, there would be a higher immunity to both differential- and common-mode noise but the shifting threshold would reduce the data rate capability. The effect of hysteresis on data rate is the opposite of peaking.

3.2   Parameter Characterization

As seen in Section 3.1, a variety of input, output, and coupled characteristics are possible, depending on the choices of input and output devices and the manner in which they are optically coupled. Figure 3.2-1 illustrates a type of optoisolator with respect to which all the important parameters, analog as well as digital, can be visualized and described. Listed in order of their importance:

1.   Isolation (Common Mode Rejection, CMR)

2.   Insulation (maximum V_{I-O})

3.   Speed (modulation bandwidth, propagation delay)

4.   Reverse coupling (ground looping)

5.   Forward coupling (Current Transfer Ratio, CTR; fan-out)

3.2.1   Isolation

The fundamental purpose of an isolator, whether optically, electrically, or magnetically coupled, is to enhance, in the output, the ratio of differential-mode to common-mode signals. Optical coupling is superior to electric or magnetic coupling because the photons that carry the differential mode signal do not carry any charge or require a magnetic flux to support their movement. Thus, the only means by which the common-mode signal can appear in the output are by:

(a)   modulating the input current and

(b)   stray capacitive coupling.

Means (a) is not really a property of the optoisolator and can be eliminated by impedance balancing, as seen in Figure 3.2-1, where R_{P1} and R_{P2} can be selected or adjusted to make $\partial I_F / \partial e_{CM} = 0$. Isolation characterization is therefore examined with respect to (b) stray capacitive coupling.

Analog isolation is seen simply as the ratio of the relative effects of differential-mode voltage, e_{DM}, and common-mode voltage, e_{CM}, on output current, I_C, expressed as the Common-Mode Rejection Ratio, CMRR:

$$CMRR = \left(\frac{\partial I_C / \partial e_{DM}}{\partial I_C / \partial e_{DM}} \right) \left(\frac{e_{DM}}{e_{CM}} \right) \qquad (3.2.1\text{-}1)$$

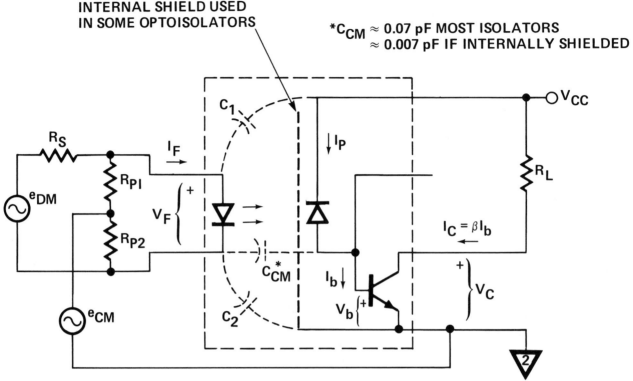

INTERNAL SHIELD USED
IN SOME OPTOISOLATORS

*$C_{CM} \approx 0.07$ pF MOST ISOLATORS
≈ 0.007 pF IF INTERNALLY SHIELDED

- ISOLATION:

    ANALOG: CMRR $\triangleq \dfrac{\partial I_C / \partial e_{DM}}{\partial I_C / \partial e_{CM}}$ CMR = 20 $\log_{10}$ (CMRR)

    DIGITAL: MAX. TOLERABLE COM. MODE FOR PROPER DIGITAL OUTPUT
    CMRV $\triangleq$ MAX. TOLERABLE e_{CM} CMTR $\triangleq$ MAX. TOLERABLE de_{cm}/dt
    CM_H: CMRV or CMTR for $V_C \geq V_{OH}$ (INPUT DIODE OFF)
    CM_L: CMRV or CMTR for $V_C \leq V_{OL}$ (INPUT DIODE ON)

- INSULATION: $V_{I-O} \triangleq e_{CM}$ ABOVE WHICH DAMAGE MAY OCCUR.

- SPEED:
    ANALOG: 3 dB BANDWIDTH FOR dV_C / de_{CM}
    DIGITAL: DELAY IN PROPAGATION OF CHANGE IN LOGIC STATE, $e_{DM} \rightarrow V_C$

- REVERSE COUPLING: $C_{I-O} = C_1 + C_2 + C_{CM}$

- FORWARD COUPLING: CURRENT TRANSFER RATIO, CTR $\triangleq \dfrac{I_C}{I_F} \times 100\%$

Figure 3.2-1 Important Optoisolator Parameters; Illustration and
Brief Description

CMRR is often expressed in dB, in which case it is called the Common Mode Rejection, CMR:

$$\text{CMR} = 20 \log_{10} (\text{CMRR}) \qquad (3.2.1-2)$$

CMR cannot be specified without reference to the input circuit design. This is seen by analyzing equation 3.2.1-1:

$$\frac{\partial I_C}{\partial e_{DM}} = \beta \frac{\partial I_b}{\partial e_{DM}} = \beta \frac{\partial I_p}{\partial e_{DM}} = \beta \frac{\partial I_p}{\partial I_F} \frac{\partial I_F}{\partial e_{DM}} \qquad (3.2.1-3)$$

$$= \beta \frac{\partial I_p}{\partial I_F} \left(R_S + \frac{dV_F}{dI_F} \right)^{-1}$$

$$\frac{\partial I_C}{\partial e_{CM}} = \beta \frac{I_b}{\partial e_{CM}} = \beta \frac{I}{\partial e_{CM}} \, \partial (C_{CM} \frac{de_{CM}}{dt}) \qquad (3.2.1-4)$$

$$= \beta \, C_{CM} \, (2 \pi f_{CM})$$

Taking the ratio of equation 3.2.1-3 to equation 3.2.1-4 gives

$$\text{CMRR} = \frac{\beta(I_p/I_F)}{\beta \, (2\pi f_{CM} \, C_{CM}) \, (R_S + \frac{dV_F}{dI_F})} \qquad (3.2.1-5)$$

The photocurrent-to-input current ratio (I_P/I_F) and common-mode coupling capacitance, C_{CM}, are properties of the optoisolator, however, CMRR can be made arbitrarily high by making R_s and dV_F/dI_F small. Note also that CMRR decreases as the frequency, f_{CM}, of the common mode signal rises. There is a limit to this effect. The cutoff frequency of the amplifier limits response to both e_{DM} and e_{CM} making the ratio unimportant. Because of the non-isolator variables affecting CMR, any meaningful description of this parameter must give conditions for I_F, R_s and f_{CM}. Notice that CMRR is independent of the amplifier gain, but does depend on the diode-to-diode current transfer ratio, I_P/I_F. This ratio is fairly constant: $I_P/I_F \approx 0.0015$ for all HP optoisolators and is very nearly the same also for non-HP types in which the "photodiode" is the base-to-collector junction of a phototransistor. Notice also that C_{CM} is only a small part of the total input-to-output capacitance, which is given in data sheets as C_{I-O}. C_{CM} also is fairly constant; $C_{CM} \approx 0.07$ pF in most HP isolators, but in the internally shielded types $C_{CM} \approx 0.007$ pF.

At present, the internal shield is offered in only the very high speed optoisolator, 5082-4361, in which the broader bandwidth would, without the shield, allow CMRV to dip lower as in Figure 3.2.1-1, or shift downward and leftward the curve in Figure 3.2.1-2.

For the unshielded single-transistor analog types, CMR can be improved by adding a neutralizing capacitor between the collector and either of the input pins. The value of this capacitor should be $\beta \times C_{CM} \approx 7$ pF. Neutralization can also be used with dual isolators of the single-transistor type, providing the neutralizing capacitor couples each collector to its corresponding input diode. Obviously, the neutralizing capacitor must have a voltage rating compatible with the application.

Digital operation requires that the output remain in proper logic state despite interference from e_{CM}. The extent to which such interference is tolerable can be described in either of two ways. For sinusoidal e_{CM}, there is a maximum tolerable amplitude, called the Common Mode Rejection Voltage, CMRV, which, if exceeded, will cause the output to change. However, a single value of CMRV does not describe e_{CM} tolerance. Since e_{CM} is capacitively coupled, CMRV varies inversely with frequency, as in Figure 3.2.1-1, up to the amplifier cutoff frequency, beyond which the slope is positive. For non-sinusoidal e_{CM}, i.e. transients, it is more convenient to describe the maximum tolerable rate of change, $\partial e_{CM}/\partial t$, in terms of Common Mode Transient Rejection, CMTR. Just as for CMRV, a single value of CMTR does not describe $\partial e_{CM}/\partial t$ tolerance. The abscissa over which CMTR is examined may be either the amplitude, e_{CM}, or the duration, t_{TR}, of the transient, as in Figure 3.2.1-2 because:

$$e_{CM} = e'_{CM} \times t_{TR} \qquad (3.2.1-6)$$

Observations have shown that e'_{CM} is very nearly a hyperbolic function of e_{CM} according to:

$$(e_{CM} - e_{CMO}) (e'_{CM} - e'_{CMO}) \approx \text{CONSTANT} \quad (3.2.1-7)$$

The "CONSTANT" is a function of the gain and speed of the amplifier in the isolator output. The asymptotes, e_{CMO} and e'_{CMO}, describe what is intuitively clear. That is, there is some transient excursion amplitude, e_{CMO} which is small enough that no matter how high the rate of rise may be, the isolator output will remain in its proper logic state. Similarly, for a sufficiently small rate of rise e'_{CMO}, the excursion amplitude e_{CM} is not limited (except by the insulation, Section 3.2.2). These asymptotes can be described in terms of current and voltage increments Δi_b and Δv_b at the amplifier input. To some extent, therefore, there are some circuit design choices that affect the isolator's defense against common mode transients.

Notice in Figures 3.2.1-1 and 3.2.1-2 that different curves are shown, CM_H and CM_L, representing common mode voltage tolerance with the output in logic HIGH state and LOW state, respectively. In general, reducing the value of R_L (see Figure 3.2-1) raises CM_H and lowers CM_L. Raising the level of input current, I_F, raises CM_L. For sinusoidal common mode voltage, polarity is of no concern because the rate of rise equals the rate of fall. For non-sinusoidal transients, it is worth noting that if the output is in a logic low state, a positively sloped transient has no effect. Thus in Figure 3.2.1-2, the CM_H curve applies only to positive

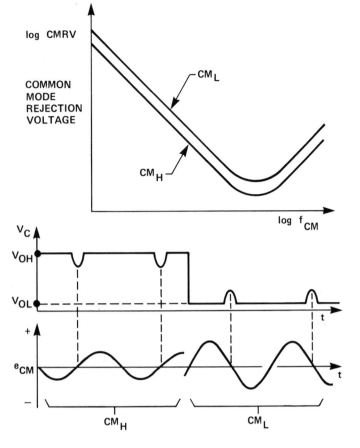

Figure 3.1.1—1 Sinusoidal Common Mode Voltage Rejection Property, CMRV.

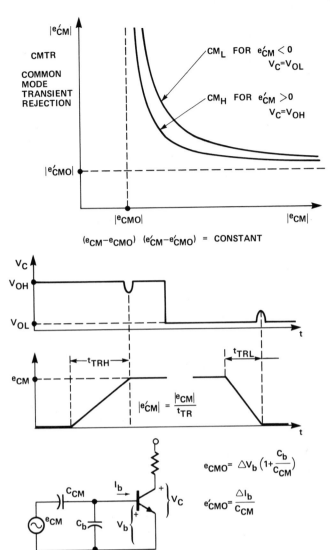

$$(e_{CM}-e_{CMO}) \ (e'_{CM}-e'_{CMO}) = CONSTANT$$

$$|e'_{CM}| = \frac{|e_{CM}|}{t_{TR}}$$

$$e_{CMO} = \Delta V_b \left(1+\frac{C_b}{C_{CM}}\right)$$

$$e'_{CMO} = \frac{\Delta I_b}{C_{CM}}$$

Figure 3.2.1-2   Transient Common Mode Voltage Rejection Property, CMTR.

transients. Conversely, in similar fashion, the CM_L curve applies only to negative transients. Therefore, if observation or other evidence suggests that transients predominantly have a higher rate of change in one direction than in the other (e.g. a sawtooth), R_L and I_F can be adjusted, as above, to selectively improve CM_H and CM_L. Such unequal rates of change are common in circuits having diodes or other nonlinear elements.

The curves in Figures 3.2.1-1 and 3.2.1-2 characterize only the static immunity to common mode interference. Common mode sinusoids and transients can also affect the dynamic performance adversely. For example, a transient occuring during a differential-mode logic transition can affect the rate of rise or fall in the output circuit, thus preventing a short bit from causing a proper logic transition at the output. Also, high frequency common mode interference can cause more than one output transition to occur during a single transition of the differential-mode signal. (See Section 3.6.2 and 3.6.6 for defensive techniques.)

Characterization of dynamic immunity to any common-mode interference can be done quite simply with an up/down decade counter as in Figure 3.2.1-3. With a reference data stream at the "up" input and the data

transmitted through the isolator at the "down" input, the counter should only toggle between two adjacent states -- either 4, 5, 4, 5 or 5, 6, 5, 6. Preset at 5 re-sets both R-S latches, causing LEDs 2 and 4 to glow. Should the isolator fail occasionally to make an output transition, the count will advance until a "carry" output sets the "carry" latch. Should extraneous transitions occur, the counter will decrement until a "borrow" output sets the "borrow" latch. Perfect operation is indicated as long as LEDs 1 and 3 remain off in the presence of e_{CM}.

CAUTION: getting a clean stream of reference data to the "up" input in the presence of e_{CM} *may pose enough difficulty to require extreme patience and heroic measures, especially if very high data rates are involved.*

The scheme shown in Figure 3.2.1-4 has the best chance of yielding valid results. Since the differential count is all that matters, the use of two $\div$ x counters allows the reference

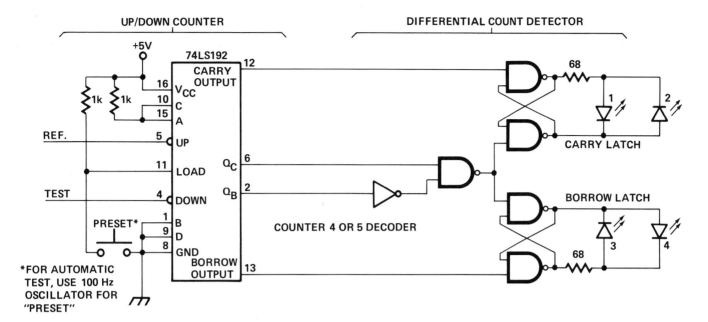

UP/DOWN COUNTER

DIFFERENTIAL COUNT DETECTOR

COUNTER 4 OR 5 DECODER

*FOR AUTOMATIC TEST, USE 100 Hz OSCILLATOR FOR "PRESET"

CONDITION		INDICATOR NUMBER (LED)				DESCRIPTION OF CONDITION
		1	2	3	4	
PASS	A	OFF	ON	OFF	ON	PERFECT OPERATION
FAIL	B	ON	ON	OFF	ON	TEST INPUT MISSING SOME COUNTS
	C	OFF	ON	ON	ON	EXTRANEOUS COUNTS AT TEST INPUT
	D	ON	ON	ON	ON	HIGH-RATE MIXTURE, CONDITIONS B&C
	E	ON	OFF	ON	OFF	PROBABLY OK – CHECK PRESET
	F	FLICKER	FLICKER	OFF	ON	OCCASIONAL CONDITION B
	G	OFF	ON	FLICKER	FLICKER	OCCASIONAL CONDITION C
	H	FLICKER	FLICKER	FLICKER	FLICKER	LOW-RATE MIXTURE, CONDITIONS B&C

Figure 3.2.1-3   Up/Down Counter, Differential Count Detector Sense Transmission Error.

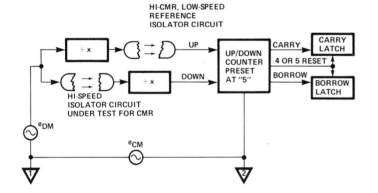

Figure 3.2.1-4   "Heroic Measures" to Assure Clean Reference Count in Dynamic CMR Test.

isolator to operate at a lower data rate so speed can be traded for better CMR. The isolator circuit under test must operate at the full data rate in the presence of the same interference. If the test is to be done with e_{DM} being a randomly coded data stream, the two ÷ x counters should be pre-set at the same count (e.g. 0000) with e_{DM} at a fixed logic state. For most consistent results, inverters should be used, as needed, so that at "preset" all counter inputs are "zero" if they are negative-edge triggered or "one" if they are positive-edge triggered.

3.2.2   Insulation

A common, but usually erroneous, assumption is that insulation can be operated at any voltage up to that at which it breaks down. Many designers are unaware of corona and its effect on insulation. Of those who are aware of corona, some believe it occurs only at exposed terminals.

Corona (also known as "partial discharge") can occur within insulation materials, particularly those having an abundance of "microvoids". Due to inhomogenous electric fields within the material, the local field across a "microvoid" can rise to a level at which there is a local breakdown resulting in a partial discharge. This partial discharge at one microvoid shifts the field so it builds up across some other microvoid. Each partial discharge causes ions that locally degrade the insulation. The cumulative effect over long periods of time (weeks or months) is to lower the terminal-to-terminal breakdown voltage.

There is a voltage, called Corona Inception Voltage, CIV, above which corona can be observed using specialized equipment capable of responding to each partial discharge. As the test voltage is raised above CIV, there is an increase in the rate of occurrence of these partial discharges. Correspondingly, there is an increase in the rate at which the insulation is degraded.

To maintain a high quality of insulation, HP optoisolators are given a special treatment called "backfilling". After molding, they are placed in a silicone oil bath in a chamber which is evacuated until the gases from the microvoids have escaped. Then the chamber is pressurized, forcing silicone oil to fill the microvoids. With this treatment, CIV is believed to be well above the rated V_{I-O}, but 100% production testing for corona is not practical; with the methods and equipment now available, it is too time consuming.

However, to insure rejection of devices with initially defective insulation, there is 100% production testing with 3000 Vdc applied for 5 seconds at 45% relative humidity, rejecting for leakage greater than $1.0 \mu A$.

The production test assures that each part meets the Underwriters Laboratories requirements for 220 Vac, 50/60 Hz operation. The UL formula for prescribing the test voltage:

line voltage = 220 Vrms
times 2 = 440 Vrms
plus 1000 Vrms = 1440 Vrms 50/60 Hz for 1 minute
or plus 20% = 1728 Vrms 50/60 Hz for 1 second
or times $\sqrt{2}$ = 2444 Vdc for 1 second

Except for those in hermetic packaging, all HP optoisolators easily meet UL requirements for status as a "Recognized Component" under File No. E55361.

Another important property of insulation is its leakage resistance. In HP optoisolators R_{I-O} is typically 10^{12} ohms. This appears to be an unreasonable claim when compared with the $1.0 \mu A$ leakage allowed in the 3000 Vdc test. Here again, it is a matter of practicality. With 3000 Vdc applied to 10^{12} ohms, the resulting current is only 3 nA. Precise observation of so small a current in a 5 second test is impractical, so the $R_{I-O} = 10^{12}$ ohms in given as typical, with periodic verification done on randomly selected samples. Both kinds of testing (R_{I-O} and V_{I-O}) are performed with all input leads connected to one side of the test voltage and all output leads to the other side -- a two-terminal test.

3.2.3   Speed of Response

For both analog and digital operation, the speed of response is highly dependent on the circuit as well as on characteristics of the optoisolator. In both analog and digital modes, the speed of response can be enhanced by use of feedback and peaking. These techniques are

discussed in Section 3.5.6 for analog operation, and in Section 3.6.3 for digital operation. In this section, discussion centers on isolator properties affecting speed of response, and how the speed is characterized.

Analog operation requires the isolator to operate in its "active region", i.e., with the output collector neither cut off nor saturated over the required excursion range. In the circuit of Figure 3.2-1, e_{DM}, R_s, R_{PI}, and R_2 should be selected for $I_F = (I_{Fdc} \pm \Delta I_F)$ such that $0 \ll V_C \ll V_{CC}$ between the excursion limits of I_F. Speed of response is then characterized either in terms of 10%-90% rise time if ΔI_F is a step function, or in terms of 3 dB bandwidth if ΔI_F is a sinusoid.

In the isolator itself, the principal bandwidth-limiting elements are the photoemitter, the amplifying transistor, and the capacitance of the photodiode. In the photodiode, the photoelectrons are created within a few picoseconds after photons enter. The resulting photocurrent flows to the base with a rise time constant as seen at the base. The base time constant depends not only on the transistor in the isolator but also on the circuit used with the transistor. If the transistor is operated common-emitter as in Figure 3.2-1, then the base time constant:

$$\tau_B = R_B (C_{PD} + C_{BC}) + \beta R_L C_{BC} \qquad (3.2.3\text{-}1)$$

where:

$C_{PD} \approx$ **10 pF is the photodiode capacitance**

$C_{BC} =$ **Base-to-collector capacitance = 0.5 pF plus stray external capacitance between collector and base connections**

$R_L =$ **Load resistance**

$R_B =$ **Dynamic resistance to ground at the base**

If no external resistance is added to bypass the base, then R_B is just the dynamic resistance at the base:

$$(3.2.3\text{-}2)$$

$$R_B = \frac{\partial V_B}{\partial I_b} = \frac{25\,mV}{I_b} = \beta \left(\frac{25\,mV}{I_C}\right) = \beta \left(\frac{25\,mV}{\frac{CTR}{100\%} \times I_{Fdc}}\right)$$

where CTR = Current Transfer Ratio (see Section 3.2.5). Then in equation 3.2.3-1

$$(3.2.3\text{-}3)$$

$$\tau_B = \beta \left(\frac{25\,mV}{\frac{CTR}{100\%} \times I_{Fdc}}\right) (C_{PD} + C_{BC}) + R_L C_{BC}$$

Substituting typical values, CTR = 20% and I_{Fdc} = 15 mA yields

$$(3.2.3\text{-}4)$$

$$\tau_B = \beta C_{BC} [R_L + 8.33\Omega (1 + \frac{C_{PD}}{C_{BC}})] = \beta C_{BC} [R_L + 175\Omega]$$

Equation 3.2.3-4 shows the importance of a low value of C_{BC}. It also shows that the base time constant is limited by R_L only if $R_L > 175\Omega$. Compare this with a phototransistor for which C_{BC} (≈ 20 pF) is larger by 40 times; also, $C_{PD} = 0$ and the base time constant is limited by $R_L > 8.33\Omega$.

Although this is not recommended for high speed operation with HP optoisolators, the transistor can be operated "common collector", as in Figure 3.2.3-1. This, incidentally, IS the recommended circuit for optimizing the speed of a phototransistor. Here the base-to-collector capacitance is not Miller-effect multiplied so the base time constant is:

$$\tau_B = (C_{PD} + C_{BC}) [R_B + (\beta + 1) R_L] \qquad (3.2.3-5)$$

Assuming the same values for CTR and I_F in the expression of equation 3.2.3-2 for R_B, equation 3.2.3-5 becomes:

$$\tau_B = (\beta + 1) (C_{PD} + C_{BC}) [R_L + \frac{R_B}{\beta + 1}] \qquad (3.2.3-6)$$

$$= (\beta + 1) (C_{PD} + C_{BC}) [R_L + \frac{8.33\Omega}{1 + 1/\beta}]$$

$$= (\beta + 1) (C_{PD} + C_{BC}) [R_L + 8.25\Omega]$$

Notice now that the base time constant is limited by $R_L > 8.25\Omega$ regardless of the relative values of C_{PD} and C_{BC}. Comparing equation 3.2.3-4 and equation 3.2.3-6, it is clear that if $C_{BC} \ll C_{PD}$ as in HP isolators, the common-emitter circuit is preferred for superior bandwidth, whereas for phototransistor types the common-collector circuit is better, particularly if the gain requirement makes $R_L \gg 8.25\Omega$.

Although in HP optoisolators the photoemitter is not usually the limiting element in speed of response, it should be mentioned. When it is current-source driven, the step response (10%-90% rise/fall time) is approximately 20 ns. This corresponds to a 3 dB bandwidth greater than 15 MHz.

For a given junction area and optical port, the remaining chip design parameters that raise the efficiency also lower the speed. This is because the photons are produced by electron-hole recombination. The photon emission rate is, therefore, proportional to the recombination rate, which is proportional to minority carrier density, which in turn is proportional to junction charge, Q_j. Thus:

$$\phi_e = \text{radiant flux} \propto Q_j \qquad (3.2.3-7)$$

Since Q_j is the product of forward current I_F, and minority carrier lifetime, τ_j:

$$\phi_e \propto Q_j = I_F \tau_j \qquad (3.2.3-8)$$

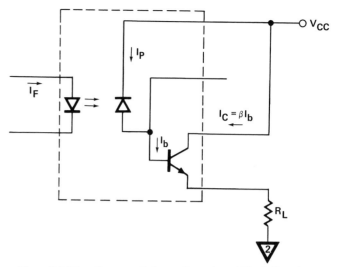

Figure 3.2.3-1 Common-Collector Operation of Single-Transistor Optoisolator.

It is clear that, for a fixed forward current, the speed (inverse of lifetime) can be increased only at the expense of efficiency.

If the photoemitter is the speed limitation in a particular circuit, some improvement can be obtained by peaking. In analog operation, peaking the photoemitter drive can also compensate for speed deficiency in the output circuit.

Digital operation requires the isolator to switch from one logic state to another. For the low state I_F (Figure 3.2-1) must be large enough to bring V_C well below some defined threshold, and for the high state I_F must be low enough to allow V_C to rise well above that threshold. Speed of response is therefore defined with respect to the time required for V_C to reach the threshold in response to a change of logic state at the input, i.e., switching I_F from one state to the other. Characterization of speed is either in terms of propagation delay or data rate.

Propagation delay is the time required for a change of logic state to propagate through the isolator and cause a change of logic state in its load. t_{PHL} is the propagation delay in causing the isolator output to drop from the high state to a specified threshold. t_{PLH} is the propagation delay in causing the output to rise from the low state to the threshold. t_{PHL} and t_{PLH} are shown in Figure 3.2.3-2.

While isolator characteristics (CTR, β, C_{PD}, C_{BC}, etc.) influence t_{PHL} and t_{PLH}, propagation delay is also influenced by the circuit. Raising I_F reduces t_{PHL} by causing a large collector current, but raises t_{PLH} by causing the output to be more deeply saturated and thus increasing the "storage" time. Raising the value of R_L also reduces t_{PHL} by reducing the current it sources to the V_C node but unfortunately this also raises t_{PLH} by reducing the current available to pull V_C up again when the input logic state is

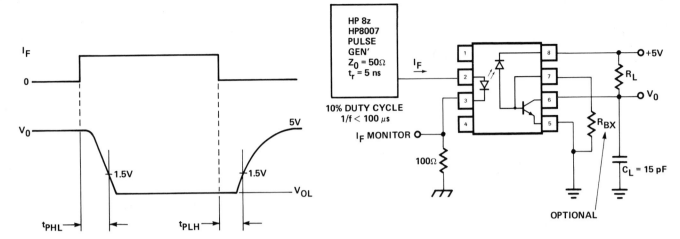

Figure 3.2.3-2   Propagation Delays, t_{PHL} and t_{PLH}; Definition and Measurement.

changed. A resistor, R_{BX} (Figure 3.2.3-2) in parallel with the base-to-emitter junction aids in turning off the transistor to reduce t_{PLH}; adding R_{BX} will require additional input current to keep the output current as high -- and t_{PHL} as low -- as it was before R_{BX} was added. The input current increment, I_{FX}, is required to make up for the current bypassed by R_{BX}, so its value depends on what type of isolator is used. As a general rule:

for single-transistor types:

$$I_{FX} = \left(\frac{V_{be}}{R_{BX}}\right)\left(\frac{I_F}{I_P}\right); I_{FX} \text{ (mA)} \approx \frac{700(V)}{R_{BX} \text{ (k}\Omega)} \qquad (3.2.3-9)$$

for split-darlington types:

$$(3.2.3-10)$$

$$I_{FX} = \frac{V_{be}}{R_{BX}} \quad \frac{I_F}{I_P} \quad \frac{I}{\beta 2} \; ; \; I_{FX} \text{ (mA)} \approx \frac{7(V)}{R_{BX} \text{ (k}\Omega)}$$

Making I_F very large, then adding R_{BX} reduces both t_{PHL} and t_{PLH}.

Data rate characterization of speed can be directly related to t_{PHL} and t_{PLH}, as shown in Figure 3.2.3-3. Data rate is defined for NRZ (non return to zero) data as the maximum rate for error-free transmission of a random data pattern. Random data implies that following an arbitrarily long string of consecutive ones or zeroes, there may be a string of ones and zeroes alternating at the maximum rate. The minimum duration of the zero or one level is the reciprocal of the maximum data rate in bits per second (b/s) whereas the maximum duration is arbitrary.

The pattern in Figure 3.2.3-3a produces the response in Figure 3.2.3-3b when observed with a 'scope triggered from the slow clock (t_2). If the scope is triggered at every step of

the fast clock (t_1), the result is the classical "eye" picture in Figure 3.2.3-3c. Data are recoverable as long as the "eye" is "open" above and below the threshold for a period of time greater than the "set-up" time required before clocking. Good practice would be to design for an "open eye" duration about twice the "set-up" time. Applying this rule gives the maximum data rate:

$$f_{NRZ(MAX)} \leqslant \frac{1}{t_{SET\text{-}UP} + t_{P(MAX)}} \qquad (3.2.3-11)$$

where $t_{P(MAX)}$ is either t_{PHL} or t_{PLH} -- whichever is the greater. The technique applies also to system characterization; that is, for the entire system (drivers, lines, wiring, etc.) a $t_{PHL(SYS)}$ and $t_{PLH(SYS)}$ can be observed and applied in equation 3.2.3-11.

With RZ (return to zero) data, or other self-clocking patterns, the concept of random data does not apply. Each bit interval includes a period of time at the high state and a period of time at the low state, as in Figure 3.2.3-4. With such a pattern, the longest time spent in either logic state cannot exceed one bit interval and the shortest is never less than half a bit interval.

Characterization of the RZ data rate in terms of t_{PHL} or t_{PLH} is valid, indeed equivalent, for systems with linear transient response and balanced threshold. However, optoisolators are usually non-linear. Furthermore, a transmission line long enough to affect the rise time will exhibit non-linear response. The characterization of RZ data rate should therefore be done with RZ data. A safe rule in judging the data rate that can be expected is:

$$f_{RZ(MAX)} < \frac{1}{2\,t_{P(MAX)}} \qquad (3.2.3-12)$$

where $t_{P(MAX)}$ is either t_{PLH} or t_{PHL}, whichever is the greater.

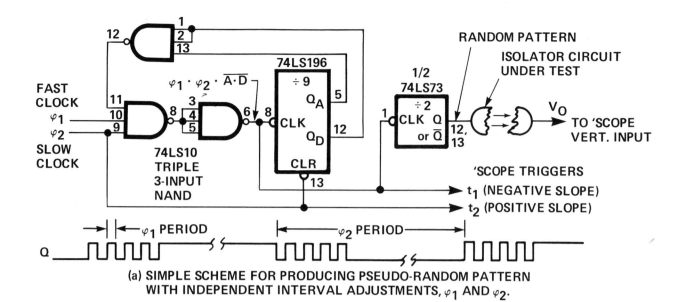

(a) SIMPLE SCHEME FOR PRODUCING PSEUDO-RANDOM PATTERN
WITH INDEPENDENT INTERVAL ADJUSTMENTS, φ_1 AND φ_2.

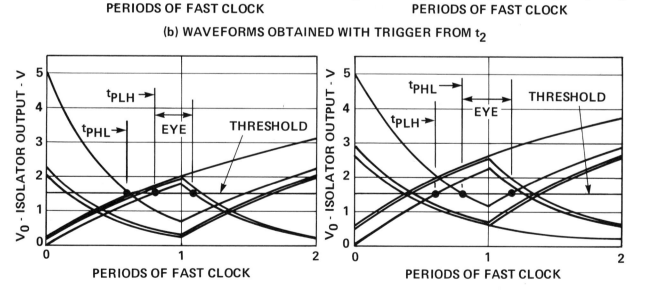

(b) WAVEFORMS OBTAINED WITH TRIGGER FROM t_2

(c) CLASSICAL "EYE" PICTURE OBTAINED WITH TRIGGER FROM t_1

Figure 3.2.3-3   Pseudo-Random Code Generator and Observation
of NRZ Data Rate.

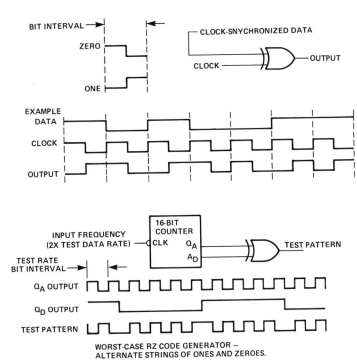

Figure 3.2.3-4 RZ Data Waveforms and Worst-Case Code Pattern Generator

3.2.4 Reverse Coupling

An important consideration in many applications of isolators is the degree to which they prevent the flow of ground loop current. Indeed, this is often the basis for use of *optically* coupled isolators in preference to other kinds. Ground loop current can be troublesome not only in the module generating the signal to be transmitted, but also in the module receiving the signal. As seen in Figure 3.2-10, voltage drops in connecting wires can cause annoying offsets.

Optoisolators virtually eliminate dc ground loops ($R_{I-O} \approx 10^{12}$ ohms), but they do allow some ac ground loop current. Internal capacitance between the input and output circuits permit ac ground loop current according to

$$i_{GL} = C_{I-O} \frac{de_{CM}}{dt} \qquad (3.2.4-1)$$

Note that the capacitance here is not merely the 0.07 pF of C_{CM} in Figure 3.2-1. The internally shielded optoisolators have a lower C_{CM}, but C_{I-O} is the same in both kinds. There are no circuit tricks that can be done to reduce C_{I-O} or its effects.

In applications where the $C_{I-O} \approx 1$ pF of most optoisolators is intolerably high, the only recourse is the selection of a type that has a larger physical separation between the input and output circuits. Such types are usually more costly because they require lenses or fiber optics to obtain adequate optical coupling between widely separated circuits.

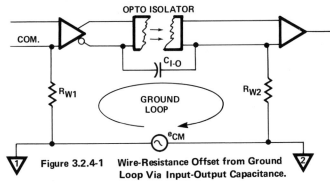

Figure 3.2.4-1 Wire-Resistance Offset from Ground Loop Via Input-Output Capacitance.

With $C_{I-O} = 1$ pF, the ground loop current that flows from a 60 Hz e_{CM} is only 533 pA per volt rms. Even with 220 V rms, the ground loop current is less than 120 nA. If several isolators are used, however, or if e_{CM} has a high frequency or voltage, the resulting ground loop current should be considered.

3.2.5 CTR (or Gain)

With respect to analog types of optoisolators (Figure 3.2-1), the gain is simply defined as the ratio of the output current to the input current. This property is called the Current Transfer Ratio (CTR) and is usually given in percent. If an analog optoisolator is to be used in a digital application, it is worth noting that the CTR is given for a very low collector voltage. This assures a designer that for the specified level of input current, the collector voltage will be adequately low if the current available from the load (I_C in Figure 3.2-1) does not exceed I_F x CTR $\div$ 100%.

For optoisolators intended mainly for digital applications, the CTR may be given differently. Rather than being described as ratio of output/input current, it may be described as the "fan out" capability relative to some particular logic family. For example, specification of a "fan out" of 8 TTL gates means that the output of the optoisolator can "sink" a current of 8 x 1.6 mA while maintaining an adequately low output voltage. Where an external pullup resistor is used, the current available from the pullup resistor must be considered in the design.

A more common way of specifying CTR for a digital application is to give the maximum value of the output voltage for given levels of input and output currents. This voltage, V_{OL}, is the output voltage at logic low. For example, if $V_{OL} = 0.4$V max at $I_F = 1.6$ mA and $I_O = 4.8$ mA, this means that with $V_O = 0.4$V, the minimum CTR is 300% and as long as the available load current does not exceed 4.8 mA, the output voltage will be less than 0.4V with an input current of 1.6 mA or more.

CTR is not constant for all levels of input current. This is due partly to the supralinearity of the photoemitter, but a larger portion of this variation is due to the change in the gain, h_{fe}, of the output amplifier, especially if it has more than one transistor. This variation is described by giving either the output current or the CTR as a function of input current.

3.14

3.3 Isolator Packaging

3.3.1 Packaging of Plastic DIP Isolators

Figure 3.3.1-1 shows the mechanical construction of Hewlett-Packard's 8 pin optically coupled isolators. The GaAsP emitter and silicon detector are die attached and wire bonded to separate four-pin lead frames. They are then covered with an inert silicone junction coating. An insulating film is sandwiched between the emitter and detector lead frames. Teflon FEP film is used as the insulating medium between the lead frames. This insulating film assures the excellent low leakage, high voltage insulation between input and output. Finally, the entire assembly is encapsulated in epoxy to insure package integrity.

The same 8 pin package can also be used for dual channel isolators. Mechanical construction of a dual is similar to the construction of a single channel isolator. Due to pin limitations on the dual, V_{CC} and GND of each detector are normally connected together and brought out to two of the four output pins. The remaining two output pins are used for the output of each detector. This configuration retains the high speed performance obtained by biasing the photodiode at a constant reverse voltage. Since the base of the output transistor is unconnected, the base to collector capacitance is lowered and the dual isolator has a somewhat shorter propagation delay than the corresponding single channel isolator. However, circuits that require feedback applied to the base or an external resistor from base to emitter to improve speed, cannot use an 8 pin dual channel isolator. The pinouts of Hewlett-Packard's 8 pin single and dual channel isolators are shown in Figure 3.3.1-2.

3.3.2 Packaging of High Reliability Isolators

Some applications require higher reliability than can be obtained with a plastic encapsulated opto isolator. In general, if an application requires integrated circuits in ceramic packages and metal can transistors for higher reliability in severe environments, then a hermetically packaged isolator should be used. Presently, hermetic isolators are either packaged in a metal can or in a 16-pin ceramic package. Figure 3.3.2-1 shows the mechanical construction of Hewlett-Packard's 16-pin hermetic optically coupled isolator. Silicon detectors are die attached and wire bonded inside a 16 pin ceramic package. GaAsP emitters are die attached and wire bonded to a separate ceramic insert. Solder preforms are applied between the ceramic package and the insert is then soldered in place. Next, the assembly is potted with transparent insulating material to improve optical coupling and electrical insulation between the emitter and detector. Finally, a metal lid is attached to the package to insure a hermetic seal. The finished optically coupled isolator can withstand storage

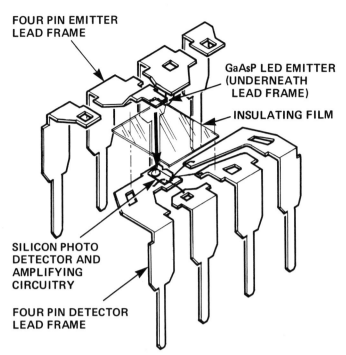

EXPLODED VIEW OF HP OPTICALLY COUPLED ISOLATOR

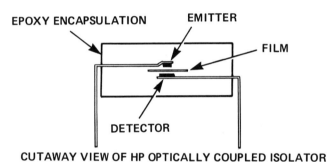

CUTAWAY VIEW OF HP OPTICALLY COUPLED ISOLATOR

Figure 3.3.1-1   Mechanical Construction of Hewlett Packard's Plastic Encapsulated Opto Isolators.

temperatures from -65°C to +150°C, operating temperature extremes from -55°C to +125°C and 98% relative humidity at 65°C without failure.

These high reliability isolators are also available with additional high reliablity screening for military applications. Hewlett Packard has established two standard high reliability test programs, patterned after MIL-M-38510, Class B. These programs are known as the TX and TXB programs.

3.3.3 Compatibility of Six and Eight Pin Isolators

In many respects, six and eight pin optically coupled isolators have direct pin compatibility. Figure 3.3.3-1 illustrates this compatibility for six pin phototransistor or photodarlington isolators and for eight pin photodiode-transistor or photodiode-split darlington isolators. The eight pin package allows a separate V_{CC} bias supply that provides high speed operation and further circuit versatility while still maintaining direct mechanical compatibility.

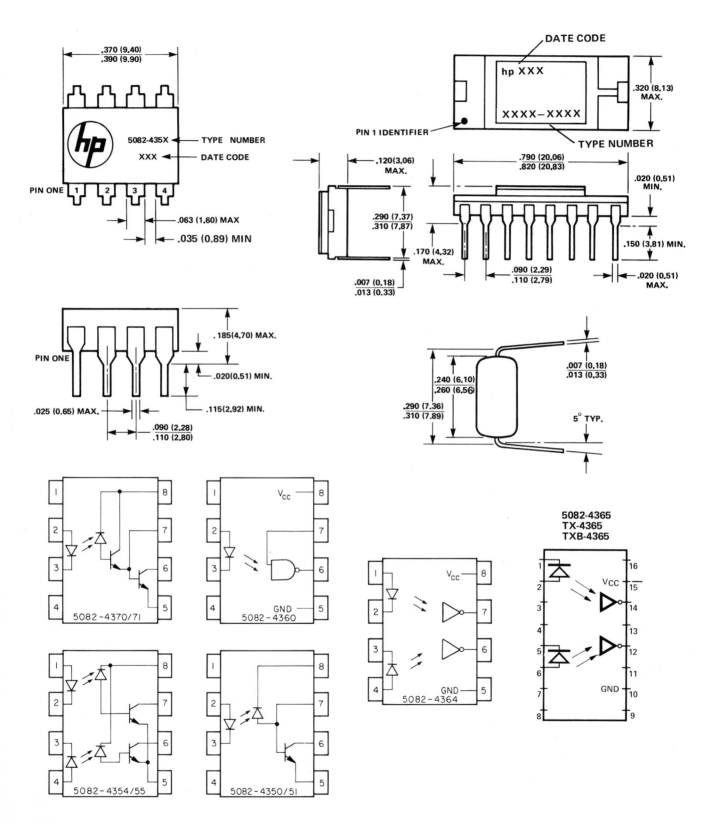

Figure 3.3.1-2   Outline Drawings and Pinouts of Hewlett Packard's
                  Opto Isolators.

3.16

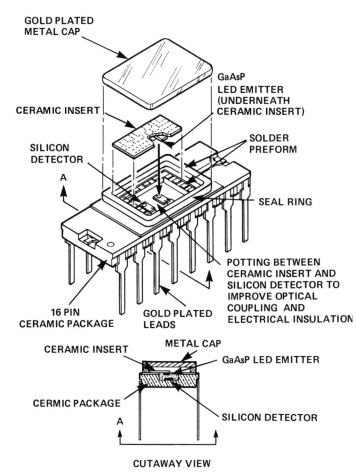

GOLD PLATED METAL CAP

GaAsP LED EMITTER (UNDERNEATH CERAMIC INSERT)

CERAMIC INSERT

SILICON DETECTOR

A

SOLDER PREFORM

SEAL RING

POTTING BETWEEN CERAMIC INSERT AND SILICON DETECTOR TO IMPROVE OPTICAL COUPLING AND ELECTRICAL INSULATION

16 PIN CERAMIC PACKAGE

GOLD PLATED LEADS

CERAMIC INSERT

METAL CAP

GaAsP LED EMITTER

CERMIC PACKAGE

SILICON DETECTOR

A

CUTAWAY VIEW

Figure 3.3.2-1 Mechanical Construction of Hermetic Opto Isolator Manufactured By Hewlett Packard.

The only extra cost of this interchangeability is an allowance for two additional holes on the printed circuit. These holes are used for pins 1 and 8 of the eight pin isolator and left open for a six pin isolator. To provide proper bias for the photodiode, pin 8 should be connected to V_{CC}. Since six pin isolators have an overall length of 8.89 mm (.350 inches) and eight pin isolators have an overall length of 9.91 mm (.390 inches), both isolators require 10.16 mm (.400 inch) spacing on standard 2.54 mm. (.100 inch) printed circuit pad spacing. Using dual channel eight pin isolators, four isolator channels can be realized in the same space as one 16 pin integrated circuit.

This direct pin compatibility holds for all of Hewlett-Packard's eight pin single channel optically coupled isolators. The 5082-4350 family and the 5082-4370 family offer switching speed advantages over phototransistor and photodarlington isolators. The split darlington configuration of the 5082-4370 family allows the output transistor to saturate as low as .1V for improved noise margin with TTL circuitry. For a V_{CC} of 5 volts, the 5082-4360 family is also directly pin compatible. The 5082-4360 family has extremely fast switching speeds, low input current requirements and a TTL compatible enable input.

3.3.4 Layout Considerations for Optically Coupled Isolators

Optically coupled isolators have excellent common mode rejection characteristics. However, for optimum common mode rejection, special considerations should be given to the circuit layout. Stray capacitance between the input circuit and the output circuit should be minimized. This can be done by physically separating all input and output circuitry on the circuitboard. For line receiver applications, the shielded cable should also be dressed properly to minimize stray capacitance. Since the base of the output transistor is especially susceptible to common mode transients, a special ground trace under the isolator can serve as a shield. To further minimize capacitive coupling into the base, pin 7 can be clipped off at the side of the package. A simpler approach in an application requiring more than one isolator is to use dual packages, because the base is unconnected.

Circuit layout can also effect propagation delays through the optically coupled isolators. Since most opto isolators have an open collector output, excessive shunt capacitance on pin 6 can limit rise and fall times of the isolator. For this reason, the logic gate being driven by the isolator should be as close to the isolator as possible and the pullup resistor mounted in close proximity to the isolator. An external resistor can be connected between pin 5 and pin 7 to reduce t_{PLH} of the 5082-4370 isolator. If this technique is used, capacitance between pin 5 and pin 6 should be minimized because of the "Miller" effect. Figure 3.3.4-1 illustrates some of these circuit layout considerations.

3.3.5 Bypass Capacitor Requirements

Some isolators such as the 5082-4360 series family require an external bypass capacitor to prevent internal oscillations. An isolator that is not functioning properly due to internal oscillations exhibits the following symptoms: the isolator has an extremely low current transfer ratio which requires an excessively high input LED current to cause the output collector to saturate. If noise is coupled into the isolator from its load, the isolator may have multiple transitions for a single input pulse. To prevent these internal oscillations, a low inductance .01 μF ceramic capacitor should be placed between pins 5 and 8 as close to the device as possible. For optimum results, the bypass capacitor should be connected in such a manner as to minimize the coupling of noise generated by the isolator load into the V_{CC} and ground connections of the isolator. This can be accomplished by using separate V_{CC} and ground lines for the isolator or isolators than those used by the digital logic. A second technique to minimize noise coupling is to lay out the printed circuit with a topology that connects the V_{CC} and ground for the digital logic between the power supply and the V_{CC} and ground of the opto isolator. These techniques

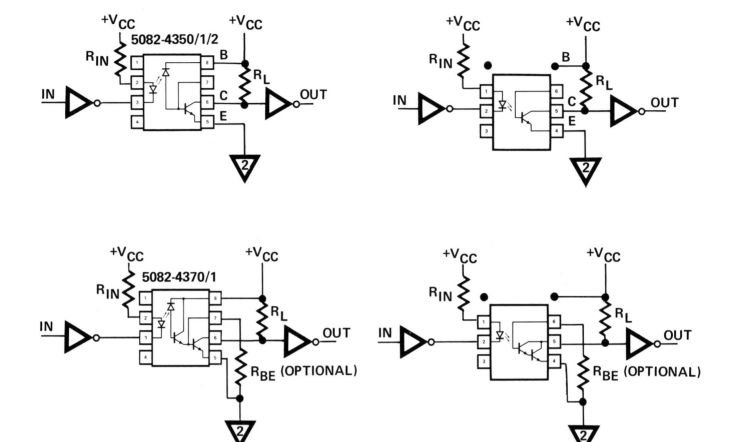

SIX AND EIGHT PIN OPTICALLY COUPLED ISOLATORS ARE DIRECTLY
PIN COMPATIBLE WHEN TWO EXTRA HOLES AND SHORT V_{CC} TRACE
ARE ADDED TO PRINTED CIRCUIT BOARD.

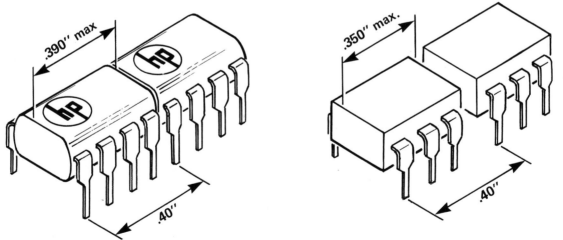

MOUNTED ON .1 INCH CENTERS, SIX PIN OPTO ISOLATORS REQUIRE
THE SAME SPACE AS EIGHT PIN OPTO ISOLATORS.

Figure 3.3.3-1   Pinout Compatibility and End Stackability of 6
and 8 Pin Isolators.

3.18

- KEEP INPUT CIRCUITRY SEPARATED FROM OUTPUT CIRCUITRY.
- ADD SPECIAL GROUND TRACE UNDERNEATH ISOLATOR.
- KEEP OUTPUT LEADS TO INTEGRATED CIRCUIT AND LOAD RESISTOR AS SHORT AS POSSIBLE.
- IF EXTERNAL RESISTOR FROM PIN 5 TO PIN 7 IS USED, KEEP LEADS AS SHORT AS POSSIBLE AND MINIMIZE CAPACITIVE COUPLING TO PIN 6.

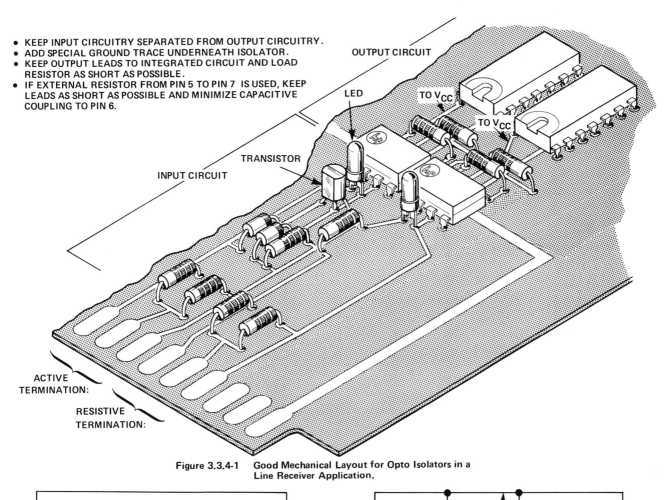

Figure 3.3.4-1   Good Mechanical Layout for Opto Isolators in a Line Receiver Application.

A.

BEST - ISOLATOR IS CONNECTED BY SEPARATE CONNECTIONS THAN THOSE USED FOR COMMON V_{CC} AND GROUND TRACES.

C.

NOT RECOMMENDED - CURRENT SPIKES GENERATED BY SWITCHING OF LOAD CAUSE NOISE INTO ISOLATOR BECAUSE OF RESISTANCE AND INDUCTANCE OF V_{CC} AND GROUND TRACES.

B.

GOOD - LOAD BEING DRIVEN BY ISOLATOR IS BETWEEN POWER SUPPLY AND ISOLATOR.

D.

NOT RECOMMENDED - SAME REASON AS CONFIGURATION C.

Figure 3.3.5-1   Recommended Placement of Bypass Capacitor for Optimum Results.

3.19

are illustrated in Figure 3.3.5-1. When several isolators are used on a single printed circuit board, a separate bypass capacitor should be connected across pins 5 and 8 of each isolator. These same bypassing techniques also apply to the 5082-4365 isolator with the exception that V_{CC} and ground are on pins 15 and 10. When several isolators are end stacked, a very clean mechanical layout can be accomplished by connecting the pullup resistor to the V_{CC} of an adjacent isolator.

In noisy environments, bypassing may also be required for the 5082-4350 series and for the 5082-4370 series isolators to prevent V_{CC} transients from being coupled into the photodiode. The voltage transients can be large enough to cause multiple transitions of the output while the collector is switching from one state to another. When extremely high common mode transients are present, sufficient energy might be coupled into the photodiode and overstress or cause a catastrophic failure of the detector. A resistor connected between V_{CC} and pin 8 of the isolator will limit these current surges to safe levels but will not affect the normal operation of the opto isolator. This technique is illustrated in Figure 3.6.2-4.

3.4   CTR Degradation

3.4.1   Introduction — Optocouplers Aging Problem

A persistent, and sometimes crucial, concern of designers using optocouplers is that of the current transfer ratio, CTR, changing with time. The CTR is defined as the ratio of the output current, I_o, of the optocoupler divided by the input current, I_F, to the light emitting diode expressed as a percentage value at a specified input current. The resulting optocoupler's gain change, ΔCTR^+, with time is referred to as CTR degradation. This change, or degradation, must be accounted for if long, functional lifetime of a system is to be guaranteed.

A number of different sources for this degradation will be explained in the next section, but numerous studies have demonstrated that the predominant factor for degradation is reduction of the total photon flux being emitted from the LED, which, in turn, reduces the device's CTR. This degradation occurs to some extent in all optocouplers.

$$^+\Delta CTR = CTR_{final} - CTR_{initial} \qquad (3.4.1\text{-}1)$$

3.4.2   Causes

The main cause for CTR degradation is the reduction in efficiency of the light emitting diode within the opto-coupler. Its quantum efficiency, η, defined as the total photons per electron of input current, decreases with time at a constant current. The LED current is comprised

primarily of two components, a diffusion current component, and a space-charge recombination current:

$$(3.4.2\text{-}1)$$

$$I_F(V_F) = \underbrace{A\,e^{qV_F/kT}}_{\text{Diffusion}} + \underbrace{B\,e^{qV_F/2kT}}_{\text{Space-Charge Recombination}}$$

where A and B are independent of V_F, q is electron charge, k is Boltzmann's constant, T is temperature in degrees Kelvin, and V_F is the forward voltage across the light emitting diode.

The diffusion current component is the important radiative current and the non-radiative current is the space-charge recombination current. Over time, at fixed V_F, the total current increases through an increase in the value of B. From another point of view, with fixed total current, if the space-charge recombination current increases, due to an increase in the value of B, then the diffusion current, the radiative component, will decrease. The specific reasons for this increase in the space-charge recombination current component with time are not fully understood.

The reduction in light output through an increase in the proportion of recombination current at a specific I_F is due to both the junction current density, J, and junction temperature, T_J. In any particular optocoupler, the emitter current density will be a function of not only the required current necessary to produce the desired output, but also of the junction geometry and of the resistivity of both the P and N regions of the diode. For this reason, it is important not to operate a coupler at a current in excess of the manufacturer's maximum ratings. The junction temperature is a function of the coupler packaging, power dissipation and ambient temperature. As with current density, high T_J will promote a more rapid increase in the proportion of recombination current.

The junction and IC detector temperature of Hewlett-Packard optocouplers can be calculated from the following expressions:

$$(3.4.2\text{-}2)$$

$$T_J = T_A + \theta_{JA}(V_F I_F) + \theta_{D\text{-}E}(V_o I_o + V_{cc} I_{cc})$$

$$(3.4.2\text{-}3)$$

$$T_D = T_A + \theta_{E\text{-}D}(V_F I_F) + \theta_{DA}(V_o I_o + V_{cc} I_{cc})$$

where the T_J is the junction temperature of the LED emitter, T_D is the junction temperature of the detector

This study is based on a total of 640 optocouplers of the 6N135 type (Figure 3.4.2-1) with 700 nm GaAs$_{.7}$P$_{.3}$ LEDs from twenty different epitaxial growth lots representing a range of n-type doping and radiance. The 6N135 allows access to measurement of the emitter degradation via the relative percentage change in photodiode current, $\Delta I_P/I_P$, as well as output amplifier β change. Stress currents of I_{FS} = .6, 7.5, 25 and 40 mA were applied to different groups of optocouplers, and at each measurement time of t = 0, 24, 168, 1000, 2000, 4000 and 10,000 hours, measurement currents of I_{FM} = .5, 1.6, 7.5, 25 and 40 mA were used to determine the CTR.

The important results to be noted are the following. First, a factor of major significance in the study of CTR degradation is the ΔCTR varies as a function of the ratio of $I_{FS}/I_{FM} \equiv R$. Large values of R will result in greater CTR degradation than at lower R values with the same magnitude of I_{FS}. However, knowledge of the ratio of I_{FS}/I_{FM} alone does not give a complete picture of degradation because ΔCTR is also dependent upon the absolute magnitude of the stress current, $|I_{FS}|$. The following data will allow the derivation of the necessary equations with which to predict ΔCTR as a function of I_{FS}, I_{FM} and time.

Figure 3.4.3-1 displays the mean and mean plus 2σ values of emitter degradation versus R for 1K, 4K, and 10K hours at 25°C. Accelerated degradation can be seen at larger R values.

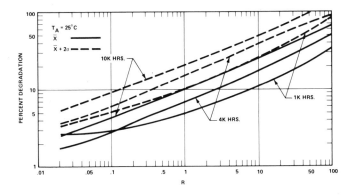

Figure 3.4.3-1. Emitter Degradation vs. R (Ratio of Stress Current to Measurement Current) for 1k, 4k, and 10k Hours. Mean, Mean + 2σ Distribution, T_A = 25°C.

The data of Figure 3.4.3-1 can be replotted to illustrate the percentage degradation versus time as a function of R. Figure 3.4.3-2 illustrates the mean and mean plus 2σ distribution with R = 1 and 50.

From this curve, a useful expression which relates the average degradation in emitter efficiency to time is obtained for the mean or mean plus 2σ distributions. [The symbol "D" will refer to CTR degradation due solely to emitter degradation, $\Delta\eta/\eta$, whereas ΔCTR/CTR will refer to total CTR degradation as expressed in Equation (3.4.2-5)].

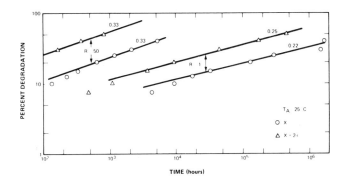

Figure 3.4.3-2. Degradation vs. Time at R = 1 and R = 50 for Mean, Mean + 2σ Distributions, T_A = 25°C.

$$(3.4.3\text{-}1)$$

$$D_{\bar{x}} \text{ or } D_{\bar{x}+2\sigma} \equiv \frac{-\Delta I_P}{I_P} = A_0 R^\alpha t^{n(R)} \text{ for } I_{FS} = \overline{I}_{FS} \text{ in } \%$$

where t is in 10^3 hours and A_0 and α differ for mean or mean plus 2σ. Equation (3.4.3-1) represents an average degradation corresponding to a specific R, t, and an average stress current $\overline{I}_{FS}$. A knowledge of $\overline{I}_{FS}$ and the actual device operating stress I_{FS} can be utilized to correct $\overline{D}$ to reflect the absolute magnitude of I_{FS}. This will be shown in the development of Equations (3.4.3-6) and (3.4.3-8). The data shows that $\overline{I}_{FS}$ increases with R and can be represented as follows:

$$(3.4.3\text{-}2)$$

$$\overline{I}_{FS}(R) = 14.13 + 9.06 \log_{10}R \quad , \quad T_A = 25°C$$

$$(3.4.3\text{-}3)$$

$$\overline{I}_{FS}(R) = 10.5 + 5.76 \log_{10}R \quad , \quad T_A = 85°C$$

These equations are obtained from averaged degradation data versus I_{FS} at different measurement times.

The expression for n(R) was found to obey the relationship

$$n(R) = .0475 \log_{10}R + .25 \qquad (3.4.3\text{-}4)$$

A_0 and α were determined from degradation data versus R and are found in Figure 3.4.5-1, "Matrix of Coefficients."

IC, T_A is ambient temperature, and the thermal resistances are the emitter junction to ambient, θ_{JA} = 370°C/W = θ_{DA} detector to ambient, and the detector to emitter thermal resistance is $\theta_{D\text{-}E}$ = 170°C/W = $\theta_{E\text{-}D}$. V_F, I_F are the forward LED voltage and current; V_O, I_O are the output stage voltage, and current and V_{cc}, I_{cc} are the power supply voltage and current to the device. In general, it is desirable to maintain $T_J \leqslant 125°C$.

A useful model can be constructed to describe the basic optocoupler's parameters which are capable of influencing the current transfer ratio. The 6N135 optocoupler, Figure 3.4.2-1, is the simplest device and one which is easily accessible for needed parameter measurements. However, any optocoupler can be modeled in this fashion within its linear region. Figure 3.4.2-1 shows the system block diagram which yields the relationship of input current, I_F, to output current, I_O. The resulting expression for CTR is:

$$CTR = \frac{I_O}{I_F}(100\%) = K R \eta(I_F,t) \beta(I_P,t) \qquad (3.4.2-4)$$

where K represents the total transmission factor of the optical path, generally considered a constant as is R, the responsivity of the photodetector, defined in terms of electrons of photocurrent per photon. η is the quantum efficiency of the emitter defined as the photons emitted per electron of input current and depends upon the level of input current, I_F, and upon time. Finally, β is the gain of the output amplifier and is dependent upon I_P, the photocurrent, and time. Temperature variations would, of course, cause changes in η, β as well.

From Equation (3.4.2-4), a normalized change in CTR, at constant I_F, can be expressed as:

$$\qquad\qquad\qquad\qquad\qquad\qquad (3.4.2-5)$$

$$\frac{\Delta CTR}{CTR} = \left(\frac{\Delta\eta}{\eta}\right)_{I_F} + \left(\frac{\Delta\eta}{\eta}\right)_{I_F}\left(\frac{\partial\ln\beta}{\partial\ln I_P}\right)_t + \left(\frac{\Delta\beta}{\beta}\right)_{I_P}$$

The first term, $\Delta\eta/\eta$, represents the major contribution to ΔCTR due to the relative emitter efficiency change; generally, over time, $\Delta\eta$ is negative. This change is strongly related to the input current level, I_F, as discussed earlier and more elaboration will be given later. The second term, $(\Delta\eta/\eta)I_F$ $(\partial\ln\beta/\partial\ln I_P)_t$, represents a second order effect of a shift, positive or negative, in the operating point of the output amplifier as the emitter efficiency changes. The third term, $(\Delta\beta/\beta)I_P$, is a generally negligible effect which represents a positive or negative change in the output transistor gain over time. The parameters K and R are considered constants in this model.

3.4.3   Degradation Model

In this section, an extensive test program conducted at Hewlett-Packard to characterize the CTR degradation of optocouplers is discussed. The development which will follow is mainly of interest to those concerned with reliability and quality assurance. From the basic data, the CTR degradation equations will be developed in order to predict the percentage change in CTR with time. Complete data and analysis of CTR degradation will be found in an internal Hewlett-Packard report.

Equation (3.4.3-1) gives a direct relationship between the average degradation, $\overline{D}$, and time. As mentioned earlier, the magnitude of the stress current also determines the amount of degradation. In order to allow for the effect of $|I_{FS}|$, empirical observations were made on D at different I_{FS} and at different times for several values of R. The dependence of degradation on stress current is linear

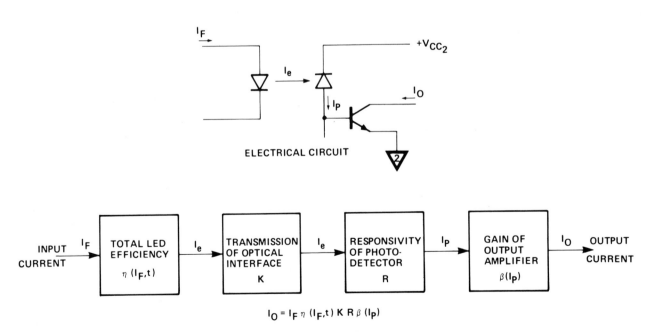

$$I_O = I_F \eta (I_F,t) K R \beta (I_P)$$

Figure 3.4.2-1. System Model for an Optocoupler

up to $I_{FS} = 40$ mA, for all values of R. From these observations, the average rate of change, or slope, S(R,t), of degradation D with I_{FS} over time was found to behave in the following fashion for any R:

$$S \equiv \frac{\partial D}{\partial I_{FS}} = \alpha(R) \log_{10}t + \beta(R) \quad \%/mA \qquad (3.4.3\text{-}5)$$

where t is in 10^3 hours, the coefficients $\alpha(R)$ and $\beta(R)$ can be found on Figure 3.4.5-1.

Along with Equation (3.4.3-5), the mean distribution degradation, $D_{\overline{x}}$, can be estimated for any specific stress current, I_{FS}, ratio R, and time t via the subsequent expression:

$$D_{\overline{x}} = \overline{D}_{\overline{x}} + S \, [I_{FS} - \overline{I}_{FS}] \quad \% \qquad (3.4.3\text{-}6)$$

or substituting Equation (3.4.3-1),

$$D_{\overline{x}} = A_o R^\alpha t^{n(R)} + S \, [I_{FS} - \overline{I}_{FS}] \quad \% \qquad (3.4.3\text{-}7)$$

where, again, D_x is the average degradation at time t, in units of 10^3 hours, corresponding to a stress current, $\overline{I}_{FS}$, given by Equations (3.4.3-2) and (3.4.3-3); I_{FS} is the actual stress current and R = I_{FS}/I_{FM}; S is the expression (3.4.3-5) for the change of slope of D versus I_{FS} with time; n(R) is a power of t, given by Equation(3.4.3-4), and A_o, α are found in Figure 3.4.5-1.

Equation (3.4.3-7) gives the mean distribution degradation by using a degradation value, $\overline{D}$ (first term), corresponding to the ratio of I_{FS}/I_{FM}, or a stress current, $\overline{I}_{FS}$, and then applying a correction quantity (second term) to $\overline{D}$ due to the magnitude of the actual stress current, I_{FS}, yielding the actual degradation D.

The expression for the mean + 2σ distribution degradation, $D_{\overline{x} + 2\sigma}$, (worst case) is almost of the same form as Equation (3.4.3-7). The dissimilarity arises from the fact that the standard deviation, σ, is dependent upon the stress current, I_{FS}, the ratio R, and upon time. This complex dependency was analytically deduced from the data to be the following expression:

$$D_{\overline{x} + 2\sigma} = \overline{D}_{\overline{x} + 2\sigma} + [S + 2P] \, [I_{FS} - \overline{I}_{FS}] \quad \% \quad (3.4.3\text{-}8)$$

or substituting Equation (3.4.3-1)

$$D_{\overline{x} + 2\sigma} = A_o R^\alpha t^{n(R)} + [S + 2P] \, [I_{FS} - \overline{I}_{FS}] \quad (3.4.3\text{-}9)$$

where $D_{\overline{x} + 2\sigma}$ is the degradation for $\overline{x} + 2\sigma$ distribution corresponding to the stress current $\overline{I}_{FS}$, Equations (3.4.3-2) and (3.4.3-3). A_o and α are found in Figure 3.4.5-1 under the $\overline{x} + 2\sigma$ category. S [Equation (3.4.3-5)] represents the slope to correct for actual I_{FS} versus $\overline{I}_{FS}$ current levels, and

P [Equation 3.4.3-10)] is the new term which is a slope to correct for the σ variation with I_{FS}, R and t. The coefficients $\gamma(R), \delta(R)$ in P are found in Figure 3.4.5-1.

$$P = \gamma(R) \log_{10}t + \delta(R) \quad \%/mA \qquad (3.4.3\text{-}10)$$

where t is in 10^3 hours.

The degradation Equations (3.4.3-6) and (3.4.3-8) are considered accurate for the ranges of $I_{FS} \leqslant 40$ mA and R $\leqslant$ 20; outside this range the model does not predict degradation as well. Hence, check to see if I_{FS} and R satisfy the above conditions. If I_{FS} or R exceed these limits, prediction of D will be, in general, greater than the actual degradation due to large values for S and P which do not reflect actual S and P. If $\overline{I}_{FS}$ is approximately equal to the actual I_{FS}, then the second term in the degradation equations need not be determined. Otherwise, the second term needs to be determined to obtain true emitter degradation, D. If $\overline{I}_{FS} < I_{FS}$, then the degradation, D, will be less than the degradation, $\overline{D}$, corresponding to $\overline{I}_{FS}$, and vice versa when $\overline{I}_{FS} > I_{FS}$. A quick and coarse estimate for degradation $\overline{D}$ can be obtained by using $\overline{D} = A_o R^\alpha t^{n(R)}$ for a specific R with approximate values for $\alpha \approx 0.4$ and $n \approx 0.3$. Figure 3.4.3-3 represents plots of Equations (3.4.3-6) and (3.4.3-8) for R = 1 and $\overline{I}_{FS} = 1.6$, 6.3, and 16mA at both $T_A = 25°C$ and $T_A = 85°C$. These plots are very useful in making a quick approximation of D for the specific conditions for which the plots have been made. These conditions represent the recommended operating conditions for the three HP optocoupler families.

This discussion of reliability data and its interpretation with model equations is qualified to specific optocouplers, 6N135 and 6N138, where continuous LED operation was maintained, and extrapolation of data for times beyond 10,000 hours is assumed to be valid. Different types of LEDs or preparation processes may produce different results than those presented in this section. These expressions only incorporate the first order effect, emitter degradation $\Delta\eta/\eta$, whereas comments about higher order effects upon total CTR degradation will be given in the following section. With these expressions for degradation, accelerated testing may be accomplished by employing large values of R. Such testing can provide a means by which to determine acceptable emitter lots for optocoupler fabrication, acceptable degradation performed for lot selection, or predict functional lifetime expectancy for optocouplers under specific operational conditions.

An important point to note is that the total operational life of an optocoupler is greater than the worst case mean plus 2σ distribution implies. Specifically, the worst case degradation given in Figures 3.4.3-3a (25°C) and 3.4.3-3b (85°C) are for the continuous operation of the 6N135 optocoupler. The actual lifetime for an optocoupler is greater than Figures 3.4.3-7and 3.4.3-9would indicate since

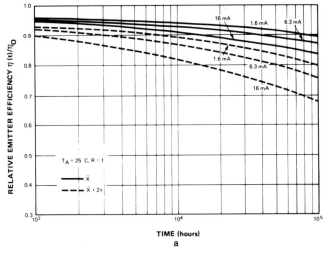

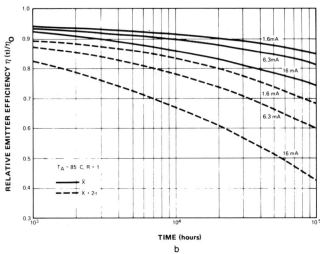

$$t_{\substack{continuous \\ lifetime}} = \begin{bmatrix} t_{\substack{system \\ lifetime}} \end{bmatrix}\begin{bmatrix} Data\ Duty \\ Factor \end{bmatrix}\begin{bmatrix} System\ Use \\ Data\ Factor \end{bmatrix}$$

Another equally important point to observe is that of the worst case conditions under which the optocoupler is used. As will be illustrated in the design examples, the worst possible combination of variations in V_{cc1}, V_{cc2}, R_{in}, CTR, R_L, I_{IL}, and temperature still result in the optocoupler functioning over an extended length of time (10^5 hours) for a particular maximum allowable degradation. However, the likelihood of seven parameters all deviating in their worst directions at the same time is extremely remote. A thorough statistical error accumulation analysis would illustrate that this worst-worst case is not a representative situation from which to design.

3.4.4   Higher Order Effects

The first order effect of emitter degradation, $\Delta\eta/\eta$, has a pronounced influence upon the ΔCTR as explained in the previous sections; however, consideration of higher order effects is important as well.

Consider the second term in Equation (3.4.2-5), $(\Delta\eta/\eta)I_F(\partial\ln\beta/\partial\ln I_P)_t$, the emitter degradation part has been explained; however, $(\partial\ln\beta/\partial\ln I_P)_t$ represents a shift in the operating point of the output amplifier of an optocoupler. The term $(\partial\ln\beta/\partial\ln I_P)$ can be rewritten as $(1/2.3\beta)(\partial\beta/\partial\log_{10}I_P)$ which is more convenient to use with the accompanying typical curves of β versus $\log_{10}I_P$ for the two optocouplers 6N135 and 6N138, given in Figure 3.4.4-1a.

If the operating photocurrent, I_P, is to the right of the maximum β point of either curve, then with reduced emitter efficiency over time, I_P will decrease, but the increasing β will tend to compensate for this degradation. However, if the operating I_P is to the left of the maximum β and then I_P decreases, the β change will accentuate the emitter's degradation, yielding a larger CTR loss. The magnitude of the contributions of $\partial\ln\beta/\partial\ln I_P$ to overall CTR degradation can be illustrated by the following examples.

Consider a 6N138 optocoupler of Figure 3.4.4-1c operating at its recommended $I_F = 1.6$ mA which corresponds to an $I_P \approx 1.6\mu$A. (An I_F to I_P relationship for Hewlett-Packard optocouplers is 1 mA input current yields approximately 1μA of photodiode current.) At $I_P = 1.6\mu$A, the slope of the $V_{CE} = 5$V curve is equal to -15,000 and the gain is $\beta = 26,000$; hence, $\partial\ln\beta/\partial\ln I_P \approx$ -0.25. If, for instance, the emitter degradation $\Delta\eta/\eta$ is -10%, then the second order term would improve the overall CTR degradation, i.e.,

(3.4.4-1)

$$\frac{\Delta CTR}{CTR} = \left(\frac{\Delta\eta}{\eta}\right) + \left(\frac{\Delta\eta}{\eta}\right)\left(\frac{\partial\ln\beta}{\partial\ln I_P}\right) + \ldots = -10\% + 2.5\% = -7.5\%$$

Figure 3.4.3-3. Calculated Curves of Relative Emitter Efficiency vs. Time for R = 1: $I_{FS} = I_{FM}$ = 1.6, 6.3, and 16 mA Which are Recommended I_F for 6N138, 6N137, and 6N135 Optocouplers Respectively. Mean, Mean + 2σ Distributions. a) $T_A = 25°$C, b) $T_A = 85°$C.

the majority of units will be centered around the mean distribution lifetime. Secondly, the optocoupler which is operated at some signal duty factor less than 100%, for example 50%, would increase the optocoupler's life by a factor of two. Third, the fact that an optocoupler is used within equipment which may have a typical 2000 hours per year (8 hours/day — 5 days/week — 50 weeks/year) instrument or system operating time, could expect to increase the optocoupler's life by another factor of 4.4 in terms of years of useful life. The appropriate operating time considerations will vary depending upon the designer's knowledge of the system in which the optocoupler will be used. The operating lifetime of an optocoupler can be expressed, for a maximum allowable degradation at a particular I_{FS}, by using Figures 3.4.3-7 and 3.4.3-9 and the following expression for $t_{continuous\ lifetime}$:

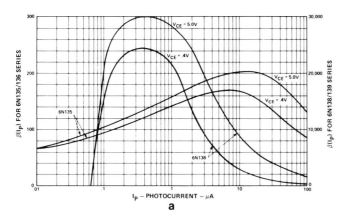

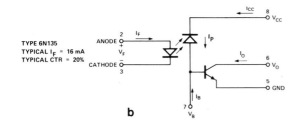

a

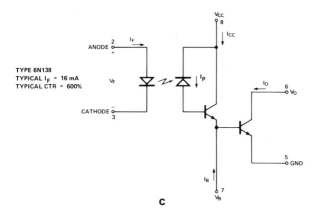

TYPE 6N135
TYPICAL I_F = 16 mA
TYPICAL CTR = 20%

b

TYPE 6N138
TYPICAL I_F = 16 mA
TYPICAL CTR = 600%

c

Figure 3.4.4-1 a) DC Current Gain, β, vs. Photocurrent, I_P, for 6N135 and 6N138 Optocouplers. Current Diagrams and Typical Values of I_F and CTR for Hewlett-Packard Optocouplers, b) 6N135, c) 6N138.

This improvement is what was expected while operating on the right side of the β maximum. In fact, with an I_F = 4 mA or $I_P \approx 4\mu A$, the term $\partial ln\beta/\partial lnI_P$ = -0.8, and again, if $\Delta\eta/\eta$ = -10%, the resulting $\Delta CTR/CTR$ = -2%, nearly cancelling the emitter's degradation.

With the 6N135 optocoupler, Figure 3.4.4-1b, operating at I_F = 10 mA, or $I_P \approx 10\mu A$, which corresponds to the maximum β point on the V_{CE} = .4V curve, the slope is zero and the total CTR degradation is basically the emitter's degradation.

Another subtle effect is seen from the third term in Equation (3.4.2-5), $(\Delta\beta/\beta)I_P$, over time. At constant I_P, β can increase or decrease by a few percent over 10,000 hours. This change is so small that the third term is generally neglected.

For the optocouplers containing an output amplifier, such as the 6N137, which switches abruptly about a particular threshold input current, the actual emitter degradation can be determined from Equations (3.4.3-6) and (3.4.3-8). An appropriate $I_{F_{initial}}$ can be determined to provide for adequate guard band current which will allow the optocoupler emitter to degrade while maintaining sufficient I_P to switch the amplifier. An actual design procedure to determine the needed $I_{F_{initial}}$ for proper operation of Hewlett-Packard optocouplers is given in the design examples section.

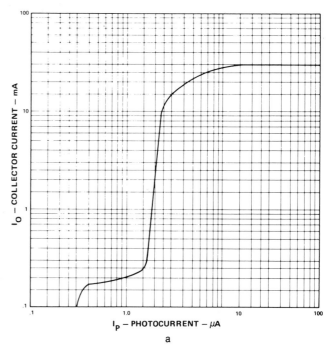

a

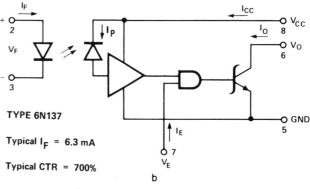

TYPE 6N137

Typical I_F = 6.3 mA

Typical CTR = 700%

b

Figure 3.4.4-2. a) Output Current, I_o, vs. Photocurrent, I_P, for 6N137 Optocoupler.

b) Circuit Diagram and Typical Values of I_F and CTR for 6N137 Optocoupler.

3.4.5 Procedure for Calculation of CTR Degradation

1. Specify I_{FS}, I_{FM}

2. Determine $R = I_{FS}/I_{FM} \leq 20$

   $I_{FS} \leq 40$ mA

   Degradation Model Equations (3.4.3-6) and (3.4.3-8) Valid

3. First Approximation of Degradation

   $$\overline{D}_{\overline{x}} = A_o R^{\alpha} t^n \quad (\%) \qquad \text{with } \alpha \approx .4, A_o \text{ (Figure 3.4.5-1)}$$
   or
   $$\overline{x} + 2\sigma \qquad\qquad\qquad n \approx .3, t \text{ in } 10^3 \text{ hours}$$
   $$(D \text{ corresponds to } I_{FS})$$

4. Calculate $\overline{I}_{FS} = \begin{cases} 14.13 + 9.06 \log_{10}R \text{ @ } 25°C & \text{Equation (3.4.3-2)} \\ 10.5 + 5.76 \log_{10}R \text{ @ } 85°C & \text{Equation (3.4.3-3)} \end{cases}$

   If $I_{FS} \approx \overline{I}_{FS}$, Step 6 and the second terms in Equations (3.4.3-6) and (3.4.3-8) do not need to be calculated.

5. Calculate $n(R) = .0475 \log_{10}R + .25$

6. Calculate $S = \alpha(R) \log_{10}t + \beta(R)$ $\qquad \alpha(R), \beta(R)$ $\qquad$ Figure 3.4.5-1

   $P = \gamma(R) \log_{10}t + \delta(R)$ $\qquad \gamma(R), \delta(R)$ $\qquad$ t in 10^3 hours

7. Calculate Mean, Mean + 2σ Degradation

   $$D_{\overline{x}} = A_o R^{\alpha} t^{n(R)} + S \, [I_{FS} - \overline{I}_{FS}] \qquad\qquad \% \qquad \text{Equation (3.4.3-6)}$$

   $$D_{\overline{x} + 2\sigma} = A_o R^{\alpha} t^{n(R)} + [S + 2P] \, [I_{FS} - \overline{I}_{FS}] \qquad \% \qquad \text{Equation (3.4.3-8)}$$

   (A_o, α via Figure 3.4.5-1, t in 10^3 hours)

8. For Second Order Effect, Determine Slope

   $$\frac{\partial \ln\beta}{\partial \ln I_P} = \frac{1}{2.3\beta} \frac{\partial \beta}{\partial \log_{10}I_P}$$

   Figure 3.4.4-1a – typical curves with an approximation for HP optocouplers of $I_F = 1$ mA yields $I_P \approx 1\mu A$

9a. Total CTR Degradation for Mean Distribution

   $$\frac{\Delta CTR}{CTR} = D_{\overline{x}} + D_{\overline{x}} \frac{\partial \ln\beta}{\partial \ln I_P}$$

9b. Total CTR Degradation for Mean + 2σ Distribution

   $$\frac{\Delta CTR}{CTR} = D_{\overline{x} + 2\sigma} + D_{\overline{x} + 2\sigma} \frac{\partial \ln\beta}{\partial \ln I_P}$$

3.26

	25°C		85°C			
	$\bar{X}$	$\bar{X}+2\sigma$	$\bar{X}$		$\bar{X}+2\sigma$	
			R < 6	6 ≤ R	R < 8	8 ≤ R
A_o	4.95	9.7	6.8	5.0	15.0	11.0
α	.388	.428	.302	.467	.284	.430

	25°C		85°C	
	R ≤ 1	R ≥ 1	R ≤ 1	R ≥ 1
$\alpha(R)$	$.19\,R^{.052}$	$.19\,R^{.32}$	$.32\,R^{.08}$	$.32\,R^{.30}$
$\beta(R)$	$.055$	$.055\,R^{.68}$	$.11\,R^{.25}$	$.11\,R^{.65}$

	25°C	85°C
$\gamma(R)$	$.063\,R^{.30}$	$.154\,R^{.26}$
$\delta(R)$	$.081\,R^{.38}$	$.196\,R^{.39}$

Figure 3.4.5-1. Matrix of Coefficients.

3.4.6 Practical Application

A very common application of an optocoupler is to function as the interfacing element between digital logic. In this section, the designer will be shown an approach which will insure the initial and long term performance of such an interface, and take into account the practical aspects of the system that surrounds it. These system elements include the data rate, the logic families being interfaced, the variations of the power supply, the tolerances of the components used, the operational temperature range, and lastly the expected lifetime of the system.

The system data speed can be considered as the primary selection criteria for selecting a specific optocoupler family. Figure 3.4.6-2 lists the ranges of data rates for four Hewlett-Packard optocoupler families when driven at specified LED input current, I_F. With this table, and the knowledge of the system data rate requirements, it is possible to select an optimum coupler.

An example of an optocoupler interconnecting two logic gates is shown in Figure 3.4.6-1. A logic low level is insured

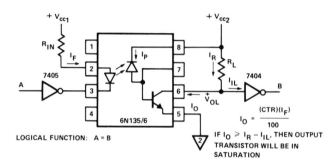

$$I_O = \frac{(CTR)(I_F)}{100}$$

LOGICAL FUNCTION: A = B

IF $I_O \geq I_R - I_{IL}$, THEN OUTPUT TRANSISTOR WILL BE IN SATURATION

Figure 3.4.6-1. Typical Digital Interface Using an Optocoupler.

FAMILY	RZ DATA RATE BITS/S	INPUT CURRENT – I_F						
		.5mA	1.0mA	1.6mA	7.5mA	10mA	12mA	16mA
6N135/6 SINGLE TRANSISTOR	MIN							333k
	TYP							2M
6N138/9 SPLIT DARLINGTON	MIN	12k		22k			125k	
	TYP	100k		200k			840k	
4N45/6 DARLINGTON	MIN					1.8k		
	TYP	640				6.5k		
6N137 OPTICALLY COUPLED GATE	MIN				6.7M			
	TYP					10M		

Figure 3.4.6-2. Optocoupler Data Rates Specifications.

3.27

FAMILY		% CTR @ I_F = (mA)						TEMP °C	V_{OL}
		.5	1.0	1.6	5	10	16		
SINGLE TRANSISTOR	6N135						7	25	0.4
	6N136						19		
SPLIT DARLINGTON	6N138		300					0–70	0.4
	6N139	400	500					0–70	0.4
DARLINGTON	4N45		250			200		0–70	1.0
	4N46	350	500			200		0–70	1.0
OPTICALLY COUPLED GATE	6N137				400			0–70	0.6

Figure 3.4.6-3. Optocoupler CTR (MIN).

when the saturated output sinking current, I_O, is greater than the combined sourcing currents of the pull-up resistor, and the logic low input current, I_{IL}, of the interconnecting gate. Using the coupler specifications selected from Figure 3.4.6-2 and the corresponding CTR (MIN) from Figure 3.4.6-3, it is possible to determine from Equation (3.4.6-1) the minimum initial value of I_F for the coupler. The design criteria is that $I_O \geq I_{IL} + I_R$ for the V_{IL} specified in Figure 3.4.6-4.

$$I_F = \frac{I_o \times 100}{CTR(MIN)} \qquad (3.4.6\text{-}1)$$

$$R_{in} = \frac{V_{cc1} - V_F - V_{OL}}{I_F} \qquad (3.4.6\text{-}2)$$

$$I_{F(MIN)} = \frac{V_{cc1(MIN)} - V_{F(MAX)} - V_{OL}}{R_{in(MAX)}} \qquad (3.4.6\text{-}3)$$

$$I_{F(MAX)} = \frac{V_{cc1(MAX)} - V_{F(MIN)} - V_{OL}}{R_{in(MIN)}} \qquad (3.4.6\text{-}4)$$

Using Equation (3.4.6-2), the typical value of R_{in} can be calculated for the selected I_F and the logic low output voltage, V_{OL}, of the driving gate. The V_{OL} of the logic family is given in Figure 3.4.6-4. The next step is to determine the worst case value of the LED input current, I_F, resulting from the tolerance variations of the LED

current limiting resistor, R_{in}, and the power supply voltage, V_{cc1}. The conditions of $I_{F(MIN)}$ and the initial CTR (MIN) are then used to determine the initial worst case value of $I_{O(MIN)}$. Conversely, the worst case CTR degradation will occur when the LED is stressed at $I_{F(MAX)}$ conditions; thus, $I_{F(MAX)}$ will be used to determine the worst case degradation of the optocoupler performance. Using the maximum V_{cc1} and the minimum R_{in} will accomplish this worst case calculation, as shown in Equation (3.4.6-4).

TTL FAMILY	I_{IL}	V_{IL}	I_{IH}	V_{IH}	I_{OL}	V_{OL}	I_{OH}	V_{OH}
74S	−2 mA	.8V	50 µA	2V	20 mA	.5V	−1000 µA	2.7V
74H	−2 mA	.8V	50 µA	2V	20 mA	.4V	− 500 µA	2.4V
74	−1.6 mA	.8V	40 µA	2V	16 mA	.4V	− 400 µA	2.4V
74LS	−.36 mA	.8V	20 µA	2V	8 mA	.5V	− 400 µA	2.7V
74L	−.18 mA	.7V	10 µA	2V	3.6 mA	.4V	− 200 µA	2.4V

Figure 3.4.6-4. Logic Interface Parameters.

The change in CTR from the initial value at time $t = 0$ to a final value at some later time can be compensated by choosing a value of R_L which is consistent with $I_{o(MIN)} - mI_{IL}$ at the end of system life. Equation (3.4.6-5) describes this worst case calculation.

$$(3.4.6\text{-}5)$$

$$R_{L(MIN)} \geq \frac{V_{cc2(MAX)} - V_{OL}}{\left[\dfrac{I_{F(MIN)} \cdot CTR(MIN) \cdot \left(1 - \left(\dfrac{D_{\bar{x}+2\sigma}}{100}\right)\right)}{100} - mI_{IL} \right]}$$

$D_{x+2\sigma}$ = worst case CTR degradation

3.28

The selection of the maximum value of R_L is also of important in that its value insures that the collector is pulled up to the logic one voltage conditions, V_{IH}, under the conditions of maximum I_{OH} of the coupler, and the I_{IH} of the interconnecting gate.

$$\text{(3.4.6-6)}$$

$$R_{L\ (MAX)} \leqslant \frac{V_{cc2\ (MIN)} - V_{IH}}{I_{OH\ (MAX)} + m\ I_{IH}}$$

The selection of the value of R_L between the boundaries of $R_{L\ (MIN)}$, and $R_{L\ (MAX)}$ has certain trade offs. As in any open collector logic system, T_{PLH} increases with increasing R_L. Conversely, as R_L is increased above $R_{L_{MIN}}$, a larger guardband between I_{oMIN} and $I_{IL} + I_R$ is achieved. Engineering judgement should be employed here to achieve the optimum trade off for desired performance.

Using the coefficient table, Figure 3.4.5-1, and Equations (3.4.3-6) and (3.4.3-8), the following examples are developed to demonstrate the methods of optocoupler system design in the presence of the mean and mean plus two sigma CTR degradation.

3.4.6-1   Example 1.

System Specifications

Data Rate	20 k bit NRZ
Logic Family	Standard TTL
Power Supply 1 & 2	5V ± 5
Component Tolerances	± 5%
Temperature Range	0 – 70°C
Expected System Lifetime	350 k hr (40 yr) at 50% system use time and 50% Data Duty Factor

Interface Specifications

Coupler 6N139

$CTR\ (MIN) = 500\%\ @\ I_F = 1.6\ mA$
$V_{OL\ (MAX)} = .4V\ @\ I_F = 1.6\ mA$
$I_{OH\ (MAX)} = 250\mu A\ @\ V_{cc2} = 7V$
$V_{F\ (MAX)} = 1.7V\ @\ I_F = 1.6\ mA$
$V_{F\ (MIN)} = 1.4V\ @\ I_F = 1.6\ mA$
$V_{F\ (TYP)} = 1.6V\ @\ I_F = 1.6\ mA$

Logic Standard TTL

I_{IL}	= 1.6 mA	I_{IH}	= 40μA
V_{IL}	= .8V	V_{IH}	= 2V
I_{OL}	= 16 mA	I_{OH}	= 400μA
V_{OL}	= .4V	V_{OH}	= 2.4V

Step 1. $R_{in\ (TYP)}$

$$R_{in} = \frac{V_{cc1\ (TYP)} - V_F\ (TYP) - V_{OL}}{I_F\ (TYP)} \qquad \text{(3.4.6.1-1)}$$

$$R_{in} = \frac{5.0 - 1.6 - .4}{1.6 \times 10^{-3}} = 1.87k\Omega,\ \text{select}\ 1.8k\Omega\ \pm 5\%$$
$$R_{(MIN)} = 1710\Omega$$
$$R_{(MAX)} = 1890\Omega$$

Step 2. $I_{F\ (MAX)}$

$$I_{F\ (MIN)} = \frac{V_{cc1\ (MIN)} - V_{F\ (MAX)} - V_{OL}}{R_{in\ (MAX)}} \qquad \text{(3.4.6.1-2)}$$

$$I_{F\ (MIN)} = \frac{4.75 - 1.7 - .4}{1890\Omega} = 1.4\ mA$$

Step 3. $I_{F\ (MAX)}$

$$I_{F\ (MAX)} = \frac{V_{cc1\ (MAX)} - V_{F\ (MIN)} - V_{OL}}{R_{in\ (MIN)}} \qquad \text{(3.4.6.1-3)}$$

$$I_{F\ (MAX)} = \frac{5.25 - 1.4 - .4}{1710\Omega} = 2.02\ mA$$

Step 4.   Determine continuous operation time for LED emitter.

$$t_{\substack{continuous \\ lifetime}} = \left[t_{\substack{system \\ lifetime}}\right]\left[\substack{Data\ Duty \\ Factor}\right]\left[\substack{System\ Use \\ Duty\ Factor}\right]$$

$$= (40\ yr \times 8.76\ k\ hr/yr)(50\%)(50\%)$$

$$t_{\substack{continuous \\ lifetime}} = 87.60K\ hr$$

Step 5.   Obtain the mean and mean + 2σ CTR degradation at $I_{F\ (MAX)}$ and $t_{continuous\ lifetime}$ either as an approximation from Figure 3.4.3-3 or by calculations as shown below.

Step 5a.   Determine $D_{\bar{x}}$

$$D_{\bar{x}} = A_o t^{.25} + S\ [I_{FS} - \bar{I}_{FS}] \qquad \text{(3.4.6.1-4)}$$

$$D_{\overline{x}} = 4.95t_{(k\ hr)}{}^{.25} + [.186 \log t_{(k\ hr)} + .055]$$

$$[I_{F\ (MAX)} - 14.13\ mA]$$

$$D_{\overline{x}} = 4.95\ (87.6)^{.25} + (.186 \log 87.6 + .055)$$

$$(2.02\ mA - 14.13\ mA)$$

$$D_{\overline{x}} = 10.10\%\ \text{for 40 yr system operation}$$

Step 5b. Determine $\overline{D}_{x} + 2\sigma$

$$D_{\overline{x} + 2\sigma} = A_{o}t^{.25} + [S + 2P]\ [I_{FS} + \overline{I}_{FS}] \qquad (3.4.6.1\text{-}5)$$

$$D_{\overline{x} + 2\sigma} = 9.7t_{(k\ hr)}{}^{.25} + [2\ (.063 \log t_{(k\ hr)} + .081)$$

$$+ (.186 \log t_{(k\ hr)} + .055)]$$

$$\times\ [I_{F\ (MAX)} - 14.13\ mA]$$

$$D_{\overline{x} + 2\sigma} = 9.7\ (87.6)^{.25} + [2\ (.063 \log 87.6 + .081)$$

$$+ (.186 \log 87.6 + .055)]$$

$$\times\ [2.02\ mA - 14.13\ mA]$$

$$D_{\overline{x} + 2\sigma} = 19.71\%$$

Step 6.   Guardband the worst case value of CTR degradation.

It is often desirable to add some additional operating margin over and above conditions dictated by simple worst case analysis. The use of engineering judgement to increase the worst possible CTR degradation by an additional 5% margin would insure that the entire distribution would fall within the analysis. Thus,

$$D_{\overline{x} + 2\sigma} + 5\% = 24.71\%$$

Step 7.   Selecting $R_{L\ (MIN)}$ for guardbanded worst case

$$D_{\overline{x} + 2\sigma} + 5\% \qquad ,\ m = 1$$

$$(3.4.6.1\text{-}6)$$

$$R_{L(MIN)} \geqslant \cfrac{V_{cc2\ (MAX)} - V_{OL}}{\cfrac{I_{F(MIN)} \cdot CTR_{(MIN)} \cdot 1 - \left(\cfrac{D_{\overline{x}+2\sigma} + 5\%}{100}\right)}{100} - mI_{IL}}$$

$$R_{L(MIN)} \geqslant \cfrac{5.25 - .4}{\cfrac{1.4 \times 10^{-3} \cdot 500\% \cdot 1 - \left(\cfrac{24.71\%}{100}\right)}{100} - 1 \cdot 1.6\ mA}$$

$$R_{L\ (MIN)} = 1.32k\Omega$$

Step 8.   Select $R_{L\ (MAX)}$

$$R_{L\ (MAX)} \leqslant \cfrac{V_{cc2\ (MAX)} - V_{OL}}{I_{OH\ (MAX)} + mI_{IH}} \qquad (3.4.6.1\text{-}7)$$

$$R_{L\ (MAX)} \leqslant \cfrac{4.75 - 2.4}{250\mu A + 40\mu A} = 8.1k$$

The range of R_{L} is from $1.32k\Omega$ to $8.1k\Omega$. It is desirable to select a pull-up resistor which optimizes both speed performance and additional I_{O} guardband. This criteria leads to a tradeoff between a value close to $R_{L\ (MIN)}$ for speed performance and one boardering near $R_{L\ (MAX)}$ for I_{O} guardbanding. In this design example, the system's lifetime has a higher priority than does the moderate speed performance demanded from the optocoupler. An R_{L} of $3.3k\Omega \pm 5\%$ is selected under this condition.

An additional guardband of 5% was added to the worst case $D_{\overline{x} + 2\sigma}$ CTR degradation guardband to insure that even a greater percentage of the distribution would be accounted for. The actual percentage difference between $I_{OL\ (MAX)}$ and $I_{O\ (MIN)}$ at the end of system life is shown below:

$$(3.4.6.1\text{-}8)$$

$$I_{O\ (MIN)} = \cfrac{CTR_{(MIN)} \cdot I_{F\ (MIN)} \cdot 1 - \left(\cfrac{\overline{D}_{\overline{x} + 2\sigma}}{100}\right)}{100}$$

$$(3.4.6.1\text{-}9)$$

$$I_{OL\ (MAX)} = \cfrac{V_{cc2\ (MAX)} - V_{OL}}{R_{L\ (TYP - 5\%)}} + m|I_{IL}|$$

$$\%\ \text{Guardband} = \left[1 - \cfrac{I_{OL\ (MAX)}}{I_{O\ (MIN)}}\right] \times 100 \qquad (3.4.6.1\text{-}10)$$

For the example shown, the additional end of system life I_{O} guardband results from the selection of an R_{L} greater than the $R_{L\ (MIN)}$ as shown in Steps 9, 10, and 11.

Step 9. $I_{O\ (MIN)}$ at end of system life

$$I_{O\ (MIN)} = \cfrac{500\% \cdot 1.4\ mA \cdot \left(1 - \cfrac{19.17\%}{100}\right)}{100} = 5.65\ mA$$

Step 10. $I_{OL \, (MAX)}$ **for worst case of** $I_{R \, (MAX)} + I_{IL}$

$$(3.4.6.1\text{-}11)$$

$$I_{OL \, (MAX)} = \frac{5.25 - .4}{3.13k\Omega} + 1.6 \text{ mA} = 3.14 \text{ mA}$$

Step 11. % Guardband

$$\% = 1 - \frac{3.14 \text{ mA}}{5.65 \text{ mA}} \quad 100 = 44.4\% \qquad (3.4.6.1\text{-}12)$$

Thus, this circuit interface design offers an additional 44.4% I_O guardband beyond the 19.71% required to compensate for the CTR change caused by 86.7k hr of continuous operation at an $I_{F \, (MAX)}$ of 2 mA. This extra guardband results from having chosen an $R_L = 3.3k$ rather than the lowest allowable value of R_L plus the engineering guardband chosen in Step 6.

3.4.6.2 Example 2.

System Specifications

Data Rate	250K bit NRZ
Logic Family	TTL to LSTTL
Power Supply 1 and 2	5V ± 5%
Component Tolerance	± 5%
Temperature Range	25°C
Expected System Lifetime	175 k hr (20 yr) at 50% System Use Time and 50% Data Duty Factor

Interface Conditions

Coupler 6N136

$CTR_{(MIN)}$ = 19% @ I_F = 16 mA
V_{OL} = .4V
I_{OH} = 500 nA @ V_{cc2} = 5.0V
$V_{F(TYP)}$ = 1.6V @ I_F = 16 mA
$V_{F(MIN)}$ = 1.5V @ I_F = 16 mA
$V_{F(MAX)}$ = 1.7V @ I_F = 16 mA

Logic LSTTL

I_{IL}	= .36 mA	I_{OL}	= 8 mA
V_{IL}	= .8V	V_{OL}	= .5V
I_{IH}	= 40µA	I_{OH}	= 400µA
V_{IH}	= 2V	V_{OH}	= 2.7V

Again using Figure 3.4.5-1, the data rate dictates the use of a 6N136 at an $I_{F \, (TYP)}$ of 16 mA. Using the same 12 step

worst case analysis, it is possible to determine the values of R_{in}, R_L and the degree of guardbanding of I_O at end of system lifetime.

Step 1. R_{in} = 187Ω, select 180Ω ± 5%
$R_{L \, (MIN)}$ = 179Ω
$R_{L \, (MAX)}$ = 189Ω

Step 2. $I_{F \, (MIN)}$ = 14.02 mA

Step 3. $I_{F \, (MAX)}$ = 19 mA

Step 4. System Lifetime

t = 43.8k hr

Step 5. $D_{\overline{x}}$ and $D_{\overline{x} + 2\sigma}$ for $I_{F \, (MAX)}$ of 19 mA

by calculation or from Figure 3.4.3-3

$D_{\overline{x}}$ = 14.5% $\qquad$ 43.8k hr

$D_{\overline{x} + 2\sigma}$ = 28.5% $\qquad$ continuous lifetime

Step 6. Engineering Guardband of 5%,

$D_{\overline{x} + 2\sigma} + 5\%$ = 33.5%

Step 7. R_L selection with guardbanding of $D_{\overline{x} + 2\sigma} + 5\%$

$R_{L \, (MIN)}$ = 3.44kΩ

Step 8. $R_{L \, (MAX)}$ = 50kΩ

Step 9. $R_{L \, (TYP)}$ = 5.1kΩ ± 5%, $R_{L \, (TYP \, - \, 5\%)}$

= 4.84kΩ, $R_{L \, (MAX \, + \, 5\%)}$

= 5.35kΩ

Step 10. End of System Life $I_{O \, (MIN)}$

$I_{O \, (MIN)}$ = 1.5 mA

Step 11. $I_{OL \, (MAX)}$ = 1.36 mA

Step 12. Engineering % Guardband of $I_{O \, (MIN)}$ = 9.3%

3.4.6.3 Example 3.

If a particular design requirements specifies a maximum tolerable degradation over a system lifetime, the optimum value of $I_{F \, (TYP)}$ can be obtained from Figure 3.4.6.3-1. For example, if a maximum acceptable degradation, $D_{\overline{x} + 2\sigma}$, is 40%, and a continuous operation of 400k hr is

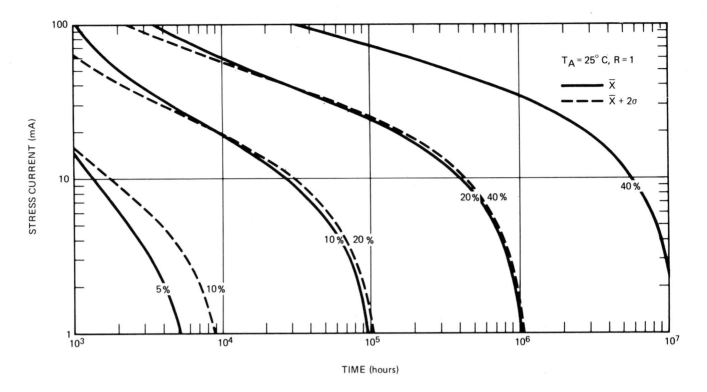

Figure 3.4.6.3-1. Stress Current (I_{FS}) vs. Time vs. % Degradation.

desired, this curve specifies that $I_{F\ (TYP)}$ should be less than or equal to 10 mA. A 400k hr continuous operation with 100% system duty factor as might be encountered in telephone switching equipment is equivalent to 45 years of system lifetime.

If a 6N139 split Darlington were used to interface an LSTTL logic gate with the system specifications stated, a collector pull-up resistor of as low as 160Ω could be used. If an R_L of 1k were selected, this optocoupler would offer an additional end of life guardband of 81.8%. This worst case analysis points out that with the knowledge of selecting proper values of R_L, the CTR performance of the coupler far exceeds the normal MTBF requirements for most commercial electronic systems.

3.4.6.4 Consideration of the Optically Coupled Gate

System data speed requirements in the multi-megabit range can also be communicated through an optocoupler. The first three coupler families listed in Figure 3.4.6-2 are not applicable in these very high speed data interface applications; however, the optically coupled gate, 6N137, will function to speeds of up to 10 MHz. This type of coupler differs in operation from the single transistor and Darlington style units in that it exhibits a non-linear transfer relationship of I_F to I_O. This is shown in Figure 3.4.6.4-1. The relationship is described as a minimum threshold of LED input current, $I_{F\ th}$ which is required to cause the output transistor to sink the current supplied by

the pull-up resistor and interconnected gate. As the LED degrades, the effect is that a larger value of $I_{F\ th}$ is required to create the same detector photodiode current necessary to switch the output gate.

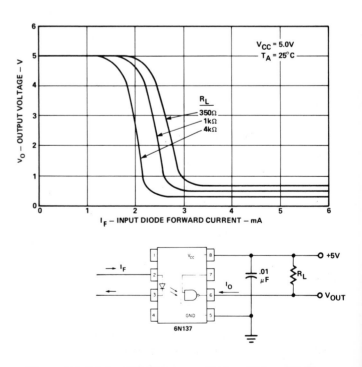

Figure 3.4.6.4-1. 6N137 Input — Output Characteristics.

In the previous interface examples, the worst case analysis and guardbanding is based on the output collector current, I_O. With the optically coupled gate, worst case guardbanding is concerned with the selection of the initial value of the I_F, which at end of system lifetime will generate the necessary threshold photocurrent demanded by the gate's amplifier to change state.

The calculation of the required I_F to allow for worst case LED degradation is approached by guardbanding the guaranteed minimum isolator input current, I_{FH}, for a specified I_{OL} and V_{OL} interface. Equation (3.4.6.4-1) shows the relationship of the I_P to I_F for this coupler.

$$I_P \alpha (I_F)^n \quad \text{, where } 1.1 \leqslant n \leqslant 1.3 \qquad (3.4.6.4\text{-}1)$$

Using the concept that the guardbanding of the initial value of I_F will result in a similarly guardbanded I_P, the relationship presented in Equation (3.4.6.4-2) results:

$$\left[1 - \frac{D_{\bar{x} + 2\sigma}}{100} \right] = \left[\frac{I_{PH}}{I_P} \right] = \left[\frac{I_{FH}}{I_F} \right]^n \qquad (3.4.6.4\text{-}2)$$

$$I_F = \frac{I_{FH}}{\left[1 - \frac{D_{\bar{x} + 2\sigma}}{100} \right]^n} \qquad (3.4.6.4\text{-}3)$$

The previous interface example showed that the first term of the $D_{\bar{x} + 2\sigma}$ equation dominated the magnitude of the worst case degradation. This term, $A_o R^\alpha t^{n(R)}$, i.e., $(9.7 \cdot t_{(k\,hr)}{}^{.25})$, does not contain an I_F current dependent term; thus, an approximation of the worst case LED degradation can be made that relates to the system's lifetime. This initial value of $D_{\bar{x} + 2\sigma}$ can be used in Equation (3.4.6.4-3) to calculate the initial value of the I_F. With this initial I_F, a more accurate degradation value can be calculated using Equation (3.4.6.1-5). This procedure results in an iterative process to zero in on a value of I_F that will insure reliable operation.

The following example will illustrate this approach.

3.4.6.5   Example 4.

System Specifications

Data Rate	6 MHz NRZ
Logic Family	LSTTL to TTL
Power Supply 1 and 2	5V ± 5%
Component Tolerance	± 5%
Temperature Range	0 – 70°C
Expected System Lifetime	203k hr (23 yr) at 50% System Use Time and 50% Data Duty Factor

Step 1. Determine the continuous operation time for LED emitter

$$t_{\substack{\text{continuous} \\ \text{lifetime}}} = \left[\begin{matrix} t_{\text{system}} \\ \text{lifetime} \end{matrix} \right] \left[\begin{matrix} \text{Data Duty} \\ \text{Factor} \end{matrix} \right] \left[\begin{matrix} \text{System Use} \\ \text{Factor} \end{matrix} \right]$$

$$= \left[23\ \text{yr} \cdot 8.76\text{k hr/yr} \right] \left[50\% \right] \left[50\% \right]$$

$$= 50.3\text{k hr}$$

Step 2. Calculate the worst case LED degradation

$$D_{\bar{x} + 2\sigma} \approx 9.7\ t_{(k\,hr)}{}^{.25}$$

$$D_{\bar{x} + 2\sigma} \approx 9.7\ (50.3)^{.25}$$

$$D_{\bar{x} + 2\sigma} \approx 26\%$$

Step 3. Calculate the first approximation of guardbanded I_F, n = 1.2

$$I_F = \frac{I_{FH}}{\left[1 - \frac{(\approx D_{\bar{x} + 2\sigma})}{100} \right]^{1/n}} = \frac{5\ \text{mA}}{.78} = 6.41\ \text{mA}$$

Step 4. Calculate input resistor R_{in}

$$R_{in} \leqslant \frac{V_{cc1\,(MIN)} - V_{F\,(MAX)} - V_{OL}}{I_F}$$

$$R_{in} \leqslant \frac{4.75 - 1.7 - .4}{.00641}$$

$$R_{in} \leqslant 413\Omega \text{ select } R_{in} = 390\Omega \ \pm 5\%$$

Rin (MAX)

$$R_{in\,(MAX)} = 409\Omega$$

$$R_{in\,(MIN)} = 370\Omega$$

Step 5. Calculate the $I_{F\,(MAX)}$

$$I_{F\,(MAX)} = \frac{V_{cc1\,(MAX)} - V_{F\,(MIN)} - V_{OL}}{R_{in\,(MIN)}}$$

$$I_F = \frac{5.25 - 1.4 - .4}{370}$$

$$I_F = 9.32 \text{ mA}$$

Step 6. Calculate the worst case $D_{\overline{x} + 2\sigma}$ for I_F (MAX)

$$D_{\overline{x} + 2\sigma} = 25.8\% + .747 \, (9.32 \text{ mA} - 14.13 \text{ mA})$$

$$D_{\overline{x} + 2\sigma} = 22.2\%$$

Step 7. Calculate the new minimum required I_F at end of life based on degradation found in Step 6.

$$I_{F(EOL)} = \frac{I_{FH}}{\left[1 - \frac{22.2}{100}\right]^{1/1.2}} = \frac{5}{.81} = 6.16 \text{ mA}$$

Step 8. Calculate I_F (MIN)

$$I_{F \text{ (MIN)}} = \frac{V_{cc1 \text{ (MIN)}} - V_{F \text{ (MAX)}} - V_{OL}}{R_{in \text{ (MAX)}}}$$

$$I_{F \text{ (MIN)}} = \frac{4.75 - 1.7 - .4}{409}$$

$$I_{F \text{ (MIN)}} = 6.47 \text{ mA}$$

Step 9. $R_{L \text{ (MIN)}}$, m = 1

$$R_{L \text{ (MIN)}} = \frac{V_{cc2 \text{ (MAX)}} - V_{OL}}{I_{OL \text{ (MIN)}} - m I_{IL}}$$

$$= \frac{5.25 - .6}{.016 - .0016}$$

$$R_{L \text{ (MIN)}} = 332\Omega$$

Step 10. $R_{L \text{ (MAX)}}$, m = 1

$$R_{L \text{ (MAX)}} = \frac{V_{cc2 \text{ (MAX)}} - V_{OH}}{I_{OH \text{ (MAX)}} + m I_{IH}}$$

$$R_{L \text{ (MAX)}} = \frac{4.75 - 2.4}{250\mu A + 40\mu A}$$

$$R_{L \text{ (MAX)}} = 8.1 k\Omega$$

Step 11. Minimum % Emitter Degradation Guardband

$$\%_{\text{(MIN)}} = \left[1 - \frac{I_{F \text{ (EOL)}}}{I_{F \text{ (MIN)}}} \cdot 100\right] \qquad (3.4.6.5\text{-}1)$$

$$4.8\% = \left[1 - \frac{6.16 \text{ mA}}{6.47 \text{ mA}} \cdot 100\right]$$

where I_F (EOL) represents the switching threshold at the end of life.

Step 12. Maximum % Emitter Degradation Guardband

$$\%_{\text{(MAX)}} = \left[1 - \frac{I_{F \text{ (EOL)}}}{I_{F \text{ (MAX)}}}\right] 100 \qquad (3.4.6.5\text{-}2)$$

$$34\% = \left[1 - \frac{6.16 \text{ mA}}{9.32 \text{ mA}}\right] 100$$

The conclusions that are to be drawn from this analysis are that as long as the $I_{F \text{ (MAX)}}$ is less than $I_{\overline{FS}} = 14.13$ mA, the worst-worst case CTR degradation may be calculated using only the first term, $A_o R^{\alpha} t^{n(R)}$, of the $D_{\overline{x} + 2\sigma}$ case. In the example presented, 26% degradation was determined from the first term, and when the more accurate calculation using Equation (3.4.5.1-5) was used, a 22% degradation resulted. The end of life I_F guardband may be calculated using Equations (3.4.6.5-1) and (3.4.6.5-2). Using Equation (3.4.6.5-1), the minimum guardband is 5.7%, and with Equation (3.4.6.5-2), the maximum guardband is 35%.

3.5 Analog Applications of Optically Coupled Isolators

3.5.1 Introduction

Optically coupled isolators are useful for applications where analog or DC signals need to be transferred between two isolated systems in the presence of a large potential difference or induced noise. They are an inexpensive way to eliminate the shock hazard between an input transducer and an output circuit. Optically coupled isolators can also reduce the common mode noise generated between an isolated input circuit and an output circuit. Potential applications include those in which large transformers, expensive instrumentation amplifiers, or complicated A/D conversion schemes have been used. Examples include sensing circuits (thermocouples, transducers ...), patient monitoring equipment, power supply feedback, high voltage current monitoring, and audio or video amplifiers. In many applications, the isolator can transmit the analog signal directly. However, in applications where very high linearity and stability are critical, the analog signal can be converted into a digital form and then isolated. Overall circuit parameters like linearity bandwidth, and stability determine which approach is best.

3.5.2 Analog Model for an Optically Coupled Isolator

For an optically coupled isolator with a photodiode detector, or with an integrated photodiode and transistor detector such as the 5082-4350 family, the output current for a wide range of input currents can be expressed as:

$$I_O = K(I_F/I_{F'})^n \qquad (3.5.2\text{-}1)$$

where I_F is the input LED current, $I_{F'}$ is the input LED current where K is measured, K is the output current when $I_F = I_{F'}$, and n is the slope of I_O vs. I_F on logarithmic coordinates. For a photodiode opto isolator, I_O is the current flowing into the cathode of the zero or reverse biased photodiode. For an integrated photodiode and transistor isolator, I_O is the collector current of the output transistor when biased in the active mode.

If n is equal to one, then the input to output transfer function of the opto isolator is linear. For most optically coupled isolators, n is not equal to one. For the 5082-4350 isolator, n varies from approximately 2 at very low inputs currents to 1 or less at higher input currents. Typical I_O vs. I_F characteristics for the 5082-4350 type opto isolator are shown in Figure 3.5.2-1.

3.5.3 Types of Analog Circuits

There are several analog techniques that use optically coupled isolators to isolate an analog signal. The Servo and Differential techniques are dc coupled isolated amplifiers that use two opto isolators for improved linearity and temperature stability. The servo linearizer forces the input current of one optically coupled isolator to track the input current of a second opto isolator by servo action. If $n_1 \approx n_2$ over the excursion range of the input current of the isolators, then the non linearities will cancel and the overall transfer function will be linear. The differential linearizer causes the input current of one opto isolator to increase in response to an input signal while the input current of the second optically coupled isolator decreases by an equal amount. If $n_1 \approx n_2$ over the excursion range of

the input currents of the isolators, then a gain increment in the first opto isolator will be approximately balanced by a gain decrement in the second opto isolator and the overall transfer function will be linear. For AC coupled applications, reasonable linearity can be obtained with a single optically coupled isolator. The opto isolator is biased at higher levels of input LED current where the ratio of incremental photodiode current to incremental LED current ($\partial I_P/\partial I_F$) is more nearly constant. For the 5082-4350 series of opto isolators, this occurs at input currents greater than 15 mA, as shown by Figure 3.5.3-1. The input current of the optically coupled isolator is modulated by the input signal and the modulation of the photodiode current is detected and amplified. The linearity of this type of circuit is determined by the amount by which incremental CTR ($\partial I_P/\partial I_F$) changes over the range over which the input LED current is modulated. For many applications, at least one of these three techniques can be used. Table 3.5.3-1 compares the advantages and disadvantages of each of these analog techniques.

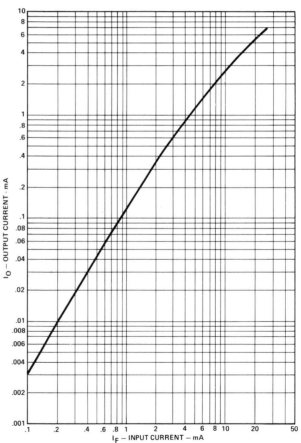

Figure 3.5.2-1   Output Current vs. Input Current for the 5082-4350, 4351, 4352, 4354, and 4355 Optically Coupled Isolators.

3.5.4 Servo Isolation Amplifier

The servo isolation amplifier shown in Figure 3.5.4-1 operates on the principle that two optically coupled isolators will track each other if their gain changes by the same amount over some operating region. U_2 compares the outputs of each opto isolator and forces I_{F2} through the second isolator to be proportional to I_{F1} through the first isolator. The constant current sources bias each input LED at 3 mA quiescent current. R_1 has been selected so that I_{F1} varies from 2 mA to 4 mA as V_{IN} varies from -5V to +5V. R_1 can be selected to accomodate any desired input range.

The zero adjustment potentiometer must have sufficient dynamic range to compensate for a worst case spread of opto isolator current transfer ratios (the values of $K/I_{F'}$ x 100%) at the input quiescent current of 3 mA. Then with V_{IN} at some value, R_4 can be adjusted for a gain of one. This potentiometer only requires sufficient dynamic range to compensate for I_{CC1} not being equal to to I_{CC2}. If

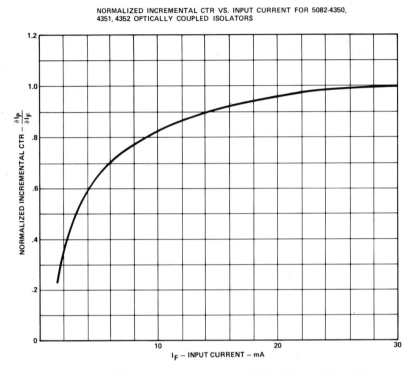

NORMALIZED INCREMENTAL CTR VS. INPUT CURRENT FOR 5082-4350, 4351, 4352 OPTICALLY COUPLED ISOLATORS

Figure 3.5.3-1. Normalized Incremental CTR vs. Input Current for 5082-4350, 4351, 4352 Optically Coupled Isolators.

	SERVO	DIFFERENTIAL	AC COUPLED
Number of Opto Isolators per Channel:	2: 5082-4350/1/2 OR 1: 5082-4354/5	2: 5082-4350/1/2 OR 1: 5082-4354/5	1: 5082-4350/1/2
Overall Linearity:	.5%	1%	1%
Input Current Range:	2–4 mA	2–4 mA	15–25 mA
Frequency Response:	DC	DC	AC
Bandwidth:	100 KHz	1 MHz	10 MHz
Temperature Stability — Offset:	±.01%/°C	±.04%/°C	Not Applicable
Temperature Stability — Gain:	−.03%/°C	−.4%/°C*	−.5%/°C*
Common Mode Rejection	>46 dB @ 1 KHz	>70 dB @ 1 KHz	>22 dB @ 1 MHz

*can be improved with additional thermistor

Table 3.5.3-1 Comparison of the Typical Characteristics of Servo, Differential, and ac Coupled Isolated Amplifiers.

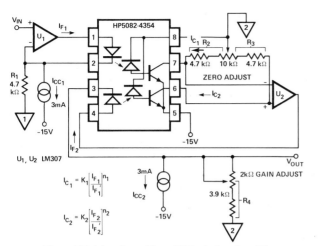

$$I_{C_1} = K_1 \left[\frac{I_{F_1}}{I_{F'_1}}\right]^{n_1}$$

$$I_{C_2} = K_2 \left[\frac{I_{F_2}}{I_{F'_2}}\right]^{n_2}$$

Figure 3.5.4-1   Servo Type DC Isolation Amplifier.

TABLE 3.5.4-1 Typical Performance for the Servo Linearized DC Amplifier.

I_{CC1} is equal to I_{CC2}, R_4 can be replaced with a single resistor equal to R_1. The transfer function of the servo isolation amplifier is:

$$V_{OUT} = R_4 \left[(I_{F'2}) \left(\frac{K_1 R_2 (I_{CC1})^{n_1}}{K_2 R_3 (I_{F'1})^{n_1}} \right)^{1/n_2} \right. \tag{3.5.4-1}$$

$$\left. \times \left(1 + \frac{V_{IN}}{R_1 I_{CC1}} \right)^{n_1/n_2} - I_{CC2} \right]$$

After zero adjustment, this transfer function reduces to:

$$V_{OUT} = R_4 I_{CC2} [(1+x)^n - 1] \text{ where } x = \frac{V_{IN}}{R_1 I_{CC1}}, n = \frac{n_1}{n_2} \tag{3.5.4-2}$$

Using a binomial expansion, V_{OUT} can be written as:

$$V_{OUT} = R_4 I_{CC2} \left[nx + \frac{n(n-1)x^2}{2!} + \frac{n(n-1)(n-2)x^3}{3!} + \ldots \right] \tag{3.5.4-3}$$

The linearity error in the transfer function when $n_1 \neq n_2$ can be written as:

$$\frac{\text{linearity error}}{\text{desired signal}} = \frac{(1+x)^n - nx - 1}{nx} \tag{3.5.4-4}$$

For example, if $|x| \leq .35$, $n_1 = 1.9$ and $n_2 = 1.8$, then the error is 1% of the desired signal. Overall linearity can be improved by reducing $|x|$ or by matching the optically coupled isolators for n. The current transfer ratio (K/$I_{F'}$ x 100%) has no effect on overall linearity.

While temperature stability is also dependent on the stability of current sources and resistors, changes in current transfer ratio (the value of K/$I_{F'}$, x 100%) of the optically coupled isolators will have negligable effect on overall gain and offset as long as the ratio of K_1 to K_2 remains constant. Typical characteristics of the servo isolator amplifier (Figure 3.5.4-1) are given in Table 3.5.4-1.

3.5.5   Differential Isolation Amplifier

The differential isolation amplifier shown in Figure 3.5.5-1 operates on the principle that an operating region exists where a gain increment in one optically coupled isolator can be approximately balanced by a gain decrement in a second opto isolator. As I_{F1} increases in the first opto isolator due to changes in V_{IN}, I_{F2} in the second opto isolator decreases by an equal amount. If $n_1 \approx n_2$, then the gain increment caused by increases in I_{F1} will be approximately balanced by the gain decrement caused by decreases in I_{F2}. The constant current source biases each input LED at 3 mA quiescent current. R_1 and R_2 are selected so that I_F varies from 2 mA to 4 mA as V_{IN} varies from -5V to +5V. R_1 and R_2 can be selected to accomodate any desired input range. At the output, U_3 and U_4 force V_{OUT} to be proportional to the difference between I_{C1} and I_{C2}.

$$V_{OUT} = R_5 [(R_3/R_4) I_{C1} - I_{C2}] \tag{3.5.5-1}$$

The zero adjustment potentiometer, R_3, allows V_{OUT} to be set to zero when V_{IN} is equal to zero. The gain adjustment potentiometer, R_5, allows the overall gain to be set to one. R_3 and R_5 must have sufficient dynamic range to compensate for a worst case spread of opto isolator current transfer ratios (the values of K/$I_{F'}$ x 100%) at the input quiescent current of 3 mA. The transfer function of the differential isolation amplifier is:

$$V_{OUT} = R_5 \left[\left(\frac{K_1 R_3}{R_4} \right) \left(\frac{I_{CC}}{2 I_{F'1}} \right)^{n_1} \left(1 + \frac{V_{IN}}{R I_{CC}} \right)^{n_1} \right. \tag{3.5.5-2}$$

$$\left. - K_2 \left(\frac{I_{CC}}{2 I_{F'2}} \right)^{n_2} \left(1 - \frac{V_{IN}}{R I_{CC}} \right)^{n_2} \right]$$

$$\text{if } R \equiv R_1 \equiv R_2$$

After zero adjustment, this transfer function reduces to:

$$V_{OUT} = R_5 K' \left[\left(1 + \frac{V_{IN}}{R I_{CC}} \right)^{n_1} - \left(1 - \frac{V_{IN}}{R I_{CC}} \right)^{n_2} \right] \tag{3.5.5-3}$$

$$\text{where } K' = \frac{K_1 R_3}{R_4} \left(\frac{I_{CC}}{2 I_{F'1}} \right)^{n_1} = K_2 \left(\frac{I_{CC}}{2 I_{F'2}} \right)^{n_2}$$

Using a binomial expansion, V_{OUT} can be written as:

$$(3.5.5\text{-}4)$$

$$V_{OUT} = R_5 K' \left[(n_1+n_2)x + \frac{[n_1(n_1-1)-n_2(n_2-1)] \, x^2}{2!} + \frac{n_1(n_1-1)(n_2-2)-n_2(n_2-1)(n_2-2) \, x^3}{3!} + \ldots \right]$$

where $x = \dfrac{V_{IN}}{R \, I_{CC}}$

If both isolators have a square law response, i.e., $n_1 = n_2 = 2$, then all non linear terms will cancel and V_{OUT} will be proportional to V_{IN}. However, if $n_1 = n_2$, then all even order terms will cancel, and the total linearity error can be reduced below 1%. The linearity error in the transfer function for any values of n can be written as:

$$(3.5.5\text{-}5)$$

$$\frac{\text{linearity error}}{\text{desired signal}} = \frac{(1+x)^{n_1} - (1-x)^{n_2} - (n_1+n_2) \, x}{(n_1+n_2) \, x}$$

For example, if $|x| \leqslant .35$, $n_1 = 1.9$ and $n_2 = 1.8$, then the linearity error is 1.5% of the desired signal. Overall linearity can be improved by reducing $|x|$ or by matching the optically coupled isolators for n. The current transfer ratio has no effect on linearity.

While temperature stability is also dependent on the stability of current sources and resistors, changes in current transfer ratio of the optically coupled isolators over temperature will cause a change in gain of the circuit. This change in gain can be compensated with a thermistor in either the input or output circuit. For example, if R_5 is replaced by a positive TC thermistor, or R_1 and R_2 are replaced with negative TC thermistors, then the gain will tend to increase as temperature goes up. Zero offset over temperature will remain stable as long as the ratio of K_1 to K_2 remains constant. Typical characteristics of the differential isolation amplifier (Figure 3.5.5-1) are given in Table 3.5.5-1.

3.5.6 AC Coupled Isolation Amplifier

The AC coupled isolation amplifier shown in Figure 3.5.6-1 operates on the principle that a single optically coupled isolator can be biased in a region where incremental CTR $(\partial I_P/\partial I_F)$ is constant. Q_1 is biased by R_1, R_2 and R_3 for a quiescent collector current of 20 mA. R_3 allows I_F to vary from 15 mA to 25 mA for a 1V peak-to-peak input signal. Under these operating conditions, the 5082-4351 operates in a region of almost constant incremental CTR as shown by Figure 3.5.3-1. The varying photon flux is detected by the photodiode and amplified by Q_2 and Q_3. Q_2 and Q_3 form a cascade amplifier with feedback applied by R_4 and R_6. I_3 is selected to allow Q_3 to operate at its maximum gain bandwidth product. R_6 is selected as V_{be}/I_3 and R_7 is selected to allow maximum excursions of V_{OUT} without clipping: $R_7 = (V_{CC} - V_{be} - V_{ce\,sat})/(2I_3)$. R_5 provides dc bias for Q_3. Closed loop gain of the output amplifier can be adjusted with the gain adjustment potentiometer, R_4. Linearity can be improved at the expense of signal to noise ratio by reducing the excursions of I_F. This can be accomplished by increasing R_3 and adding a resistor from the collector of Q_1 to ground to obtain the desired quiescent collector current of 20 mA. This circuit has no feedback around the optically coupled isolator, so any parameter that causes the incremental CTR $(\partial I_P/\partial I_F)$ to vary will cause a change in gain of the circuit. Since the quantum efficiency of the input LED varies with temperature, the $\partial I_P/\partial I_F$ and thus the overall gain will also vary with temperature. However, a thermistor can be used in the output amplifier to compensate for this change in gain. For example, if R_7 is replaced by a positive TC thermistor, or R_6 is replaced by a negative TC thermistor, then the gain will tend to increase as temperature goes up. The transfer function of the AC coupled isolation amplifier is:

$$\frac{V_{OUT}}{V_{IN}} \cong \left[\frac{\partial I_P}{\partial I_F} \right] \left[\frac{1}{R_3} \right] \left[\frac{R_4 R_7}{R_6} \right] \qquad (3.5.6\text{-}1)$$

Typical characteristics of the AC coupled isolation amplifier (Figure 3.5.6-1) are given in Table 3.5.6-1:

3% linearity for 10V p-p dynamic range
Unity voltage gain
25 kHz bandwidth (limited by U_1, U_2, U_3, U_4)
Gain drift: $-.4\%/^\circ C$
Offset drift: ± 4 mV/$^\circ$C
Common mode rejection: 70 dB at 1 kHz
3000V DC insulation

Table 3.5.5-1

Typical Performance of the Differential Linearized DC Amplifier

---

2% linearity over 1V p-p dynamic range
Unity voltage gain
10 MHz bandwidth
Gain drift: $-.6\%/^\circ C$
Common mode rejection: 22 dB at 1 MHz
3000V DC insulation

Table 3.5.6-1

Typical Performance of the Wide Bandwidth AC Amplifier

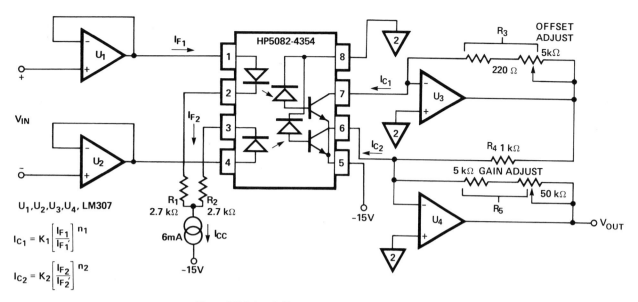

$$I_{C_1} = K_1 \left[\frac{I_{F_1}}{I_{F_1'}} \right]^{n_1}$$

$$I_{C_2} = K_2 \left[\frac{I_{F_2}}{I_{F_2'}} \right]^{n_2}$$

Figure 3.5.5-1   Differential Dc Isolation Amplifier.

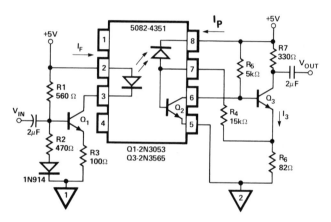

Figure 3.5.6-1   Ac Coupled Isolation Amplifier.

3.5.7   Digital Isolation Techniques

The Servo, Differential, and AC coupled analog circuits offer a simple and low cost means to optically isolate an analog signal. However, the smallest linearity error that can be achieved with these techniques is limited to about .5% to 1%. Moreover, stability of gain and offset over temperature may not be adequate for some applications. For applications that require higher linearity or better stability than can be achieved with analog techniques, isolation can be realized with digital techniques. A second reason for using digital isolation techniques is that in many applications, the analog signal must be converted into a digital format anyway before it can be processed by a microprocessor or by a digital output device. Optically coupled isolators can then provide isolation between the analog input device and the digital output device and at the same time provide all the advantages of digital isolation techniques.

With digital isolation techniques, the analog signal is converted into some type of digital code that can be transmitted through an optically coupled isolator. At the output, the digital code is converted back into analog form, or used directly by digital processing circuitry. Since the isolator is used only as a high speed switch, linearity, stability and bandwidth are determined by the conversion accuracy, temperature stability and conversion time of the A to D and D to A converters. Selection of the proper opto isolator is determined by such factors as maximum data rate, input current limitations, and proper logic interface at the output.

3.5.7.1 Isolated Analog to Digital Techniques

These techniques convert the analog signal into a digital signal through an isolated interface but leave the digital signal in a form that can be used by some type of digital processor. The A to D converter technique and variable pulse width monostable multivibrator technique illustrate two ways in which this analog to digital conversion can be accomplished. The circuits that are shown illustrate the use of optically coupled isolators to provide isolation between analog and digital circuitry and how high speed isolators can reduce systems cost.

The most commonly used digital technique to isolate an analog signal is the use of an A to D converter with optically coupled isolators providing isolation of the digital signals. This technique is outlined in Figure 3.5.7.1-1. The digital information can be transmitted through the isolators in either serial or parallel format, depending on the outputs available on the A to D converter. Serial transmission is especially practical for 8, 10 and 12 bit A to D converters because it replaces one isolator per bit with one or two high

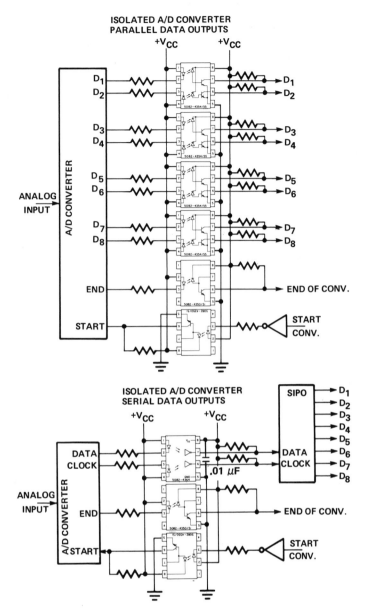

ISOLATED A/D CONVERTER
PARALLEL DATA OUTPUTS

ISOLATED A/D CONVERTER
SERIAL DATA OUTPUTS

Figure 3.5.7.1-1 Opto Isolators Provide Isolation for A/D Converters. High Speed Isolators can Reduce Cost by Transmitting Data in Serial Form.

speed isolators. If serial outputs are not available from the A to D converter, parallel outputs can be converted into serial format with a PISO shift register. The A to D converter digital isolation technique is especially useful where the digital information will be used as data for some type of digital system such as a microprocessor or LED display. A to D converters are available with straight binary, offset binary, two's complement binary, or binary coded decimal outputs.

A second technique that can be used to convert an analog signal into digital form while maintaining electrical isolation between the analog signal and the digital circuitry is the circuit shown in Figure 3.5.7.1-2. Whenever the monostable

multivibrator is triggered by a negative going pulse at TRIGGER, Q_0 goes high and stays high for a time proportional to the magnitude of V_{IN}. The time during which Q_0 is high is measured and used to provide a digital representation of V_{IN} to the output circuit. Q_0 can be used to gate an oscillator into a counter. After Q_0 goes low, the contents of the counter can be displayed by an LED display or read into a microprocessor. This concept is illustrated by Figure 3.5.7.1-2a. The counter can also be implemented in microprocessor software. With this approach, TRIGGER is attached to one bit of an output port of the microprocessor and Q_0 is attached to one bit of an input port. For this example, Q_0 is attached to D_7 to simplify the software implementation. With the program illustrated in Figure 3.5.7.1-2b, register HL will contain a binary number proportional to V_{IN} after the conversion is completed. The total time, t, that has elapsed will be equal to (29)(HL)(clock rate of microprocessor). While this approach uses less hardware than the first approach, the microprocessor must wait until the conversion is completed. Commercially available timers, such as the 555 timer, can be used to implement this technique at minimal cost.

3.5.7.2 Isolated Analog to Digital to Analog Techniques

These techniques perform the same type of transfer functions as the strictly analog servo, differential and AC coupled circuits described earlier. The difference between these techniques and the analog circuits discussed earlier is that the optically coupled isolators are used in a digital fashion. Thus, these circuits are more immune to CTR degradation and some types of common mode transients. These techniques are useful where some type of high stability isolated instrumentation amplifier is required. The pulse width modulator and the voltage to frequency converter illustrate two ways in which this analog to digital to analog conversion can be accomplished.

The circuit shown in Figure 3.5.7.2-1 shows a pulse width modulated scheme to isolate an analog signal. The oscillator operates at a fixed frequency, f, and provides a continuous trigger to the monostable multivibrator. Once triggered, the monostable multivibrator gives an output pulse proportional to the value of V_{IN}. The output of the monostable multivibrator is a square wave of frequency f but with a duty factor proportional to V_{IN}. The maximum frequency of the oscillator is determined by the required linearity of the circuit and the propagation delay of the optically coupled isolator.

$$(t_{max} - t_{min})(\text{required linearity}) \geqslant |t_{PLH} - t_{PHL}| \qquad (3.5.7.2-1)$$

For example, if f is 1 MHz, and the monostable multivibrator can vary the duty factor from 10% to 90%, then $(t_{max} - t_{min}) = 800$ nS. If the isolator is a 5082-4360, then $|t_{PLH} - t_{PHL}|$ varies from about 10 nS to 60 nS depending on drive currents, and temperature. Thus, the

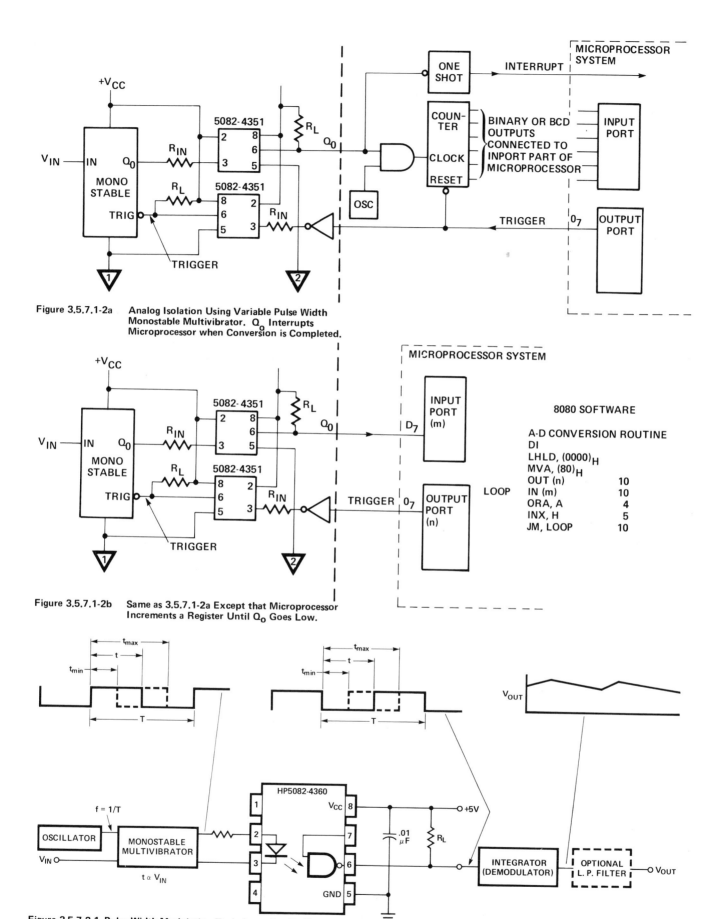

Figure 3.5.7.1-2a Analog Isolation Using Variable Pulse Width
Monostable Multivibrator. Q_o Interrupts
Microprocessor when Conversion is Completed.

8080 SOFTWARE

A-D CONVERSION ROUTINE
DI
LHLD, (0000)$_H$
MVA, (80)$_H$

OUT (n)	10
IN (m)	10
ORA, A	4
INX, H	5
JM, LOOP	10

Figure 3.5.7.1-2b Same as 3.5.7.1-2a Except that Microprocessor
Increments a Register Until Q_o Goes Low.

Figure 3.5.7.2-1 Pulse Width Modulation Techniques Can Be Used
to Isolate an Analog Signal.

3.41

worst case linearity error would be <7.5%. At 100 KHz, the worst case linearity error would be reduced to <.75%, assuming that the monostable multivibrator and output circuits had perfect linearity. At the output, the squarewave would be integrated or demodulated such that V_{OUT} would be proportional to the duty factor of the squarewave. An additional low pass filter would help to reduce the ripple at the output frequency f.

Figure 3.5.7.2-2 shows a voltage to frequency conversion scheme to isolate an analog signal. The voltage to frequency converter gives an output frequency that is proportional to V_{IN}. The maximum frequency that can be transmitted through the isolator is limited by the maximum data rate of the isolator. At the output, the frequency is converted back into a voltage. The overall circuit linearity is dependent only on the linearity of the voltage to frequency and frequency to voltage converters. Another modification on this technique is frequency modulation about a carrier frequency, f_0. Here, V_{IN} modulates f_0 by $\pm\Delta f$ depending on the amplitude of V_{IN}. At the output, V_{OUT} is reconstructed with a phase locked loop or similar circuit.

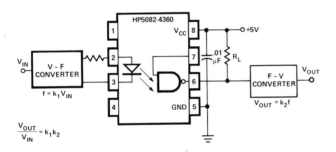

Figure 3.5.7.2-2 Voltage to Frequency Conversion Techniques Can Be Used to Isolate an Analog Signal.

3.6 Digital Applications

While there are many different digital applications of optoisolators, there are some input design considerations they all have in common. These considerations are:

a) the "off" state: input diode forward current very small, zero, or negative (reverse biased)

b) the "on" state: input diode forward current at a steady, adequate value.

c) threshold: a transitional but definable level of input diode forward current.

For (a) and (c), the only consideration usually given is that they are unimportant, but this is not always true. For (b), the "worst case" limits are usually narrower than for (a) and (c), so (b) receives more attention.

In the "off" state, the voltage of the input diode is usually zero or negative. It may, however, be slightly positive. It is worth noting that at forward voltage as high as 1.2 volts, the forward current is often neglible (<10μA). The "off" state consideration may also have some relationship to the threshold condition that should be considered.

For the "on" state and threshold condition, the consideration is illustrated in Figure 3.6-1. In analog types of optoisolators, the output current is a relatively smooth function of input current, the ratio being the CTR (Current Transfer Ratio, see Section 3.2.5). As indicated in Figure 3.6-1(a) the required minimum level of forward current, I_F, for a proper "on" state is related to the output circuit. A good practice is to design the drive circuit so that I_F is at the level for which data sheet specs are given, then adjust the load accordingly, allowing an appropriate guardband for CTR degradation. Note also that there are maximum limits on I_F for both DC and pulsed operation.

The threshold condition can also be defined, with reference to Figure 3.6-1(a), as the threshold input current, I_{Fth}, at which the output voltage, V_O, is at the threshold level, V_{Oth}, for the particular logic type used in the output circuit. For example, with TTL logic, $V_{Oth} \approx 2V_{be} \approx 1.5V$. Because of the influence of temperature, CTR degradation, and unit-to-unit variability, I_{Fth} is not precise enough to describe the threshold for a comparator--type application unless the reference current is much greater than I_{Fth}. Nevertheless, I_{Fth} can have some influence on the propagation delay through the isolator, and should therefore be considered.

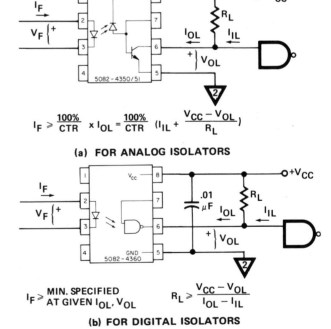

Figure 3.6-1 Input and Output Considerations for Digitally Operated Optoisolators.

In digital types of optoisolators, the threshold and "on" state input current levels are not subject to load adjustment. The output current is not a smooth function of input current. An example is shown in Figure 3.6-1(b). In the 5082-4360 type of optoisolator the threshold may be anywhere between the limits specified for I_{OH} and I_{OL}. There is, however, a specified level of I_F above which the output transistor will be capable of sinking a specified current. Guardbanding the load to allow for CTR degradation is not useful. Good practice is to design the drive circuit to provide a level of I_F slightly higher than the specified minimum. (This technique may also be applied to the circuit of Figure 3.6-1(a).) This form of guardbanding for CTR degradation narrows the spread between the minimum I_F required and the maximum data sheet rating. Techniques are given in Section 3.6.1.2 for dealing with a narrow spread.

3.6.1 Line Receivers

When digital data is transmitted over any appreciable length of transmission line (even less than a meter), there arises a possibility of ground shift, ground looping, etc. Optoisolators can reduce the amount of ground loop current and the effects of the resulting common mode voltage. They are, therefore, very useful as line receivers. This section discusses the fundamentals of designing optoisolators into line receivers.

Most line drivers are capable of sourcing a line voltage higher than the minimum (≈ 1.5V) needed to turn "on" an optoisolator. They usually also deliver a line current greater than the value of I_F as determined with reference to Figure 3.6-1. It is possible, of course, to design a line driver that will source the proper current and no more, but that is usually not good practice, as will be described later. Good practice is to drive the line with all the voltage and current available, then deal with the excess in the design of the termination of the line.

In the design of the termination, the "off" state is usually ignored unless pre-bias is used, and this is discussed in Section 3.6.3. There are usually **three objectives** in the design of the termination:

(A) proper "on" state I_F
(B) threshold level, I_{Fth}
(C) reflections due to impedance mismatch.

3.6.1.1 Resistive Terminations

As seen earlier, objective (A) is mandatory; objectives (B) and (C) are discretionary. If discretion allows (B) and (C) to be neglected, the termination may be as simple as in Figure 3.6.1-1 requiring only a single resistor. In most cases, the

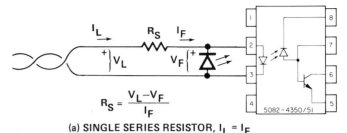

$$R_S = \frac{V_L - V_F}{I_F}$$

(a) SINGLE SERIES RESISTOR, $I_L = I_F$

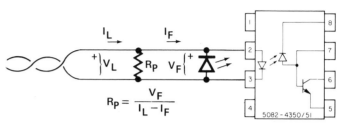

$$R_P = \frac{V_F}{I_L - I_F}$$

(b) SINGLE PARALLEL RESISTOR, $V_L = V_F$
NOT RECOMMENDED IF $I_F \ll I_L$

Figure 3.6.1-1   Single-Resistor Terminations using Optoisolators as Line Receivers.

series resistor termination of Figure 3.6.1-1(a) would be used because it will accomodate a broader range of driver and line resistance variables. It is slower than the single parallel resistor termination of Figure 3.6.1-1(b) because the input diode is driven from a higher impedance. Slowness can be remedied with a peaking capacitor in parallel with R_s; if peaking capacitance is applied, the anti-parallel diode should be used (even if the driver is polarity *non*-reversing) to allow the peaking capacitor to charge and discharge the maximum amount.

The anti-parallel LED in Figure 3.6.1-1 is recommended also when polarity-reversing drive is used. It should be a GaAsP/GaAs device (e.g. HP type 5082-4850), or the input diode of another isolator of the same type to provide a balanced output.

When, in addition to proper I_F, consideration must be given to either threshold current, I_{Fth}, or to line reflections, an additional resistor provides one additional degree of freedom. A two-resistor termination can accomodate the additional objective of either setting the I_{Fth} level or of approximate impedance matching but not both. Approximate impedance matching will be described first because it is then easier to describe the condition for threshold setting.

First of all, the nature of the reflection problem must be understood. This can best be explained with reference to a "design" having large, oscillatory reflections, illustrated in Figure 3.6.1-2. (This "design", incidentally is not recommended; it is given here only as an example of what can take place.) When the driver makes a logic transition from "low" to "high", the source line is changed so rapidly

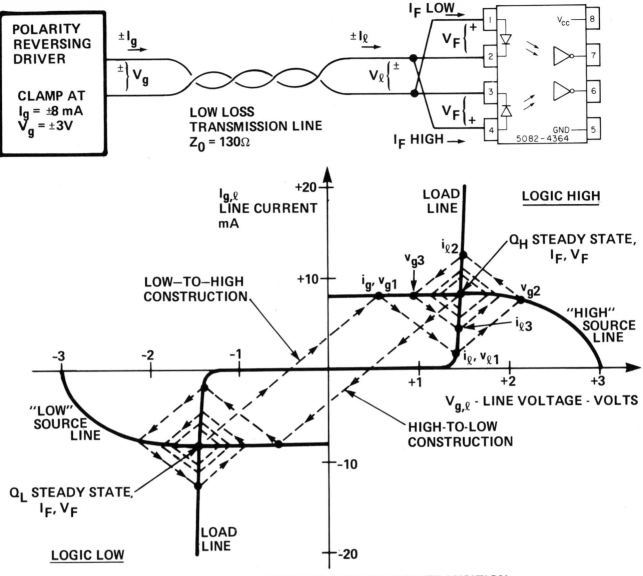

REFLECTION CONSTRUCTION FOR LOW-TO-HIGH TRANSITION
(ASSUME QUIESCENT LOW INITIAL CONDITION, Q_L)

1. FROM Q_L CONSTRUCT A LINE WITH SLOPE $\partial V/\partial I = +Z_0$ TO WHERE IT INTERSECTS THE <u>SOURCE</u> LINE FOR LOGIC HIGH AT POINT i_g, v_{g1}. THIS IS THE INITIAL STEP CURRENT IN THE LINE.

2. FROM i_g, v_{g1}, CONSTRUCT A LINE WITH SLOPE $\partial V/\partial I = -Z_0$ TO WHERE IT INTERSECTS THE <u>LOAD</u> LINE FOR LOGIC HIGH AT POINT i_ℓ, $v_{\ell 1}$. THIS IS THE INITIAL STEP CURRENT IN THE LOAD.

3. FROM i_ℓ, $v_{\ell 1}$ CONSTRUCT $+Z_0$ LINE TO v_{g2} ON SOURCE LINE.

4. FROM v_{g2} CONSTRUCT $-Z_0$ LINE TO $i_{\ell 2}$ ON LOAD LINE.

5. ETC., ETC., UNTIL CONSTRUCTION CONVERGES AT Q_H.

GENERAL RULE: $+Z_0$ LINES TO SOURCE LINE, $-Z_0$ LINES TO LOAD LINE.

Figure 3.6.1-2   Graphical Analysis of Reflections with Non-Linear Line Driver & Load

as to be regarded as a step function. At this instant, the load on the driver is just the transmission line with dynamic resistance Z_0 and initial conditions of voltage and current corresponding to point Q_L. Thus the initial voltage and current in the line at the driver end are found as the intersection of the "high" source line with a line through Q_L and having a POSITIVE Z_0 slope. This point is labelled, i_g, v_{g1}, and this is the voltage and current that will propagate toward the load. Upon reaching the termination, the transmission line is now regarded as a source having Z_0 internal resistance with point i_g, v_{g1} as its initial condition. It is, therefore, represented as a line passing through i_g, v_{g1} with a NEGATIVE Z_0 slope, and its intersection with the load line gives the first voltage and current, $i_\ell, v_{\ell 1}$, at the termination (load). This initial value, $i_\ell, v_{\ell 1}$, will now travel back to the driver where it presents to the driver a load of dynamic resistance Z_0 and initial conditions $i_\ell, v_{\ell 1}$. This is represented by a line through $i_\ell, v_{\ell 1}$ with a POSITIVE Z_0 slope and, intersecting the source line gives a voltage v_{g2}, the second terminal voltage at the driver. The driver terminal voltage, v_{g2}, now travels back to the load, again along a NEGATIVE Z_0 slope, to intersect the load line at $i_{\ell 2}$, the second load terminal current. Then a POSITIVE Z_0 slope from $i_{\ell 2}$ intersects the source line at v_{g3} and a NEGATIVE Z_0 slope through v_{g3} intersects the load line at $i_{\ell 3}$. The construction is repeated until it converges at Q_H. Notice that the transmission line is alternately represented as a Z_0 load (POSITIVE Z_0 slope to intersect source line) and as a Z_0 source (NEGATIVE Z_0 slope to intersect load line).

This graphical method does not account for lossiness in the line, or for transient capacitive or inductive effects at the terminals. Nevertheless, it is a powerful aid in visualizing the effects of any load line or source line adjustments made by adding or changing resistances. Consider, for example, the addition of a series resistor, R_s, at the load in Figure 3.6.1-2. This would be represented as a positive slope of R_s starting from ≈ 1.5 volts at zero current. This would have brought $i_{\ell 2}$ much closer to v_{g2} without having changed the current at Q_H by very much. Consider also the case of a purely resistive Z_0 load; Q_L and Q_H would then be joined by a $+Z_0$ sloped line and there would be no reflections.

Such reflectionless design is not possible with optoisolator terminations because the impedance lowers as the voltage rises from below the turn-on voltage of the input diode to above turn-on. The best that can be done is to make sure that whatever reflections there are permit a monotonic approach to convergence at the quiescent point (Q_L or Q_H), rather than the oscillatory convergence seen in Figure 3.6.1-2. Monotonic convergence means that each successive step at the termination lies *between* the last step and the quiescent point -- not beyond the quiescent point. The problem with oscillatory convergence is that the first step may be high enough to initiate turn-on, while the next step

might be below turn-on, etc. Thus, in response to a single-edge logic transition there may be one or more extraneous pulses, i.e., one or more extra edges resulting from one edge.

In working out a design, there are three cases to consider:

CASE 1: quiescent voltage-to-current ratio is equal to Z_0

CASE 2: dynamic load resistance never greater than Z_0

CASE 3: dynamic load resistance never less than Z_0

The optimal termination lies somewhere between CASE 2 and CASE 3, with CASE 1 being a good starting point. Graphical construction of two-resistor terminations are shown in Figures 3.6.1-3 and 3.6.1-4. They are called the low-threshold (LT) and high threshold (HT) circuits with reference to the line voltage at which turn-on begins. They have the same relative merits, respectively, as the series and parallel one-resistor terminations of Figure 3.6.1-1.

Applying CASE 1, then doing a reflection construction by the method outlined in Figure 3.6.1-2 suggests that CASE 1 would be an optimal design, but while this is nearly true, it neglects the fact that turn-on of the input diode does not happen instantly. The first load voltage point must lie on an extension of the dynamic resistance below turn-on. This is why CASE 1 is a good first approximation, but upward or downward adjustment of the dynamic resistance above and below turn-on might be necessary. Direction of adjustment depends on the dynamic resistance of the driver in the neighborhood of the intersection of its source line with the Z_0 line. If its dynamic resistance here exceeds Z_0 adjust toward CASE 3, but if less than Z_0 adjust toward CASE 2. Going all the way to the CASE 2 or CASE 3 limits might be too far because they may place the quiescent point in a region where the dynamic resistance of the source line is radically different from what it is at its intersection with the Z_0 line.

If the source is very non-linear, a good approach is to select as the quiescent condition that point on the source line where the dynamic resistance is Z_0 ($-Z_0$ slope along source line). Then use general construction to find the resistor values for either the LT or the HT termination.

For line lengths greater than 50 meters, reflections are unlikely to be a problem, so attention can be focused on making a threshold selection.

There are THREE SITUATIONS for which threshold adjustment makes any sense:

(a)   to establish the threshold for a comparator-type application with slowly varying line drive.

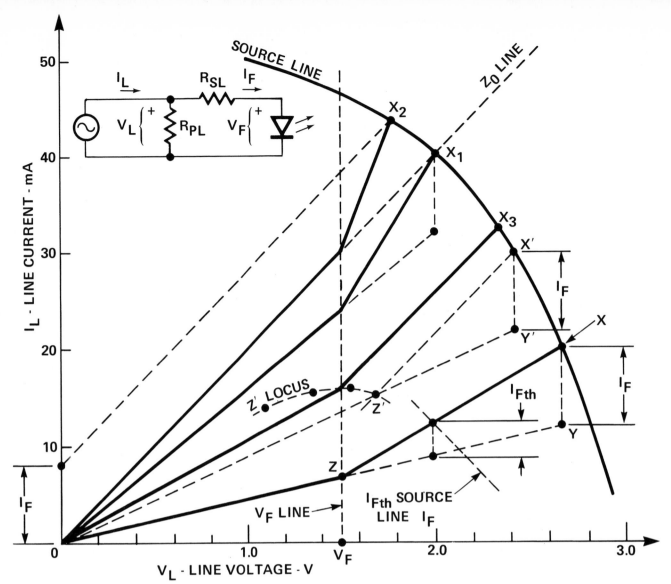

GENERAL CONSTRUCTION FOR ANY POINT, X, ON SOURCE LINE:

FROM POINT Y AT A CURRENT I_F BELOW X, DRAW LINE TO ORIGIN 0, INTERSECTING V_F LINE AT Z, THEN DRAW ZX.

$$\text{SLOPE OF OZ:} \quad \frac{\partial I_L}{\partial V_L} = \frac{1}{R_{PL}} \qquad \text{SLOPE OF ZX:} \quad \frac{\partial I_L}{\partial V_L} = \frac{1}{R_{PL}} + \frac{1}{R_{SL}}$$

CASE 1 CONSTRUCTION - DRAW Z_0 LINE TO INTERSECT SOURCE LINE AT X_1, THEN PROCEED AS FOR POINT X TO FINE R_{PL}, R_{SL}.

$$\frac{V_L}{I_L} = Z_0$$

CASE 2 CONSTRUCTION - FROM A POINT I_F ABOVE 0, DRAW LINE PARALLEL TO Z_0 INTERSECTING SOURCE LINE AT X_2, THEN PROCEED AS FOR X TO FIND R_{PL}, R_{SL}.

$$\frac{\partial V_L}{\partial I_L} \leqslant Z_0$$

CASE 3 CONSTRUCTION - CONSTRUCT A Z' LOCUS OF INTERSECTIONS OBTAINED FROM SEVERAL POINTS X' ON SOURCE LINE: THROUGH EACH X' DRAW A LINE SLOPED $1/Z_0$ TO INTERSECT AT Z' WITH LINE OY', WHERE Y' IS AT A CURRENT I_F BELOW X'. X_3 IS THE POINT X' FOR WHICH THE Z' LOCUS INTERSECTS THE V_F LINE. FROM X_3 PROCEED AS FOR X TO FIND R_{PL}, R_{SL}.

$$\frac{\partial V_L}{\partial I_L} = Z_0$$

Figure 3.6.1-3   Graphical Design of Low-Threshold Two-Resistor Termination.

3.46

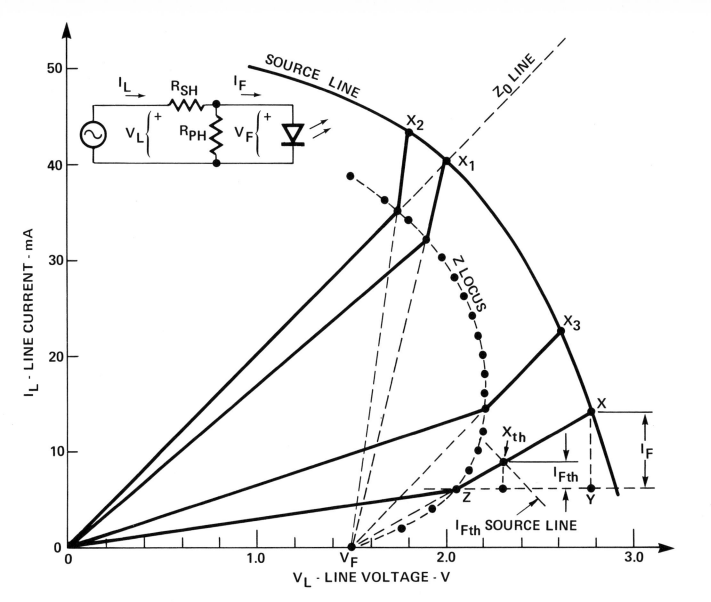

GENERAL CONSTRUCTION FOR ANY POINT X ON SOURCE LINE:

A LINE FROM X TO V_F IS INTERSECTED AT Z BY A CONSTANT-CURRENT LINE THROUGH Y AT A CURRENT I_F BELOW X.

SLOPE OF OZ: $\dfrac{\partial V_L}{\partial I} = R_{SH} + R_{PH}$ SLOPE OF ZX: $\dfrac{\partial V_L}{\partial I_L} = R_{SH}$

CASE 1 CONSTRUCTION -   DRAW Z_0 LINE TO INTERSECT SOURCE LINE AT X_1, THEN PROCEED AS FOR POINT X TO FIND R_{PH}, R_{SH}.

$\dfrac{V_L}{I_L} = Z_0$

CASE 2 CONSTRUCTION -   CONSTRUCT Z LOCUS (OF TURN-ON VOLTAGES), FOR SEVERAL POINTS X, USING "GENERAL CONSTRUCTION". FROM V_F DRAW A LINE THROUGH THE INTERSECTION OF THE Z LOCUS AND THE Z_0 LINE TO INTERSECT SOURCE LINE AT X_2, THEN PROCEED AS FOR POINT X TO FIND R_{PH}, R_{SH}.

$\dfrac{\partial V_L}{\partial I_L} \leqslant Z_0$

CASE 3 CONSTRUCTION -   FROM V_F DRAW A LINE SLOPED $1/Z_0$ TO INTERSECT SOURCE LINE AT X_3, THEN PROCEED AS FOR POINT X TO FIND R_{PH}, R_{SH}.

$\dfrac{\partial V_L}{\partial I_L} \geqslant Z_0$

Figure 3.6.1-4   Graphical Design of High-Threshold Two-Resistor Termination.

(b)   to balance delays when a single isolator is used with polarity *non*-reversing drive.

(c)   to optimize data rate with either single or dual isolators and polarity reversing drive.

The last one (c) is easy: make the threshold as low as possible. This is best done with an active voltage-clamp termination, as will be seen later. In designing for (a) or (b), it is first necessary to determine what the SOURCE LINE is for I_{Fth}.

The most precise determination of the I_{Fth} source line is by transient measurement of the V-I characteristics at the termination for the instant at which threshold current is desired. This is usually more labor than is justifiable by the results because any solution will yield only approximate resistor values and subsequent tweaking will be needed for either (a) or (b) above.

The next best is to make shrewd assumptions based on known system characteristics. Some good possibilities:

1.   If the driver is back-matched (impedance = Z_0) construct the I_{Fth} SOURCE LINE parallel to the steady-state source line -- spaced halfway between zero and steady-state for situation (b), and spaced at whatever is appropriate for (a).

2.   If the line is long enough that the round trip travel time of a pulse exceeds t_{PHL} or t_{PLH}, then, for (b) construct the first two legs of the reflection diagram (Figure 3.6.1-2) starting from $V_L, I_L = 0$. If the second leg (NEGATIVE Z_0 slope through i_g, v_{g1}) falls below the steady-state source line, take this as the I_{Fth} source line. If it falls above the steady-state source line, forget it.

3.   For either (a) or (b), assume that the driver voltage and current rise proportionately and construct the I_{Fth} source line as the locus of points obtained by taking points on the steady state source line and reducing the voltage and current by the same proportion -- by 0.5 for (b), and by whatever is appropriate for (a).

Next, for points on the steady-state source line perform the general construction (Figure 3.6.1-3 or 3.6.1-4) to find the point at which the desired values of I_F and I_{Fth} are obtained. With a backmatched source and assumption No. 1, the solution can be found analytically by substituting I_{Fth} and E_{th} for I_F and E in the circuit equation (E and E_{th} are intercepts of the steady-state and I_{Fth} source lines at $I_L = 0$). These substitutions yield a pair of simultaneous equations to be solved for the resistor values. For the LT circuit, use:

$$E = I_F Z_0 + (V_F + I_F R_{SL}) \left(1 + \frac{Z_o}{R_{PL}}\right) \qquad (3.6.1\text{-}1)$$

For the HT circuit use:

$$E = (R_{SH} + Z_o) \left(I_F + \frac{V_F}{R_{PH}}\right) \qquad (3.6.1\text{-}2)$$

To satisfy all three of the design objectives requires a three-resistor termination in either a "π" or "T" configuration. This is a rather complicated procedure, and for some driver/load combinations, solutions do not exist. The subject is discussed in EDN, in a five-part article: Feb. 5, 20, Mar. 5, 20, and April 5, 1976.

3.6.1.2 Active Terminations

Unlike resistive terminations, active terminations can respond to a variety of driver and line conditions to deliver the proper I_F to the optoisolator input, provide a low threshold, and absorb re-reflections from a mismatched driver. They should, therefore, be considered for the following situations:

- Busing or current looping where changing the number of stations affects the current available to each station.
- Power supply fluctuation at the driver.
- Design flexibility requirement; same termination to be used with any of several types of drivers.
- Accomodation of variation in line characteristics or length, where line resistance is significant.
- Temperature change on long lines ($\Delta R/R \approx 0.4\%/°C$ for copper wire).
- Data rate enhancement where a long or lossy line degrades the rise time at the termination.
- Driver mismatch causing reflections.

Active terminations are simply current regulators for the optoisolator input diode. There are two classes: current-clamp regulators allowing terminal voltage to rise, and voltage-clamp regulators allowing terminal current to rise. Whereas for resistive terminations, there are three design objectives to consider, with active terminations there is really only one: proper steady-state I_F. So far as threshold is concerned, a voltage-clamp regulator allows it to be lower than any resistive termination can permit without allowing I_F to rise excessively. Reflections are of no concern unless there are other stations on the line that might suffer -- the active termination absorbs the current/voltage fluctuations caused by reflections. If the active termination causes reflection problems for other stations, these can be reduced by addition of series or shunt resistance.

Current-clamp circuits are shown in Figure 3.6.1-5, and the operating principles for polarity non-reversing drive are seen in the V-I characteristics. Note that isolator input current does not flow until $V_L > (V_F + V_{be})$, and beyond that, there is some time delay in turning on Q2. Lowering R2 to speed up the Q2 turn-on might cause unsatisfactory regulation of I_F in the one-port circuit, but is no problem with the two-port. The main advantage of the one-port over the two-port is the simplicity of converting it to polarity-reversing drive if $|V_L| > (4 V_{be} + V_F)$ *(if this were not so, a current-clamp regulator would be a poor choice anyway)* -- a diode bridge allows the same regulator to limit current in both directions.

Voltage Clamp circuits are shown in Figure 3.6.1-6 for polarity non-reversing drive, along with V-I characteristics. Circuits for polarity-reversing drive are shown in Figure 3.6.1-7. Note that turn-on begins at V_F, and fully regulated current is obtained with $V_L = (V_F + V_{be})$ for either polarity-reversing or polarity non-reversing drive. Compare this with the current-clamp circuits requiring $V_L = (V_F + 2V_{be})$ for polarity non-reversing and $V_L = (V_F + 3V_{be})$ or $(V_F + 4V_{be})$ for polarity-reversing drive.

Note also that the regulator does not become active until Q1 turns on. That is, there is no transistor turn-on delay of isolator input current (as for Q2 in the current-clamp circuits). Moreover, since the turn-on of Q1 lags the flow of line current, there will be a momentary surge of line current into the isolator input diode. A similar lag in turn-off of Q1 hastens the turn-off of the optoisolator. *This peaking effect enhances the data rate especially where long or lossy lines cause long rise-time at the termination.*

The terminal voltage compatibility makes the voltage-clamp regulator ideal for busing. Although slight differences in V_F, V_{be}, or β might cause one regulator to operate at higher current than others in parallel with it, the voltage will be the same on all, so each isolator input diode will have its own proper I_F. That is, isolators requiring high I_F can be terminated in parallel with low-I_F types, providing, of course, the line driver can supply the total current needed.

In current looping, the build-up of terminal voltages around the loop limits the number of stations that can be operated. Also, as stations are added or removed, there may be a considerable change in loop current. For these reasons, the low operating voltage and broad range of regulated operation makes the voltage-clamp regulator a good choice.

3.6.2 Common Mode Rejection (CMR) Enhancement

Common mode interference is so named because it appears as a voltage which is, as referred to output ground, *common*

to both input terminals of the optoisolator. It is labelled e_{CM} in Figure 3.6.2-1. The desired signal appears *differentially* and is therefore called the differential mode signal, e_{DM}.

Characterization of the basic optoisolator CMR properties are discussed in Section 3.2.1. Attention here is given to various circuit techniques for improving CMR over that of the basic isolator.

e_{CM} comes about in either of two ways: "induced", represented by e_1 in Figure 3.6.2-1, or "conducted", represented by e_2. In some situations, e_2 is inherent to the system; for example, if the isolator is used to control a module floating at some large voltage with respect to driver ground, this large voltage is represented by e_2. Another example of conducted interference is that which comes through the interwinding capacitance of a power transformer. e_1 represents the voltage between the transmission line and ground or between the line and some parallel conductor. Although shown as capacitively coupled in Figure 3.6.2-1, e_1 could also be magnetically induced, as might occur in industrial machine control applications in which control and power lines are close together over appreciable distances.

If e_1 is more tightly coupled to one side of the line than to the other, the difference can cause a substantial voltage difference between the two lines. Twisted pair lines are used to balance the common mode coupling, and a shield over the twisted pair aids in balancing induced interference even if the shield floats; that is, no shield connection at either end. If the shield is to be connected anywhere, it should be to a point at the junction of two resistors connected in series across the receiving end of the transmission line. The only exception to this rule occurs when the shield is used as a third conductor to pre-bias a balanced split-phase receiver, as in Figure 3.6.3-3. A further aid in balancing out e_1 is to have the internal driver impedance balanced to driver ground. The effect of unbalanced induced interference can be somewhat compensated by adjustment of the resistors from each side of the line to the shield at the receiving end.

Even with e_1 balanced, it is still present on both terminals at the receiver, and, along with e_2, comprises e_{CM}, which can be capacitively coupled to the amplifier on the output side of the isolator, as discussed in Section 3.2.1. There are a number of techniques for dealing with e_{CM}:

- Neutralization
- Balanced differential amplification
- Amplifier de-sensitization
- Selective flip-flop output
- Exclusive-OR flip-flop
- Use of high-CMR devices

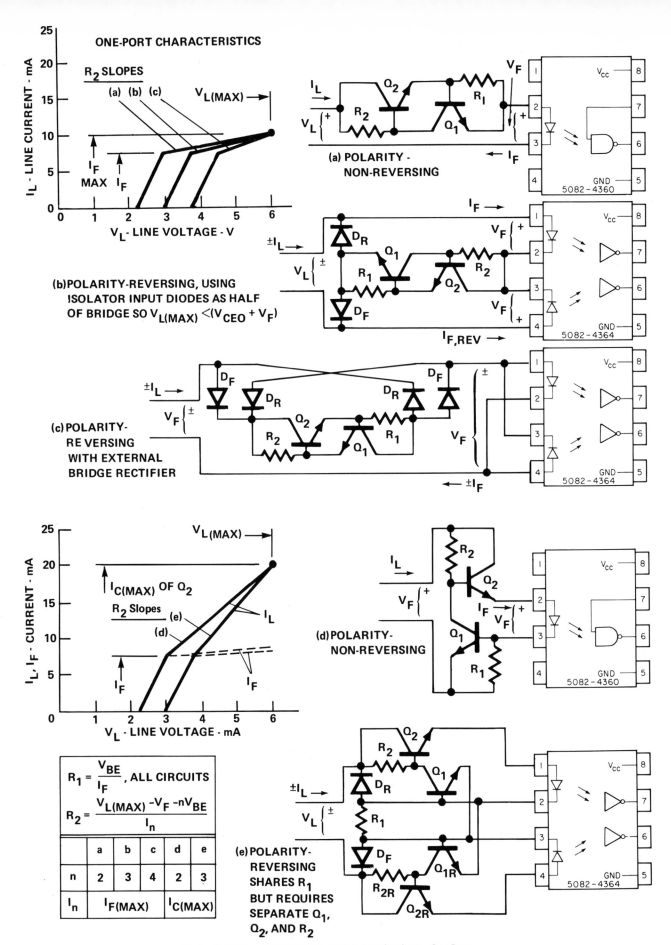

$$R_1 = \frac{V_{BE}}{I_F}, \text{ ALL CIRCUITS}$$

$$R_2 = \frac{V_{L(MAX)} - V_F - nV_{BE}}{I_n}$$

	a	b	c	d	e
n	2	3	4	2	3
I_n	$I_{F(MAX)}$			$I_{C(MAX)}$	

Figure 3.6.1-5   Current-Clamp Active Terminations: One-Port (a, b, c) and Two-Port (d, e).

3.50

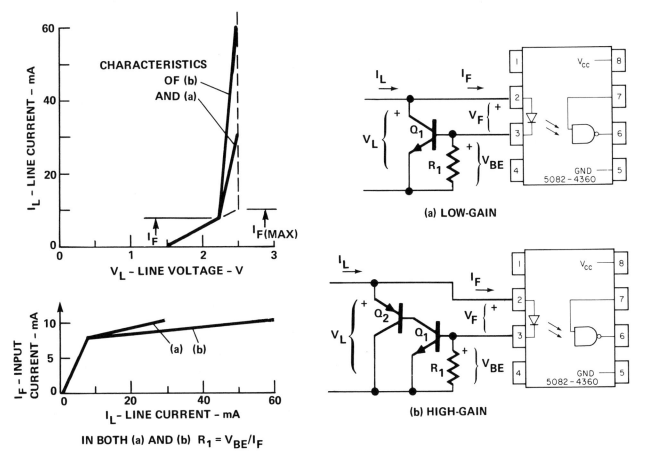

IN BOTH (a) AND (b) $R_1 = V_{BE}/I_F$

Figure 3.6.1-6   Voltage-Clamp Active Terminations for Polarity-Non-Reversing Drive.

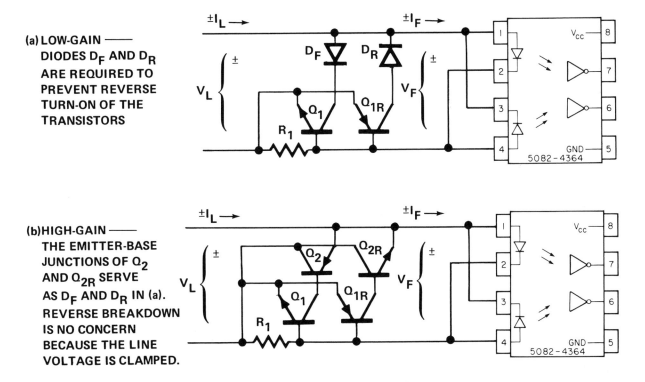

(a) LOW-GAIN — DIODES D_F AND D_R ARE REQUIRED TO PREVENT REVERSE TURN-ON OF THE TRANSISTORS

(b) HIGH-GAIN — THE EMITTER-BASE JUNCTIONS OF Q_2 AND Q_{2R} SERVE AS D_F AND D_R IN (a). REVERSE BREAKDOWN IS NO CONCERN BECAUSE THE LINE VOLTAGE IS CLAMPED.

Figure 3.6.1-7   Voltage-Clamp Active Terminations for Polarity-Reversing Drive.

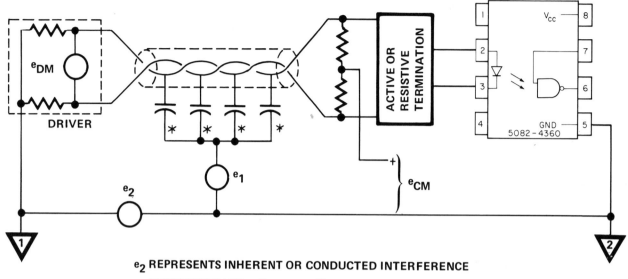

e_2 REPRESENTS INHERENT OR CONDUCTED INTERFERENCE

e_1 REPRESENTS EMI (ELECTRO-MAGNETICALLY INDUCED) INTERFERENCE

$e_{CM} = (e_1 + e_2)$ = COMMON MODE INTERFERENCE

e_{DM} = DIFFERENTIAL MODE SIGNAL

Figure 3.6.2-1   Common Mode Interference; Schematic Representation.

Neutralization can be used with optoisolators having a single-transistor amplifier operated common-emitter. Also, it is effective only while the transistor is active -- not saturated or cut-off. However, since it is while the transistor is active that e_{CM} is most troublesome, it is a technique worth considering. It consists simply of a neutralizing capacitor, C_N in Figure 3.6.2-2(a) that sources a current $C_N (de_{CM}/dt)$ which is opposite in polarity to the collector current $\beta\, C_{CM} (de_{CM}/dt)$ caused by e_{CM}. If $C_N = \beta\, C_{CM}$, these opposing currents will be balanced and the collector voltage will not be affected by e_{CM}. Even if C_N does not precisely neutralize e_{CM}, its effect is beneficial. *CAUTION: C_N must have a voltage rating high enough to accomodate the maximum voltage that e_{CM} might attain.*

Balanced differential amplification also works only if the isolator output amplifiers are active. It is clear in Figure 3.6.2-3 that if both isolators are driven to cutoff or saturation by an e_{CM} transient, they cannot maintain a differential response to e_{DM}.

Amplifier de-sensitization is perhaps the simplest defense to raise against e_{CM}. The base bypass resistor, R_{BE} in Figure 3.6.2-4 reduces the impedance in which the current flows that is coupled via C_{CM} by any e_{CM} transient. Thus a larger e_{CM} transient can be tolerated. It also reduces the amplifier's sensitivity to photocurrent resulting from e_{DM}, and therefore, requires a higher current in the input diode

for a proper "on" state. The base bypass can be used with either single-transistor or split-darlington types of amplifiers and offers the additional benefit of higher data rate capability (see Figure 3.6.3-5). R_{SK} is recommended where there is a risk that a body charged with static electricity may be discharged with all or part of the surge current passing through C_{CM}. Being amplified by the first transistor, such surge current can then be destructive.

A selective flip-flop output circuit can take advantage of a situation in which the e_{CM} transients have a higher rate of change in one direction than in the other, such as a sawtooth. It can also take advantage of a circuit with $CM_H > CM_L$ or vice versa (see Section 3.2.1). Since a NAND flip-flop can tolerate having both inputs high, it should be used where the likelihood is greater for both isolator outputs to be high (due to e_{CM} transient) than is the likelihood that both outputs will be low. Conversely, a NOR flip-flop can tolerate having both inputs low, so the rule is reversed. These rules are summarized in Figure 3.6.2-5.

Exclusive-OR flip-flop, whether of NOR or NAND construction can tolerate either both inputs high or both inputs low without either of its outputs changing state. The outputs can change state only in response to a change in differential input, so it has **infinite** common mode rejection for a static condition of e_{DM} in either logic state. Note in

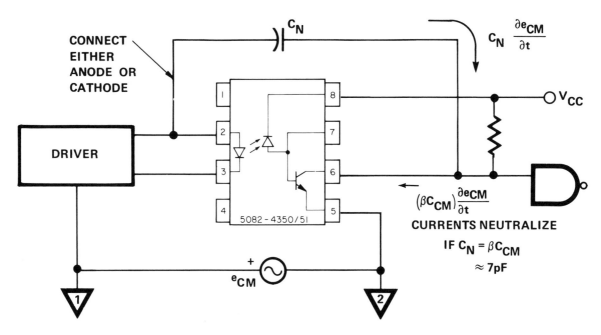

(a) BASIC SCHEME FOR NEUTRALIZATION. NEUTRALIZING
CAPACITOR MUST HAVE ADEQUATE VOLTAGE RATING.

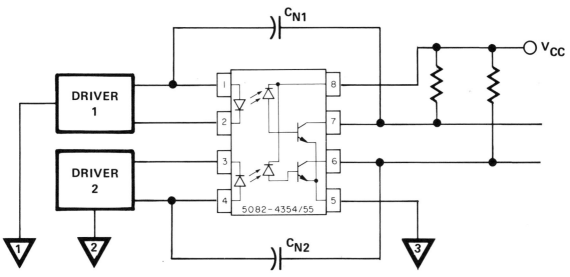

(b) NEUTRALIZATION APPLIED TO OPTOISOLATOR WITH DUAL
SINGLE-TRANSISTOR AMPLIFIERS. SEPARATE NEUTRALIZING
CAPACITORS ARE NEEDED FOR EACH COLLECTOR REQUIRING
NEUTRALIZATION. INPUT DRIVERS NEED NOT HAVE A COMMON
GROUND REFERENCE.

Figure 3.6.2-2   Neutralization of Common-Mode Interference with
External Capacitor.

3.53

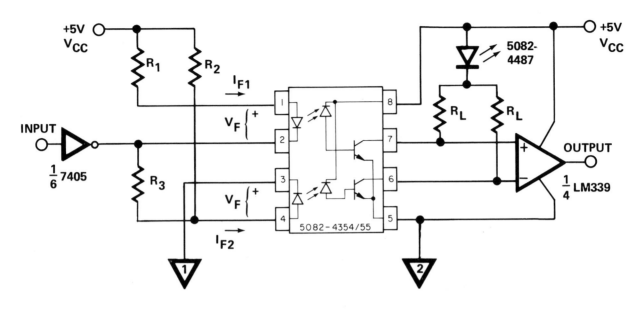

INPUT	I_{F1}	I_{F2}	OUTPUT
0	I_L	I_H	0
1	I_H	I_L	1

INPUT	I_{F1}	I_{F2}
0	$\dfrac{V_{CC} - 2V_F}{R_1 + R_3}$	$\dfrac{V_{CC} - 2V_F}{R_1 + R_3} + \dfrac{V_{CC} - V_F}{R_2}$
1	$\dfrac{V_{CC} - V_F}{R_1}$	$\dfrac{V_{CC} - V_F}{R_2} - \dfrac{V_F}{R_3}$

RESULTS OBTAINED WITH
$R_1 = R_3 = 180,\ R_2 = 270$

INPUT	I_{F1} (mA)	I_{F2} (mA)
0	5.6	18.5
1	19.4	4.6

REQUIRED RATIO OF RESISTORS :

$$R_1 = R_3 = \frac{R_2}{2\left(1 - \dfrac{V_F}{V_{CC}}\right)}$$

$$\frac{I_H}{I_L} = 2\left(\frac{V_{CC} - V_F}{V_{CC} - 2V_F}\right) \qquad R_1 \geqslant \frac{V_{CC} - V_F}{I_H}$$

$$R_L \geqslant \frac{V_{CC} - V_F}{I_{C(MAX)}} \approx 470$$

IT IS NOT NECESSARY TO HAVE PRECISE BALANCE OF ON- AND OFF-STATE INPUT CURRENTS AS LONG AS THE RATIO OF THE LOWER HIGH TO THE HIGHER LOW EXCEEDS THE RATIO OF THE HIGHER CTR TO THE LOWER CTR. SUCH UNBALANCE CAN ALSO BE COMPENSATED BY ADJUSTMENT OF R_L.

Figure 3.6.2-3   Active Differential Drive for CMR Enhancement and High Data Rate.

REDUCE R_{BE}, R_L FOR: $\left\{\begin{array}{l}\text{LOWER AMPLIFIER GAIN}\\\text{HIGHER CMR}\\\text{LOWER CTR}\end{array}\right.$

R_{SK} RESTRICTS SURGE; DOES NOT HELP OR HURT CMR OR CTR. RECOMMENDED VALUE: $R_{SK} = \dfrac{1\ \text{VOLT}}{.15 \times I_F}$

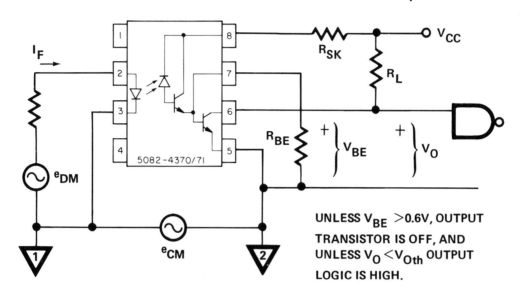

UNLESS $V_{BE} > 0.6V$, OUTPUT TRANSISTOR IS OFF, AND UNLESS $V_O < V_{Oth}$ OUTPUT LOGIC IS HIGH.

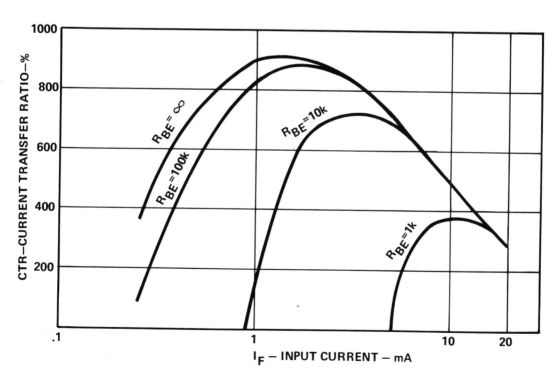

Figure 3.6.2-4   Amplifier Desensitization for CMR Enhancement by Lowering CTR.

3.55

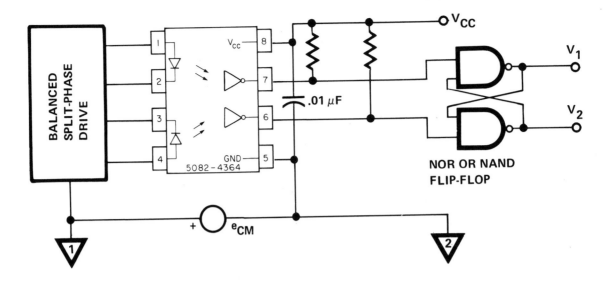

CONSIDERATIONS AFFECTING CHOICE OF NOR OR NAND

WITH RESPECT TO:	CMR		ISOLATOR PROPAGATION DELAY
	ISOLATOR PROPERTIES	e_{CM} WAVEFORM	
PREFER NAND	$CM_L < CM_H$		$t_{PHL} > t_{PLH}$
PREFER NOR	$CM_L > CM_H$		$t_{PHL} < t_{PLH}$

Figure 3.6.2-5   Selective Flip-Flop Output for CMR Enhancement and Edge Sharpening.

Figure 3.6.2-6 that the exclusive -OR flip-flop requires the output of each isolator to drive two gate inputs. If the isolator cannot handle two inputs, a buffer inverter can be inserted at no loss of CMR.

CAUTION: although the exclusive-OR flip-flop has infinite static immunity to e_{CM} its *dynamic immunity* is **not** *infinite.* If a common mode transient capable of holding both isolator outputs either high or low should persist throughout the duration of a differential mode pulse, the pulse will escape detection. Other than this, the worst a common mode transient can do to the exclusive-OR flip-flop is to advance or retard the timing of the flip-flop output transition, i.e., cause jitter of the edge.

High CMR devices are those in which the ratio of optical coupling to capacitive coupling has been made higher than that of the *usual* "sandwich"-type optoisolators. One such technique is the use of a conductive transparent screen over the detector (see Section 3.2.1). Another is the use of lenses or fiber-optics to couple the optical signal efficiently over longer distances, thereby physically separating the detector from the input and reducing capacitive coupling without a severe penalty in optical coupling. Such arrangements are discussed further in Section 2.5.5.

3.6.3   Data Rate Enhancement

In addition to the t_{PHL} and t_{PLH} data rate restrictions imposed by the optoisolator there are system limitations. Some of these can be optimized and others can be compensated. In *compensating,* the usual tradeoffs are reduction in differential-mode noise margin or common-mode rejection. This is especially true when single-ended (rather than balanced split-phase) optoisolators are used because the compensation reduces the amount of differential-mode voltage change needed to cause the isolator output circuit to change logic state. The circuit is therefore more vulnerable to differential-mode noise and to common-mode noise which has been partly differentialized by unbalanced impedance.

System optimization consists of reducing the degradation in the rise time of step changes in the differential mode signal, e_{DM}. If such degradation is due to stray wiring capacitance or inductance the remedies include reducing the impedance of the e_{DM} signal source, enlarging the wires or circuit board traces, and dressing leads properly.

The potential data rate devastation of a transmission line is often underestimated. This is because its transient response

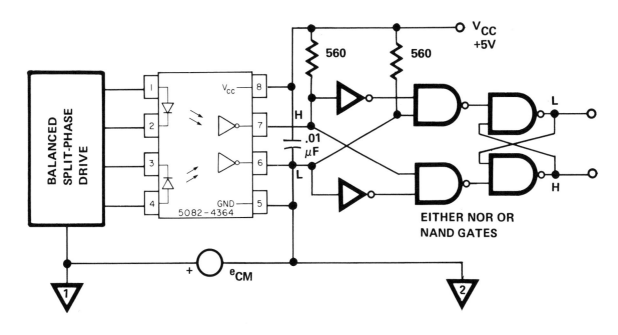

Figure 3.6.2-6   Exclusive-Or Flip-Flop for Infinite CMR Under
                  Steady State Conditions.

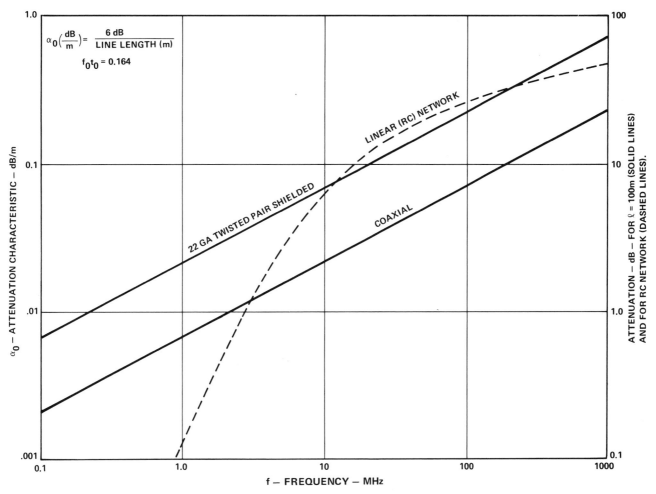

Figure 3.6.3-1   Transmission Line Attenuation as a Function of
                  Frequency.

3.57

is not the same as the familiar RC-type transient. A good estimate of transmission line performance, and appropriate design choices, can be made using the relationships described in Figures 3.6.3-1 and 3.6.3-2.

Due to skin effect, most transmission lines have an attenuation vs. frequency that varies as the square root of frequency over the frequency range of the solid lines in Figure 3.6.3-1. If a curve is not available, it is simple enough to generate one by measuring the relative output voltage, V_1 and V_2, from two lengths, ℓ_1 and ℓ_2, of cable at some particular frequency, f. Then on coordinates of log α_O vs. log f, plot the point obtained by measurement at f:

$$\alpha_O \ (f) = \frac{[20 \log (V_2/V_1)]}{[\ell_1 - \ell_2]} \quad \left(\frac{dB}{m}\right) \qquad (3.6.3\text{-}1)$$

Through this point, draw a line sloping upward at half a decade of α_O per decade of f. The lengths of line used in the measurement need not be the same as required for the system -- sample lengths will do, but the longer the samples, the greater the precision. If the generator and load impedances match the characteristic impedance of the line, a single length, ℓ_1, can be used; then substitute $\ell_2 = 0$ and V_2 = generator voltage (with matched load connected) in equation 3.6.3-1. The frequency used should be in the range of 1 to 10 MHz.

Response of a transmission line of length, ℓ, to a step input is found as follows:

1. Compute α_O by dividing the line length, ℓ, into 6 dB:

$$\alpha_O \left(\frac{dB}{m}\right) = \frac{6 \ dB}{\ell(m)} \qquad (3.6.3\text{-}2)$$

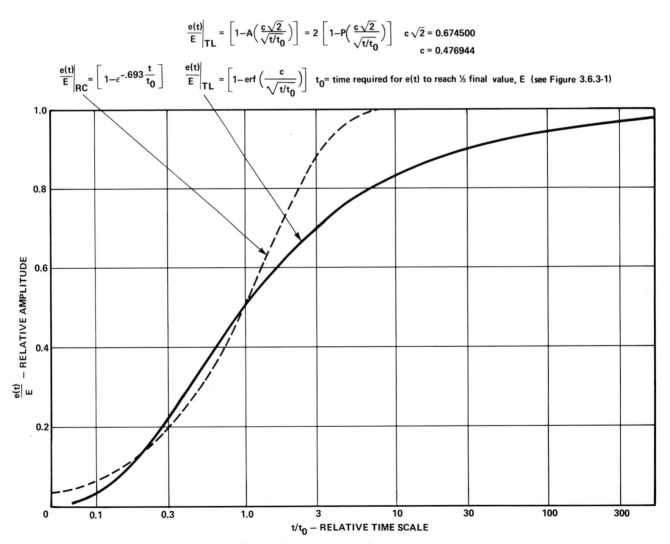

$$\left.\frac{e(t)}{E}\right|_{TL} = \left[1 - A\left(\frac{c\sqrt{2}}{\sqrt{t/t_0}}\right)\right] = 2\left[1 - P\left(\frac{c\sqrt{2}}{\sqrt{t/t_0}}\right)\right] \qquad c\sqrt{2} = 0.674500$$
$$c = 0.476944$$

$$\left.\frac{e(t)}{E}\right|_{RC} = \left[1 - \epsilon^{-.693 \frac{t}{t_0}}\right] \qquad \left.\frac{e(t)}{E}\right|_{TL} = \left[1 - \text{erf}\left(\frac{c}{\sqrt{t/t_0}}\right)\right] \qquad t_0 = \text{time required for e(t) to reach } \frac{1}{2} \text{ final value, E (see Figure 3.6.3-1)}$$

Figure 3.6.3-2   Transmission Line Response to a Step Change at the Input.

2. With α_o from Step 1, enter Figure 3.6.3-1 and find f_o at the intersection of α_o with the curve for the selected cable. (f_o is the 6-dB frequency for the length, ℓ, and could be obtained by the measurement procedure described above.)

3. Compute t_o from the relationship:

$$f_o\, t_o = 0.164 \qquad\qquad (3.6.3\text{-}3)$$

4. Step response, excluding travel time, is found from the normalized curve (solid line) in Figure 3.6.3-2, where t_o is the time required for the transient to change through half its asymptotic value. The logic delay, t_{TL}, imposed by the transmission line is then:

$$t_{TL} = t_o \times (t_{th}/t_o) \qquad\qquad (3.6.3\text{-}4)$$

where (t_{th}/t_o) is the value of (t/t_o) at which

$$\qquad\qquad\qquad\qquad\qquad (3.6.3\text{-}5)$$

$$\frac{e(t)}{E} = \frac{E_{th} - E_o}{E_{as} - E_o} = \frac{\Delta E}{E}$$

where: E_o = initial line voltage
E_{as} = asymptotic voltage
E_{th} = voltage where switching occurs

EXAMPLE: For $\ell = 100$m of twisted pair shielded cable,

Step 1 gives $\alpha_o = 0.06$

Step 2 gives $f_o = 8.5$ MHz (typical curve, Figure 3.6.3-1)

Step 4 if $\Delta E/E = 0.7$, $(t_{th}/t_o) = 3$
$t_{TL} = 19.3$ ns x $3 = 57.9$ ns

Note that if $\Delta E/E$ had been 0.8, $t_{TL} = 19.3$ ns $\times$ 7 = 135.1 ns, which is more than twice the time delay for only a small change of threshold. The importance of threshold in dealing with transmission line effects can hardly be given enough emphasis.

Shown for comparison in Figure 3.6.3-1 is a linear (RC) transient which also rises to half its final value at $t = t_o$, but note that it attains 90% in just $3.3 \times t_o$ while the transmission line transient requires $\approx 30 \times t_o$. The attenuation vs. frequency of the same RC network is shown in Figure 3.6.3-1 for comparison with the performance of 100m of twisted pair shielded cable -- scales at right side. The normalized transmission line transient can be plotted either from tables of *erf* (x) or from tables of the area under the "normal curve of error", A(x):

$$A(x) = \frac{1}{\sqrt{2\pi}} \int_{-x}^{x} \epsilon^{-\frac{1}{2} t^2}\, dt \qquad\qquad (3.6.3\text{-}6)$$

$$= \frac{2}{\sqrt{2\pi}} \int_{0}^{x} \epsilon^{-\frac{1}{2} t^2}\, dt$$

so:

$$\qquad\qquad\qquad\qquad\qquad (3.6.3\text{-}7)$$

$$A(x\sqrt{2}) = \frac{2}{\sqrt{\pi}} \int_{0}^{x} \epsilon^{-t^2}\, dt$$

$$= erf\,(x)$$

Some math tables give the area A(x) between the limits -x and +x; others the area P(x) for -∞ to +x; still others give the area between +x and +∞. Since the normal curve is symmetrical and the total area, -∞ to +∞, is just 1.0, any of these tables can be used. Jahnke and Emde (4th Edition, p.24) gives *erf* (x) with adequate precision. A great deal of precision is not worthwhile because the results of the procedure given here are only estimates to be used for avoiding serious problems with data rate limitation.

The procedure can also be used in reverse:

4. Given threshold, initial, and asymptotic voltages, use equations 3.6.3-4 and 3.6.3-5 and Figure 3.6.3-2 to find a value of t_o for tolerable delay, t_{TL}.

3. Find f_o from t_o using equation 3.6.3-3.

2. Enter Figure 3.6.3-1 with f_o to where it intersects the cable curve at α_o.

1. From equation 3.6.3-2, compute the maximum length of cable that can be used. If that is too short, calculate α_o from equation 3.6.3-2 for the desired length; then the intersection of this value of α_o with f_o from Step 3 gives a point on Figure 3.6.3-1 below which the cable curve must lie. A proper cable may thus be defined.

A handy relationship to bear in mind is that t_o **increases as the square of the length of the cable.**

The curve of Figure 3.6.3-2 can be used even if the line is mismatched at both ends. During the transient, the transmission line, as seen at the termination, is a source with impedance Z_0 and open-circuit (Thevenin) voltage changing in the manner described by Figure 3.6.3-2. The voltages, E_o, E_{as}, E_{th} to be used in equation 3.6.3-5 are found most easily by graphical construction, as shown in Figure 3.6.3-3. Worst-case delays occur under initial conditions as determined by the quiescent point (Q_L or Q_H). For the transient going from Q_L toward Q_H, a line with NEGATIVE Z_0 slope through Q_L intersects the $I_L = 0$ line at E_o. (NOTE: Because $I_L = 0$ at Q_L, the NEGATIVE Z_0 line has zero length in this example. For a non-zero example, see Figure 3.6.3-4b.) To find E_{as} begin with a POSITIVE Z_0 sloped line through Q_L to intersect the SOURCE LINE for Q_H (not necessarily at Q_H); from this intersection, a line with NEGATIVE Z_0 slope intersects the $I_L = 0$ line at E_{as}. E_{th} is a property of the load characteristic upon which there is a point, X_{th}, defining threshold conditions for the termination. Thus a line with NEGATIVE Z_0 slope through X_{th} intersects the $I_L = 0$ line at E_{th}. For a transient going from Q_H to Q_L, construction is symmetrical with a balanced polarity-reversing system; with polarity non-reversing, construction is not symmetrical but the rules above are the same for finding E_o, E_{as}, and E_{th}.

Compensation for either linear or transmission line transients can be done by peaking or by threshold-lowering, or both.

Without peaking or threshold-lowering the lowest possible threshold voltage at the terminals of an isolator input circuit are:

$$V_{Lth} = V_F + I_{Fth} R_S \qquad (3.6.3-8)$$

where the series resistance may be either R_S (Figure 3.6.1-1) or R_{SL} (Figure 3.6.1-3) or r_1 (Figures 3.6.1-6,7). V_{Lth} is the voltage at the point X_{th} in Figure 3.6.3-3. By prebiasing the input diode, this threshold can be lowered. As long as the prebias voltage is below 1.2V, the forward current in the isolator input diode will be less than $10 \, \mu A$.

The effects of such prebias on the logic delay, where the isolator terminates a transmission line, are shown in Figure 3.6.3-3. The technique is useful also where no transmission line is involved, but transient delays are present.

With a single-ended system, the threshold may already be near optimum, as seen in Figure 3.6.3-3(a). The net effect of threshold adjustment is determined with reference to Figure 3.6.3-2 for a transmission line transient -- in other situations, the effects can be found with reference to the particular kind of transient involved, e.g. RC. In general, threshold adjustment in a single-ended circuit shortens delay in one direction while lengthening it in the other, so it can be used to compensate for unbalanced t_{PLH} and t_{PHL} in the isolator.

With a balanced, polarity reversing system, the optimum threshold is at $V_L = 0$, so the objective is to make it as low as possible. The effect on data rate can be profound. In Figure 3.6.3-3(b), the one-volt prebias changes the delay from $\approx 12 \times t_o$ to $\approx 2.4 \times t_o$. If t_o is only 20 ns, this changes the delay from 240 ns down to 48 ns and raises the NRZ data rate from less than 3.5 Mb/s to approximately 10 Mb/s.

Peaking is another form of compensation. The usual technique is to place a capacitor across the series current-limiting resistor. Then when the driver changes state, the voltage on the capacitor, for as long as it persists, augments the source. With respect to the circuit in Figure 3.6.3-4(a), the peaking can be analyzed with or without transmission line effect. The construction at the left shows the "temporary" load line caused by the voltage, V_p, on the peaking capacitor. As the source line changes from Q_L to Q_H, turn-off begins immediately in one input diode and requires only a $2 \times V_F$ change of source voltage for turn-on to begin in the other. If V_p persists while the source line changes all the way from Q_L to Q_H, there will be a surge of forward current in the diode being turned on. As described in Section 3.2.3, this will enhance the data rate. **Caution:** Peaking must not allow I_{SURGE} to exceed the amplitude or duration permitted by maximum ratings. If the source or transmission line impedance is too low, peaking within limits permitted can be done with bypassing around only part of R_S. In general, the rule for construction of the "temporary" load line is to represent only that portion of the load which is NOT bypassed, and shift it by the voltage on the peaking capacitor.

If peaking is applied to a single-ended polarity-non-reversing termination, the construction rule is the same as above. It should be apparent, in construction, that an anti-parallel diode is needed across the isolator input diode to allow C_p to charge and discharge, even though the drive polarity does not reverse.

A safe, but slightly less effective peaking technique for polarity-reversing balanced drive is shown in Figure 3.6.3-4(b). With respect to steady state conditions (Q_L, Q_H) the series resistor limits input current. In the transient state, the diode being turned off has its charge transferred to the diode being turned on. Until the initially-on diode is discharged, the voltage on it augments the line-to-line drive. Threshold is very near zero, being offset only by the series resistance, R_D, of the initially-on diode. Discharge time is very short (≈ 20 ns), and if the drive transient does not rise through $\Delta E/E$ within this time, the threshold shifts over to the X_{th} point on the steady state load line. This form of peaking is, therefore, effective only for short lines or very low-loss lines, having $t_o < 20$ ns (see Figure 3.6.3-1).

Safe peaking is also obtained with the voltage-clamp regulators of Figure 3.6.1-6,7. Until the regulator transistor is turned on, current through the input diode of the isolator is limited only by r_1 and Z_o. The effect can be enhanced

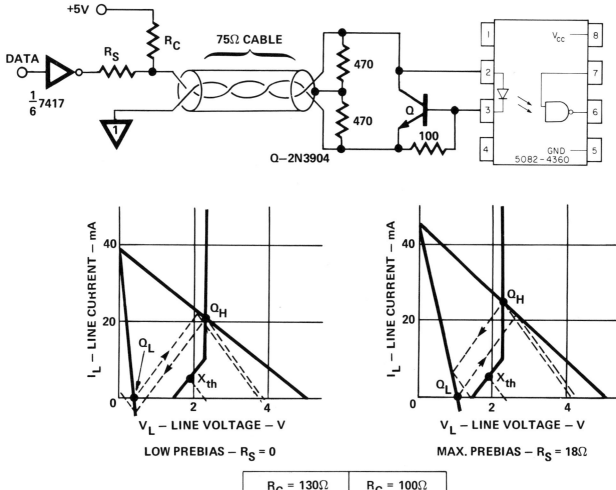

*TAKING ACCOUNT OF INTERNAL RESISTANCE OF GATE	$R_C = 130\Omega$		$R_C = 100\Omega$		
	$R_S* = 0 + 10*$		$R_S* = 18 + 10*$		
	L → H	H → L	L → H	H → L	
E_o	.4	3.85	1.05	4.15	$\frac{\Delta E}{E} = \frac{E_{th} - E_o}{E_{as} - E_o}$
E_{as}	3.75	.15	4.25	1.40	
E_{th}	2.3	2.3	2.3	2.3	
$\Delta E/E$	.57	.42	.39	.67	

THRESHOLD ADJUSTMENTS FOR DATA RATE ENHANCEMENT OR DELAY BALANCING WITH SINGLE-ENDED RECEIVERS — POLARITY NON-REVERSING.

Figure 3.6.3-3a Threshold Adjustment (PRE-BIAS) for Data Rate
Enhancement; Polarity Non-Reversing.

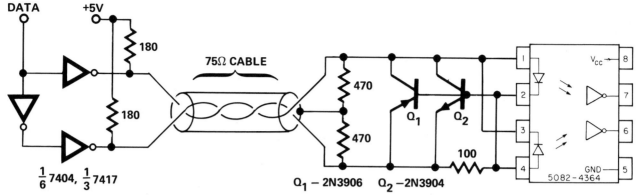

NO PREBIAS – SHIELD BALANCED FOR OPTIMAL CMR AT THE
EXPENSE OF DATA RATE.

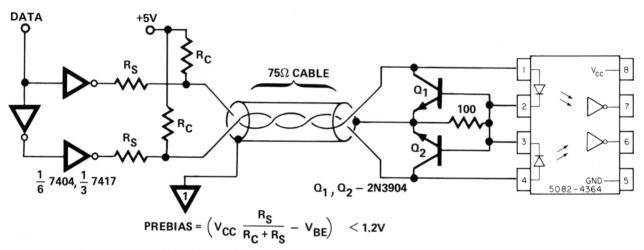

$$\text{PREBIAS} = \left(V_{CC} \frac{R_S}{R_C + R_S} - V_{BE} \right) < 1.2V$$

SHIELD RETURNS CURRENT OF "ON" SIDE, ALLOWS "OFF" SIDE TO BE PREBIASED.

ALTERNATIVE USE OF SHIELD WITH SPLIT-PHASE POLARITY-REVERSING DRIVE
AND BALANCED DUAL-ISOLATOR LOAD – EITHER CMR OPTIMIZATION OR DATA
RATE ENHANCEMENT.

Figure 3.6.3-3b Threshold Adjustment (PRE-BIAS) for Data Rate
Enhancement; Polarity Reversing

with a capacitor bypass across all or part of R_1. This also gives more consistent peaking because it is less dependent on the transistor turn-on time.

CAUTION: Both peaking and threshold-lowering tend to reduce CMR. If common-mode interference includes large, rapidly changing transients, the CMR enhancement techniques of Section 3.6.2 should be applied. For balanced terminations, the exclusive-OR flip-flop of Figure 3.6.2-6 is recommended. Single-ended terminations may require internally shielded optoisolators.

Output circuitry may also require attention in data rate optimization. With very high speed optoisolators the pullup resistor should be made as small as possible, consistent with current sinking capability of the optoisolator output and current sinking requirements of the logic inputs being driven.

Slower types of optoisolators can be made faster by "swamping" the base of the output transistor. With a resistor, R_{BE}, connected from the base to the emitter, the logic delay, especially t_{PLH}, is greatly enhanced by the discharge path that R_{BE} provides for the base. The gain (CTR), however, is reduced, especially for low input currents. It is therefore necessary to raise the input current to obtain proper operation. Lowering R_{BE}, at any particular level of input current, reduces t_{PLH} but raises t_{PHL}, as seen in Figure 3.6.3-5. Thus at any level of input current there is a value of R_{BE} at which t_{PLH} and t_{PHL} are approximately balanced. The optimum value of R_{BE} depends also on the value of the pullup resistor. Figure 3.6.3-5 shows recommended values of R_{BE} and the results to be expected as a function of input current.

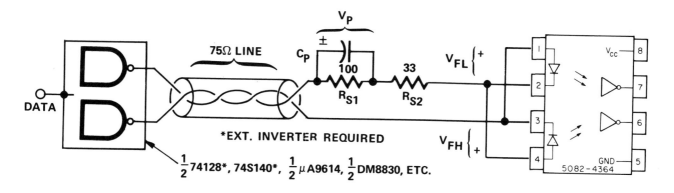

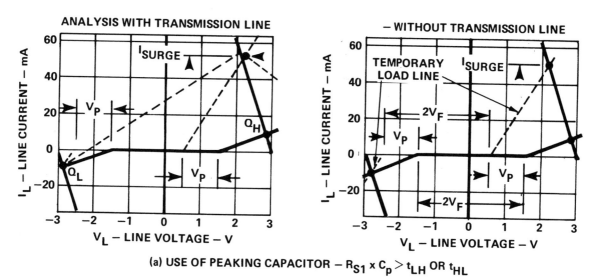

(a) USE OF PEAKING CAPACITOR — $R_{S1} \times C_p > t_{LH}$ OR t_{HL}

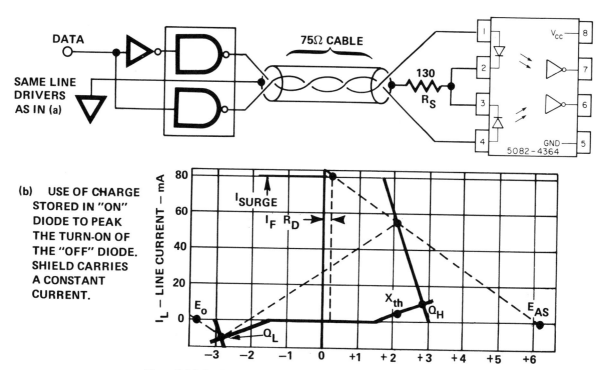

(b) USE OF CHARGE STORED IN "ON" DIODE TO PEAK THE TURN-ON OF THE "OFF" DIODE. SHIELD CARRIES A CONSTANT CURRENT.

Figure 3.6.3-4   Peaking Techniques with Split-Phase Polarity-Reversing Line Drive.

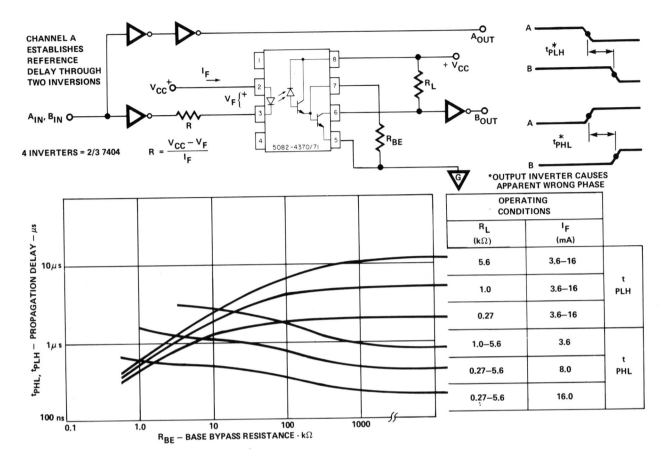

CHANNEL A
ESTABLISHES
REFERENCE
DELAY THROUGH
TWO INVERSIONS

4 INVERTERS = 2/3 7404

$$R = \frac{V_{CC} - V_F}{I_F}$$

5082-4370/71

*OUTPUT INVERTER CAUSES APPARENT WRONG PHASE

OPERATING CONDITIONS		
R_L (kΩ)	I_F (mA)	
5.6	3.6—16	t_{PLH}
1.0	3.6—16	
0.27	3.6—16	
1.0—5.6	3.6	t_{PHL}
0.27—5.6	8.0	
0.27—5.6	16.0	

Figure 3.6.3-5 Base Bypass Resistor, R_{BE}, Effect on Propagation Delay.

3.6.4 Party Line Operation (Bussing, Current Looping)

Some situations require two or more stations to operate from a common transmission line. There may be only one station capable of transmitting while all others on the line are receiving; this is called a simplex system. If two stations are each capable of transmitting to the other, it is called a duplex system. The more general case, where any one of several stations is capable of transmitting to all others, as well as receiving, is called a multiplex system.

RS-232 is a popular form of simplex bussing, documented by EIA as RS-232-C. The EIA specifications do not require ground separation, however, optoisolators are often used as extra protection with lines longer than specified in RS-232C. The EIA receiver input specifications are completely met using the termination in Figure 3.6.1-1(a) with $R_s \geq 3,000$ ohms as per RS-232C. Since RS-232-C allows a line voltage as low as 3 volts, the isolator input is required to operate with I_F as low as 0.5 mA. Also, RS-232-C limits the maximum line voltage excursions to ±25V, so the power into the input diode cannot exceed $25^2/(4 \times 3k) = 52$ mW, and the antiparallel LED, while recommended, is not essential.

RS-422 and **RS-423** are EIA documents for simplex data transmission at data rates higher than that specified in RS-232-C. RS-422 is for line voltage balanced with respect to driver ground and RS-423 is for unbalanced (single-ended), but both specify polarity-reversing line drive. Here, also, ground separation is not required, but optoisolators of adequate speed of response can be used. The RS-422/423 receiver specifications include a zero threshold voltage, so the specification cannot be completely met by any optoisolator due to the V_F threshold for turn-on. However, the RS-422/423 driver specifications are compatible with the termination in Figure 3.6.1-7 if R_1 is 90 to 100 ohms.

Wire cost in long simplex runs can be reduced using the scheme in Figure 3.6.4-1 to transmit both clock and data along the same line. To avoid distortion of the reconstructed clock pulse, the data transitions at the input should be permitted only while the clock is low. By using a pair of high-gain optoisolators at the output, the buffer inverters can be omitted. Omitting the buffers will invert the data but the clock output will be the same.

3.64

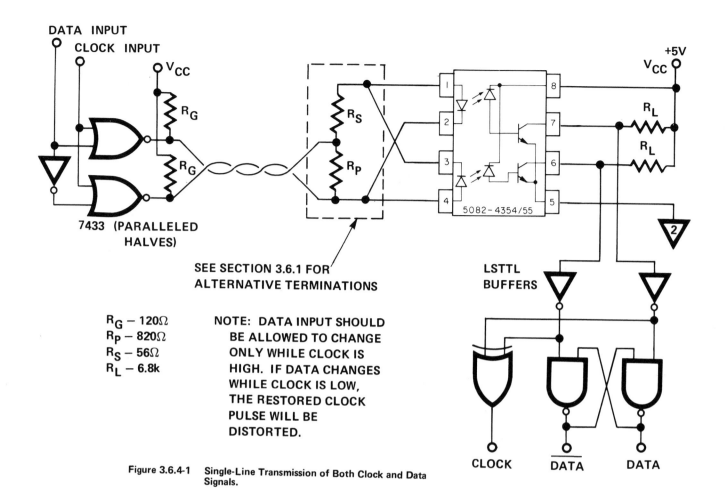

RG – 120Ω
RP – 820Ω
RS – 56Ω
RL – 6.8k

NOTE: DATA INPUT SHOULD
BE ALLOWED TO CHANGE
ONLY WHILE CLOCK IS
HIGH. IF DATA CHANGES
WHILE CLOCK IS LOW,
THE RESTORED CLOCK
PULSE WILL BE
DISTORTED.

Figure 3.6.4-1   Single-Line Transmission of Both Clock and Data
Signals.

Duplex data transmission between two modules is usually done along a single line, with each module normally in the "receive" mode. If ground separation between the modules is required, a pair of optoisolators can be connected as shown in Figure 3.6.4-2(a) to establish two-way data transmission. In the quiescent state, i.e., if there are no externally applied gates pulling P_1 or P_2 low, they will be pulled up by pullup resistors and both isolators will be off. Applying a zero at P_1 turns on U_{12} causing a zero to appear at P_2; the output of N_2 remains low, and U_{21} remains off, so when the zero is removed from P_1 the circuit returns to its quiescent state. Similarly, if a zero is applied at P_2, U_{21} goes on and U_{12} remains off. With both isolators "off" in the quiescent state, loss of power in either module does not affect the other. Use of optoisolators with output current sinking capability high enough to accomodate more than one gate loading permits dropping the buffer inverters (B_1, B_2) and leads to the modification in Figure 3.6.4-2(b). Operation is similar, except that both isolators are "on" in the quiescent state so loss of power in one module causes a zero at P in the other. By driving the cathodes of the input diodes, rather than the anodes, NOR gates may be used in place of OR gates.

Multiplexing is commonly done by a "wired-or" bus as in Figure 3.6.4-3. Any station can drop a zero on the line, causing a zero to appear at all other stations. Obviously, such a system requires some protocol to establish transmission priorities. "Wired-or" bussing requires the logic family in any station to be compatible with that of other stations. Moreover, the common ground, shared by all stations, may allow ground loop interference and here optoisolators offer a solution by permitting ground separation at any or all of the stations. Optoisolators also allow use of any logic family in any station, with the isolator's "line" side performing the interface.

Each station to be ground-separated requires one optoisolator to receive and one to transmit. The "receive" isolator must be capable of operating with low enough input current that the required number of stations can be served, while the "transmit" isolator must be capable of sinking enough current to cause a low at all the other stations.

3.65

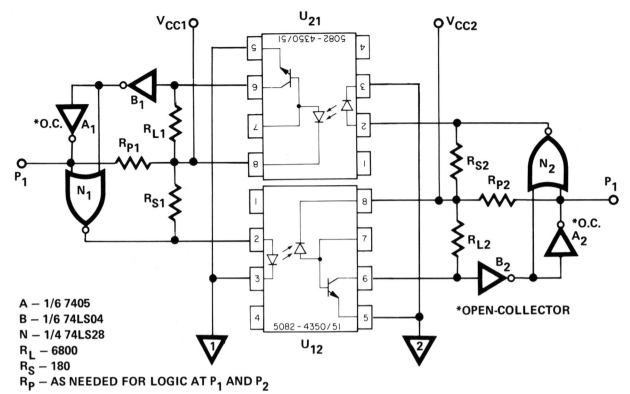

A – 1/6 7405
B – 1/6 74LS04
N – 1/4 74LS28
R_L – 6800
R_S – 180
R_P – AS NEEDED FOR LOGIC AT P_1 AND P_2

*OPEN-COLLECTOR

(a) MEDIUM-SPEED DUPLEX ISOLATOR CIRCUIT

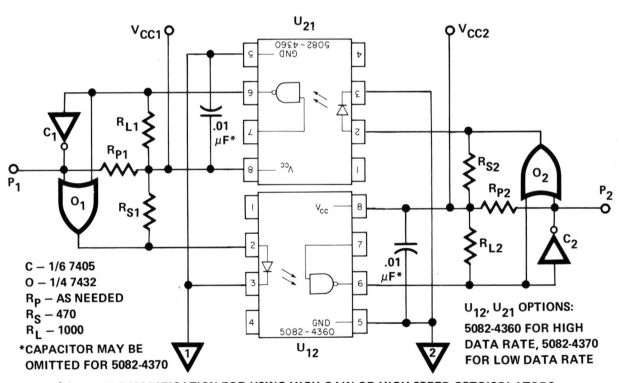

C – 1/6 7405
O – 1/4 7432
R_P – AS NEEDED
R_S – 470
R_L – 1000

*CAPACITOR MAY BE
OMITTED FOR 5082-4370

U_{12}, U_{21} OPTIONS:
5082-4360 FOR HIGH
DATA RATE, 5082-4370
FOR LOW DATA RATE

(b) CIRCUIT MODIFICATION FOR USING HIGH GAIN OR HIGH SPEED OPTOISOLATORS

Figure 3.6.4-2   Duplex Data Transmission with Optoisolators for
Ground Separation.

3.66

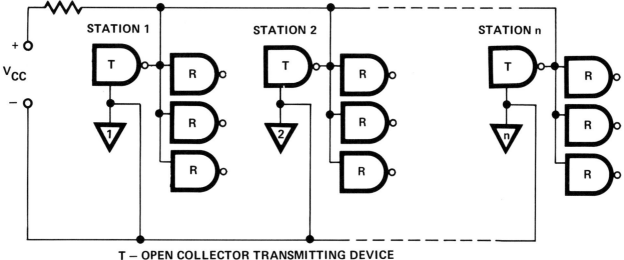

T – OPEN COLLECTOR TRANSMITTING DEVICE
R – RECEIVER – POSSIBLY MORE THAN ONE PER STATION

Figure 3.6.4-3 **Wired-OR Bus Multiplexing Requiring Common Ground for all Stations.**

In the circuit of Figure 3.6.4-4(a), the buffer inverters match the isolator input/output characteristics to whatever logic family is used in the module. A low applied to P_T turns on the "transmit" isolator, causing the line voltage to drop. Whenever the line voltage drops low enough to approach the threshold voltage of the "receive" isolator, it will turn off and cause P_R to go low. Also, a low would appear at the P_R terminal of each optoisolator receiver on the line. The optional diodes D_3 and D_2 enhance the data rate by keeping the line voltage from dropping very far below the "receive" threshold. An additional advantage is that D_3 will glow to indicate which, of the several stations on the bus, is the one from which the "zero" is originating.

To have in each station a single terminal at which data can be received as well as transmitted requires some kind of anti-lock logic. For example, if in Figure 3.6.4-4(a), the terminals P_R and P_T were to be connected together, the entire system would lock low. Figure 3.6.4-4(b) shows one form of anti-lock logic; in quiescent state the line is high so V_O is low and P_{TR} is high. In the receiving mode, as the line goes from high to low, V_O goes from low to high, and P_{TR} goes from high to low, but the output of the NOR gate *remains low* because V_O is applied directly to one of its inputs, and inverted to the other. In transmitting, when a zero is *applied* at P_{TR}, since V_O is already a zero, the output of the NOR gate will rise and allow the "transmit" isolator to turn on, and its output will drop the line voltage. This will turn off the "receive" isolator, but V_O cannot rise because the Open-Collector (O.C.) anti-lock inverter, B, is holding V_O low. With the input to inverter A held low, P_{TR} will go to a high when the zero ceases to be applied.

In general, the principle of anti-lock logic is to have within the station a means to:

(a)   hold the "transmit" isolator off during reception

(b)   allow the "transmit" isolator to turn on when a zero is applied at P_{TR}

(c)   make P_{TR} return to a high when the zero ceases to be applied.

Requirements (a) and (b) are met with O.C. inverters A followed by OR-type logic (NOR or OR), however, implied in (b) is a means to hold the input of inverter A at a low when a low is applied at P_{TR}, and this would then also take care of requirement (c). Inverter B does this; note that without inverter B, applying a low at P_{TR} would initiate a low line which would raise V_O and cause the line to rise again -- oscillation would result as long as P_{TR} is held low. The O.C. inverter B works because the isolator output is also O.C. If it were not, i.e., if the isolator had active pullup, the anti-lock loop would require a gate, rather than the "wired-or" situation at the output of the isolator.

Alternatives to the use of inverter B are shown in Figure 3.6.4-4(c) and (d). In (c) the "receive" isolator's output transistor is held "on" with base current supplied from the cathode of the "transmit" isolator's input diode. This reduces component count, but the data rate capability is reduced because of the capacitance of the input diode being added at the base. In (d) the "receive" isolator is held "on" by running the entire line source current through the input diode. This reduces component count still further (eliminates D3) and has a higher data rate capability than (c), but the large current forced into the transmit isolator's input will cause a higher rate of CTR degradation, especially if the "transmit" mode occurs frequently. Both (c) and (d) will have a longer delay than (b) in recovering

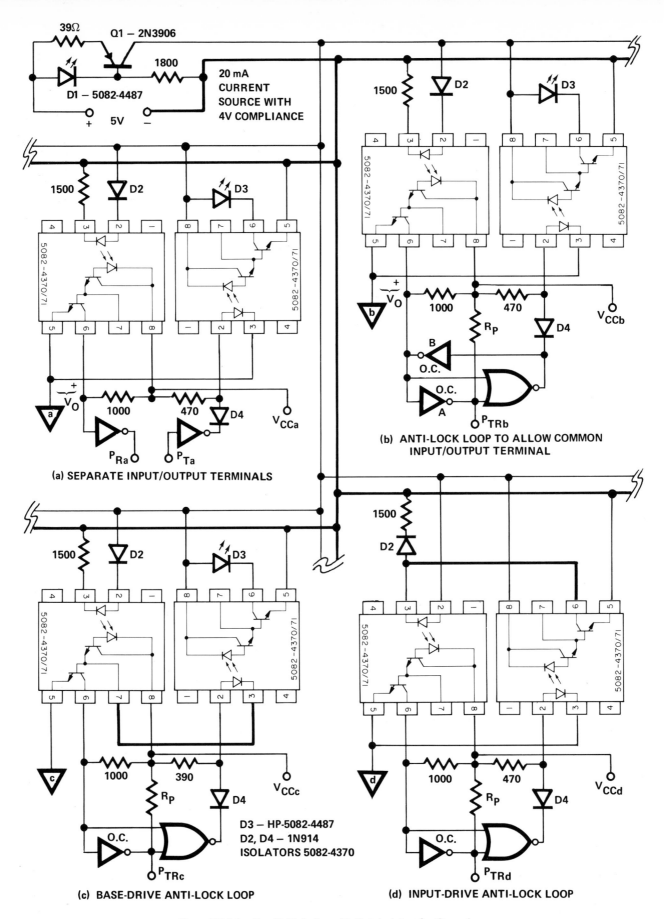

Figure 3.6.4-4   Bus Multiplexing with Optoisolators for Ground Separation of Stations.

3.68

from the "transmit" mode to the "receive" mode, due to the higher base current in the isolator output transistor during the "transmit" mode.

Other means of changing from "receive" mode to "transmit" mode are discussed in Section 3.6.6 describing the use of optoisolators with microprocessors.

Current-looping is another form of data transmission which also allows multiplex as well as simplex operation. The basic difference between current-loop multiplexing and bus multiplexing is that in a current loop each transmitting station transmits by interrupting, rather than shunting, the line current. Thus, as seen in Figure 3.6.4-5(a) and (b), the "transmit" isolator is turned "off", rather than "on" when a zero is applied at P_T (or P_{TR}).

When the "transmit" isolator is "off" the terminal voltage, V_T, rises to a value, V_{TL}, which depends on how many stations are in the loop. Worst case would be with only one station in the loop, in which case the full supply voltage would appear across the terminals of the "transmit" isolator's output. If the loop supply voltage exceeds the isolator output voltage rating, a buffer switch can be used. By using the isolator to bypass the buffer switch base current, the voltage across the isolator output never exceeds the base-to-emitter voltage of the buffer switch. Also, when the "transmit" isolator is "on", loop current may exceed the rating of the "receive" isolator input diode. A shunt regulator, such as that of Figure 3.6.1-6, can be used to bypass line current in excess of that required by the isolator.

Terminal voltage for the high state, V_{TH} limits the number of stations the loop can accomodate. It is, therefore, desirable to make it as low as possible. If separate buffer switch and input-current regulator are used, the least possible terminal voltage would be the sum of $(V_F + V_{be})$ for the regulator and $(2V_{be})$ for the buffer, and would require four transistors, two of which must carry full line current. By combining the buffer/regulator functions as in Figure 3.6.4-5(c), two transistors can be eliminated.

V_{TH} may be as low as $(V_F + V_{be} + V_{SAT,Q2}) \approx 2.5V$ but may also be somewhat higher, depending on the β of Q2, as seen in Figure 3.6.4-5(c). Such dependence on β can be eliminated by using a darlington in place of Q2, making $V_{TH} \approx 3.0V$, but not dependent on β.

Notice that with the shunt-operated buffer in Figure 3.6.4-5(c), the logic relationship between the input of the "transmit" isolator and loop is inverted from what it is in (a) or (b). Thus, the anti-lock loop of Figure 3.6.4-4(b) or (c), but not (d), can be used.

3.6.5 Telephone Circuit Applications

Any apparatus connected to telephone lines must meet two requirements. The primary requirement is that it must respond to the desired signal only. The second is that it must not interfere with the normal function of the line. Optoisolators are a nearly ideal solution to the second requirement, and proper circuit design can satisfy the first.

Ring detection requires the circuit to respond to ring signals only. These can occur over a fairly broad range of frequencies. Simple peak detection of the ac ring signal is unsatisfactory because there may be other high-amplitude voltage variations on the line, such as dialing "spikes". When a ring occurs, there are several cycles in succession at a higher repetition rate than dialing spikes. In Figure 3.6.5-1, the input diode of the optoisolator and the antiparallel diode allow ac current to flow in the coupling capacitor, C1, with each half cycle causing a peak current of about 0.5 mA in the input diode. The resulting current spikes are integrated by C2 in the output circuit. After a few cycles of the ring signal, the charge on C2 has changed far enough to turn on the output transistor. Dial spikes are ignored by this circuit. Loading of the telephone line cannot be worse than the 200k established by the series resistors. At 20 Hz, it is more nearly 450k, and at dc it is limited only by capacitor leakage.

On/off hook detection requires a means to sense current in the line, flowing in a particular direction. With no interference a single optoisolator will do, as in Figure 3.6.5-2(a). If transient common mode current is present, a pair of optoisolators with OR (or NOR) logic can be used, as in Figure 3.6.5-2(b). Common mode current in either direction can only turn on one of the two isolators -- a differential current is required to turn on both and get an output. If the common mode current is not transient, but a persistent dc level, it may prevent response of one side or the other to differential mode current. In such a situation, the circuit of Figure 3.6.5-2(c)can be used. Here the common mode current, I_{CM}, decrements the input current in one isolator as it increments the input current in the other. The isolators should be a type that is approximately linear and the threshold set at a level requiring the combined collector current resulting from differential current, I_{DM}, to obtain an output. This circuit can respond only to I_{DM} as long as $|I_{CM}| < I_{DM}$. In the LED/resistor network, the purpose of the resistor is to adjust each isolator channel to have the same ratio of collector current to line current. The LED is there to make this ratio nearly constant, down to very low current levels, without requiring a large voltage drop across the resistors.

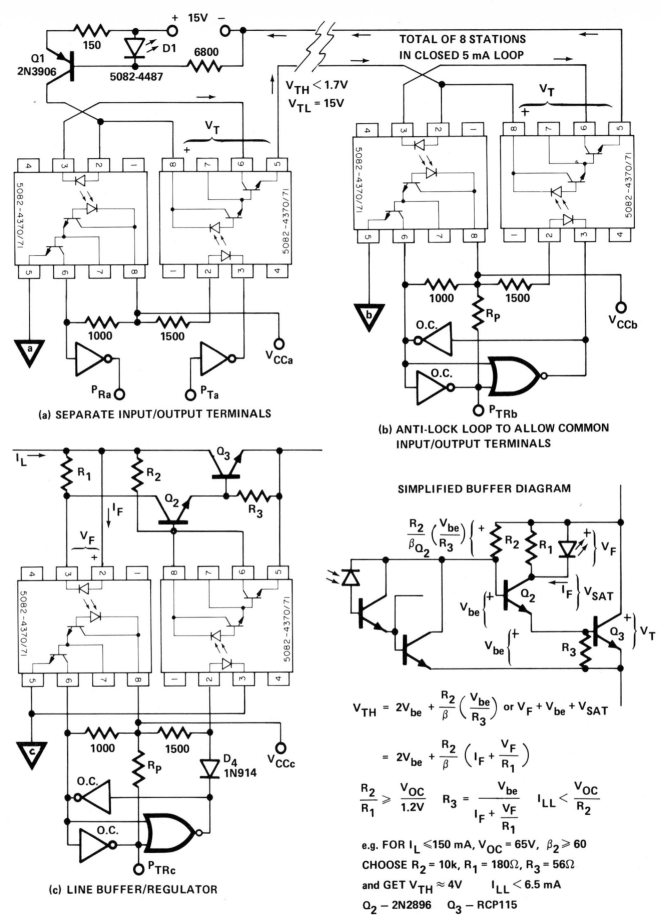

(a) SEPARATE INPUT/OUTPUT TERMINALS

(b) ANTI-LOCK LOOP TO ALLOW COMMON INPUT/OUTPUT TERMINALS

(c) LINE BUFFER/REGULATOR

SIMPLIFIED BUFFER DIAGRAM

$$V_{TH} = 2V_{be} + \frac{R_2}{\beta}\left(\frac{V_{be}}{R_3}\right) \text{ or } V_F + V_{be} + V_{SAT}$$

$$= 2V_{be} + \frac{R_2}{\beta}\left(I_F + \frac{V_F}{R_1}\right)$$

$$\frac{R_2}{R_1} \geqslant \frac{V_{OC}}{1.2V} \qquad R_3 = \frac{V_{be}}{I_F + \frac{V_F}{R_1}} \qquad I_{LL} < \frac{V_{OC}}{R_2}$$

e.g. FOR $I_L \leqslant 150$ mA, $V_{OC} = 65$V, $\beta_2 \geqslant 60$
CHOOSE $R_2 = 10$k, $R_1 = 180\Omega$, $R_3 = 56\Omega$
and GET $V_{TH} \approx 4$V $\qquad I_{LL} < 6.5$ mA
$Q_2 - 2N2896 \qquad Q_3 - RCP115$

Figure 3.6.4-5 Current-Loop with Optoisolators to Accommodate Loop Voltage Drops.

3.70

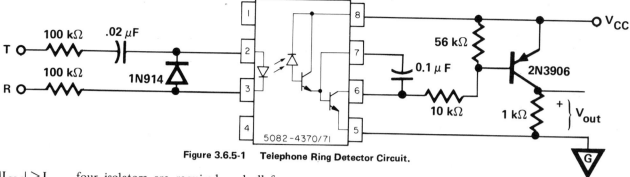

Figure 3.6.5-1   Telephone Ring Detector Circuit.

If $|I_{CM}| > I_{DM}$, four isolators are required, and all four must be adjusted to the *same ratio* of *collector current* to *line current,* as shown in Figure 3.6.5-2(d). In this scheme, I_{CM} of either polarity causes current both into and out of the collector current summing node, leaving its voltage changed only by the amount of unbalance in the collector-current-to-line-current ratio. I_{DM} of either polarity unbalances the output causing the summing node voltage to move up or down according to the polarity of I_{DM}. For the on/off hook application, only R_{L1} and the "-I_{DM}" comparator are needed, but for general application, this scheme, by using two load resistors and two comparators can not only detect I_{DM}, but also its polarity.

3.6.6   Microprocessor Applications

Ground looping in microprocessors has been blamed for everything from giving free "games" to blowing the entire circuit. Certainly there are static electricity hazards to circuits. These can be relieved by using optoisolators to open the ground loops, with little or no impairment of the system function. The slight additional expense is minuscule compared with the cost of troubleshooting.

The hazards usually arise where peripheral hardware (memory, I/O units, etc.) are connected with long bus runs, but can also exist in some intra-modular situations. The standard bus interface devices do offer some CMR protection, but are themselves vulnerable to large ground loop surges.

An arrangement whereby a peripheral unit can exchange information with the data bus and yet be isolated is shown symbolically in Figure 3.6.6-1 and schematic details are in Figure 3.6.6-2. As the truth table shows, the direction of data flow is basically controlled by the Receive Enable (RE) and Transmit Enable (TE) functions. With RE low, the R_n outputs remain high regardless of what B_n is; with TE low, the B_n outputs are "open collector" so the T_n inputs cannot control B_n. With both RE and TE low, the bus side of the interface unit is entirely "open collector"; not only are the outputs "open collector", but the inputs also present only "open collector" loading, thus making the current sinking efforts of other bus drivers more effective. The schematic of Figure 3.6.6-2 shows why.

When RE is high, the Q_{RE} transistors are switched on and are capable of sourcing 2 mA to any of the four isolator inputs facing the bus side; so whenever B_n is pulled low (by some other transmitter on the bus) the isolator at B_n is turned on and R_n goes low. Thus $R_n = B_n$.

When TE is high, four of the 20 mA current sources at Q_{TE} source current to the bus lines, while the fifth energizes the V_{CC} terminals of the four "transmit" isolators; so whenever T_n goes low, the corresponding isolator is turned on and B_n goes low. Thus $B_n = T_n$.

RE and TE are permitted to be simultaneously high, but if R_n is connected to T_n, the line will lock low. If it is desirable to have R_n and T_n connected so as to have a common receive/transmit terminal on the I/O side, the lock-low situation can be prevented by using an anti-lock loop, such as described in Section 3.6.4. A more common way is to use the logic driving RE and TE, shown in Figures 3.6.6-1 and 3.6.6-2. When $\overline{CS}$ (Chip Select) is high, both RE and TE are low. With $\overline{CS}$ low, either RE **or** TE is high, depending on whether R/T (Receive/Transmit) is high or low. If R/T is high, only RE can go high, and since TE is low, a low at T_n cannot be transmitted to B_n, so the line will not lock low with R_n connected to T_n. If R/T is low, only TE can go high; B_n will then go low when T_n goes low, but with RE low, Q_{RE} is off and the output transistor of the "receive" isolator remains off, so again the line will not lock low if R_n is connected to T_n.

The electrical ports on the bus side of the optoisolator interface are all compatible with the usual logic levels required at inputs of those modules not requiring isolation. The bus-side inputs require that non-isolated (as well as isolated) drivers sink no more than 2 mA, because that is all the Q_{RE} sources make available, and as little as 1 mA will do, depending on what logic current the R_n output must sink. At the 2 mA level of input current, the $\overline{R}_n$ output can handle TTL.

It is usually not necessary to install optoisolator interfacing as extensively as Figure 3.6.6-3 suggests, but it can be done. If only one or two of the modules is likely to suffer (or cause) difficulty, it is usually necessary only to isolate those

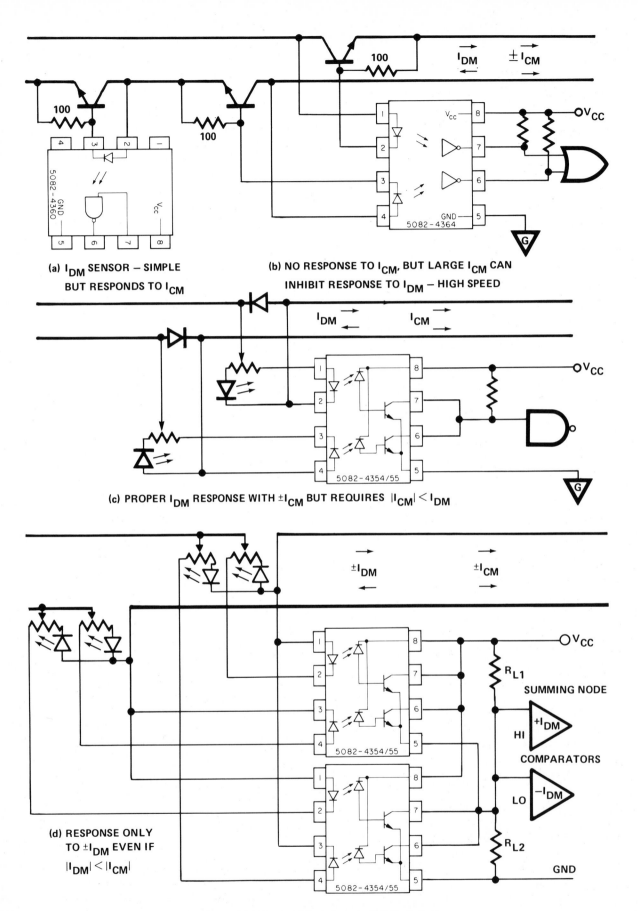

(a) I_{DM} SENSOR – SIMPLE
BUT RESPONDS TO I_{CM}

(b) NO RESPONSE TO I_{CM}, BUT LARGE I_{CM} CAN
INHIBIT RESPONSE TO I_{DM} – HIGH SPEED

(c) PROPER I_{DM} RESPONSE WITH $\pm I_{CM}$ BUT REQUIRES $|I_{CM}| < I_{DM}$

(d) RESPONSE ONLY
TO $\pm I_{DM}$ EVEN IF
$|I_{DM}| < |I_{CM}|$

Figure 3.6.5-2.   Telephone Line On/Off Hook Detectors.

3.72

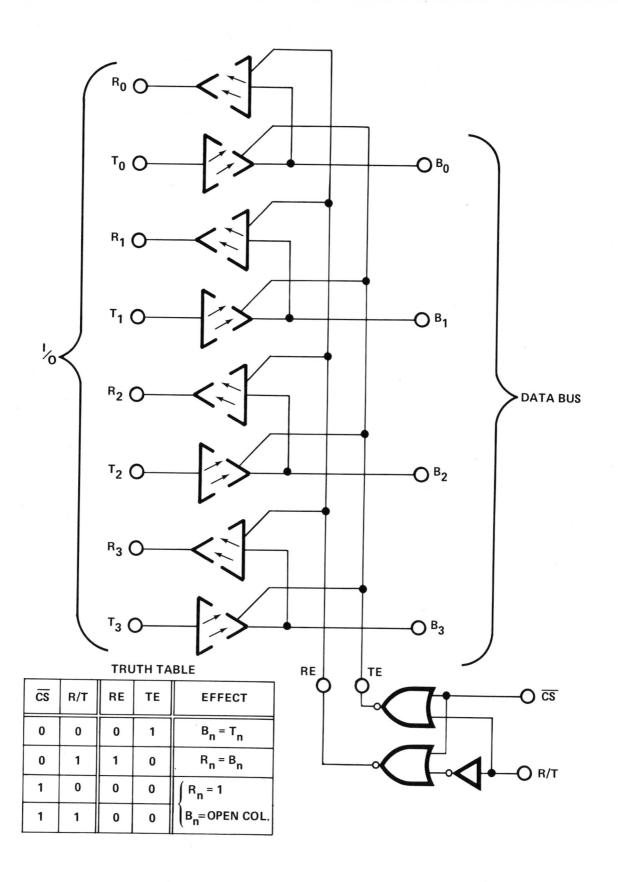

$\overline{CS}$	R/T	RE	TE	EFFECT
0	0	0	1	$B_n = T_n$
0	1	1	0	$R_n = B_n$
1	0	0	0	$\begin{cases} R_n = 1 \\ B_n = \text{OPEN COL.} \end{cases}$
1	1	0	0	

TRUTH TABLE

Figure 3.6.6-1   Symbolic Representation of Isolated Bus-I/O
Interface Circuit.

3.73

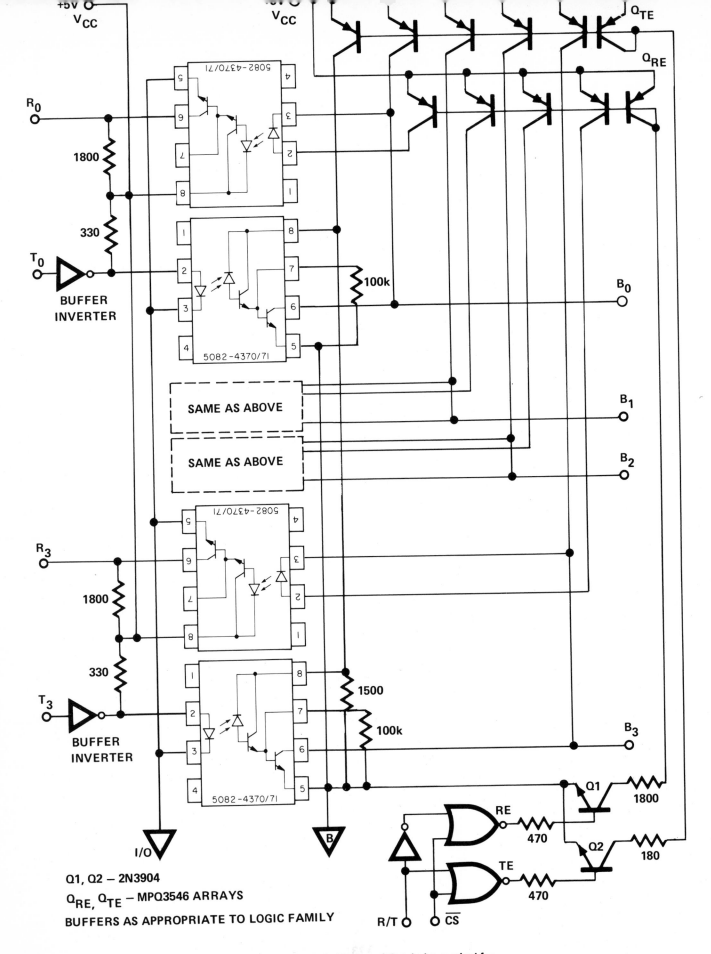

Figure 3.6.6-2   Schematic Diagram of Optoisolators wired for
Bus-I/O Interface Circuit.

3.74

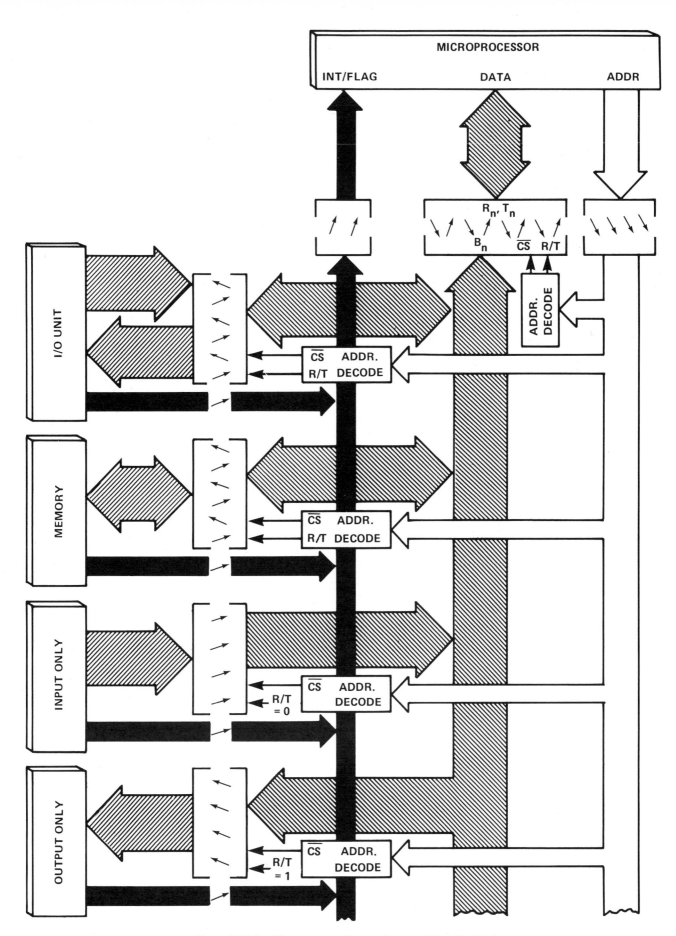

Figure 3.6.6-3   Microprocessor System; Suggested Use of Isolated
Bus-I/O Interface Circuit.

3.75

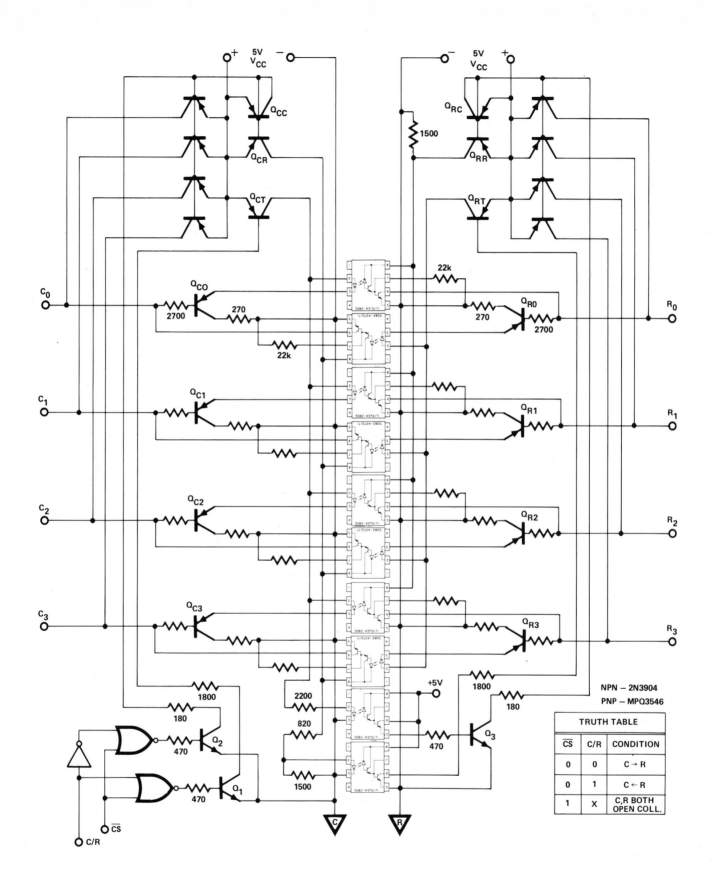

Figure 3.6.6-4 Symmetrical Four-Bit Data Bus Isolator.

NPN — 2N3904
PNP — MPQ3546

	TRUTH TABLE	
$\overline{CS}$	C/R	CONDITION
0	0	C → R
0	1	C ← R
1	X	C,R BOTH OPEN COLL.

3.76

particular units. If isolation is used on *any* of the wires connecting such a particular unit, then, for best isolation, *all* wires to that unit should be isolated. Any un-isolated wires allow a ground loop closure that can affect those which are isolated.

The microprocessor can also be isolated from the data bus it is controlling, as in Figure 3.6.6-3, top right. The reason the microprocessor appears to be addressing itself is that directional control of the flow of data (to or from the bus) is through inputs on the bus side of the optoisolator interface. In most cases, the $\overline{CS}$ input of the microprocessor isolator would be left low and its "address decoder" would only decode direction information from appropriate lines in the address bus. If for any reason, modules are connected on the bus lines between the microprocessor and the microprocessor isolator, care should be taken to make sure their electrical requirements are compatible with the R_n, T_n terminal characteristics. Such caution is especially required if the realization in Figure 3.6.6-2 is used because it is not symmetrical. That is, the R_n outputs do not have as large a current sinking capability as the B_n outputs.

A symmetrical interface isolator circuit is shown in Figure 3.6.6-4. Achieving symmetry requires the buffer transistor (Q_{Cn}, Q_{Rn}) to obtain 10mA of isolator input current from less than 1mA of current being sunk at the input (base). If there is assurance of at least a 10mA current sinking capability from any other drivers on the line, then Q_{Cn} and/or Q_{Rn} can be omitted, and the cathode of the isolator input diode would be connected through 270 ohms to the collector of the output isolator. This change would still leave an open-collector condition whenever $\overline{CS}$ is high.

The symmetrical interface isolator is especially useful when the data bus must serve several units that do not require isolation as well as some that do require isolation from the microprocessor but not from each other. The alternative to using the symmetrical interface isolator would be to use an interface isolator, such as the one in Figure 3.6.6-2 at each of the units requiring isolation, and this would require many more parts than symmetrical isolation.

Section 4
Photodiodes

4.0 PHOTODIODES

This section deals with the bare fundamentals of photodiode design and construction and with the basic characteristics of PIN photodiodes. Amplifier configurations are described for linear and logarithmic response to optical signals. Also given are circuits and suggested applications for utilizing the performance features of PIN photodiodes.

4.1 Theory and Characterization

4.1.1 Photodiode Design and Construction

When a photon is absorbed in a semiconductor, an electron-hole pair is formed. Photocurrent results when the photon-generated electron-hole pairs are separated, electrons going to the N side, holes to the P side.

Separation of a photon-generated electron-hole pair is more likely to occur when the pair is formed in a region of the semiconductor where there is an electric field (see Figure 4.1.1-1). The alternative to separation is for the electron-hole pair to simply recombine, thereby causing no charge displacement and thus no contribution to photocurrent.

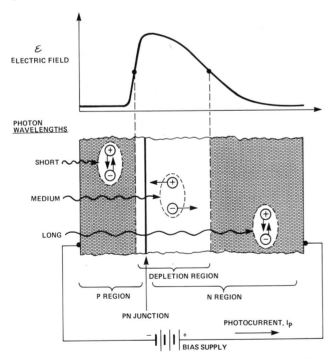

Figure 4.1.1-1   P-N Photodiode Junction; Diagram of Internal Field Effect on Detection.

Electric field distribution in a semiconductor diode is not uniform. In the regions of the P-type diffusion (front) and N-type diffusion (back) the field is much weaker than it is in the region between, known as the depletion region. For best performance, a photodiode should be made to allow the largest possible number of photons to be absorbed in the depletion region. That is, the photons should not be absorbed until they have penetrated as far as the depletion region, and should be absorbed before penetrating beyond the depletion region.

The depth to which a photon penetrates before it is absorbed is a function of the photon wavelength. Short wavelength photons are absorbed near the surface while those of longer wavelength may penetrate the entire thickness of the crystal. For this reason, if a photodiode is to have a broad spectral response (respond well to a broad spectrum of wavelengths) it should have a very thin P-layer to allow penetration of short wavelength photons, as well as a thick depletion region to maximize photocurrent from long wavelength photons.

The thickness of the depletion region depends on the resistivity of the region to be depleted and on the reverse bias.

A depletion region exists even if no reverse bias is applied. This is due to the "built-in" field produced by diffusion of minority carriers across the junction. Reverse bias aids the built-in field and expands the depletion region.

The extent of the depletion region at any voltage is larger in devices made with higher resistivity at the junction; but low resistivity is required at both surfaces for making ohmic contact to the device. P-N photodiodes, such as solar cells, are made with P diffusion into N-type material of low resistivity. In P-N photodiodes, a thin P diffusion allows good short-wavelength response, but a relatively high reverse bias is required to extend the depletion to the depth required for good long-wavelength response. A deep P diffusion degrades the short-wavelength response but lowers the bias required for good response at longer wavelengths.

Optimization of both short- and long-wavelength response at low reverse bias requires a P-I-N, rather than a P-N diode structure. A PIN diode has a thin P-type diffusion in the front and an N-type diffusion into the back of a wafer of very high resistivity silicon (see Figure 4.1.1-2). The high resistivity material between the P-type and N-type diffusions is called the intrinsic region, or I-layer.

In Hewlett-Packard PIN photodiodes, the I-layer has a resistivity so high that even at zero bias, the depletion region extends from the P-layer to approximately halfway through the I-layer. With as little as 5 volts reverse bias, depletion is extended all the way to the N-layer; this is called the "punch-through" voltage. Since breakdown voltage is over 200V, it is often desirable to operate at reverse voltages well above punch-through so as to keep the I-layer fully depleted even at high flux levels. This insures best linearity and speed of response.

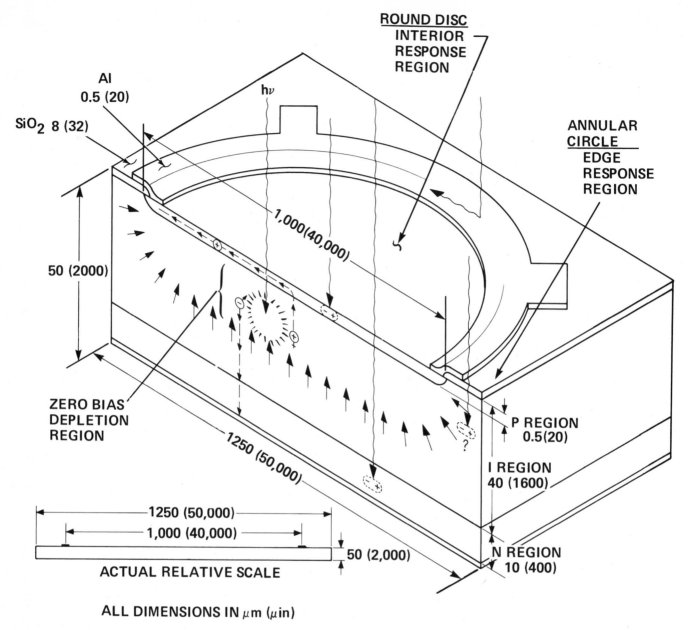

ALL DIMENSIONS IN μm (μin)

Figure 4.1.1-2   P-I-N Photodiode; Isometric Cutaway Distorted to Clarify Main Features.

As reverse voltage is increased, the depletion region expands sideways, extending beyond the contact rim with as little as 20 volts applied. This permits photons incident at the edge outside the ring to enter and be absorbed in the depletion region without having to penetrate the P-layer. Since the silicon dioxide surface passivation is very transparent, even to very short (UV) wavelengths, the quantum efficiency of edge response can actually exceed unity (see Figure 4.1.2-1). Quantum efficiency exceeding unity is possible because the shorter wavelength photons have energies greater than twice the bandgap of silicon.

Virtually any kind of semiconductor junction exhibits photoresponse, but most devices are designed and packaged

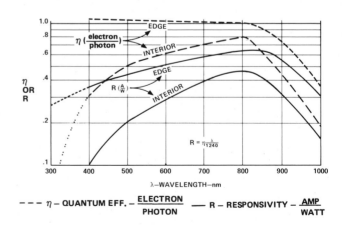

Figure 4.1.2-1   Spectral Response of Interior and Edge Regions of HP P-I-N Photodiode.

to prevent radiant flux from interfering with their intended function. LEDs are packaged to permit radiation of flux from their junctions, so their junctions are easily exposed to radiation. Of course, they are not designed for optimal performance as photodiodes; nevertheless, they perform well enough to be useful in many applications. Their spectral response peak is at a wavelength much shorter than a silicon photodiode spectral peak, so, in the absence of filtering, they give a closer approximation to photopic (human visual) response.

4.1.2 Photodiode Characterization

A figure of merit for photodiode response can be described by the quantum efficiency. Ideally, each photon (or quantum of energy) should cause a contribution of one electron to the stream of photocurrent. *Quantum efficiency*, η_q, is therefore dimensioned as "electrons per photon".

For most engineering, a more familiar performance parameter is the *flux responsivity*, R_ϕ, which takes account of the photon energy. It is the ratio of photocurrent to spot flux:

$$R_\phi = \eta_q \frac{\lambda}{1240} = \frac{I_P}{\phi_e} \qquad (4.1.2\text{-}1)$$

where: R_ϕ = flux responsivity in amps per watt

   η_q = quantum efficiency in electrons per photon

   λ = photon wavelength in nanometers

   I_P = photocurrent in amperes

   ϕ_e = radiant flux in watts

In Figure 4.1.2-1 the spectral quantum efficiency is shown with dashed lines and the responsivity with solid lines. These same values apply to all Hewlett-Packard PIN photodiodes, regardless of their size or lens magnification. This is because responsivity is area-independent, being defined for an incrementally small spot. Note that there is a substantial difference between the responsivity of the interior region and that of the edge region (see also Figure 4.1.1-2). Actually, the edge response is obtained only when reverse bias is applied. Interior response is nearly independent of reverse bias at wavelengths shorter than peak; at longer wavelengths interior responsivity does increase slightly when reverse bias is applied — depending on how much reverse bias is applied and how high the flux level is. At reverse bias greater than punch through, the variation of interior responsivity with reverse bias is nearly zero for moderate flux level.

Another handy performance parameter is the incidence response, R_E, which takes account of the photosensitive area (or apparent area, for photodiodes with magnifying lenses). It is the ratio of photocurrent to incidance:

$$R_E = \frac{I_P}{E_e} = \int [R_\phi (A_D)] \, d A_D \approx R_\phi \, A_D \qquad (4.1.2\text{-}2)$$

where: R_E = incidance response in amps per watt per square millimeter*

   E_e = radiant incidance in watts per square millimeter*

   A_D = effective photosensitive area in square millimeters*

   I_P, R_ϕ are defined in equation 4.1.2-1

*The SI units of area are square millimeters and square meters but square centimeters are still frequently used in describing incidance.

Incidance response describes the photodiode performance when it is floodlighted (uniform incidance over the entire device) so that edge effects are included along with interior response. The incidance response varies with reverse bias as the depletion region spreads out from the P-diffusion area. Consequently, the description of incidance response as the product of responsivity times area is a useful approximation only. The perimeter-to-area ratio decreases as photodiode area is increased, so the approximation improves as the area is increased. With very large reverse bias (>100V) the entire surface of the chip can be photosensitive — that is, the depletion region can be extended all the way from the edge of the P-diffusion to the edge of the chip.

Speed of response also depends on what areas of the photodiode are irradiated and how much reverse bias is applied, as seen in Figure 4.1.2-2. The speed observed also depends on the load resistance. Photocurrent begins to flow just a few picoseconds after flux is applied, but there is junction capacitance, package capacitance, and stray wiring capacitance to be charged. The rise/fall time constant therefore depends largely on load resistance unless the load resistance is so small that the internal resistance of the photodiode limits the speed. The internal resistance is mainly the sheet resistance of the very thin P-diffusion. With flux applied at the very center of the interior region, the resulting photocurrent encounters maximum sheet resistance and this may be as much as 50 ohms. Flux applied to the interior at portions closer to the aluminum contact ring produces photocurrent that encounters a lower sheet resistance.

Low noise is another benefit resulting from the extremely high resistivity of the I-layer in Hewlett-Packard PIN photodiodes. Diode-noise-limited operation can be achieved over a modulation frequency range extending from dc to more than 10 KHz, as seen in Figure 4.1.2-3. The horizontal dashed line is the noise calculated from the shot noise formula:

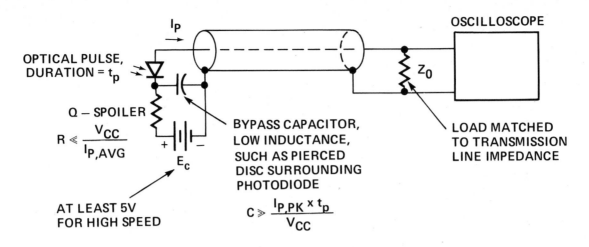

OPTICAL PULSE, DURATION = t_p

I_p

Q — SPOILER

$R \ll \dfrac{V_{CC}}{I_{P,AVG}}$

E_c

AT LEAST 5V FOR HIGH SPEED

BYPASS CAPACITOR, LOW INDUCTANCE, SUCH AS PIERCED DISC SURROUNDING PHOTODIODE

$C \gg \dfrac{I_{P,PK} \times t_p}{V_{CC}}$

OSCILLOSCOPE

Z_0

LOAD MATCHED TO TRANSMISSION LINE IMPEDANCE

(a) CIRCUIT ARRANGEMENT FOR SPEED OF RESPONSE OBSERVATION

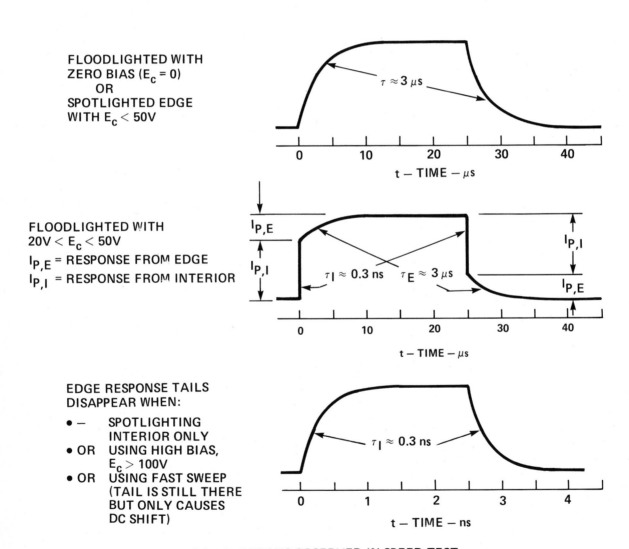

FLOODLIGHTED WITH ZERO BIAS ($E_c = 0$) OR SPOTLIGHTED EDGE WITH $E_c < 50V$

$\tau \approx 3\,\mu s$

0   10   20   30   40

t — TIME — μs

FLOODLIGHTED WITH $20V < E_c < 50V$

$I_{P,E}$ = RESPONSE FROM EDGE

$I_{P,I}$ = RESPONSE FROM INTERIOR

$I_{P,E}$

$I_{P,I}$

$\tau I \approx 0.3\,ns$ $\tau E \approx 3\,\mu s$

$I_{P,I}$

$I_{P,E}$

0   10   20   30   40

t — TIME — μs

EDGE RESPONSE TAILS DISAPPEAR WHEN:

- — SPOTLIGHTING INTERIOR ONLY
- OR USING HIGH BIAS, $E_c > 100V$
- OR USING FAST SWEEP (TAIL IS STILL THERE BUT ONLY CAUSES DC SHIFT)

$\tau I \approx 0.3\,ns$

0   1   2   3   4

t — TIME — ns

(b) WAVEFORMS OBSERVED IN SPEED TEST

Figure 4.1.2-2   Speed of Response Observation Apparatus and Waveforms.

4.4

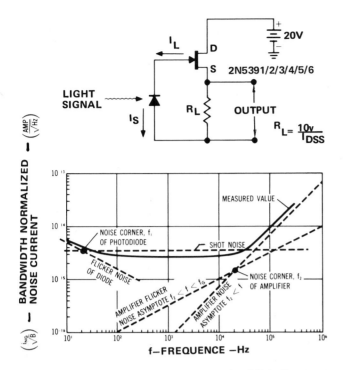

Figure 4.1.2-3   Modulation-Bandwidth-Normalized Noise Spectrum Referred to Amplifier Input.

$$\frac{I_{N,SHOT}}{\sqrt{B}} = \sqrt{2 q I_{dc}} = 17.9 \sqrt{I_{dc} (nA)} \left(\frac{fA}{\sqrt{Hz}}\right) \qquad (4.1.2-3)$$

where: $I_N/\sqrt{B}$ = bandwidth normalized noise current in femtoamps per root hertz

q = electron charge = 1.602×10^{-19} coulombs

I_{dc} = dc current flowing in the photodiode in nanoamps -- usually taken as total dark current.

The noise current calculated from equation 4.1.2-3 is usually greater than the measured noise at frequencies above the flicker noise corner, f_N. This is because the dc current used in the formula actually consists of two components: leakage current and junction current. Only junction current causes full shot noise; the only noise arising from leakage current is the thermal noise of the leakage resistance and the flicker noise. The two components are difficult to distinguish, so a worst case value is obtained by applying the shot noise formula (equation 4.1.2-3) to the entire dark current.

In a frequency band extending from f_1 at the low end to f_2 at the high end, the total noise current is:

$$i_N (f_2, f_1) = i_{NO} \sqrt{(f_2 - f_1) + f_n \ln (f_2/f_1)} \qquad (4.1.2-4)$$

where: $i_N (f_2, f_1)$ = total noise current in amps

i_{NO} = bandwidth normalized noise current in amps per root hertz from equation 4.1.2-3

f_2, f_1 = upper and lower 3 dB frequency in hertz

f_N = flicker noise corner in hertz, typically less than 20 Hz

Zero bias noise is just the thermal noise in the dynamic resistance of the photodiode at zero bias. According to the thermal noise formula:

$$I_{N, THERM} = \sqrt{\frac{4 k T}{R_{DO}}} \qquad (4.1.2-5)$$

$$= \frac{4}{\sqrt{R_{DO} (G\Omega)}} \left(\frac{fA}{\sqrt{Hz}}\right)$$

where: $I_N/\sqrt{B}$ = bandwidth normalized noise current in femtoamps per root hertz

k = Boltzmann's constant 1.38×10^{-23} joules per degree Kelvin

T = absolute temperature, taken as $290°$ K

R_{DO} = photodiode dynamic resistance at zero bias in gigaohms

The zero bias dynamic resistance is difficult to measure directly, but can be calculated from:

$$R_{DO} (G\,\Omega) = \left[\frac{k T}{q} (V)\right] \frac{1}{I_S (nA)} \qquad (4.1.2-6)$$

where: I_S = "reverse saturation current", measured as in Figure 4.2.2-2, in nanoamps

$\frac{kT}{q} \approx 0.025V$, defined in equations 4.1.2-3 and 4.1.2-5

Inserting R_{DO} from equation 4.2.1-6 into equation 4.2.1-5 leads to the interesting result:

$$I_{N, THERM} = \sqrt{4 q I_S} \qquad (4.1.2-7$$

$$= 25.3 \sqrt{I_S (nA)} (fA/\sqrt{Hz})$$

Comparing this with equation 4.1.2-3 suggests that a low leakage diode would be noisier at zero bias — and this is true. However, at zero bias there is no flicker noise, so this mode is usually preferred for low noise operation. Because there is no flicker noise, the noise in a bandwidth, $B = (f_2 - f_1)$ is that found from equation 4.1.2-4 with $f_N = 0$.

When signal flux is applied to a photodiode, the resulting photocurrent produces full shot noise. (This sometimes comes as a surprise when the photodiode is used with signals of low modulation depth.) Nevertheless, signal-to-noise ratio is defined as the ratio of the photocurrent when signal is applied to the noise current when there is no signal (i.e., dark):

$$\frac{S}{N} \triangleq \frac{I_P}{I_N} = \frac{\varphi \times R_\phi}{I_N} \tag{4.1.2-8}$$

The **Noise Equivalent Power NEP** is defined as the signal flux level for which S/N = 1.0 for B = 1 Hz:

$$NEP \triangleq \frac{I_N/\sqrt{B}}{R_\phi} \left(\frac{fW}{\sqrt{Hz}}\right) \tag{4.1.2-9}$$

Because NEP varies inversely as the responsivity, reversal of the log scale of NEP in Figure 4.1.2-4 gives the log of relative spectral response.

Under floodlight conditions, the modulation-bandwidth-normalized signal/noise ratio is:

$$\frac{S}{N} = \frac{E_e R_\phi A_D}{I_N/\sqrt{B}} \tag{4.1.2-10}$$

The S/N ratio increases as the square root of the photodiode area, A_D, because the signal rises linearly as the area while noise current varies only as the square root of the area. A figure of merit called **Detectivity, D*** (DEE-STAR) characterizes the area normalized quality of a photodiode surface by normalizing the S/N ratio with respect to the square root of the area and the incidance (flux per unit area). Thus from equation 4.1.2-10:

$$D^* \triangleq \frac{S/N}{E_e\sqrt{A}} = \frac{R_\phi \sqrt{A_D}}{I_N/\sqrt{B}} \tag{4.1.2-11}$$

Then from equation 4.1.2-9, the relationship between D* and NEP is derived:

$$D^* = \frac{\sqrt{A_D}}{NEP} \tag{4.1.2-12}$$

The usual units for D* are centimeters root hertz per watt.

Because they both depend on responsivity, which varies with wavelength; and on noise, which varies with modulation frequency and bandwidth, NEP and D* are both usually given with parenthetical conditions, e.g.: D* $(\lambda, f, \Delta f)$ or NEP $(\lambda, f, \Delta f)$.

4.2   Photodiode Operation

4.2.1   Circuit Model

The simplest model of a photodiode is just an ordinary diode having in parallel with it a current source as in Figure 4.2.1-1. The magnitude of this current source is proportional to the radiant flux being detected by the photodiode. The polarity of this photocurrent is from cathode to anode. It is thus apparent that if zero external

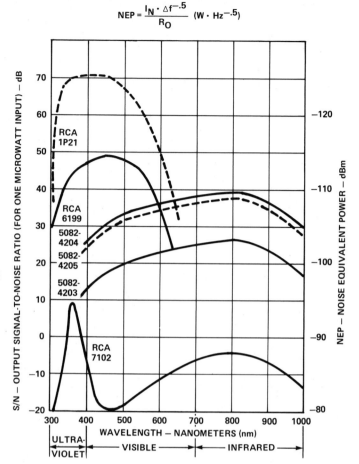

$$NEP = \frac{I_N \cdot \Delta f^{-.5}}{R_O} \quad (W \cdot Hz^{-.5})$$

Figure 4.1.2-4   Optical Spectrum of Noise Equivalent Power (NEP) for HP P-I-N Photodiodes.

bias is applied, the photocurrent will cause the anode to become positive with respect to the cathode. Part of the photocurrent will flow back through the photodiode, and part will flow in the load resistance. If the load resistance is open or extremely high, most of the photocurrent flows in the forward direction through the diode. This may seem paradoxical, but such a model does describe the results obtained in an open circuit.

Operation with zero bias is called the photovoltaic mode because the photodiode is actually generating the load voltage. Photovoltaic operation can be either linear or logarithmic depending upon the value of the load resistance. Logarithmic operation is obtained if the load resistance is very high ($>10^{11}$ ohms). Linear operation is obtained if the load resistance is very low with respect to the dynamic resistance of the photodiode. The upper limit of linear zero-bias operation is at $V_L \approx 100$ mV depending on the precision of the linearity requirement. With higher values of R_L, sensitivity can be increased for detecting very low level signals, but the dynamic range of linear response is decreased. The maximum practical value of R_L ranges from 25MΩ for large area photodiodes to 550MΩ for the smaller devices.

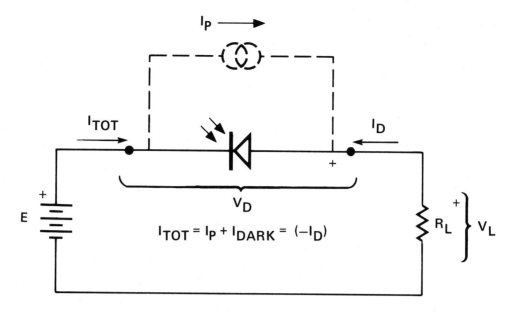

(a) CIRCUIT MODEL OF A PIN PHOTODIODE

$$I_{TOT} = I_P + I_{DARK} = (-I_D)$$

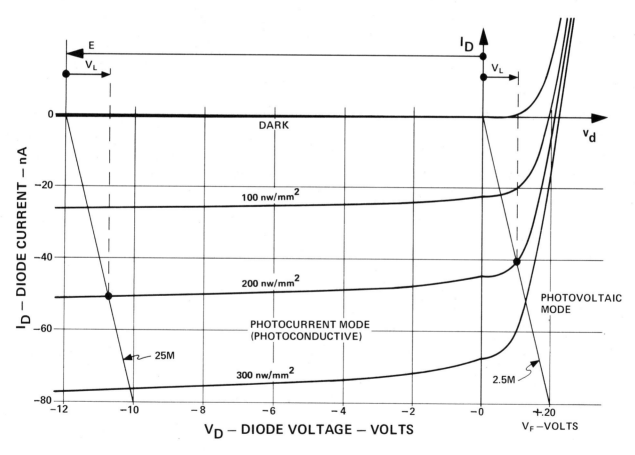

(b) VOLTAGE/CURRENT CHARACTERISTICS OF A PIN PHOTODIODE (NOTE CHANGE OF VOLTAGE SCALE THROUGH ZERO)

Figure 4.2.1-1   Electrical Characteristics of HP P-I-N Photodiode, 5082-4207.

Operation with reverse bias is called the photocurrent or the photoconductive mode. As compared with photovoltaic mode, the photocurrent mode offers:

(1)   higher speed
(2)   better stability
(3)   larger dynamic range
(4)   lower temperature coefficient
(5)   improved long-wavelength response over the interior region
(6)   short-wavelength (ultraviolet) response in the edge region.

The main drawback in photocurrent-mode operation is the flow of dark current, due to the reverse bias. Dark current is that which flows when no radiant flux is applied to the photodiode. Dark current flowing in R_L produces an offset voltage that varies exponentially with temperature. There is also likely to be some flicker noise due to reverse bias. Since offset and flicker noise disappear when zero bias is used, the photovoltaic mode is usually preferred unless the application requires one of the six advantages enumerated above.

4.2.2   Basic Amplifier Arrangements

For linear operation, the photodiode should be operated with as small a load resistance as possible. Figure 4.2.2-1 shows the recommended amplifier arrangement. The negative-going input is at virtual ground; the dynamic resistance seen there by the photodiode is R_1 divided by loop gain. If the op-amp has extremely high input resistance, loop gain is very nearly the forward gain of the op-amp. R_2 can be omitted if the photocurrent is reasonably high — its purpose is only to balance off the effect of offset current. As shown, the output voltage will rise in response to the optical signal. If it is preferable to have the output drop in response to optical input, then *both* the photodiode *and* E_c should be reversed. E_c may, of course, be zero. Speed of response is usually limited by the time constant of R_1 with its own capacitance, so it is improved by using a string of two or more resistors in place of a single R_1.

Logarithmic operation requires the highest possible load resistance — at least $10G\Omega$. With an FET-input op-amp, this is easily achieved as in Figure 4.2.2-2. If the offset current of the amplifier poses a problem, a resistor can be added between the positive- and negative-going inputs. Its value should not be less than $10G\Omega$ divided by loop gain. If the amplifier has a very high input resistance, loop gain is equal to the forward gain of the amplifier divided by $(1 + R_2/R_1)$ so making $R_2 = 0$ allows the smallest possible resistance between the inputs. The speed of response of this amplifier will be very low, with a time constant $\tau \approx 0.1s$. If high

speed logarithmic operation is required, it is best to use the linear amplifier of Figure 4.2.2-1 followed by a logarithmic converter.

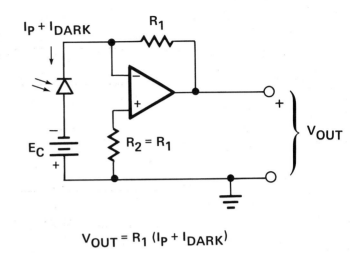

$$V_{OUT} = R_1 (I_P + I_{DARK})$$

Figure 4.2.2-1   Linear Response; Photodiode and Amplifier Circuit Arrangement.

4.2.3 | Suggested Applications

PIN photodiodes are extremely stable, have a zero temperature coefficient for $\lambda < 800nm$, and operate linearly over 100 dB with <1% distortion. The success achieved with a servo system such as that in Figure 4.2.3-1 is limited more by mechanical stability of the components than by the photodiodes. The loop consisting of amplifier A_1 with optical feedback from D_1 to D_2 stabilizes the intensity I_{e2} and makes it linearly proprotional to V_{REF}. If the beam splitter is stable, then the ratio of I_{e3} to I_{e2} is

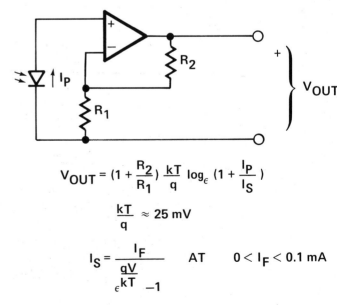

$$V_{OUT} = \left(1 + \frac{R_2}{R_1}\right) \frac{kT}{q} \log_\epsilon \left(1 + \frac{I_P}{I_S}\right)$$

$$\frac{kT}{q} \approx 25\ mV$$

$$I_S = \frac{I_F}{\epsilon^{\frac{qV}{kT}} - 1} \qquad AT \qquad 0 < I_F < 0.1\ mA$$

Figure 4.2.2-2   Logarithmic Response; Photodiode and Amplifier Circuit Arrangement.

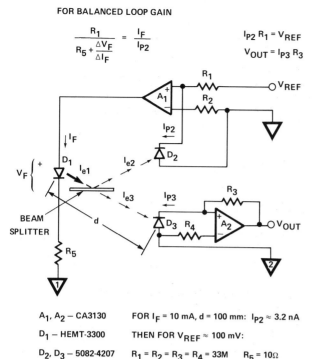

FOR BALANCED LOOP GAIN

$$\frac{R_1}{R_5 + \frac{\Delta V_F}{\Delta I_F}} = \frac{I_F}{I_{P2}}$$

$$I_{P2} R_1 = V_{REF}$$

$$V_{OUT} = I_{P3} R_3$$

A_1, A_2 – CA3130   FOR I_F = 10 mA, d = 100 mm: $I_{P2} \approx 3.2$ nA

D_1 – HEMT-3300   THEN FOR $V_{REF} \approx 100$ mV:

D_2, D_3 – 5082-4207 $R_1 = R_2 = R_3 = R_4 = 33M$ $R_5 = 10\Omega$

Figure 4.2.3-1   Servo-Controlled Intensity for Densitometer or Linear Coupler.

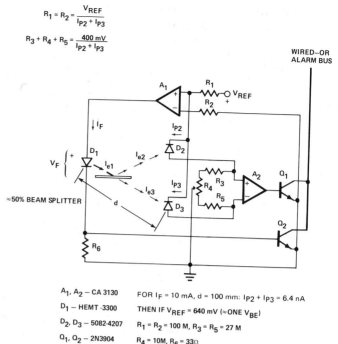

$$R_1 = R_2 = \frac{V_{REF}}{I_{P2} + I_{P3}}$$

$$R_3 + R_4 + R_5 = \frac{400 \text{ mV}}{I_{P2} + I_{P3}}$$

A_1, A_2 – CA 3130   FOR I_F = 10 mA, d = 100 mm: $I_{P2} + I_{P3}$ = 6.4 nA

D_1 – HEMT-3300   THEN IF V_{REF} = 640 mV ($\approx$ ONE V_{BE})

D_2, D_3 – 5082-4207 $R_1 = R_2 = 100$ M, $R_3 = R_5 = 27$ M

Q_1, Q_2 – 2N3904 $R_4 = 10M$, $R_6 = 33\Omega$

Figure 4.2.3-2   Sum-&-Difference Amplifiers Improve Sensitivity in Obscuration Alarm.

constant and V_{OUT} is linearly proportional to V_{REF}. This remains true even if D_1 degrades — the A_1, D_1, D_2 servo would simply force more current to D_1 to compensate the degradation. In this mode, since ground 2 need not be the same as ground 1, it is a linear optical coupler (optoisolator). With I_{e3} held stable, the incidance at D_3 would have a linear relationship to the transmittance of any material inserted between the beam splitter and D_3, making a stable optical transmissometer. If material is instead inserted between the beam splitter and D_2, I_{P3} would rise in proportion to the attenuation. Then if, instead of a linear amplifier, A_2 were a log amplifier (Figure 4.2.2-2) V_{OUT} would be logarithmically related to the attenuation of I_{e2}, thus making an optical densitometer.

The very high dynamic resistance of PIN photodiodes raises some interesting amplifier possibilities. Sum-and-difference amplification, for example, is vastly simplified because the diode impedance is high enough to isolate the "sum" side from the "difference" side as in Figure 4.2.3.2. At the input of amplifier A_1, the photocurrents I_{P2} and I_{P3} are summed to make the loop gain that stabilizes I_{e1} greater by 6 dB than it is in Figure 4.2.2-1. The expression for the loop gain for A_1 is:

$$A_{LOOP\ 1} = \left(\frac{I_{p2} + I_{p3}}{I_F}\right)\left(\frac{R_1}{R_6 + \frac{\Delta V_F}{\Delta I_F}}\right)(A_{V1})$$

$$= 1.80 \times 10^5 \text{ or } 105.12 \text{ dB}$$

assuming that A_1 has a forward gain $A_{V1} = 10^5$. The extra 6dB loop gain is probably not needed with $A_{V1} = 10^5$, but as long as loop gain is high, the cathodes of D_2 and D_3 are at virtual ground. Consequently, the voltages at the anodes of D_2 and D_3 may be as high as 100mV and still provide linear operation. Thus the resistance from either anode to ground may be as high as 100 mV/3.2 nA = 31.3MΩ.

If A_2 is connected as a linear differential amplifier, it would be wise to keep the anode voltages below 50 mV to allow 0 to 100% excursion of I_{P2} or I_{P3}. The system can then be used to give linearly the relative transmission of the I_{e2} and I_{e3} flux paths for the full range of 0 to 100% transmittance.

As shown, however, A_2 is connected as a very sensitive obscuration alarm, with R_4 adjusting the threshold. If the obscuration should increase in the I_{e3} flux path, or decrease in the I_{e2} flux path, A_2 would turn on Q_1. Should obscuration of both flux paths increase, LOOP 1 reacts to raise I_F, and at approximately 20 mA Q_2 will turn on. Also, if the efficiency of D_1 should degrade, Q_2 will turn on, indicating that something is wrong. Notice that if obscuration increases in flux path I_{e3} there will be proportional increase of the flux in path I_{e2} to keep $I_{P2} + I_{P3}$ constant. Thus the use of sum-and-difference amplifier provides differential sensing with 6dB higher gain than would be obtained by simply stabilizing flux path I_{e2} and sensing a change in I_{e3}.

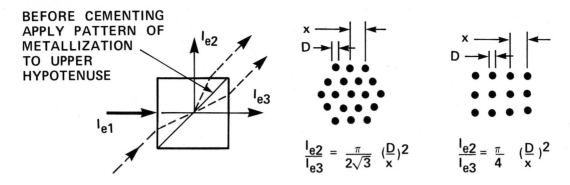

BEFORE CEMENTING APPLY PATTERN OF METALLIZATION TO UPPER HYPOTENUSE

I_{e2}

I_{e3}

I_{e1}

x

D

x

D

$$\frac{I_{e2}}{I_{e3}} = \frac{\pi}{2\sqrt{3}} \left(\frac{D}{x}\right)^2$$

$$\frac{I_{e2}}{I_{e3}} = \frac{\pi}{4} \left(\frac{D}{x}\right)^2$$

SUGGESTED METALLIZATION PATTERNS: <u>HONEYCOMB</u>   <u>SQUARE</u>

(a) BEAM SPLITTER CUBE

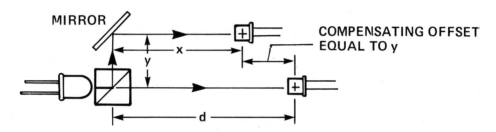

MIRROR

COMPENSATING OFFSET EQUAL TO y

x

y

d

(b) SPLITTER-TO-DETECTOR DISTANCE EQUALIZATION FOR PARALLEL BEAM OPERATION

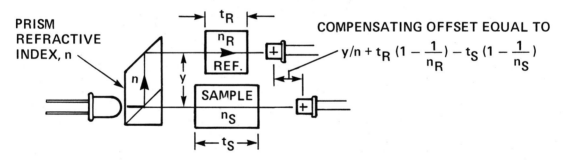

PRISM REFRACTIVE INDEX, n

t_R

n_R

REF.

SAMPLE

n_S

t_S

n

y

COMPENSATING OFFSET EQUAL TO

$$y/n + t_R \left(1 - \frac{1}{n_R}\right) - t_S \left(1 - \frac{1}{n_S}\right)$$

(c) DISTANCE EQUALIZATION ACCOUNTING FOR VARIATION OF REFRACTIVE INDEX THROUGH BOTH OPTICAL PATHS

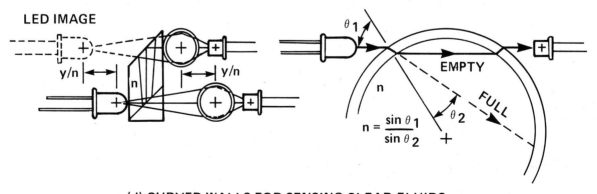

LED IMAGE

y/n

y/n

n

θ_1

EMPTY

FULL

n

θ_2

$$n = \frac{\sin \theta_1}{\sin \theta_2}$$

(d) CURVED WALLS FOR SENSING CLEAR FLUIDS

Figure 4.2.3-3   Optical Configurations for Use In Densitometer and Obscuration Alarm.

Some suggested optical path considerations are shown in Figure 4.2.3-3 to be used with the circuits of Figures 4.2.3-1,-2. Beam splitting to a very high precision can be done by applying partial metallization to the hypotenuse face of one prism, then cementing it to the hypotenuse face of another, as in Figure 4.2.3-3(a). The repetition interval of the metallization pattern should be very small to avoid interference patterns with LED details. To obtain parallel output beams, the incident beam can be applied parallel to the hypotenuse face, thus causing the split beams to emerge parallel (similar to a Dove prism), as shown by the dashed lines. Separation between the two beams may not be adequate, however, and the off-axis angle of incidence introduces "tilted-plate" astigmatism. Alternative means for obtaining parallel beams are shown in Figure 4.2.3-3(b,c) — these require distance compensation for beam offset, which (a) does not. All three, however, require compensation for sample length, as in (c).

Figure 4.2.3-3(d) shows how sample cells with curved walls can be applied with the densitometer or the obscuration detector circuits to respond to changes of index of refraction of clear fluids, as well as relative opacity of diffuse fluids.

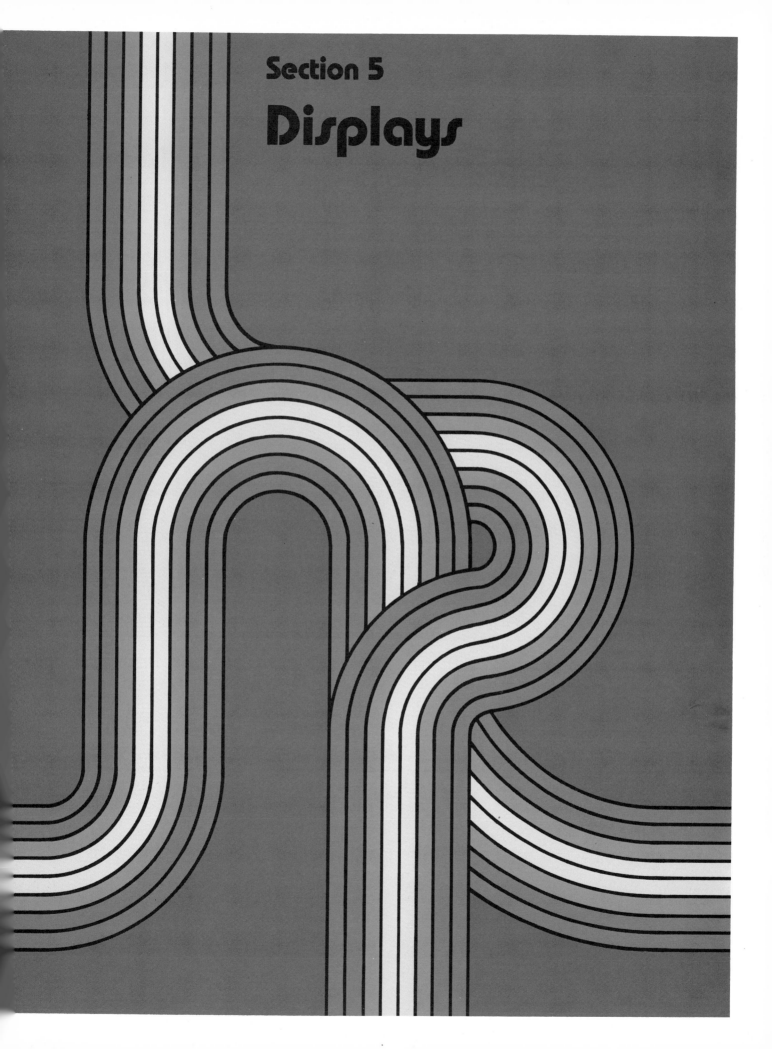

Section 5
Displays

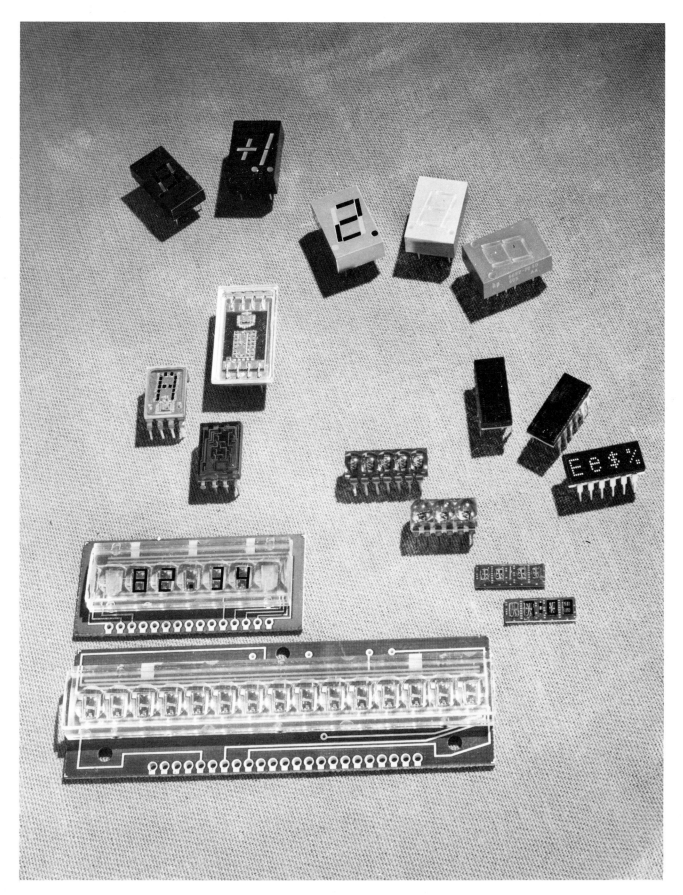

Figure 5.0.1-1 A Collection of LED Display Devices

5.1

5.0 DISPLAYS

The function of any display is to attract the attention of an observer by some recognizable arrangement of objects, illustrations, images or symbols. An electronic display has the advantage of easy control of the various light emitting or light modulating elements permitting substantial information transfer across the man-machine interface. Electronic displays may be implemented using any one of a large number of different technologies. A few of the more commonly known are:

- Incandescent
- Cathode ray tube
- Liquid crystal
- Fluorescent (operation is analogous to that of the CRT)
- Electrochromic
- Electrofluoridic
- Light emitting diode

Each of these technologies is characterized by distinct capabilities and requirements with respect to element density, color, power dissipation, packaging, unit size, drive voltages, environmental capabilities, cost, availability and other properties. The choice of any one display technology is generally a matter of optimizing some of these properties at the expense of others.

5.0.1 Types of LED Displays

The LED display technology is relatively new compared to most of the other technologies listed above. The first commercial LED products entered the marketplace in 1968. Since that time, the scope of LED display devices has expanded rapidly so that it now includes a wide variety of distinctly different products. Based on use, size, and drive requirements, these display products may be divided into four generalized categories:

1) Displays with On-Board Integrated Circuits
2) Strobable-Seven-Stretched Segment Displays
3) Magnified Monolithic Displays
4) Dot Matrix Alphanumeric Displays

Figure 5.0.1-1 depicts some of the different display products available. All of these display types have been originally developed using 655 nm red GaAsP type diode technology. Technologically, it is feasible to manufacture most of these different display categories utilizing the GaP transparent substrate colors high-efficiency red, yellow and green. However, at the present time, the availability of displays utilizing these higher technology LEDs has, for the most part, been limited to the stretched segment type

displays. The GaP transparent substrate LEDs will become available in more and more products as this technology advances.

5.0.2 Display Fonts

The most prominent feature of any display product is the physical arrangement of the display elements. This arrangement, or "font" as it is commonly termed, is important not only from the standpoint of the type information which can be transmitted but also important in that it dictates the nature and complexity of the support electronics required by the display.

Figure 5.0.2-1 depicts some of the common display fonts. Fonts A and E are by far the most common in use in LED technology. Both arrangements have some positive and negative attributes. The seven segment technique is easy to utilize from an electrical standpoint, however, it is limited to displaying numeric and a small range of alphabetic information. The 5x7 dot matrix can display a wide range of numeric, alphabetic and other characters but involves some rather expensive electronic circuitry to implement the other portions of the display subsystem. The sixteen segment approach shown in Figure C has a full alphanumeric capability but has had rather limited acceptance. This font is not capable of displaying some of the specialized characters required by sophisticated systems. Font F illustrates a 9-segment display which has a somewhat more pleasing font than a 7-segment display. It can display the same numeric information and has more alphabetic capability. The dot matrix fonts in Figures B and D are abbreviated versions of the 35 dot matrix of Figure E and are used primarily to display only numeric and hexidecimal information.

5.0.3 The Display Subsystem

The visible portion of a display (be it LED or some other technology) is only a minor portion of the electronic system necessary to convert electrically coded data into a form readily comprehended by a human viewer. Figure 5.0.3-1 is a block diagram of what may be generally termed a display subsystem. A display user can purchase one, some, or all of the functions in the block diagram as a single product. It is necessary, however, that all of the functions depicted be present in some part of any information display.

By way of example, a seven segment decoder/driver may be combined with a data latch and a seven segment diode matrix to form a display subsystem. Similarly, all four of the above functions may be purchased in a single package to perform the same end result. Often, the data latch depicted in the block diagram will be present as a part of

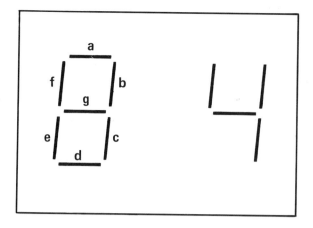

FONT A: 7—SEGMENT

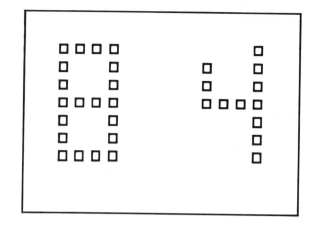

FONT B: MODIFIED 4X7 DOT MATRIX

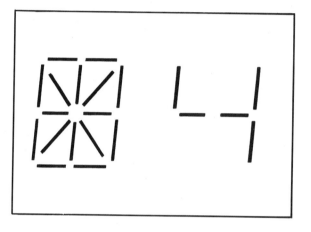

FONT C: 16—SEGMENT

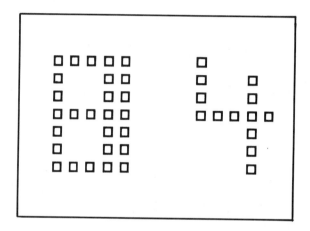

FONT D: MODIFIED 5X7 DOT MATRIX

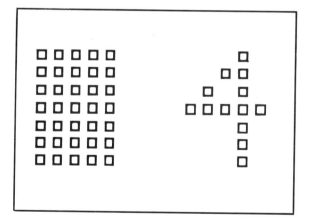

FONT E: 5X7 DOT MATRIX FONT

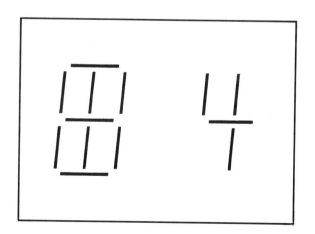

FONT F: 9—SEGMENT

Figure 5.0.2-1 Display Fonts Used in LED Displays.

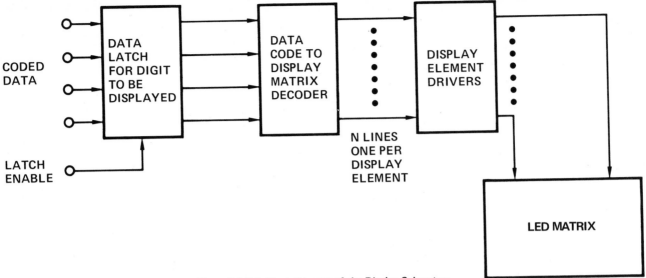

Figure 5.0.3-1 Block Diagram of the Display Subsystem.

the data source and does not have to be supplied as a portion of the display. Price, digit size, viewing conditions, power limitations, system architecture, and other considerations must be taken into account when deciding how to best design and partition a display subsystem.

5.0.4 Data Handling in Display Systems

The natural-partitioning of systems utilzing displays often leads to a situation where the display subsystem is physically separated from the data source. In these situations, it becomes necessary to transmit coded data from the data source to the display subsystem. Figure 5.0.4-1 depicts two of the most common techniques for transmitting four line BCD data plus decimal point status. The choice of technique usually depends on data source and display system requirements. For long distances, the cost of the full parallel (character parallel/bit parallel) approach can become prohibitive. If full parallel data must be converted to a character serial/bit parallel data format, the circuit depicted in Figure 5.0.4-2 may be implemented using the 3 state data bus buffers such as the National DM8095. A low true signal at the enable input of each buffer will apply the input data to the data bus. Timing of the enable signals will permit either RZ (return to zero) or NRZ (non return to zero) data formatting. In the RZ system, clocking information can be extracted at the receiving end and hence no additional transmission lines need be utilized for clocking.

5.1 Numeric Displays with an On-Board Integrated Circuit (OBIC)

For many applications, it is desirable to have the display subsystem assembled into a compact configuration and packaged separately at some distance from the data source.

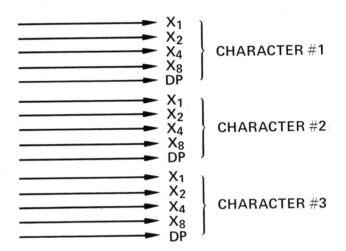

CHARACTER PARALLEL/BIT PARALLEL DATA TRANSMISSION (FULL PARALLEL)

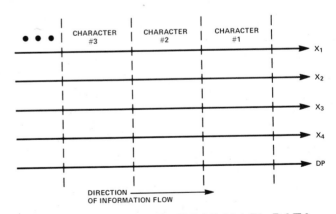

CHARACTER SERIAL/BIT PARALLEL DATA TRANSMISSION.

Figure 5.0.4-1 Two Common Techniques for Transmitting 4-Line BCD Data Plus Decimal Point Status Information.

5.4

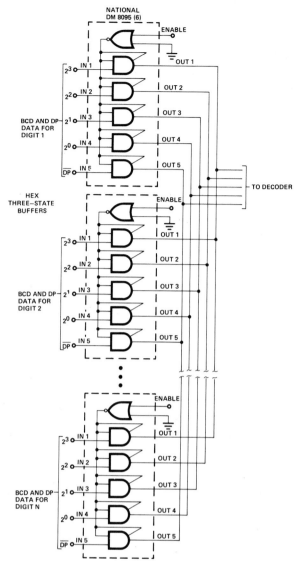

Figure 5.0.4-2 A Circuit for Character Parallel to Character Serial Data Conversion.

One possible approach is to have the display PC board contain all the necessary components that comprise the display subsystem. However, a more efficient and in many cases less costly approach is to use a display device that has an integrated circuit containing the subsystem logic functions, data latch/decoder/drivers, packaged on the same substrate as the display matrix. Such a device is termed an on-board integrated circuit (OBIC) display.

The cost per unit of an OBIC display device is more than that of a device which does not contain an integrated circuit. Even so, there are some specific advantages to an OBIC device that off-set this initial cost.

1. The pin connections between the data source and the display assembly are reduced due to the increased level of integration of the display subsystem.

2. The design of the display system is simplified and requires a minimum amount of engineering time.

3. The manufacturing costs and the time required to assemble each display system on a production basis are reduced to a minimum.

4. The reliability of the display system is significantly increased due to the reduction in circuit complexity and the reduced component count.

5. Less space is required to mount the display system assembly.

6. A pleasing font may be used, such as a modified 4x7 dot matrix, and need not conform to the font codes available with discrete commercial decoders.

The disadvantage of an OBIC display module is that as more functions are added, the module becomes unique in its configuration and a second source is usually not available.

A numeric OBIC display should provide a designer with a device that operates directly from a 5.0 volt supply, is compatible with commonly used data source logic such as TTL, has the electrical characteristic guaranteed over a wide temperature range, latches and decodes BCD data and directly drives a large pleasing character that is easily recognized at a distance.

The construction of an LED OBIC display typically begins with a ceramic substrate that has printed circuit metallization on the face and external leads brazed to the back surface. The integrated circuit and LEDs are die attached and wire bonded to the metallization. A hermetic device has a glass window covering the face of the display with a hermetic seal at the substrate rim wall-to-glass interface, as illustrated in Figure 5.1-1. A plastic device uses a coating of silicone gell to protect the wire bonds and reduce thermal stress to a minimum, then the substrate assembly is encapsulated in tinted undiffused epoxy that acts as an integral contrast filter and forms the display package.

As mentioned earlier, OBIC displays are usually unique in their design, thus the assembly technique described above will necessarily be different depending upon the package, character size and font, available functions incorporated within the integrated circuit, pin out arrangement and lead spacing. For this reason, Section 5.1.1 discusses a particular OBIC LED display family that incorporates the features described above.

5.1.1 The HP 5082-7300 OBIC Display Family

The HP 5082-7300 OBIC family offers a series of plastic or hermetic devices that display either decimal or hexidecimal numeric information. The plastic devices are epoxy encapsulated and are designed for use in consumer and commercial equipment, such as: control units for household appliances, electronic office equipment, computers and electronic measuring instruments. The hermetic devices, available in either industrial or military grade packages, are designed for use in high reliability applications involving the possible exposure to an adverse environment, such as: the instrument panels of transportation vehicles, controllers in numerical control milling machines, chemical processing equipment, military and aerospace systems. The construction of the hermetic OBIC display is pictured in Figure 5.1-1.

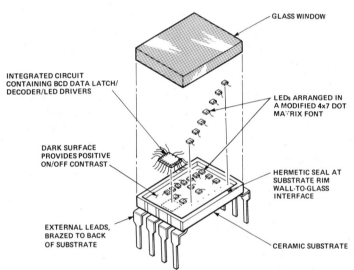

Figure 5.1-1 Construction Features of a Hermetic OBIC LED Display.

5.1.1.1 Character Font

The decimal devices are available with either right hand or left hand decimal point and display the numeric characters 0-9, and a minus sign (−). The hexidecimal devices display the characters 0-9 and A-F, and incorporate a blanking feature that allows the display to be blanked (turned off) without affecting the information stored in the data latch. A companion plus/minus one overrange display with right hand decimal point (±1.) is available in either of the hermtic or plastic packages. The character font is a 7.4mm (.29 inch) high modified 4x7 dot matrix as shown in Figure 5.1.1.1-1. The advantage of this font is that shaped characters are formed, so that a "3" does not resemble a reversed "E", a "B" looks different than an "8" and a "D" can be distinguished from a "0". The result is a font that forms characters which are easily recognized by an observer standing at a distance of 4.6 meters (15 feet).

TRUTH TABLE						
BCD INPUT				DECIMAL NUMERIC	HEXI-DECIMAL	OVER-RANGE
X8	X4	X2	X1			
L	L	L	L	0	0	±1.
L	L	L	H	1	1	
L	L	H	L	2	2	
L	L	H	H	3	3	
L	H	L	L	4	4	
L	H	L	H	5	5	
L	H	H	L	6	6	
L	H	H	H	7	7	
H	L	L	L	8	8	
H	L	L	H	9	9	
H	L	H	L	8 **2.**	A	
H	L	H	H	(Blank)	B	
H	H	L	L	(Blank)	C	
H	H	L	H	---	D	
H	H	H	L	(Blank)	E	
H	H	H	H	(Blank)	F	

NOTES:

1. H = LOGIC HIGH; L = LOGIC LOW. WITH THE ENABLE INPUT AT LOGIC HIGH CHANGES IN BCD INPUT LOGIC LEVELS HAVE NO EFFECT UPON DISPLAY MEMORY OR DISPLAYED CHARACTER.

2. TEST PATTERN FOR DECIMAL NUMERIC DEVICES.

Figure 5.1.1.1-1 Truth Table and Character Font for an OBIC LED Display.

5.1.1.2 The On-Board Integrated Circuit

The integrated circuit is composed of a 5-bit data latch for the decimal display (4-bit latch for the hexidecimal display) a decoder and LED matrix drivers. The block diagram for the integrated circuit is shown in Figure 5.1.1.2-1. On the hexidecimal devices, the decimal point is replaced by the blanking function. The data latch accepts logic high true BCD information and logic low true decimal point status. Information is loaded into the data latch when the enable input is at logic low. This information is then latched when the enable input is returned to a logic high. Changes at the data inputs will then have no effect upon the display memory or the displayed character. The blanking input controls only the LED drivers and has no effect upon the display memory. When the blanking input is at logic low, the display is on (character is illuminated) and at logic high, the display is blanked (character is not illuminated). A summary of the logic input functions is given in the truth table of Figure 5.1.1.1-1.

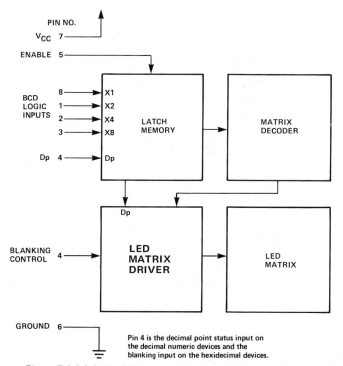

Pin 4 is the decimal point status input on the decimal numeric devices and the blanking input on the hexidecimal devices.

Figure 5.1.1.2-1 Block Diagram of the Four Subsystem Components Incorporated into an OBIC Display.

Equivalent input and LED driver circuits are shown in Figure 5.1.1.2-2. Each data input represents one TTL load. The display is blanked when a minimum threshold level of 3.5 volts is applied to the blanking input. This may be achieved by using an open collector TTL gate and a pull-up resistor. For example, (1/6) 7416 hexinverter/buffer/driver and a 120Ω pull-up resistor will provide sufficient drive to blank 12 displays. The size of the pull-up resistor required to blank a quantity of N-digits may be determined from the following formula:

$$R_{BLANK} = [V_{CC}\text{-}3.5V]/[N(1.0\ mA)] \qquad (5.1.1.2)$$

When the hexidecimal display is blanked, the power dissipation is reduced to 66% of that power dissipated when the maximum number of LED drivers are operating (displaying the Figure "B"). As a result of this, a significant power saving can be achieved by strobing the blanking input of each display. The strobing rate should be a minimum of 100 Hz or 5x the expected frequency of any mechanical vibration of the display in order to avoid flicker. Also, the time average intensity of the hexidecimal

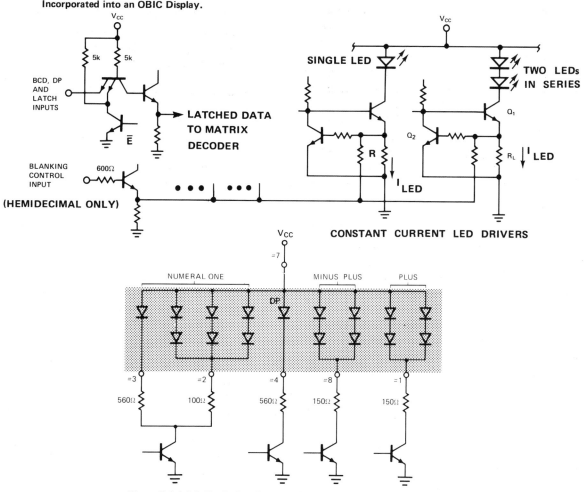

Figure 5.1.1.2-2 Equivalent Input and LED Driver Circuits for the Decimal, Hexidecimal and Overrange Displays.

display can be adjusted by blanking to comply with changing ambient lighting conditions. This is done by using pulse width modulation of the blanking control to vary the display on-time.

Constant current sources are used to drive the LED matrix. Each constant current driver is connected to either a single LED or two LEDs in series; the pattern is shown in Figure 5.1.1.2-3. A minimum V_{CC} of 4.5 volts is required to maintain sufficient compliance within the constant current drivers to produce an acceptably illuminated character. As V_{CC} falls below 4.5 volts, those drivers with two LEDs in series no longer have sufficient compliance to source current through two LED forward voltage drops, while those drivers servicing a single LED will still have sufficient compliance to source current through one LED forward voltage drop. The consequence of this may be insufficient illumination of the series diodes.

A real benefit to a designer using an OBIC display is the fact that the electrical parameters are guaranteed on the data sheet over a wide temperature range. This allows for worst case designing using known minimum and maximum parameter values at the temperature extremes. In operating a plastic OBIC device, it is the package temperature that is of concern. Therefore, the electrical parameters for the plastic OBIC devices are guaranteed over a case temperature range from -20°C to +85°C. The hermetic OBIC devices are characterized with respect to the operating ambient temperature, with the electrical parameters for the industrial devices guaranteed over the ambient temperature range from 0°C to +70°C, the same as for standard 7400 series TTL. The military grade devices are characterized over the full operating temperature range, guaranteeing to the designer the electrical parameters from -55°C to +100°C.

As is the case with the design of any data transmission system, a designer should take into account the timing requirements of both the data source and the OBIC display in order to insure that the correct information is latched and displayed. The timing requirements for this series of OBIC displays is given in Figure 5.1.1.2-4. These timing requirements are valid over the full operating temperature range for either a plastic or hermetic device. Enable rise times greater than 200 nsec may result in the latching of erroneous information. Data may be clocked into these OBIC displays at data rates up to 10 MHz.

5.1.1.3 Temperature Considerations

As indicated above, it is necessary to control the package temperature of a plastic OBIC device, as measured at the top of display pin number 3. For either a plastic or hermetic device, the primary thermal path for power

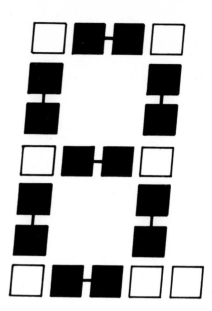

☐ = Seven constant current sources driving one LED.

▰ = Seven constant current sources driving two LEDs in series. Maximum power dissipation for the decimal numeric device occurs with nine LED drivers operating, when numeral 5 and dp are displayed. Maximum power dissipation for the hexidecimal device occurs with ten LED drivers operating, when digit B is displayed.

Figure 5.1.1.2-3 Arrangement for LED Matrix Drivers.

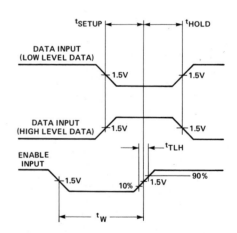

DESCRIPTION	SYMBOL	MIN.	MAX.	UNIT
Enable Pulse Width	t_W	100		nsec
Time data must be held before positive transition of enable line	t_{SETUP}	50		nsec
Time data must be held after positive transition of enable line	t_{HOLD}	50		nsec
Enable pulse rise time	t_{TLH}		200	nsec

Figure 5.1.1.2-4 Data Input and Enable Timing Requirements.

dissipation is through the device leads. The thermal resistance junction-to-lead is $15°C/W$. The maximum allowed junction temperature for a plastic OBIC device is $100°C$. Therefore, a designer should establish the thermal resistance to ambient of the display mounting structure in order to determine if the plastic OBIC devices may be operated in the maximum expected ambient temperature without heat sinking. As an example, a mounting structure consisting of DIP sockets that have been soldered onto a printed circuit board which has sufficient metallization to give a combined thermal resistance to ambient of $25°C/W$ per package, will permit the operation of plastic OBIC displays in an ambient temperature of $60°C$ without the use of external heat sinking. A mounting structure that has a thermal resistance to ambient of $35°C/W$ per package is sufficient to allow the operation of the hermetic OBIC displays in ambients up to $+100°C$, without the need of external heat sinking.

5.1.2 Intensity Control for Hexidecimal Displays Using Pulse Width Modulation

The optimal design operates a display at a light level that, when compared to the ambient or background, is bright enough to be seen without being so bright as to cause eye strain and fatigue. The fatigue becomes particularly noticeable when an operator must view an overbright display for long periods of time. An overbright display may be defined as one whose light level exceeds the optimal value by more than 10 to 1. If the ambient light conditions remain relatively constant, the selection of a good operating light level is straight forward. However, there are applications where the ambient can change by as much as 100:1, such as in the instrument panels of transportation vehicles, portable instrumentation, aircraft control tower or shipboard applications. For widely changing conditions such as these, the circuit designer may wish to add a dimming feature. Pulse width modulation can be used to control the percent time that a display is on (i.e., its "duty cycle"). By varying this duty cycle, the average light level can be varied over a wide range. Since the drive level is not changed (only the duty cycle); close matching among segments and digits is maintained even at very low light levels.

It is convenient to have the display intensity automatically track the ambient. In Figure 5.1.2-1 are two methods for pulse width modulating the displays blanking input. If a system clock is available, a photoresistor, R_x, can be used to control the duty cycle of a monostable multivibrator. The duty cycle varies from 80% in a bright ambient to 13% in a dim ambient.

If a system clock is not available, an alternative method employs a 555 timer as a free running oscillator. The photoresistor, R_x, controls the output pulse width resulting

in a 90% duty factor in a bright ambient and a 2% duty factor in a dim ambient.

The frequency at which the blanking input is strobed to turn the display off and on should be fast enough, regardless of pulse width, to insure an observer sees a continuous, flicker free, display. This minimum blanking frequency should be at least 100 Hz or 5x the vibration frequency, if mechanical vibration of the display is expected.

5.1.3 Interfacing a Microprocessor to an OBIC Numeric Display

An OBIC decimal or hexidecimal display can interface very easily to a microprocessor system. Since the OBIC numeric display contains a latch, the only extra circuitry that is required to interface to a microprocessor is some external gating. An eight bit data word can be configured as two BCD or hexidecimal characters. This allows the desired display information to be stored compactly in a RAM and easily manipulated by the microprocessor. Since the OBIC display is dc driven, the microprocessor needs to update the display only when the information needs to be changed. Thus, the OBIC display requires only a minimum amount of microprocessor time. If hexidecimal information is to be displayed, an HP 5082-7340 display or equivalent should be specified. This display has a blanking input which does not affect the latch. When decimal information is to be displayed, an HP 5082-7300 display or equivalent should be specified. This display has a latched decimal point and has the capability to display a minus sign, blank, or lamp test.

If a moveable decimal point on an OBIC decimal display or selective digit blanking on an OBIC hexidecimal display is required, some additional circuitry may be necessary. Since the OBIC hexidecimal display does not have a latched blanking control, an external latch is needed. The eight bit data word can be configured as one character (four bits) plus one additional bit for a decimal point or blanking control. An example of this technique is shown in Figure 5.1.3-1. Each digit would be addressed by a different eight bit address code. With the Intel 8080A microprocessor, an output instruction is used to load the contents of the accumulator into the specified OBIC numeric display. In this example, the lowest order four bits (D_3, D_2, D_1, D_0) are used for the numeric information and D_4 is used for blanking control or decimal point control. When it is desired to pack two hexidecimal or BCD characters in a single eight bit word, a couple of bits from the address bus can be decoded as decimal point or blanking information. An example of this technique is shown in Figure 5.1.3-2. Each pair of digits is addressed by four possible address codes: n, n+1, n+2, or n+3. The two lowest order address bus lines, A_1 and A_0 are decoded as decimal point or blanking information. The second byte of the output instruction will determine which pair of characters is to be updated and whether either digit is to be blanked or have its decimal point turned on. When the decimal point inputs or the blanking inputs do not have to be controlled by the microprocessor, only a single output address code would uniquely specify each pair of digits. This can simplfy programming.

An OBIC hexidecimal display can also be used to simplify microprocessor program debugging. Hexidecimal information is much easier to decode than binary information by the software designer. Figure 5.1.3-3 shows a complete debugging system for an Intel 8080A

microprocessor. At the beginning of each machine cycle, the 8080A microprocessor strobes eight bits of information into a status latch. These eight bits uniquely specify which one of ten machine cycles will be executed by the microprocessor. The sixteen address lines specify where the information is to be read or written. The eight data lines hold the contents of that information. If the microprocessor is operated in a single step mode, then after the end of each machine cycle, one of the ten LED lamps will indicate which machine cycle was last executed and the hexidecimal displays will indicate what was on the address and data buses during that machine cycle. In this example, a "Fetch" machine cycle was just executed and a JMP instruction was read from $(00AF)_{16}$. Other microprocessor systems can be debugged in a similar fashion, by using the appropriate control signals from the microprocessor to load the contents of the address and data buses into OBIC hexidecimal displays and decoding the other control signals to indicate what type of machine cycle was executed.

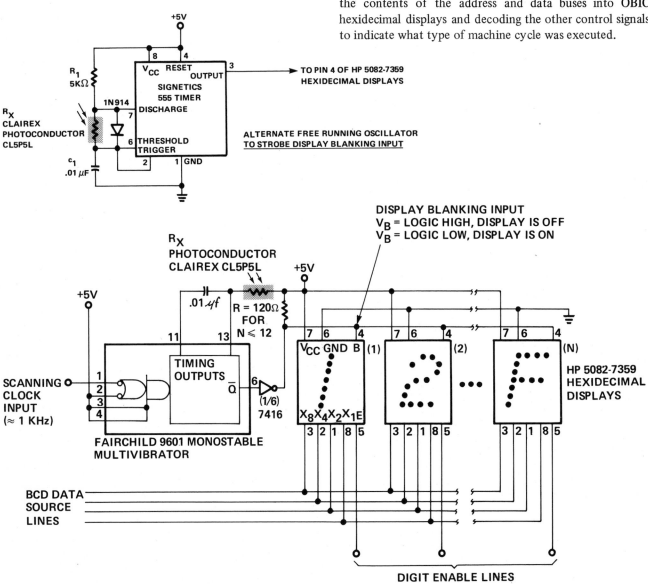

Figure 5.1.2-1 An Intensity Control Circuit for Hexidecimal Displays.

HP 5082-7340 USED AS HEXIDECIMAL OUTPUT FOR INTEL 8080A MICROPROCESSOR

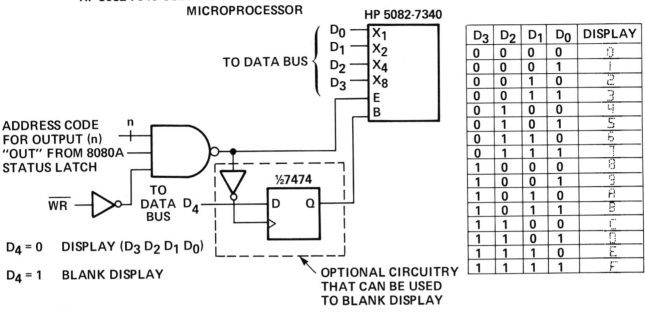

D_3	D_2	D_1	D_0	DISPLAY
0	0	0	0	0
0	0	0	1	1
0	0	1	0	2
0	0	1	1	3
0	1	0	0	4
0	1	0	1	5
0	1	1	0	6
0	1	1	1	7
1	0	0	0	8
1	0	0	1	9
1	0	1	0	A
1	0	1	1	B
1	1	0	0	C
1	1	0	1	D
1	1	1	0	E
1	1	1	1	F

$D_4 = 0$ DISPLAY ($D_3 D_2 D_1 D_0$)

$D_4 = 1$ BLANK DISPLAY

OPTIONAL CIRCUITRY THAT CAN BE USED TO BLANK DISPLAY

HP 5082-7300 USED AS DECIMAL OUTPUT FOR INTEL 8080A MICROPROCESSOR

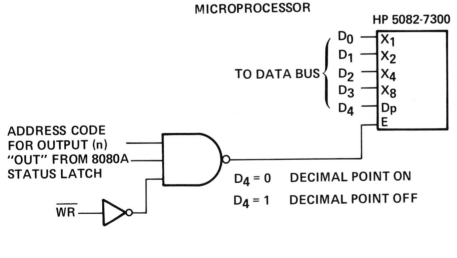

$D_4 = 0$ DECIMAL POINT ON

$D_4 = 1$ DECIMAL POINT OFF

D_3	D_2	D_1	D_0	DISPLAY
0	0	0	0	0
0	0	0	1	1
0	0	1	0	2
0	0	1	1	3
0	1	0	0	4
0	1	0	1	5
0	1	1	0	6
0	1	1	1	7
1	0	0	0	8
1	0	0	1	9
1	0	1	0	8
1	0	1	1	(BLANK)
1	1	0	0	(BLANK)
1	1	0	1	.....
1	1	1	0	(BLANK)
1	1	1	1	(BLANK)

Figure 5.1.3-1 Interfacing Obic Numeric Displays to a Microprocessor.
Five Bit Word Specifies BCD Data and Blanking/DP.

5.11

HP 5082-7340 USED AS NUMBERIC/HEXIDECIMAL OUTPUT FOR INTEL 8080A MICROPROCESSOR

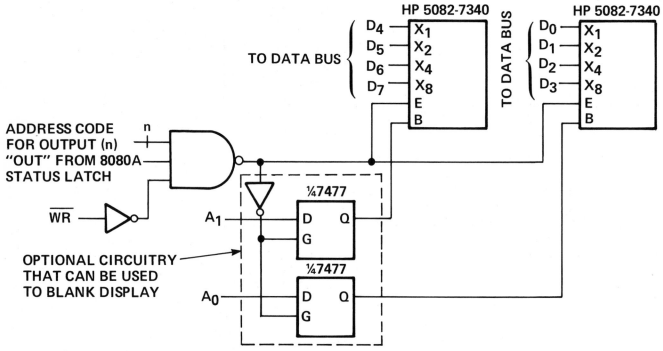

OUTPUT (n + 0) DISPLAY RESULT
OUTPUT (n + 1) BLANK LEAST SIGNIFICANT DIGIT ONLY
OUTPUT (n + 2) BLANK MOST SIGNIFICANT DIGIT ONLY
OUTPUT (n + 3) BLANK DISPLAY

HP 5082-7300 USED AS A NUMERIC DISPLAY FOR INTEL 8080A MICROPROCESSOR WITH Dp, MINUS, BLANKING, AND LAMP TEST ABILITY

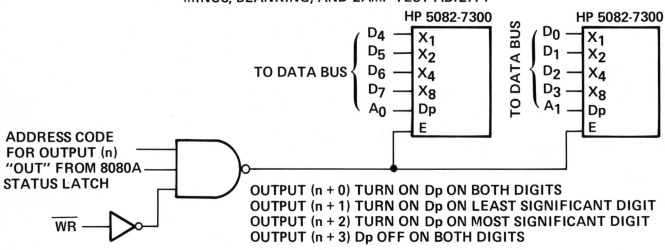

OUTPUT (n + 0) TURN ON Dp ON BOTH DIGITS
OUTPUT (n + 1) TURN ON Dp ON LEAST SIGNIFICANT DIGIT
OUTPUT (n + 2) TURN ON Dp ON MOST SIGNIFICANT DIGIT
OUTPUT (n + 3) Dp OFF ON BOTH DIGITS

Figure 5.1.3-2 Two BCD or Hexidecimal Characters can be Stored in an Eight Bit Word. Address Bus is Decoded as Blanking/DP.

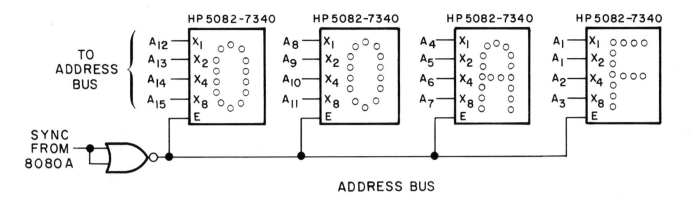

ADDRESS BUS

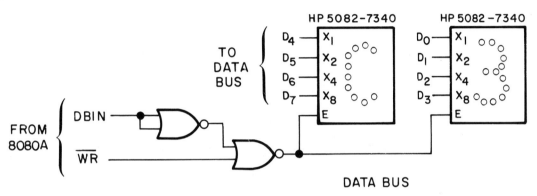

DATA BUS

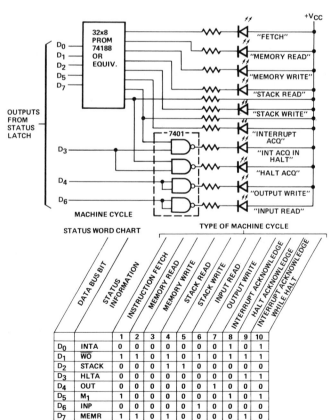

DATA BUS BIT	STATUS INFORMATION	INSTRUCTION FETCH	MEMORY READ	MEMORY WRITE	STACK READ	STACK WRITE	INPUT READ	OUTPUT WRITE	INTERRUPT ACKNOWLEDGE	HALT ACKNOWLEDGE	INTERRUPT ACKNOWLEDGE WHILE HALT
		1	2	3	4	5	6	7	8	9	10
D_0	INTA	0	0	0	0	0	0	0	1	0	1
D_1	$\overline{WO}$	1	1	0	1	0	1	0	1	1	1
D_2	STACK	0	0	0	1	1	0	0	0	0	0
D_3	HLTA	0	0	0	0	0	0	0	0	1	1
D_4	OUT	0	0	0	0	0	0	1	0	0	0
D_5	M_1	1	0	0	0	0	0	0	1	0	1
D_6	INP	0	0	0	0	0	1	0	0	0	0
D_7	MEMR	1	1	0	1	0	0	0	0	1	0

STATUS WORD CHART

TYPE OF MACHINE CYCLE

Figure 5.1.3-3   Use of LED Lamps and OBIC Numeric Displays
To Aid In Debugging A Microprocessor System.

5.2   Strobable-Seven-Stretched Segment Displays

The first seven segment LED displays to enter the marketplace were generally single digit devices with a small character height constructed on a thick-film ceramic substrate with round pins, either swaged or epoxyed into holes in the ceramic. Large rectangular slivers of GaAsP were either epoxy or eutectic die attached to the substrate in the seven segment format. Standard IC bonding techniques were then used to connect the anodes of the diodes to other traces on the substrate. The entire assembly was then encapsulated in a clear epoxy to provide protection for the bonds and chips. This construction technique requires large amounts of relatively expensive GaAsP, making the cost of the final product substantially higher than other display technologies. In addition, poor alignment of the diodes, visible metallization patterns both on the diode and on the substrate, and other problems, made most of these larger digits less than desirable in appearance by comparison to other designs.

Recently, a new stretched segment assembly technique has been developed which combines advantages of substantially lower manufacturing costs with a markedly improved display appearance. In this stretched segment design technique, a lead frame is utilized as the mechanical support for and makes electrical contact with the LED chips. Instead of encapsulating the assembly as before, a cone shaped reflecting cavity is cast inside a rectangular package above each LED, using glass filled epoxy. The top of this cavity forms the stretched segment. Light from an LED is uniformly emitted from the stretched segment resulting in a pleasing uniformly lighted rectangular area. Utilizing this technique, characters of larger size in the new high-efficiency red, yellow and green colors, as well as standard red, are assembled using LED chips an order of magnitude smaller than was used in the earlier design. Thus, the stretched segment display offers a variety of colors, sizes, improved appearance, good on/off contrast, and improved reliability at lower cost.

5.2.1   Description

The large digit LED display devices are manufactured using the concept of stretching the light from an LED by diffusion and reflection. This is accomplished by encapsulating the LED into a special cavity. This cavity is a rectangular cone with the small end down at the LED, and the large top end forming a segment of the display as shown in Figure 5.2.1-1. The area of the top end surface of the cavity may be 30 times or more the surface area of the LED. The emitted light from the LED is diffused as it passes through to the top surface of the cavity producing an evenly lighted segment. Optically, this segment will appear as an area source with a near lambertian radiation pattern.

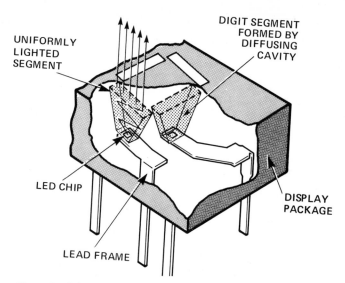

Figure 5.2.1-1   Assembly Technique of a Stretched Segment Display.

5.2.1.1 Construction

Figure 5.2.1.1-1 is a cross section of a strobable-seven-stretched-segment display. The Lexan® housing, called a "scrambler", forms the display package and contains the segment cavities. The sides of the cavities are made to have as near perfect reflection as possible to reduce light loss. The external top and side surfaces of the scrambler are coated with an epoxy paint to match the color of the LED. The colored scrambler helps provide good segment on/off contrast.

The lead frame base metal is covered with a 50 microinch copper flash and a 200 microinch silver plating. The pin size and spacing match that of a standard 14 pin DIP.

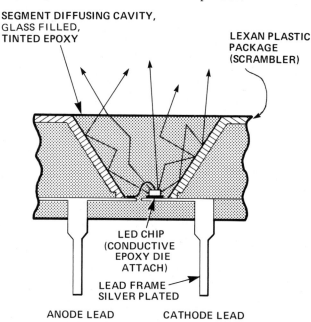

**Figure 5.2.1.1-1   Cross-Sectional Diagram of One Segment in a
Stretched Segment Display.**

The construction of a stretched segment display is as follows: the LED chip is die attached to the cathode pin of the lead frame with an electrically conductive epoxy. The top contact of the LED is wire bonded to the anode pin; a ball bond is formed on the LED top contact and a wedge bond is formed on the lead frame.

The lead frame is inserted into the scrambler and the scrambler is then filled with a glass filled tinted epoxy. This glass filled epoxy forms the rectangular cone segments, the glass acting as the light diffusing agent. The tinting in the epoxy works in conjunction with the colored scrambler to enhance segment on/off contrast.

5.2.1.2 Data Sheet Parameters

The data sheet for a stretched segment display is divided into the following basic sections:

- Package dimensions and circuit diagrams
- Absolute maximum ratings
- Electrical/optical characteristics
- Operational curves
- Operational considerations

The data sheet is configured to give a display designer as much helpful information as possible. The operational considerations gives application information covering such topics as determining V_F, selecting a set of strobed operating conditions, calculation of time average luminous intensity, contrast enhancement suggestions and post solder cleaning suggestions.

The absolute maximum ratings give those limits beyond which the device should not be operated. The device is capable of being operating at the maximum ratings, however, for best reliability it is suggested that the device be operated below the maximum ratings.

The maximum dc forward current and dc power dissipation are the maximum dc drive conditions, not temperature derated, and are used in determining maximum strobed operating conditions. These limits have been set based on a projected operating life up to 100K hours. The maximum temperature limits are based on the display package capability.

The electrical/optical characteristics list typical values for various device parameters at an ambient temperature of $T_A = 25°C$.

The operational curves are used to determine display drive conditions.

5.2.2 Determining Display Drive Conditions

The process of establishing display drive conditions involves the examination of LED junction temperature, based on derated operating currents and device thermal resistance, and the calculation of the LED forward voltage. If the display is strobed, the process also includes the determination of refresh rate, duty factor, pulse width and peak current.

5.2.2.1 Maximum DC Current (I_{DC}), Junction Temperature (T_J) and Package Thermal Resistance (θ_{JA})

The end of operating life of an LED is defined to be when the light output has degraded to 50% of its initial value. The maximum dc current is established from reliability testing results which predict an operating life of 100K hours when the LED is continuously operated at that dc current in an ambient temperature of 25°C.

To keep within the package temperature limitations, the maximum junction temperature for stretched segment displays is T_J max = 100°C. This value of T_J is based on a thermal resistance junction-to-ambient of θ_{JA} = 100°C/W per package for a device soldered into a typical PC board. One exception to this is the .3 inch common cathode display. The thermal resistance per package is 110°C/W and the T_J max = 105°C.

The junction temperature is calculated assuming all eight segment are illuminated.

$$T_J = T_A + \Delta T_J \qquad (5.2.2.1)$$
$$\Delta T_J = \theta_{JA} \text{ (8 SEGMENTS) } (P_{SEGMENT})$$
$$\Delta T_J = \theta_{JA} \text{ (8) } [I_{AVG} (V_{TURN-ON} + I_{PEAK} R_S)]$$

Example: The maximum average LED junction temperature of a high-efficiency red device, that has a maximum dynamic resistance of 33Ω, being strobed at a refresh rate of 1 kHz with a peak current of 60 mA and an average current of 12 mA in an ambient of 50°C is 84°C.

$$T_J = 50°C$$
$$\quad + (100°C/W) \text{ (8) } [.012A (1.55V + .060A (33Ω))]$$
$$T_J = 84°C$$

5.2.2.2 Forward Voltage (V_F) and LED Dynamic Series Resistance (R_s)

When calculating LED power dissipation and junction temperature, the variation in forward voltage with forward current must be taken into account. Each data sheet contains a graph of the typical variation of V_F with respect to I_F (the familiar diode curve). A linear approximation can be derived from this curve to form an equivalent circuit for an LED, see Figure 5.2.2.2-1.

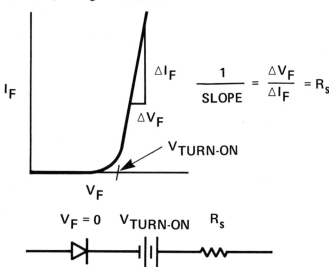

Figure 5.2.2.2-1 Equivalent Circuit for an LED.

The typical forward voltage for an LED may be scaled off of I_F vs. V_F curve or calculated. For example, the forward voltage of a standard red display may be calculated from the following formula, $V_{turn-on}$ = 1.55 volts:

$$V_F = 1.55V + I_{PEAK} R_S \qquad (5.2.2.2-1)$$
$$\text{where: } R_{S\ TYP} = 3\Omega$$
$$R_{S\ MAX} = 7\Omega$$

The forward voltage characteristics of a GaP transparent substrate device varies considerably from lot to lot. This is due to the wide range of resistivity throughout the GaP crystal and the variation in conductance of the reflective back contact. Figure 5.2.2.2-2 shows the measured forward voltage variation for the high-efficiency red display. From this kind of examination, the following table is derived:

Device	$R_{S\ MIN}$	$R_{S\ TYP}$	$R_{S\ MAX}$	$V_{TURN-ON}$
High-Efficiency Red	17Ω	21Ω	33Ω	1.55V
Yellow	15Ω	25Ω	37Ω	1.60V
Green	12Ω	19Ω	29Ω	1.75V

The maximum forward voltage for a given peak forward current may be calculated as follows:

$$V_{F\ MAX} = V_{TURN-ON} + I_{PEAK} R_{S\ MAX} \qquad (5.2.2.2-2)$$

The typical forward voltage is more accurately calculated by measuring the typical R_s value from the V_F point at I_F = 5 mA.

$$V_{F\ TYP} = V_{5mA} + R_{S\ TYP} (I_{PEAK} - 5\ mA) \qquad (5.2.2.2-3)$$

Where $V_{5\ mA}$ and R_s are obtained from the following table:

Device	V_{5mA}	$R_{S\ TYP}$
High-Efficiency Red	1.65V	21Ω
Yellow	1.75V	25Ω
Green	1.85V	19Ω

Example: For a high-efficiency red device operating at 60 mA peak, V_F = 2.81 volts:

$$V_{F\ TYP} = 1.65V + 21\Omega\ (.060-.005A) = 2.81V$$

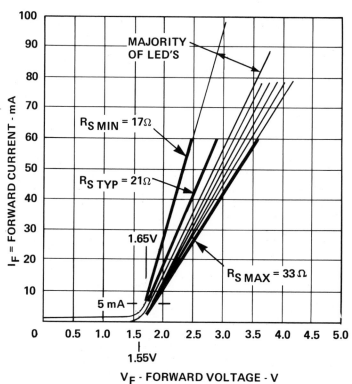

Figure 5.2.2.2-2 Measured Forward Voltage Variation for a High-Efficiency Red Display (8 Segments x 25 Devices = 200 LEDS)

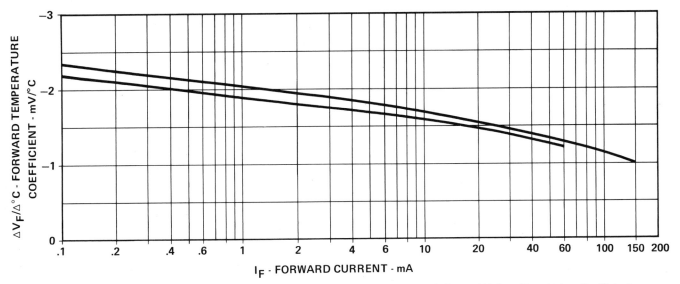

Figure 5.2.2.3-1 Forward Voltage Temperature Coefficient vs Forward Current.

5.2.2.3 Variation of Forward Voltage with Change in Temperature

The forward voltage temperature coefficient ($\Delta V_F/°C$ = -mV/°C) is given in Figure 5.2.2.3. The magnitude of this coefficient decreases in value with increasing forward current. The change in forward voltage with temperature should be taken into account when doing a worst case design.

5.2.2.4 Operational Curves for Strobing an LED Device

A power limiting criterion needs to be established in order to define the tolerable limitations with which an LED device may be operated in the strobed mode. The criterion which has been selected is the LED junction temperature. Specifically, the maximum tolerable strobe mode operating conditions are limited to a peak current (I_{PEAK}), pulse duration (t_p) and refresh rate (f) which produce a peak junction temperature ($T_{J\ PEAK}$) equal to the junction temperature obtained when the device is driven by the maximum temperature derated DC current ($I_{DC\ MAX}$). A graph of the curves which define the maximum tolerable strobe mode operating conditions for a high-efficiency red display is reproduced in Figure 5.2.2.4-1.

The thermal time constant of an LED mounted on a lead frame is about one millisecond. For this reason, refresh rates less than 1 KHz produce significant peaks and valleys in the absolute junction temperature with the average junction temperature less than the dc junction temperature. As the refresh rate increases, the average junction temperature approaches the dc junction temperature. This is illustrated by the constant duty factor line drawn across the face of Figure 5.2.2.4-1. As the refresh rate increases, the allowable operating peak current per segment also increases. It is, therefore, worthwhile to use a refresh rate of 1 KHz or faster.

5.2.2.5 Maximum DC Current ($I_{DC\ MAX}$) and Temperature Derating

The operational curves in Figure 5.2.2.4-1 are used in conjunction with the temperature derated maximum dc current to determine the maximum tolerable peak operating current per segment ($I_{PEAK\ MAX}$). The dc current derating curve for the high-efficiency red device is reproduced in Figure 5.2.2.5-1. Figure 5.2.2.5-2 shows the maximum tolerable dc junction temperature and the maximum tolerable dc power dissipation for the high-efficiency red device.

5.2.3 Sample Calculation of a Typical and Worst Case Design, Strobed Operation

Given:

1. High-efficiency red devices in a 4-digit display.

2. Maximum ambient temperature as measured between two digits just above PC board is T_A = 60°C (140°F).

3. Duty factor to be 25%.

Calculate:

1. Maximum tolerable operating conditions per segment.

2. Peak and average power dissipation.

3. Average junction temperature.

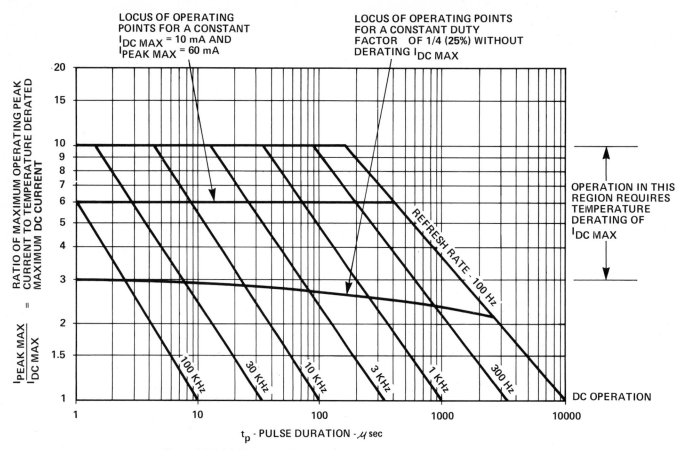

Figure 5.2.2.4-1 Maximum Tolerable Peak Current vs Pulse Duration
for a High-Efficiency Red Display.

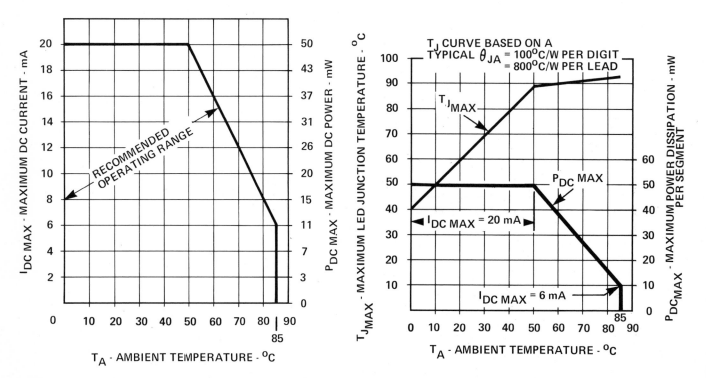

Figure 5.2.2.5-1 Maximum Allowable DC Current and DC Power
Dissipation per Segment as a Function of Ambient
Temperature for a High-Efficiency Red Display.

Figure 5.2.2.5-2 Maximum Tolerable LED Junction Temperature and
Maximum Tolerable Power Dissipation for a High-
Efficiency Red Display.

5.18

Step 1:

The duty factor is set at 1/4. It is necessary to select a refresh rate which will allow the maximum ratio of $I_{PEAK MAX}/I_{DC MAX}$ and stay within the timing constraints of the system. Refering to Figure 5.2.2.4-1, a minimum possible refresh rate of 100 Hz gives a ratio of 2.5 for a pulse duration of t_p = 2.5 msec, a refresh rate of 1 KHz gives a ratio of 2.6 for t_p = 250 μsec. and 10 KHz gives a ratio of 2.8 for t_p = 25 μsec. A good refresh rate is 3125 Hz which gives a ratio of 2.7 for t_p = 80 μsec.

In selecting a minimum refresh rate in order to avoid observable flicker, a designer should not use less than 100 Hz or less than 5X the expected vibration frequency.

Step 2:

Next determine $I_{DC MAX}$ from Figure 5.2.2.5-1, $I_{DC MAX}$ at T_A = 60°C is 16 mA.

Calculate $I_{PEAK MAX}$:

$$I_{PEAK MAX} = (2.7)(16 \text{ mA}) = 43.2 \text{ mA}$$

Determine I_{AVG} = (Duty Factor) ($I_{PEAK MAX}$):

$$I_{AVG} = (1/4)(43.2 \text{ mA}) = 10.8 \text{mA}$$

Step 3:

Now determine the power dissipation (P) and expected T_J:

$$P_{AVG SEG} = I_{AVG} V_F$$

The typical segment power may be calculated using the V_F from equation 5.2.2.2-3:

$$V_{F TYP} = 1.65V + 21\Omega (.0432 - .005A)$$
$$V_{F TYP} = 2.452V$$

The value V_F = 2.452 volts assumes T_J = 25°C. The junction temperature is above 60°C so the change in forward voltage due to junction temperature is now taken into account. This is an iterative calculation, however, one iteration is usually sufficient.

From Figure 5.2.2.3-1, $\Delta V_F/°C$ = 1.4 mV/°C for I_F = 43 mA. An approximate T_J is obtained and used to calculate an adjusted V_F:

$$T_{J APPROX} = (100°C/W)(8 \text{ SEG})(P_{SEG APPROX}) + T_A$$
$$= (100°C/W)(8 \text{ SEG})(.0108A)(2.452V) + 60°C$$
$$T_{J APPROX} = 21.2°C + 60°C = 81.2°C$$

The adjusted forward voltage is 2.373 volts:

$$\Delta V_F \approx (-.0014V/°C)(81.2 - 25°C) = -.079V$$
$$V_F = 2.452 - .079 = 2.373V$$

The typical power dissipation and typical junction temperature can now be calculated with good accuracy:

$$P_{SEG} = (.0108A)(2.373V) = .026 \text{ W/SEG}$$
$$P_{DIGIT} = (8 \text{ SEG})(.026 \text{ W/SEG}) = .208W$$
$$T_{J TYP} = (100°C/W)(.208W) + 60°C \approx 81°C$$

Now the worst case is calculated using the maximum R_s value. Equation 5.2.2.2-2 is used to calculate $V_{F MAX}$. The same iterative process, as above, is used to take into account the -1.4mV/°C change in $V_{F MAX}$.

$$V_{F MAX} = 1.55V + (.0432A)(33\Omega) = 2.976V$$
$$T_{J MAX} = (100°C/W)(8 \text{ SEG})(.0108A)(2.976V) + 60°C$$
$$T_{J MAX} = 25.7 + 60 = 85.7°C$$
$$V_{F MAX} = 2.976V - (.0014V/°C)(85.7 - 25°C) = 2.891V$$
$$P_{SEG MAX} = (.0108A)(2.891V) = .0312W$$
$$P_{DIGIT MAX} = (8 \text{ SEG})(.0312 \text{ W/SEG}) = .250W$$
$$T_{J MAX} = (100°C/W)(.250W) + 60°C$$
$$T_{J MAX} = 25 + 60 = 85°C$$

Comparing $T_{J MAX}$ and $T_{J TYPICAL}$ with the curve in Figure 5.2.2.5-2, the worst case T_J is 5°C below the limit and $T_{J TYP}$ is 9°C below the allowed limit for T_A = 60°C.

5.2.4 Relative Efficiency and Light Output

The efficiency of an LED (photons/electrons) increases with forward current until the junction reaches saturation. If the efficiency at a given current is taken as a reference, then the efficiency at other currents may be compared to the reference and a plot of relative efficiency can be generated as shown in Figure 5.2.4-1. A relative efficiency curve is actually generated by measuring the time average luminous intensity at various peak currents while maintaining a constant average current. The curve is normalized to a relative efficiency of 1.00 at that data sheet current where the minimum luminous intensity is specified.

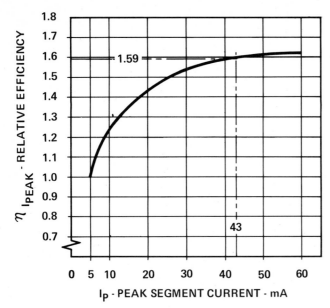

Figure 5.2.4-1 Relative Luminous Efficiency (Luminous Intensity Per Unit Current) vs Peak Current Per Segment for a High-Efficiency Red Display.

The time average luminous intensity can be calculated from the following equation:

$$(5.2.4\text{-}1)$$

$$I_v \text{ TIME AVG} = \left[\frac{I_{AVG}}{I_{AVG\,SPEC}}\right]\left[\frac{\eta_{I\,PEAK}}{\eta_{I\,PEAK\,SPEC}}\right]\left[I_{v\,SPEC}\right]$$

where:

I_v = Operating point average current

$I_{AVG\,SPEC}$ = Average current for data sheet luminous intensity value, $I_{v\,SPEC}$

$\eta_{I\,PEAK}$ = Relative efficiency at operating peak current.

$\eta_{I\,PEAK\,SPEC}$ = Relative efficiency at data sheet peak current where luminous intensity $I_{v\,SPEC}$ is specified.

$I_{v\,SPEC}$ = Data sheet luminous intensity, specified at $I_{AVG\,SPEC}$ and $I_{PEAK\,SPEC}$.

The light output of an LED also varies with temperature. As the temperature of an LED increases, the light output decreases according to the following exponential relationship:

$$I_v(T_A) = I_v(25°C)\,e^{K(T_A - 25°C)} \qquad (5.2.4\text{-}2)$$

where:

$I_v(T_A)$ = Luminous intensity at operating ambient temperature.

$I_v(25°C)$ = Luminous intensity at $T_A = 25°C$

T_A = Operating ambient temperature, °C

K = Exponent constant 1/°C, LED material dependent

e = 2.7183

LED	K
Standard Red	−.0188
High-Efficiency Red	−.0131
Yellow	−.0112
Green	−.0104

For an increase in temperature of 1°C, the light output change of the LED is as follows:

LED	+1°C Factor	% Change
Standard Red	$.9814 I_v$	−1.86
High-Efficiency Red	$.9870 I_v$	−1.30
Yellow	$.9889 I_v$	−1.11
Green	$.9897 I_v$	−1.03

Equation 5.2.4-1 applies to luminous intensity as perceived by the eye and is not applicable to radiant power output.

5.2.4.1 Sample Calculation of Time Average Luminous Intensity

The time average luminous intensity for the example in Section 5.2.3 can be calculated using the value of $I_v = 300$ mcd at $I_F = 5$ mA. The relative efficiency calculation does take into account junction temperature rise above ambient within reasonable error. The resultant $I_{v\,TIME\,AVG}$ is then corrected for the increased ambient above $T_A = 25°C$.

The relative efficiency for $I_P = 43$ mA is 1.59, as obtained from Figure 5.2.4-1, and the time average luminous intensity at $T_A = 25°C$ is calculated to be 1030 microcandelas/segment.

$$I_v \text{ TIME AVG} = \left[\frac{10.8\text{ mA}}{5\text{ mA}}\right]\left[\frac{1.59}{1}\right][300\ \mu cd]$$

$$= 1030\ \mu cd/\text{SEGMENT}$$

At $T_A = 60°C$, the light output decreases to 651 microcandelas/segment.

$$I_v(60°C) = 1030\,(.0131)\,(60 - 25°C)$$

$$I_v(60°C) = 1030\,(.6322) = 651\ \mu cd/\text{SEGMENT}$$

5.2.4.2 Digit-to-Digit and Segment-to-Segment Luminous Intensity Ratio

The luminous intensity value listed on a 7-segment display data sheet is a digit average luminous intensity for the seven segments. The ratio of the maximum I_v segment to the minimum I_v segment within a digit needs to be held to a reasonable value. Also, the digit to digit luminous intensity ratio needs to be held to a reasonable value so that digits may be stacked side by side with a pleasing appearance.

Tests have shown that when a display is lighted with a string of 8's, an observer may scrutinize the digits with a critical eye, finding all sorts of faults. However, this is not a normal mode of operation. The exact same display lighted with a string of random numbers may well be acceptable to the same observer. In fact, the observer reading the same display to obtain information during normal usage will NOT notice the minor faults he detected when scrutinizing the string of 8's. This is true when the segment-to-segment and digit-to-digit ratios are kept within certain limits.

An observer scrutinizing an LED display is able to detect a difference in light output at a threshold level of about 1.6:1. A change in light output becomes quite noticeable at the ratio of 2.3:1.

5.2.5   Driving a Seven Segment Display

A typical logic system will produce output information in the form of BCD data. This BCD data must then be converted to the 7-segment matrix code format of the display. Drivers are then required to drive each segment of the display. The two methods of driving a 7-segment display are 1) direct dc drive of a single digit and 2) strobing a string of digits.

5.2.5.1 Direct dc Driving a Display

The simplest method of driving a 7-segment display is to have one decoder/driver for each digit. This is usually cost effective up to four or five digits depending upon the trade-off between component cost and circuit complexity. Figure 5.2.5.1-1 illustrates the direct dc drive concept. Each digit has its own decoder/driver and is continuously illuminated. Though LED displays are more efficient when strobed, the advantage with direct drive is that the drivers need not handle high current levels. Also, the display circuit complexity is simple as extra timing is not required.

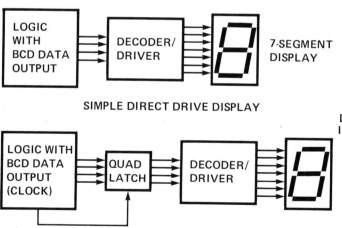

Figure 5.2.5.1-1 Block Diagram for Direct DC Drive Scheme for 7-Segment Displays.

The forward current through an LED must be limited in some fashion to obtain the desired illumination and to prevent high current damage. The two most popular methods of accomplishing these two objectives is to (1) drive the display from a decoder/driver with switching outputs that control the flow of LED current through external limiting resistors and (2) drive the display from a decoder/driver with constant current outputs. Figure 5.2.5.1-2 illustrates both methods.

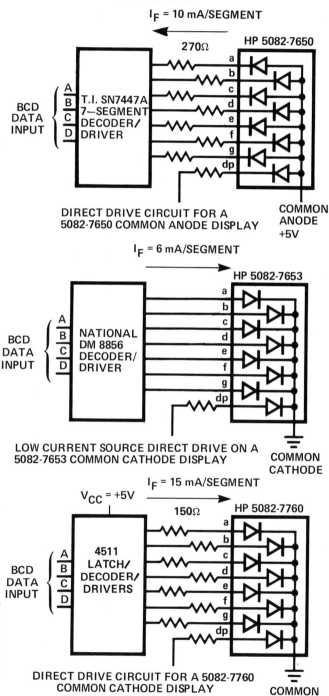

Figure 5.2.5.1-2 Example Methods for Direct DC Driving 7-Segment Displays.

When using resistor current limiting, the current limiting resistor may be determined from the following equation:

$$R = \frac{V_{CC} - V_F - V_{CE}}{I_F} \qquad (5.2.5.1\text{-}1)$$

where:

R	=	Series current limiter
V_{CC}	=	Supply voltage
V_F	=	LED forward voltage at I_F
V_{CE}	=	Saturation voltage drop across driver transistor
I_F	=	LED forward dc current

If a decoder/driver with constant current outputs is being used and the ambient temperature around the display requires dc current derating, shunt resistors at the driver outputs may be used to bleed off the excess current.

$$R_{SHUNT} = \frac{V_{OH}}{I_{OH} - I_F} \qquad (5.2.5.1\text{-}2)$$

As an example, the Fairchild 9368 7-segment latch/decoder/driver will source a maximum of 22 mA over a wide temperature range. A display is to be composed of standard red common cathode displays operating in an ambient temperature of 70°C. The data sheet derating is 0.43 mA dc/°C above T_A = 50°C. Therefore, at T_A = 70°C, $I_{DC\ MAX}$ = 16.4 mA. A worst case design requires shunting 5.6 mA from each output of the 9368. The LED forward voltage is 1.60 volts. The shunt resistor value is 286Ω:

$$R_{SHUNT} = \frac{1.60V}{(.022 - .0164)A} = 286\Omega$$

5.2.5.2 High Speed Counter Using DC Drive

Multiple function devices that will direct drive LED displays are becoming more available. An example is the Texas Instrument SN74143 BCD counter/4 bit latch/BCD to 7-segment decoder/driver. The outputs are 15 mA constant current sink. The SN74144 is the same device with a 25 mA sink capability with external current limiting resistors. A 4-digit high speed counter can be made using the SN74143 and the yellow .43 inch common anode displays as pictured in Figure 5.2.5.2-1. The 15 mA LED drive current is sufficient for the display to be read in most bright ambients.

The data strobe line may be held low to obtain a continuous display of the counter state, or strobed to latch specific counts. A decimal point input is provided in each SN74143 so that any desired decimal point position may be obtained by providing a logic high true input on the appropriate dp select line.

5.2.5.3 Concept of Strobed (Multiplexed) Operation

The most efficient method of driving an LED display is to strobe it at a high peak current and a low duty factor. This takes advantage of the increased efficiency of an LED at high peak currents and the reduction in average power dissipation when compared to dc operation. A multiplexed display design becomes cost effective when the decoder/driver and display PC board size and complexity required for dc drive become significant when compared to the timing circuitry and digit driver cost and size of PC board needed for strobed operation. The power saving that is attainable with strobed operation may show up as a need for a smaller sized power supply, contributing to cost reduction.

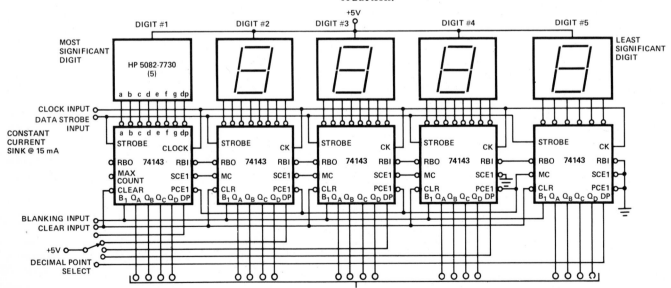

Figure 5.2.5.2-1 High Speed 5-Digit Counter Using Direct DC Driven LED Displays.

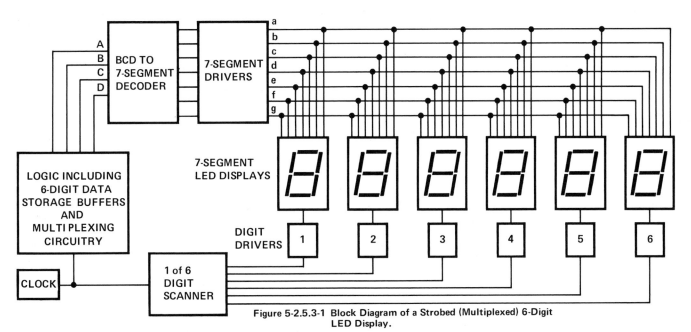

Figure 5-2.5.3-1 Block Diagram of a Strobed (Multiplexed) 6-Digit LED Display.

A strobed six digit display is illustrated in block diagram form in Figure 5.2.5.3-1. The like segments of each digit in the display are tied to a common segment bus line which is connected to one of seven segment drivers. For example, all of the "a" segments are tied common to the segment "a" driver and all of the "b" segments are tied common to the segment "b" driver. This allows one BCD to 7-segment/decoder driver to be time shared between all digits, with BCD data for each digit stored in one of six digit data storage buffers. Each digit is enabled by its own digit driver.

The operation of this multiplexing scheme is as follows: First Clock Pulse: The multiplexing logic selects digit data buffer number 1. This first digit BCD data is presented to the decoder/driver and the required segment lines are activated. The digit scanner enables only digit driver number 1 and the first number is displayed. Second Clock Pulse: The multiplexing logic selects digit data buffer number 2. This second digit BCD data is presented to the decoder/driver and again the required segment lines are activated. The digit scanner now enables only digit driver number 2 and the second number is displayed. This process is continued through digits 3, 4, 5 and 6, and then the whole process repeats.

The strobing rate to produce a flicker free display should be a minimum of 100 Hz or 5X the expect mechanical vibration. However, to optimize the overall performance of a 7-segment display, it is best to use a refresh rate of 1 KHz or faster.

When designing a multiplexed display system, timing considerations become important. The timing parameters to take into account when designing a display of a string of N-digits are:

N = quantity of digits

f = refresh rate;

t_p = on-pulse duration

t_b = blanking time between digits

The first consideration is to select a desired refresh rate. The reciprocal of the refresh rate is the refresh time period in which all N—digits are enabled, one at a time:

Refresh Period $= \tau = \dfrac{1}{f}$ (5.2.5.3-1)

The maximum allotted digit on-pulse duration is the refresh period divided by the quantity of digits to be strobed:

Maximum On-Pulse $= t_{p\ max} = \dfrac{\tau}{N}$ (5.2.5.3-2)

Figure 5.2.5.3-2 illustrates a basic timing diagram for strobing N=6 digits. Each digit is enabled in sequence without any overlapping of the digit enable time, $t_{p\ max}$. A segment line, however, may remain active if it is required in two or more successive digits. A variation of this basic timing scheme is to use Pulse Width Modulation (PWM) to vary the operating on-pulse duration, t_p, to values less than the allotted pulse duration, $t_{p\ max}$. PWM may effectively be used to vary the apparent brightness of a display to match changing ambient lighting conditions. As an example, for low level ambient conditions, t_p is reduced to dim the display. The segment peak current remains constant to take advantage of the increased LED efficiency at high peak currents, yet the average current is decreased which reduces the time average luminous intensity of the display, see equation 5.2.4-1.

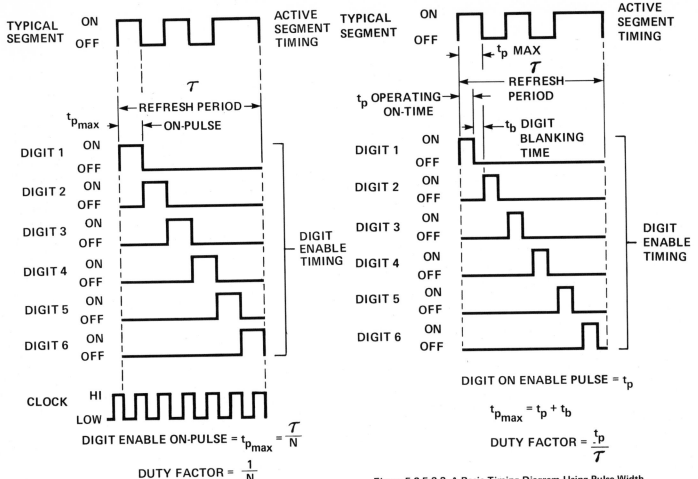

DIGIT ENABLE ON-PULSE = $t_{p_{max}} = \dfrac{\tau}{N}$

DUTY FACTOR = $\dfrac{1}{N}$

Figure 5.2.5.3-2 A Basic Timing Diagram for Strobing N=6 Digits. (The Typical Segment is Shown Active When Digits 1, 2, 5 and 6 are Enabled.)

DIGIT ON ENABLE PULSE = t_p

$t_{p_{max}} = t_p + t_b$

DUTY FACTOR = $\dfrac{t_p}{\tau}$

Figure 5.2.5.3-3 A Basic Timing Diagram Using Pulse Width Modulation to Vary Digit Enable Time, tp. The Active Segment Pulse Duration is not Modulated and Remains at tpMAX.

Figure 5.2.5.3-3 illustrates a basic timing diagram employing PWM. The digit enable on-pulse, t_p, is varied to obtain the desired display time average luminous intensity. The maximum allotted pulse duration is now composed of the digit operating on-pulse, t_p, and a blanking time, t_b:

Allotted Pulse Width = $t_{p\ max} = t_p + t_b$　　　　(5.2.5.3-3)

The duty factor is now defined as the ratio of the operating on-pulse to the refresh period:

Duty Factor (PWM) = tp/τ　　　　(5.2.5.3-4)

The active segment on-pulse duration need not be modulated and can remain at $t_{p\ max}$. It is usually sufficient and easier to modulate only the digit enable operating time, though a designer may choose to modulate either the segment on-time, digit on-time or both.

An additional use of PWM is applied when a fast strobing rate in the neighborhood of 10 KHz is used. At this fast refresh rate, the turn-off time of driving transistors may not be fast enough to prevent overlapping of digit and segment on-pulse durations. This timing overlapping is called ghosting, since a digit that should be off is still partially turned-on. A timing overlap of approximately 3% is usually sufficient to have perceptable ghosting in moderate ambient lighting conditions. To prevent this situation from occuring, PWM is employed to effect inter-digit blanking. In this mode of operation, a fixed digit (and possibly segment) blanking time, t_b, is built into the display timing to insure sufficient time for the driving transistors for one digit to turn completely off prior to turning on the driver transistors for the next digit.

The value of the segment current limiting resistors may be calculated from the generalized schematic shown in Figure 5.2.5.3-4.

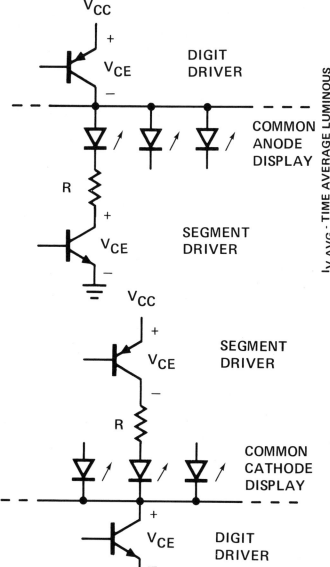

Figure 5.2.5.3-4 Generalized Drive Circuits for Strobed Operation.

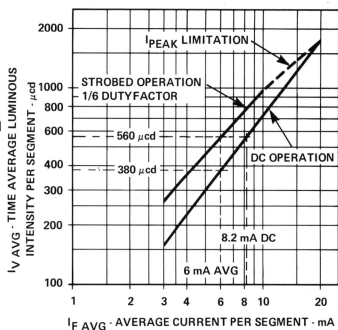

Figure 5-2.5.3-5 Comparison of Time Average Luminous Intensity as Obtained from Strobed and DC Operation, High-Efficiencly Red Display.

From Figure 5.2.5.3-4, the following equation may be used to calculate the vaue of the current limiting resistor, R:

$$R = \frac{V_{CC} - (V_F + V_{CE\ SEG} + V_{CE\ DIGIT})}{I_{PEAK}} \quad (5.2.5.3\text{-}5)$$

where:

V_{CC} = Supply voltage

V_F = LED forward voltage at I_{PEAK}

$V_{CE\ SEG}$ = Saturation voltage drop across segment driver at I_{PEAK}

$V_{CE\ DIGIT}$ = Saturation voltage drop across digit driver at (I_{PEAK} x the number of segments multipled)

I_{PEAK} = Peak LED segment current

The advantage of strobe operation over dc operation is illustrated in Figure 5.2.5.3-5. This is a graph comparing the light output obtained by strobing a 6-digit display vs. dc driving each digit individually. The design constraint is to maintain 6 mA average. At 6 mA dc the time average luminous intensity per segment is $I_{V\ AVG}$ = 380μcd. By strobing the displays at I_{PEAK} = 36 mA and a duty factor of 1/6, the light output is increased to $I_{V\ AVG}$ = 560μcd. This is an increase in light output of 1.47X over dc operation for the same average current.

At 6 mA average, the strobed operation produces an average power dissipation within the display of 10.3 mW per segment vs. 10.0 mW per segment for dc operation. To obtain the light output of 560 μcd per segment by using dc operation would require a forward current of 8.2 mA dc and a power dissipation of 14.1 mW per segment. The total power saving obtained by strobing is considerable. This power saving is realized by the reduction of the required IC components needed to drive the display. For example, in this illustration, only one decoder/driver is required for strobed operation vs. six needed for dc operation.

The human eye is a time average detector and, therefore, will easily discern the increased light output obtained by strobing.

5.2.6 Interfacing Microprocessors to Seven Segment Displays

Seven segment displays can be interfaced to a microprocessor with only a few external components. The microprocessor can be used to control the refreshing of a multiplexed display or simply to periodically up-date a dc driven display. In either case, one or more eight bit latches are required to hold the the seven segment and digit information. If the seven segment display is multiplexed, then some timing circuitry is also required to periodically request new information from the microprocessor. The timing circuitry could consist of either a monostable multivibrator or an oscillator. This circuitry would request an interrupt and use the already available hardware to handle the proper interrupt decoding. Several commercially available seven segment decoder/drivers are available that can directly drive the seven segment LED displays.

When the seven segment display is driven on a dc basis, a seven segment decoder/driver is required for each display. Figure 5.2.6-1 shows an example of a seven segment-microprocessor interface. Each display is driven by its own seven segment driver. The Fairchild 9374 has current source outputs that drive each LED segment at 15 mA dc. The decimal points are driven by a National DS8859, a hex latch with current source outputs. The Intel 8080A microprocessor updates each display, with an OUTPUT instruction. The second byte of the OUT instruction specifies an eight bit address which determines the display that is to be updated. Upon execution of the OUT instruction, the lowest four bits of the accumulator are loaded into the appropriate latched-decoder/driver. A separate OUT instruction also updates the decimal points and overflow digit. Since the upper six states of the Fairchild 9374 are decoded as (-, E, H, L, P, and blank), it is possible to blank each digit selectly under microprocessor control, indicate an overflow by a row of minus signs, and use the word "HELP" to catch someone's attention.

A microprocessor can also be used to multiplex a seven segment display. Figures 5.2.6-2 and 5.2.6-3 show two ways by which this can be accomplished. In Figure 5.2.6-2, the microprocessor outputs two bytes per digit. The first byte contains segment information for the display. Since the seven segment information is decoded by the microprocessor, the programmer can program the character font to include any special symbols that are desired. The second byte turns on the proper digit driver. At the same time, the RCA 4047 monostable multivibrator is triggered and requests another interrupt in 2 milliseconds. Upon requesting an interrupt, the interrupt circuitry (not shown) must force an RST(7) instrument to be executed by the Intel 8080A microprocessor. Then the refresh program shown in Figure 5.2.6-2b is executed. The program reads a pointer from RAM and then outputs two bytes of information to the display. The pointer is incremented by one and compared to the final address of the data file. If they are equal, the pointer is reset to the address of the first byte in the file, otherwise the pointer is set to the address of the next byte in the file. As shown, the circuit refreshes five seven segment displays on a 20% duty factor at 60 mA peak/segment. This technique can be expanded to eight displays without additional hardware. The time required to refresh the display can be determined as shown below:

$$\text{REFRESH TIME} = \frac{(165N + 7)R}{\text{MICROPROCESSOR CLOCK RATE}} \quad \text{(5.2.6-1)}$$

where N is the number of digits to be refreshed, and R is the strobing rate.

For example, suppose five digits are refreshed on a 100 Hz rate and the microprocessor uses a 2MHz clock, then 4.2% of microprocessor time is required to refresh the display. For the remaining 95.8% of the time, the microprocessor can be performing countless other tasks.

Figure 5.2.6-3 shows another technique that can be used to multiplex a seven segment display. With this technique, the microprocessor outputs only a single byte of information to the display. Four bits contain the BCD character to be displayed, one bit contains the decimal point information, and the remaining three bits determine which digit should be turned on. The Motorola MC 14511 seven segment latched decoder driver can source up to 25 mA per segment to a common cathode LED display. The high efficiency red displays that are shown can be satisfactorily operated at 25 mA/segment on a 20% duty duty factor. This power requirement is considerably below those required for standard red seven segment displays. The interrupt request circuitry is similar to the previous example. After receiving an interrupt request, the interrupt hardware must force a RST(7) instruction to be executed by the Intel 8080A microprocessor. Following the RST(7) instruction, the program shown in Figure 5.2.6-3b will be executed. This program is similar to the program shown in Figure 5.2.6-2b except that only a single byte is outputed to the display. This technique can also be expanded to eight displays without additional hardware. The time required to refresh the display can be determined by equation 5.2.6-2:

$$\text{REFRESH TIME} = \frac{(143N +7)R}{\text{MICROPROCESSOR CLOCK RATE}} \quad \text{(5.2.6-2)}$$

where N is the number of digits to be refreshed, and R is the strobing rate.

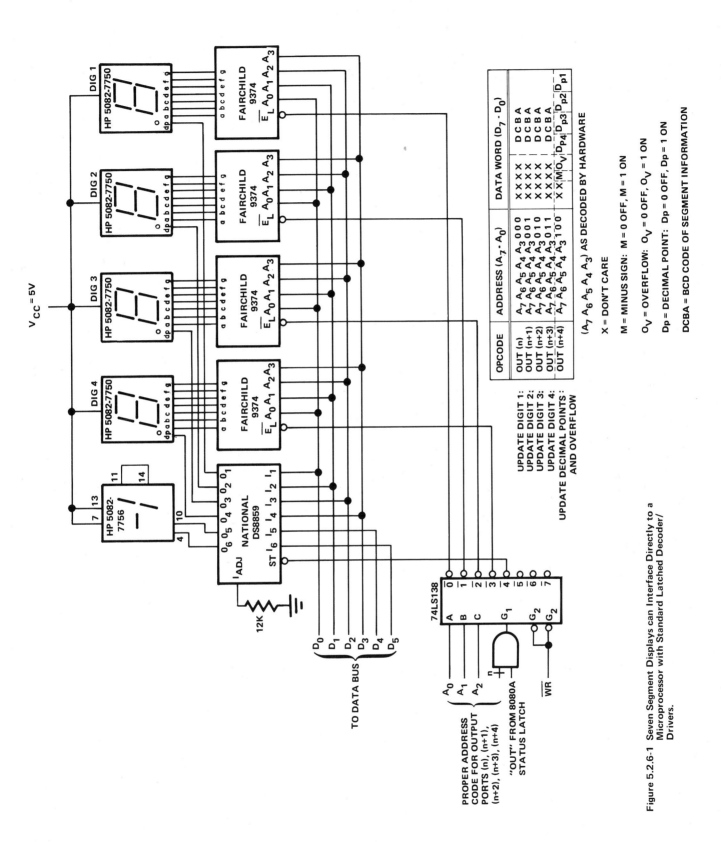

Figure 5.2.6-1 Seven Segment Displays can Interface Directly to a
Microprocessor with Standard Latched Decoder/
Drivers.

OPCODE	ADDRESS ($A_7 - A_0$)	DATA WORD ($D_7 - D_0$)
OUT (n)	$A_7\ A_6\ A_5\ A_4\ A_3\ 0\ 0\ 0$	X X X X D C B A
OUT (n+1)	$A_7\ A_6\ A_5\ A_4\ A_3\ 0\ 0\ 1$	X X X X D C B A
OUT (n+2)	$A_7\ A_6\ A_5\ A_4\ A_3\ 0\ 1\ 0$	X X X X D C B A
OUT (n+3)	$A_7\ A_6\ A_5\ A_4\ A_3\ 0\ 1\ 1$	X X X X D C B A
OUT (n+4)	$A_7\ A_6\ A_5\ A_4\ A_3\ 1\ 0\ 0$	X X M O_V D_{P4} D_{p3} D_{p2} D_{p1}

($A_7\ A_6\ A_5\ A_4\ A_3$) AS DECODED BY HARDWARE

X = DON'T CARE

M = MINUS SIGN: M = 0 OFF, M = 1 ON

O_V = OVERFLOW: O_V = 0 OFF, O_V = 1 ON

Dp = DECIMAL POINT: Dp = 0 OFF, Dp = 1 ON

DCBA = BCD CODE OF SEGMENT INFORMATION

UPDATE DIGIT 1:
UPDATE DIGIT 2:
UPDATE DIGIT 3:
UPDATE DIGIT 4:
UPDATE DECIMAL POINTS :
AND OVERFLOW

PROPER ADDRESS
CODE FOR OUTPUT
PORTS (n), (n+1),
(n+2), (n+3), (n+4)

"OUT" FROM 8080A
STATUS LATCH

TO DATA BUS

5.27

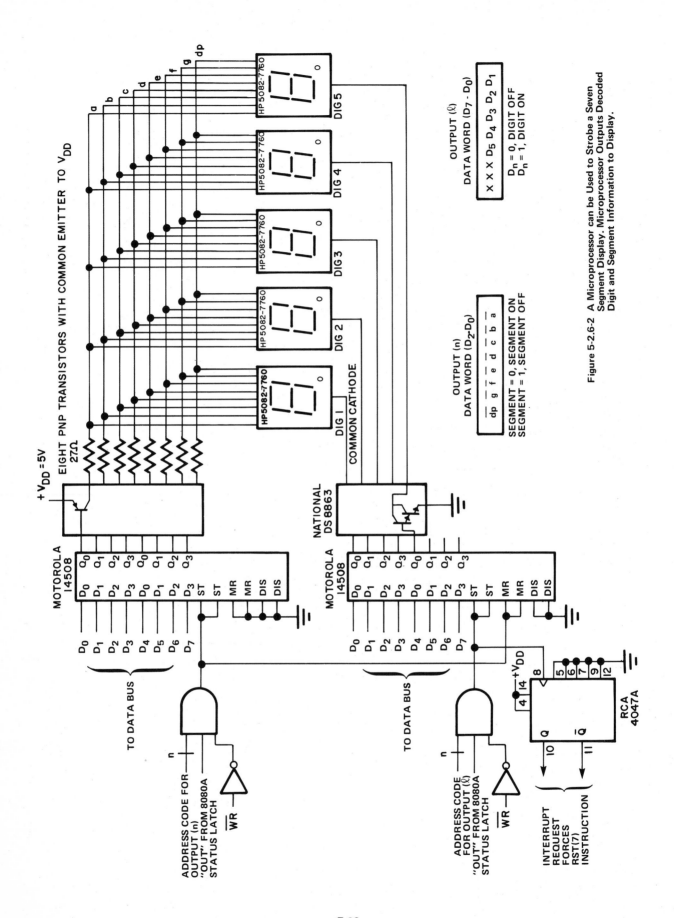

Figure 5-2.6-2 A Microprocessor can be Used to Strobe a Seven Segment Display. Microprocessor Outputs Decoded Digit and Segment Information to Display.

OUTPUT (ℓ)
DATA WORD (D₇ - D₀)

$$X\ X\ X\ D_5\ D_4\ D_3\ D_2\ D_1$$

$D_n = 0$, DIGIT OFF
$D_n = 1$, DIGIT ON

OUTPUT (n)
DATA WORD (D₇-D₀)

$$\overline{dp}\ \overline{g}\ \overline{f}\ \overline{e}\ \overline{d}\ \overline{c}\ \overline{b}\ \overline{a}$$

SEGMENT = 0, SEGMENT ON
SEGMENT = 1, SEGMENT OFF

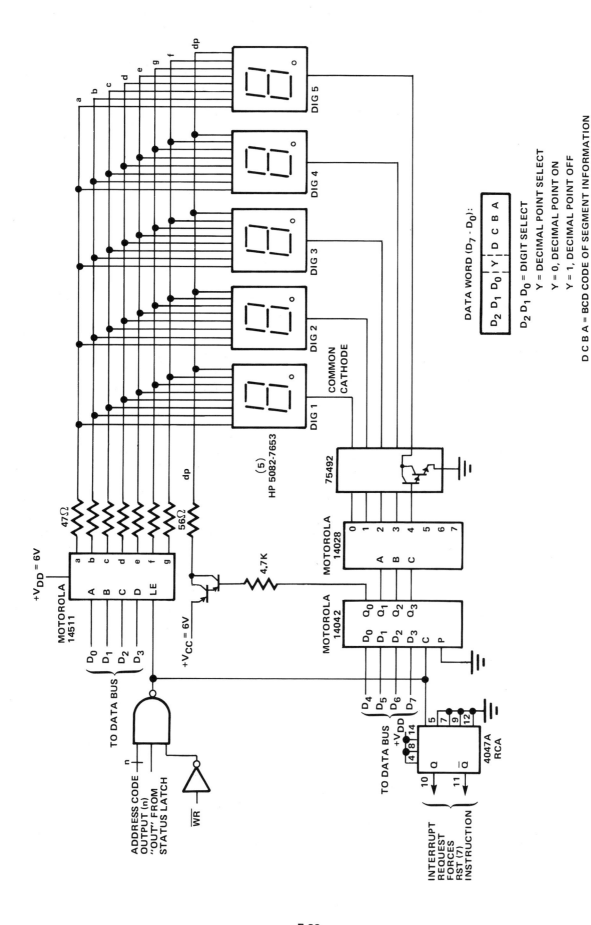

Figure 5.2.6-3a   A Microprocessor Can Be Used To Strobe a Seven Segment Display. Microprocessor Outputs a Single Byte Which Is Decoded By Display Circuitry.

5.29

ADDRESS	OP CODE	CLOCK CYCLES	COMMENTS
(0038)$_{16}$	PUSH PSW.	(11)	
	PUSH HL	(11)	
	LHLD	(16)	HL = POINTER
	A$_L$	------	
	A$_H$	------	
	MOV A, M	(7)	A = (HL)
	OUT	(10)	STORES NEW SEGMENT INFORMATION
	n	------	
	INX HL	(5)	HL = HL + 1
	MOV A, M	(7)	A = (HL)
	OUT	(10)	TURNS ON DIGIT
	I	------	DRIVER
	MOV A, L	(5)	A = L
	CPI	(7)	COMPARE L TO ADDRESS OF LAST MEMORY LOCATION
	(19)$_{16}$	------	
	JNZ LOOP	(10)	JUMP IF A ≠ (19)$_{16}$
	(ADDRESS OF	------	
	LOOP)		
	MVI, L	(7)	L = (OF)$_{16}$
	(OF)$_{16}$	------	
LOOP	INX HL	(5)	HL = HL + 1
	SHLD	(16)	POINTER = HL
	A$_L$	------	
	A$_H$	------	
	POP HL	(10)	
	POP PSW	(10)	
	EI	(4)	
	RET	(10)	

MEMORY ADDRESS	CONTENTS (D$_7$ - D$_0$)	COMMENTS
A$_H$ A$_L$	POINTER$_L$	POINTER = NEXT DIGIT TO BE DISPLAY
A$_H$ A$_L$ + 1	POINTER$_H$	
(X X 1 0)$_{16}$	dp g f e d c b a	DIGIT 1
(X X 1 1)$_{16}$	0 0 0 0 0 0 0 1	
(X X 1 2)$_{16}$	dp g f e d c b a	DIGIT 2
(X X 1 3)$_{16}$	0 0 0 0 0 0 1 0	
(X X 1 4)$_{16}$	dp g f e d c b a	DIGIT 3
(X X 1 5)$_{16}$	0 0 0 0 0 1 0 0	
(X X 1 6)$_{16}$	dp g f e d c b a	DIGIT 4
(X X 1 7)$_{16}$	0 0 0 1 0 0 0	
(X X 1 8)$_{16}$	dp g f e d c b a	DIGIT 5
(X X 1 9)$_{16}$	0 0 1 0 0 0 0	

Figure 5.2.6-2b   Refresh Program Used to Strobe the Display Shown in Figure 5.2.6-2.

ADDRESS	OP CODE	CLOCK CYCLES	COMMENTS
(0038)$_{16}$	PUSH PSW	(11)	
	PUSH HL	(11)	
	LHLD	(16)	HL = POINTER
	A$_L$	------	
	A$_H$	------	
	MOV A, M	(7)	A = (HL)
	OUT	(10)	DISPLAY IS UPDATED
	n	------	
	MOV A, L	(5)	A = L
	CPI	(7)	COMPARE L TO ADDRESS OF LAST MEMORY LOCATION
	(14)$_{16}$	------	
	JNZ LOOP	(10)	JUMP IF A ≠ (14)$_{16}$
	(ADDRESS OF	------	
	LOOP)		
	MVI, L	(7)	L = (OF)$_{16}$
	(OF)$_{16}$	------	
LOOP	INX HL	(5)	HL = HL + 1
	SHLD	(16)	POINTER = HL
	A$_L$	------	
	A$_H$	------	
	POP HL	(10)	
	POP PSW	(10)	
	EI	(4)	
	RET	(10)	

MEMORY ADDRESS	CONTENTS (D$_7$ - D$_0$)	COMMENTS
A$_H$ A$_L$	POINTER$_L$	POINTER = NEXT DIGIT TO BE DISPLAY
A$_H$ A$_L$ + 1	POINTER$_H$	
(X X 1 0)$_{16}$	0 0 0 Y	DIGIT 1
(X X 1 1)$_{16}$	0 0 1 Y	DIGIT 2
(X X 1 2)$_{16}$	0 1 0 Y BCD DATA	DIGIT 3
(X X 1 3)$_{16}$	0 1 1 Y	DIGIT 4
(X X 1 4)$_{16}$	1 0 0 Y	DIGIT 5

X = DON'T CARE
Y = DECIMAL POINT INFORMATION
    Y = 0, DECIMAL POINT ON
    Y = 1, DECIMAL POINT OFF

Figure 5.2.6-3b   Refresh Program Used to Strobe the Display Shown in Figure 5.2.6-3.

For example, 3.6% of the microprocessor's time would be required to refresh five displays at a 100 Hz refresh rate with a 2 MHz microprocessor clock. While these examples use an Intel 8080A microprocessor, these techniques can be used with other microprocessors by modifying the hardware and programs to reflect a different instruction set and different ways to output information to peripheral devices.

5.2.7   Detection and Indication of Segment Failures in Seven Segment LED Displays

The failure of a single element in a seven segment display device could result in the presentation of erroneous information which may not be detectable to the user. Such failures may merely be annoying in some equipment applications but may, on the other hand, be quite serious in the case of point of sales terminals or medical instrumentation.

Segment failures in seven segment LED displays, although rare, do have a measurable probability of occurrence. Characteristically, the failures are electrical opens or unusual degradation of light intensity. Shorted segments have such an extremely low probability of occurrence that they do not need to be considered. Fortunately, it is a rather simple matter to detect the occurrence of an open segment in a seven segment display matrix.

The circuits on the following pages depict two possible techniques for the detection and indication of such failures.

5.2.7.1 Seven-Segment Self-Test Circuit for Common Anode Displays

The circuit in Figure 5.2.7.1-1 will detect an open segment in any of the digits of a multi-digit 7 segment display. Basically, the test consists of turning on all segments for a short period not detected by the eye. Should the segment current corresponding to an "ON" segment not flow, a logic circuit blanks the entire display until reset. The portion of the circuit inside the labeled dashed line represents the failure detection system.

The scanning circuitry for the strobed display consists of a clock, a counter and a decoder. The -9601 one-shot multivibrator produces a 1 microsecond pulse out for each clock period input to the scan counter. The Q output of the one shot enables the lamp test function of the BCD to seven segment decoder driver, turning on all of the segment outputs. The status of each of the segment lines at nodes A is monitored by the 8 input -7430 NAND gate. If all segments are intact, the voltage of all the nodes A will be a logic high (V_{CC}-$V_{CE SAT}$-V_F LED) = ~2 volts. The output of the 8 input NAND will be a logic low and therefore no clock pulse will be gated through to the -7474 flip flop. If a

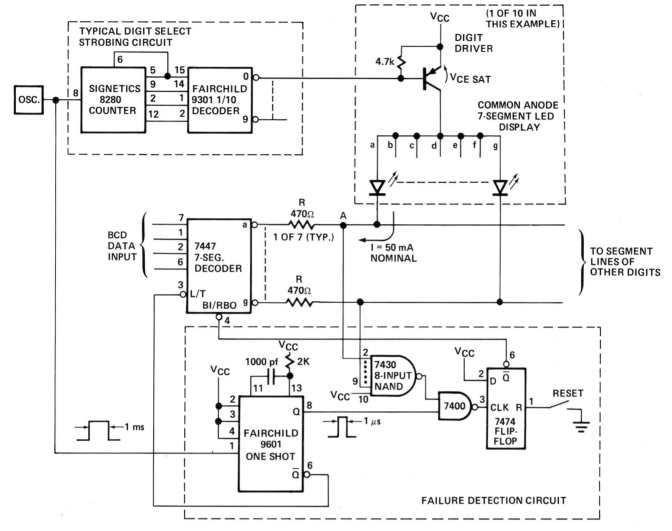

Figure 5.2.7.1-1 Self Test Circuit For 7-Segment Displays.

segment should be open, the low true output of the -7447 decoder will produce a logic 0 at the node A corresponding to the open segment. The resultant logical 1 output of the 8 input NAND will be gated through the 2 input NAND by the Q outeput of the one shot, thereby clocking the flip-flop. The 470 ohm segment current limiting resistor R is the largest value which can be used and still product a logic low at node A when using a standard TTL 8 input NAND gate.

In the circuit shown, the output of the flip-flop is used to blank the display on the occurrence of an open. The flip-flop is externally reset.

5.2.7.2 Self Test Circuit for Seven-Segment Displays and Associated Decoder Driver

The circuit in Figure 5.2.7.2-1 will detect an open segment in any of the digits of a multidigit seven segment display. Also, through the use of a redundant BCD to seven segment decoder, a malfunction of the decoder can be detected.

The circuit functions through the use of exclusive OR gating. The 8 gates labeled U_1 will only produce a logical 1 output when a segment is designated "ON" (-7447 decoder output low) and current is flowing through the segment so as to produce a logical 1 at node A and a logical 0 at node B. This is condition 1. The other two conditions are: (2) a segment turned off; nodes A and B are high, and (3) a segment is turned on by the decoder but no current flows due to an open; nodes A and B are low. Both of these conditions will result in a logical 0 at the output of gates U_1. Failure detection occurs at gates U_2. One input of each of these exclusive OR gates is the output of the gates U_1. The other input is the respective output of the -7448 BCD to 7 segment decoder. The -7448 has inverted logic outputs when compared to the -7447 (high true instead of low true). The three conditions defined above will produce the following results at the outputs of U_2.

Condition:

1.   C and D both logical 1; U_2 output logical 0.

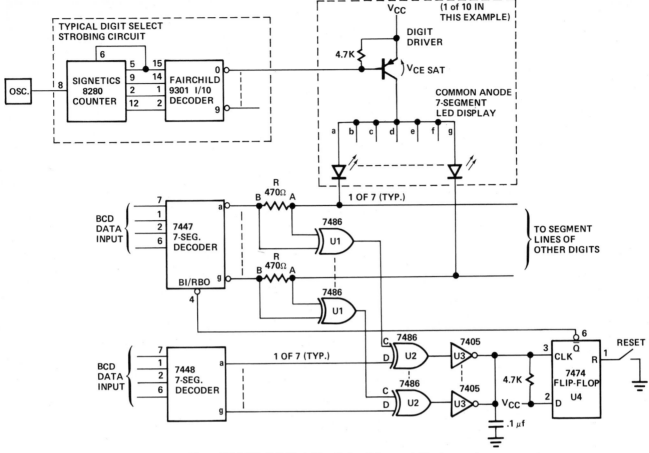

Figure 5.2.7.2-1 Self Test Circuit for 7-Segment Displays and Associated Decoder/Driver.

2.   C and D both logical 0; U_2 output logical 0.

3.   C at logical 0; D at logical 1; U_2 output at a logical 1.

Condition 3 will produce a clock input pulse to U_4, thereby setting the flip-flop; Q to logical 1.

A fourth condition will occur if either one of the decoders fails to function properly. Unless the respective segment outputs of the two decoders are always at opposite logic states, a logical 1 will be produced at the output of one of the gates U_2, causing the flip-flop to set; Q to logical 1.

In the circuit shown U_2 and U_3 are combined to form a 2 input exclusive NOR with open collector limiting resistors R to be as great as 1.1K. A capacitor is included on the clock input of U_4 to filter logic transition spikes off of the clock line allowing only a true change in the state of U_2 to clock the flip-flop. In this circuit, the Q output of the flip-flop is used to blank the display. The flip-flop must be manually reset; bringing Q to logical 0.

5.2.8.1 Suggested Drive Currents for Stretched Seven Segment Displays Used in Various Ambient Light Levels

The level of ambient light and the display contrast filter must be considered together when establishing the proper LED drive current. The LED drive current must be of sufficient magnitude to illuminate a display segment so that it can be easily recognized behind the contrast filter in the expected user ambient. As the ambient light level increases, the LED drive current must also increase to offset any masking of the display emitted light due to elevated ambient.

Table 5.2.8.1-1 lists suggested drive currents for 7.62mm (.3 inch) and 10.92mm (.43 inch) stretched seven segment displays when used in a given ambient light level with a specific filter. The currents listed should be considered as starting points and should be adjusted as necessary for each specific application. The currents listed allow the 7.62mm devices to be read from a distance of 4 meters (6 meters for the 10.92mm devices) in the corresponding ambient.

AMBIENT LIGHTING CONDITION	LED DRIVE MODE	.3 STD RED CATEGORY[2] C	.43 STD RED CATEGORY[2] B	.43 HER CATEGORY[2] AA	.43 YELLOW CATEGORY[2] B	.43 GREEN CATEGORY[2] B
DIM (HOME OR LOW LIGHT LEVEL CONTROL ROOM) (10–100 lux)	DC DRIVE	4 mA	5 mA	3 mA	5 mA	7 mA
	STROBED DRIVE[1]	3.4 mA	4.4 mA	2.1 mA	3.8 mA	4.9 mA
	FILTERS	RH 2423 P60 H1605 3MV 3M655	RH 2423 P60 H1605 3MV 3M655	P65 H1670 3M625	P27 H1720 3M590	P48 H1440 3M565
MODERATE (TYPICAL OFFICE) (100–1000 lux)	DC DRIVE	6 mA	9 mA	6 mA	11 mA	11 mA
	STROBED DRIVE[1]	5.25 mA	8.35 mA	4.5 mA	8.0 mA	7.6 mA
	FILTERS	RH2423 P60 H1605 3MV 3M655	RH2423 P60 H1605 3MV 3M655	P65 H1670 3M625	P27 H1720 3M590	P48 H1440 3M565
BRIGHT (OUTDOORS, VERY BRIGHT OFFICE) (1000–10,000 lux)	DC DRIVE	15 mA	21 mA	9 mA	16 mA	20 mA
	STROBED DRIVE[1]	14.25 mA	20.5 mA	6.25 mA (7.9 mA)	12 mA	13.5[3] mA
	FILTERS	P63 3M655L 3MNDL	P63 3M655L 3MNDL	P60 P65 3M625L 3MNDL	P27 H1720 3M590L 3MNDL	H1425 3M565L 3MNDL

NOTES: [1] FIGURE IS AVERAGE CURRENT. ASSUME $I_{pk} = 5 \times I_{AVG}$

[2] DRIVE CURRENT FOR OTHER CATEGORIES SHOULD BE VARIED BY A RATIO OF 1.5:1 PER CATEGORY WITH ADDITIONAL CORRECTION FOR RELATIVE EFFICIENCY.

[3] $I_{pk} = 4 \times I_{AVG}$

PANELGRAPHIC
P63 – DARK RED 63
P60 – RUBY RED 60
P65 – SCARLET RED 65
P27 – YELLOW 27
P48 – GREEN 48

SGL HOMALITE
H1605 – RED H100-1605
H1670 – RED H100-1670
H1720 – AMBER H100-1720
H1440 – GREEN H100-1440
H1425 – GREEN H100-1425

3M COMPANY (L = WITH LOUVERS)
3MV – VIOLET
3M655 – RED 655
3M625 – RED 625
3M590 – AMBER 590
3M565 – GREEN 565
3MND – NEUTRAL DENSITY

RH2423 = ROHM & HASS PLEXIGLAS 2423 OR 2444

Figure 5.2.8.1-1

5.3   MONOLITHIC DISPLAYS

5.3.1   Introduction

The monolithic seven segment display represents one of the largest segments of the optoelectronics industry today. Monolithic LED displays are found in many handheld calculators and digital watches. Monolithic LED displays are used because of their low cost, small size, and low power requirements. Monolithic displays have a very high sterance which allow them to be used outdoors and in other areas of high ambient lighting. While monolithic displays have generally been used for low cost applications, improvements in character size and viewing angle have also made monolithic displays suitable for portable instruments and other high quality applications. Table 5.3.1-1 compares the relative merits of a premium monolithic display to other large seven segment displays that could be used in a portable instrument application.

Monolithic displays differ from other types of LED displays in that the individual light emitting segments are formed by diffusing separate LED junctions on a single chip of GaAsP. Usually the GaAsP substrate is n doped material and each LED junction is formed by a p+ diffusion, so monolithic displays are normally of common cathode configuration. Although most monolithic displays are constructed with

5.33

	MONOLITHIC	LARGE SEVEN SEGMENT	
1) CHARACTER SIZE	4.45 mm (.175 inches)	7.62 mm (.300 inches)	10.92 mm (.430 inches)
2) DIGIT SPACING	5.84 mm (.230 inches)	10.16 mm (.400 inches)	12.70 mm (.500 inches)
3) MAXIMUM VIEWING DISTANCE	2 m	4 m	6 m
4) VIEWING ANGLE	$\pm38°$	$\pm75°$	$\pm75°$
5) TYPICAL STERANCE	300 cd/m^2 (90 fL) @2mA avg/seg 1/5 DF	120 cd/m^2 (35 fL) @15 mA avg/seg 1/5 DF	130 cd/m^2 (38 fL) @6 mA avg/seg 1/5 DF
6) POWER REQUIREMENTS	1.5—2.0 mA avg/seg	10—15 mA avg/seg (Standard Red)	4—6 mA avg/seg (High Efficiency Red)
7) DECIMAL LOCATION	RHD$_p$, CD$_p$ Colon Available	RHD$_p$, LHD$_p$	RHD$_p$, LHD$_p$
8) EASE OF USE	Single Multidigit Package Inherent Digit Alignment Inherent Luminous Intensity/Matching Must be Multiplexed Common Cathode	Single Digit Package DC or Multiplexed Operation Common Anode or Common Cathode	Single Digit Package DC or Multiplexed Operation Common Anode or Common Cathode

TABLE 5.3.1—1 COMPARISON OF MONOLITHIC AND LARGE SEVEN SEGMENT DISPLAYS

the standard seven segment format, almost any number of segments or character shapes can be diffused into the GaAsP substrate. Figure 5.3.1-1 shows the construction of a monolithic GaAsP chip. Since several segments are diffused into a single GaAsP chip, monolithic displays cannot be constructed on a GaP substrate. GaP is relatively transparent to red light, so the contrast between on and off segments would be inadequate. Because GaAsP is relatively expensive, most monolithic displays are magnified to keep chip sizes small, while attaining a viewable character size. In most cases, the monolithic display is magnified by an external lens. The resulting apparent character height is equal to the actual character height times the magnification of the lens.

5.3.2   Effect of External Lens

The external lens has a very important impact in the appearance of the final display. Besides increasing the apparent character height, the lens influences the radiation pattern, the axial luminous intensity, and the viewing angle of the display. A lens increases the axial luminous intensity by the square of the magnification:

$$I_V (MAG) = m^2 \, I_V (UNMAG) \qquad (5.3.2\text{-}1)$$

The addition of a lens may also increase the fresnel loss of the optical system. Fresnel loss, which is explained in more detail in Section 2.1.2, is the loss of light due to reflection as light passes from a medium with one index of refraction to a medium with a different index of refraction. The efficiency of light transmission between two mediums can be expressed as:

$$T = \frac{4}{2 + n_2/n_1 + n_1/n_2} \qquad (5.3.2\text{-}2)$$

where T is the transmission coefficient, n_1 is the index of refraction of the first medium, and n_2 is the index of refraction of the second medium.

Lenses for monolithic displays fall into two categories -- immersion lenses and non immersion lenses. Immersion lenses are formed by molding a lens directly over the LED chip, non immersion lenses have at least one layer of air between the LED chip and the lens assembly. The axial luminous intensity for either type of lens is equal to:

$$I_V (MAG) = \left(\prod_{i=1}^{n} T_i \right) m^2 \, I_V \qquad (5.3.2\text{-}3)$$

where $I_V (MAG)$ is the magnified axial luminous intensity, T_i is the transmission coefficient from medium i-1 to medium i, n is the total number of medium interfaces, m is the magnification of the lens, and I_V is the axial luminous intensity just beneath the surface of the LED chip.

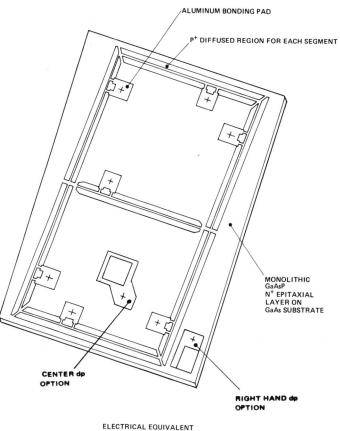

ALUMINUM BONDING PAD

P⁺ DIFFUSED REGION FOR EACH SEGMENT

MONOLITHIC GaAsP N⁺ EPITAXIAL LAYER ON GaAs SUBSTRATE

CENTER dp OPTION

RIGHT HAND dp OPTION

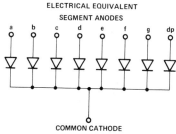

ELECTRICAL EQUIVALENT SEGMENT ANODES

a   b   c   d   e   f   g   dp

COMMON CATHODE

Figure 5.3.1-1 Construction of LED Monolithic Chip

Figure 5.3.2-1 shows examples of immersion and non immersion lenses and compares bare chip axial luminous intensity to magnified axial luminous intensity for immersion and non immersion lenses. This increase in axial luminous intensity allows a reduction in display drive currents to achieve the same luminous intensity as an unmagnified display.

The maximum off axis viewing angle at which any display can be observed is limited either by the radiation pattern of the display or by the maximum amount of distortion before the display becomes unreadable. An unmagnified LED device generally has very close to a lambertian radiation pattern, that is the luminous intensity varies as the cosine of the off axis angle:

$$I_V (\theta) = I_V \cos \theta \qquad (5.3.2\text{-}4)$$

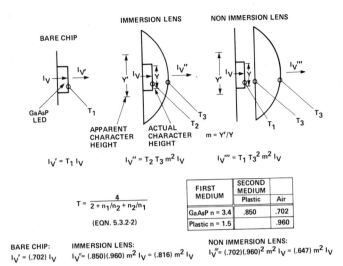

$$T = \frac{4}{2 + n_1/n_2 + n_2/n_1}$$

(EQN. 5.3.2-2)

	FIRST MEDIUM	SECOND MEDIUM	
		Plastic	Air
GaAsP n = 3.4		.850	.702
Plastic n = 1.5			.960

BARE CHIP: IMMERSION LENS: NON IMMERSION LENS:

$I_V' = (.702) I_V$ $I_V'' = (.850)(.960) m^2 I_V = (.816) m^2 I_V$ $I_V''' = (.702)(.960)^2 m^2 I_V = (.647) m^2 I_V$

Figure 5.3.2-1 Comparison of the Axial Luminous Intensity of Bare Chip, Immersion Lens, and Non-Immersion Lens Systems.

where $I_V(\theta)$ is the off axis luminous intensity, θ degrees off axis, and I_V is the axial luminous intensity.

Thus at an angle of $60°$ (total included angle of $120°$) the luminous intensity will be 50% of the axial luminous intensity. The maximum viewing angle for a display of this type is limited for a practical purposes to about $\pm75°$ because of the requirement that the display be recessed behind a filter. The radiation pattern of a magnified system may or may not be lambertian. However, in a magnified display, the viewing angle is generally limited by character distortion. The maximum viewing angle of a seven segment display can be defined as the off axis angle when the display is rotated that causes any of the segments to narrow or disappear. The maximum viewing angle can further be defined as the vertical viewing angle (segments a, d), the horizontal viewing angle (segments b, c, e, f), and the decimal point viewing angle. The maximum viewing angle is determined by magnification, type of lens, shape of lens, distance between the lens and the monolithic GaAsP chips and the spacing between characters. The calculation of maximum viewing angle is beyond the scope of this text. Figure 5.3.2-2 summarizes the relationships between magnification, luminous intensity, and viewing angle for magnified displays.

The earliest magnified monolithic displays used cast epoxy immersion lenses. The immersion lens served to increase the character height of the display, increase the axial luminous intensity and still provide a very acceptable viewing angle. Because of the relatively high cost of manufacturing this type of display, manufacturers began to experiment with other lower cost methods to manufacture a magnified monolithic display. The most commonly used technique that was developed was to die attach the monolithic GaAsP chips to some kind of substrate and then fasten an injection molded lens to the substrate. This technique was able to

reduce the manufacturing cost, improve the monolithic chip alignment, and improve the optical consistency between individual lenses in the display. The lens shape has evolved from spherical immersion lenses to plano-convex non immersion lenses, and then to positive meniscus non immersion lenses. The first non immersion lens consisted of several spherically shaped plano-convex "bubble" lenses that were molded into a linear array such that each monolithic digit was magnified by a separate "bubble" lens. Lens designers discovered that a positive meniscus lens with two spherically shaped surfaces allowed a much wider viewing angle than the simple plano-convex lens. The present state of the art is a positive meniscus lens with a spherical inner surface and an aspheric outer surface. This lens design can be optimized to allow almost as large a viewing angle as an immersion lens with the same magnification or to reduce off axis distortion of the monolithic character. An additional cylindrical magnifier can also be attached over the primary magnifying lens to further increase character height with the reduction of some viewing angle.

5.3.3 Construction of Monolithic Displays

Monolithic displays can be classified into two basic types according to whether the lens is an immersion or non immersion lens. Monolithic displays constructed with immersion lenses are manufactured by die attaching the monolithic GaAsP chips to either a ceramic substrate or to a lead frame. The die attach pad also forms the common cathode electrical connection to the monolithic chip. The aluminum contact for each segment on the monolithic chip is then wirebonded to the appropriate anode contact. A ball bond is formed on the LED top contact and a wedge bond is formed on the lead frame. The completed device is then encapsulated in epoxy. The magnifying lens is formed during encapsulation. Figure 5.3.3-1 shows the typical mechanical construction of a monolithic display constructed on a lead frame with a cast epoxy immersion lens.

Monolithic displays with non immersion lenses are usually constructed by epoxy die attaching the monolithic GaAsP chips to a special high temperature printed circuit board. The length of the printed circuit board depends on the desired number of digits and the desired digit spacing, since this type of display is generally not end stackable. Then the aluminum contact for each segment on the monolithic chip is wire bonded to the appropriate anode trace on the printed circuit board. A precision injection molded lens is then aligned and attached to the printed circuit board using holes that were drilled during fabrication of the printed circuit board. This insures accurate alignment of the lens over the LED chips. A secondary cylindrical lens can also be attached for added character height. Figure 5.3.3-2 shows the typical mechanical construction of a monolithic

EFFECTS OF MAGNIFICATION ON INTENSITY, VIEWING ANGLE, AND APPARENT SIZE

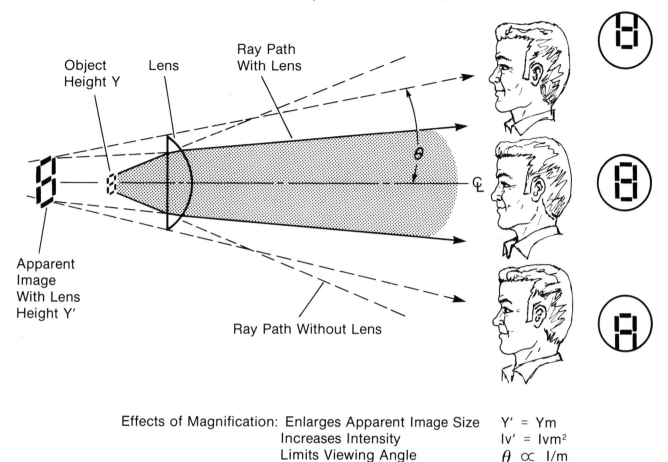

Effects of Magnification: Enlarges Apparent Image Size $\quad Y' = Ym$
Increases Intensity $\quad Iv' = Ivm^2$
Limits Viewing Angle $\quad \theta \propto I/m$

Figure 5.3.2-2 Effects of Magnification on Luminous Intensity Viewing Angle, and Apparent Character Size.

display manufactured with these techniques. Displays constructed with these techniques are commonly available in digit strings from five to fifteen digits. Table 5.3.3-1 compares the differences between these two packaging techniques with respect to manufacturing costs, ease of use by the customer, optics, and package reliability.

5.3.4 Electrical-Optical Characteristics

Because of their construction, seven segment monolithic displays are generally connected as an 8xN x-y addressable array, where N refers to the numer of digits in the display. Each of the eight anode lines addresses a different segment (a to dp) and each of the N cathode lines addresses a particular digit. This allows a five digit cluster to be constructed on a 14 pin lead frame and a fifteen digit monolithic display to be constructed with only 23 active pins. While substantially reducing the number of wires used to interconnect to the display, it also requires that the display be strobed on a 1 of N or lower duty factor.

The voltage and current characteristics of each monolithic LED are similar to the voltage and current characteristics of a single diode LED lamp. Each monolithic LED is effectively isolated from adjacent LED segments below the V_{BR} of the display. Because monolithic displays are constructed on a GaAsP substrate, they have a very low dynamic resistance in the forward region. This allows the monolithic display to be multiplexed at relatively high peak currents for higher luminous efficiency with forward voltages typically less than 1.8 volts.

Every monolithic display has upper and lower limits on the peak and average currents at which the device can be operated. The maximum peak forward current per segment has been chosen by the manufacturer to limit the maximum current density through the semiconductor junction. The average current per segment is limited by the ambient temperature and the thermal resistance of the display. In general, the junction temperature should be prevented from exceeding the maximum allowable storage temperature for

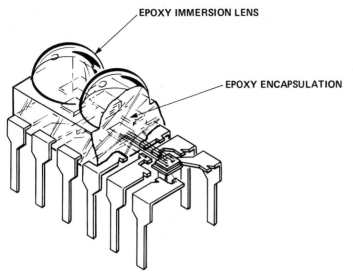

EPOXY IMMERSION LENS

EPOXY ENCAPSULATION

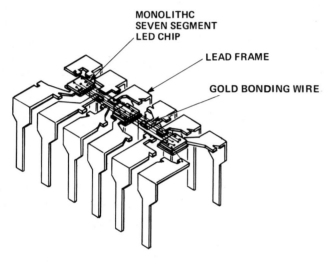

MONOLITHC
SEVEN SEGMENT
LED CHIP

LEAD FRAME

GOLD BONDING WIRE

Figure 5-3.3-1 Mechanical Construction of a Monolithic Display with an Immersion Lens.

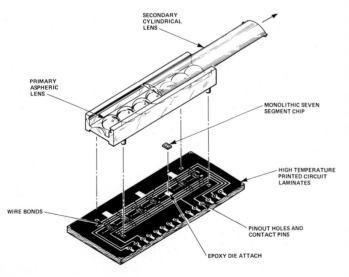

SECONDARY
CYLINDRICAL
LENS

PRIMARY
ASPHERIC
LENS

MONOLITHIC SEVEN
SEGMENT CHIP

HIGH TEMPERATURE
PRINTED CIRCUIT
LAMINATES

WIRE BONDS

PINOUT HOLES AND
CONTACT PINS

EPOXY DIE ATTACH

Figure 5-3.3-2 Mechanical Construction of a Monolithic Display Constructed on a PC Board With a Non-Immersion Lens.

the particular display. Typically monolithic displays with cast epoxy immersion lenses are limited to 100°C storage temperature and monolithic displays with non immersion lenses are limited to 85°C storage temperature. Above these temperatures, the reliability of the display may be impaired, or the lens may permanently be deformed. In most applications, monolithic displays can be driven at considerably lower average currents than this maximum average current. One exception is in high ambient applications where it is desirable to have a very high sterance display and a very dense filter to achieve an acceptable contrast ratio. The minimum peak current at which the display can be operated is determined by the junction area of each segment. Below a minimum current density, the luminous intensity matching between segments and digits may be unacceptable.

On most monolithic displays, the decimal point is located either in the center of the digit or in the lower right hand corner of the digit. For low cost applications, the right hand decimal point is recommended. The center decimal point requires a full digit location and time frame to be allocated to display the decimal. The center decimal point is an asset in quality instruments because it reduces the chances of an error caused by misreading the decimal point location. The center decimal point also allows a slightly wider viewing angle for the display since the monolithic chip can then be centered under the magnifying lens.

Most monolithic displays emit light with a peak wavelength of 655 nm. This wavelength represents the best compromise between the human eye response curve and the quantum efficiency of the GaAsP material.

Monolithic displays are usually characterized for luminous intensity under drive conditions that simulate typical usage conditions. A graph showing the relative luminous efficiency vs. peak forward current allows the designer to determine the time averaged luminous intensity for other peak and average forward currents. An example of a "relative luminous efficiency" curve is shown in Figure 5.3.4-1. The time averaged luminous intensity can be determined for any other drive condition by equation 5.3.4-1:

$$I_V \text{ TIME AVG} = \qquad (5.3.4\text{-}1)$$

$$\frac{[I_{PEAK}] \, [\text{DUTY CYCLE}] \, [\eta(I_{PEAK})] \, [I_{V \, SPEC}]}{[I_{SPEC}] \, [\text{DUTY CYCLE SPEC}] \, [\eta(I_{SPEC})]}$$

where I_{PEAK} is the desired peak current, DUTY CYCLE is the ratio of time the LED is "on" to total time, η is the relative luminous efficiency of the LED at I_{PEAK} or at I_{SPEC} and I_{SPEC} and DUTY CYCLE SPEC are the test conditions under which $I_{V \, SPEC}$ was originally characterized.

	Lead Frame Construction Cast Immersion Lens	Printed Circuit Construction Injection Molded Non-Immersion Lens
1) MANUFACTURING COST	Medium	Low
2) EASE OF USE		
a) Digit Number	2, 3, 4, 5 — end stackable (multiple clusters)	5, 8, 9, 12, 14, 15 — single module
b) Socketing	.3″ DIP	Special sockets available
c) Solderability	Wave, or Hand	Hand Solder
d) Cleaning	Vapor Degreasing	Hand Clean Only
e) Display Alignment	Clusters should be aligned	Pre-alignment included
f) Luminous Intensity	Clusters should be matched	Intensity Matching included
g) Repairability	Replace Cluster	Replace Display
3) OPTICS		
a) Viewing Angle	Good	Depends on Lens design — can be as good as Immersion Lens
b) Character Height	Varies slightly due to lens casting 2.79 mm (.11″)	Precision (may be increased by Optional Magnifier) 2.54 mm (.10″) to 4.45 mm (.175″)
4) RELIABILITY		
a) Temp Storage	-40°C to $+100^\circ$C	-20°C to $+85^\circ$C
b) Temp Cycling	Good	Very Good
c) Moisture Resistance	Good	Condensing atmosphere should be avoided
d) Mechanical Vibration	Very Good	Reasonably Good

Table 5.3.3-1 Comparison Between Monolithic Displays Constructed with Cast Immersion Lenses and Monolithic Displays with Non-Immersion Lenses.

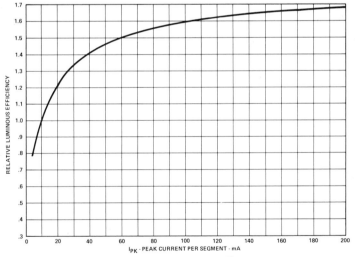

Figure 5.3.4-1 Relative Luminous Efficiency vs Peak Current Per Segment for Hewlett-Packard 5082-7265, 7275, 7285, 7295 Monolithic Displays.

For example, suppose a monolithic device is tested for luminous intensity at 10 mA peak per segment, 20% duty cycle and the typical time averaged luminous intensity is 70 μcd/segment. If the same display is strobed at 80 mA peak per segment on a 2.5% duty cycle, the typical time averaged luminous intensity will be:

$$I_V = \frac{(80)(.025)(1.55)(70)}{(10)(.20)(1.00)} \qquad (5.3.4\text{-}2)$$

$$= 108 \mu cd/segment$$

Another representation of the "relative luminous efficiency" curve is shown in Figure 5.3.4-2. In this figure, equation 5.3.4-1 and Figure 5.3.4-1 have been used to calculate typical time averaged luminous intensities for a wide variety of drive conditions. This figure graphically shows the advantage of strobing at higher peak currents to obtain a higher time averaged luminous intensity for the same average current. For example suppose a five digit display is strobed on a 20% duty cycle at 1.5 mA average current per segment (7.5 mA peak/segment), then the typical time averaged luminous intensity would be 50 μcd/segment. This represents a very acceptable luminous intensity for this type of display under ambient conditions similar to an office. If the display is strobed at higher peak currents, the average current per segment can be reduced to get the same time averaged luminous intensity or the time averaged luminous intensity can be increased. At 200 mA peak current per segment, the average current can be reduced to .82 mA per segment and still obtain the same 50 μcd/segment; or by maintaining the same average current, the time averaged luminous intensity can be increased to 90 μcd/segment.

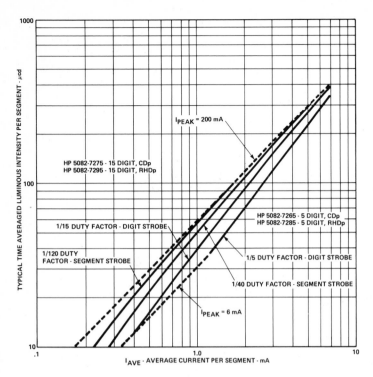

Figure 5-3.4-2 Typical Time Averaged Luminous Intensity Per Segment vs Average Current Per Segment for Hewlett-Packard 5082-7265, 7275, 7285, 7295 Monolithic Displays.

5.3.5 Driving Monolithic Displays

Monolithic displays are driven in the same way as any other common cathode strobed display. The advantage of monolithic seven segment displays is that they can be driven at considerably lower average currents than large seven segment displays for a satisfactory luminous intensity. For example, a monolithic display designed for a low cost calculator can be satisfactorily driven at average currents as low as .25 mA per segment and a somewhat larger monolithic display such as one designed for a portable instrument can be driven at average currents as low as 1 mA per segment. By comparison, a high efficiency red large seven segment display probably requires 4 to 6 mA average current per segment for a comparable luminous intensity. Since the monolithic LED display can be driven at such low average currents, it is particularly suitable for battery powered applications. Furthermore, a monolithic LED display can often be driven directly from many readily available integrated circuit devices.

Monolithic displays are commonly driven either with digit or segment strobe techniques. When a display is digit strobed, each digit is sequentially enabled and at the same time the appropriate combination of segments is turned on. For an N digit display, the maximum duty factor for a digit strobed display is 1/N. If a display is segment strobed, then each digit is sequentially enabled and while that digit is enabled, each segment is sequentially turned on. The maximum duty factor for a segment strobed system is

1/(8N). Thus for the same time averaged luminous intensity, the segment strobed display must drive each LED segment at a higher peak current where the LED has a higher quantum efficiency. This allows the display to be driven at a lower average current for the same time averaged luminous intensity. While more logic is required for a segment strobed system, segment strobing may also be preferable in custom MOS designs because the peak digit driver currents can be reduced. For example, suppose a five digit monolithic display can be described by the "relative efficiency" curve shown in Figure 5.3.4-1. The typical time averaged luminous intensity for this display is shown in Figure 5.3.4-2. When this display is digit strobed at 7.5 mA peak current per segment on a 1/5 duty factor, the typical time averaged luminous intensity is 50 μcd per segment. Each segment driver must be able to source 7.5 mA and each digit driver must be able to sink 60 mA from the display. When the display is segment strobed on a 1/40 duty factor, the same time averaged luminous intensity can be achieved at 39 mA peak current per segment. Thus each segment driver would be required to source 39 mA and each digit driver would be required to sink 39 mA from the display, a 35% reduction in average current per segment.

The primary difference between driving a monolithic display with a center decimal point and driving a monolithic display with a right hand decimal point is that the segments surrounding the center decimal point should be blanked when the decimal point is turned on. With a right hand decimal point, each digit is displayed regardless of the decimal location. Figure 5.3.5-1 shows a typical CMOS logic interface to a five digit monolithic display. The outputs of the National 74C90 counter sequentially enable each of the five digit cathodes and at the same time enable circuitry inside the digital subsystem to output the desired segment and decimal information to the display. If a display with a center decimal point is used, the proper digit can be blanked either by outputting a BCD code that is decoded as a blank state by the seven segment decoder (state 15 for the National 74C48) or using the blanking input of the seven segment decoder. A fixed decimal point can be implemented simply by connecting the Dp input to the desired cathode driver. Then whenever the selected cathode driver is on, the decimal point will be turned on. Several LSI integrated circuits are available that combine all the circuitry shown in Figure 5.3.5-1 into one or a few integrated circuit packages.

5.3.6 Interfacing Microprocessors to Monolithic Seven Segment Displays

Microprocessors can be interfaced to monolithic displays with only a few external components. In general, these external components will include one or more latches to hold the information to be displayed and additional circuitry to periodically request new information from the display. For displays of up to eight digits, either of the

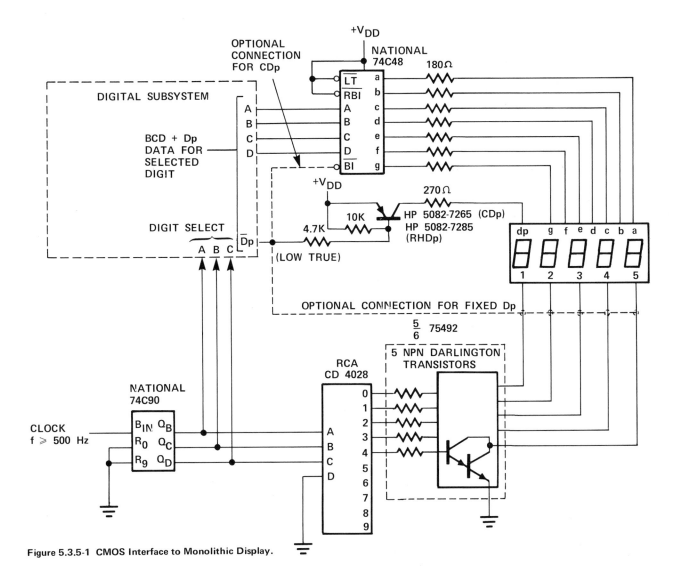

Figure 5.3.5-1 CMOS Interface to Monolithic Display.

circuits described in Figures 5.2.6-2 or 5.2.6-3 can be used. Monolithic displays of up to 16 digits can be driven from the circuit shown in Figure 5.3.6-1.

This circuit differs from any of the circuits shown in Section 5.2.6 in the way the decimal point and interrupts are handled. The microprocessor outputs only a single byte of information to the display. Four bits contain the segment and decimal point information and four bits select the proper digit enable. The upper six states (10 to 15) of the RCA CD4511 seven segment decoder are decoded as blank states. Four of these states are decoded externally to select the decimal point or minus sign. In this example, states 10 and 11 are decoded as a minus sign, states 12 and 13 are decoded as the decimal point select, and states 14 and 15 are decoded as blank states. Every 667µs, the refresh clock requests an interrupt from the Motorola 6800 microprocessor. Upon requesting an interrupt, the interrupt circuitry (not shown) must force a vectored interrupt to address "REFRESH". The interrupt request can also be decoded by a software polling program but will increase the microprocessor time that is required to service the display.

Then the refresh program shown in Figure 5.3.6-1b is executed. The program reads a pointer from RAM and then outputs the contents of the memory location specified by the pointer to the display. Then the pointer is compared to the address of the final byte of information in the data file. If they are equal, the pointer is reset to the address of the first byte of information in the data file, otherwise, the pointer is set to the address of the next byte of information in the file. Since the decimal point is decoded separately from the other seven segments, a separate time frame is required to display the decimal point. Thus, a 15 digit monolithic display with a right hand decimal point will require 16 bytes of information in the data file. The microprocessor time required to refresh any display of up to 16 digits can be determined as shown below:

(5.3.6-1)

$$\text{REFRESH TIME} = \frac{(52N + 1)R}{\text{MICROPROCESSOR OR CLOCK RATE}}$$

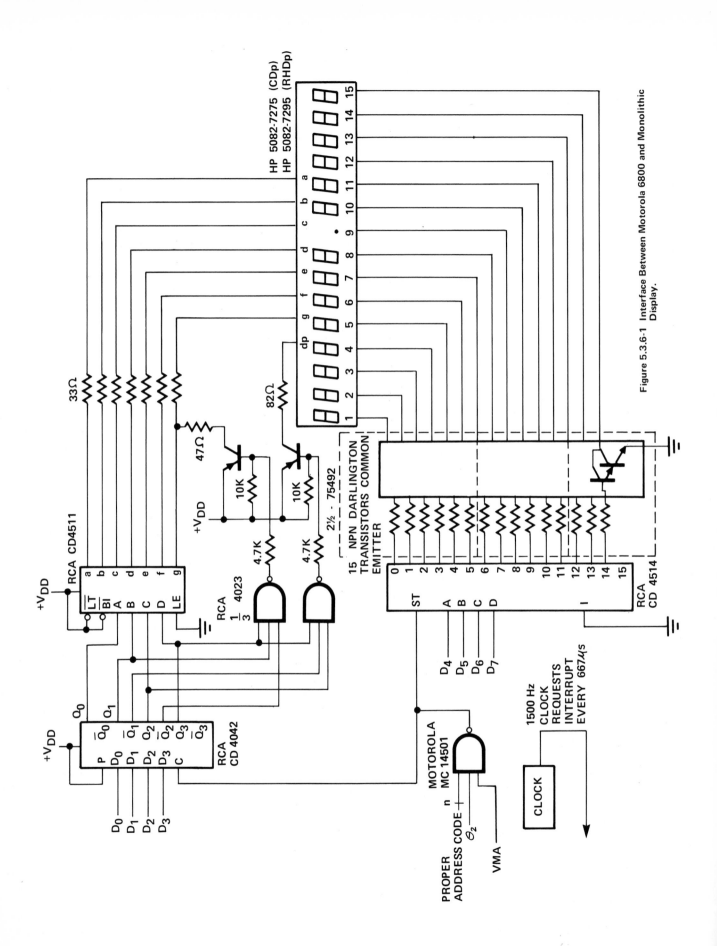

Figure 5.3.6-1 Interface Between Motorola 6800 and Monolithic Display.

5.42

ADDRESS	OP CODE	CLOCK CYCLES	COMMENTS
REFRESH	LDX	5	IX = POINTER
	A_H	----	
	A_L	----	
	L DAA, X	5	AA = (IX)
	0	----	
	STA A	5	STORES NEW SEGMENT AND DIGIT
	$DISP_H$	----	INFORMATION
	$DISP_L$	----	
	CPX #	3	IX − $(XXIE)_{16}$
	$(XX)_{16}$	----	
	$(1E)_{16}$	----	
	BNE	4	JUMP IF IX ≠ $(XXIE)_{16}$
	$(07)_{16}$	----	
	LDAA #	2	AA = $(10)_{16}$
	$(10)_{16}$	----	
	STA A	5	POINTER = $(XX10)_{16}$
	A_H	----	
	A_L + 1	----	
	CLI	2	CLEAR INT FLAG
	RTI	10	RETURN
	INC	6	POINTER = POINTER + 1
	A_H	----	
	A_L + 1	----	
	CLI	2	CLEAR INT FLAG
	RTI	10	RETURN

MEMORY ADDRESS	CONTENTS		BCD DATA TABLE	
A_H A_L	$POINTER_H$		BCD	OUTPUT
A_H A_L + 1	$POINTER_L$			
$(XX10)_{16}$	0000		0000	0
$(XX11)_{16}$	0001		0001	1
$(XX12)_{16}$	0010		0010	2
$(XX13)_{16}$	0011		0011	3
$(XX14)_{16}$	0100		0100	4
$(XX15)_{16}$	0101	BCD DATA	0101	5
$(XX16)_{16}$	0110		0110	6
$(XX17)_{16}$	0111		0111	7
$(XX18)_{16}$	1000		1000	8
$(XX19)_{16}$	1001		1001	9
$(XX1A)_{16}$	1010		1010	−
$(XX1B)_{16}$	1011		1011	−
$(XX1C)_{16}$	1100		1100	.
$(XX1D)_{16}$	1101		1101	.
$(XX1E)_{16}$	1110		1110	
			1111	

X= DON'T CARE

POINTER
(NEXT DIGIT TO BE DISPLAYED)

Figure 5.3.6-1b Refresh Program Used to Strobe the Monolithic Display Shown in Figure 5.3.6-1.

where N is the number of digits on a CDp display or one plus the number of digits on a RHDp display, and R is the strobing rate.

For example, suppose a fifteen digit RHDp display is refreshed at a 100 Hz refresh rate and the microprocessor uses a 1 MHz clock, then 8.3% of the microprocessor's time is required to service the display. For the remaining 91.7% of the time, the microprocessor can be performing other functions. While this example used a Motorola 6800 microprocessor, this technique can be used with any other microprocessor by modifying the hardware and software to reflect a different instruction set and different ways to output information to peripheral devices.

5.4 Alphanumeric Displays

Seven segment display devices are capable of transmitting only a limited number of useful different states of information. Theoretically, up to 49 different information states could be depicted utilizing the 7 segment font, however, less than 1/3 of these have any meaning in normally utilized communications context. By increasing the number of elements in a display character from 7 to 9, it becomes possible to depict in rough form most of the Roman alphabet and arabic numerals. Utilization of a 14 or 16 segment character allows better symbol differentiation and use of some special symbols. However, the lower case alphabet cannot be represented, and asthetic quality and readability are still poor. By increasing the number of elements still further, it is possible to add further refinement to the display quality. By utilizing an infinite number of elements, virtually any 2 dimensional character and character shape could be achieved. In practice, it is found that 35 elements in a 5x7 array or 63 elements in a 7x9 array offer satisfactory resolution for most alphabetic language communications. The trade-off is one of the cost of supplying and addressing the individual elements vs. the improvement gained from higher resolution. For LED alphanumeric dot matrix displays, the industry standard product is the 35 element 5x7 array.

5.4.1 The 5x7 LED Array

Figure 5.4.1-1 depicts schematically one digit of a 5x7 array. The row lines are tied to the anodes of all of the diodes of a particular row and are designated by the Roman numerials I-VII. The column lines tie together the cathodes of each diode of a particular column and are designated A-E. In multidigit arrays, the columns of digit one are designated 1A-1E, digit two 2A-2E, etc. The row lines are extended in common to all digits in an array. The effect of this type of organization is to form a 7x5N crosspoint matrix (N = number of characters) of diodes where an individual diode is energized by selecting the appropriate row and column lines. Using this technique, a 4 digit array of 140 diodes may be addressed using only 27 lines (7 row lines + 4x5 digit lines). If decoding or data storage circuits are included inside the display package, the external pin count can be even further reduced.

5.4.2 Character Generation in 5x7 Arrays

The generation of character information in a 5x7 array is somewhat more complex than the techniques utilized for 7 segment displays. In the 7 segment display, all elements of a digit can be simultaneously addressed so that the entire character is formed during one address cycle. In a 5x7 array, each character must be made up of 5 or 7 subsets of data which are presented during sequential address cycles.

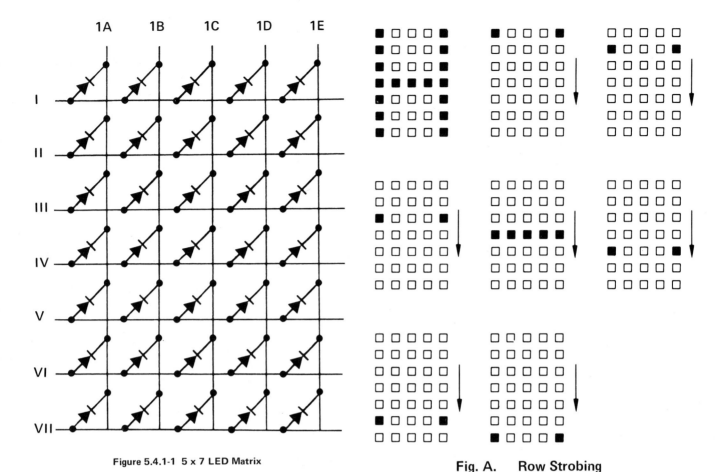

Figure 5.4.1-1 5 x 7 LED Matrix

Fig. A.   Row Strobing

By the nature of its electronic organization, then, a 5x7 display character must be operated on a strobed basis. Operating the display on this basis, though it increases the system complexity somewhat, does offer advantages which clearly outweigh the negative aspects, even for a single character display.

- A minimum number of interconnects are used to interface between the data source and the display.

- The cost of the character generation and associated drive circuitry is shared over many digits.

- Clock, timing and data storage elements required for the strobed display can be shared with other portions of the system.

There are two common methods for addressing the diodes in the 5x7 array in order to generate character information. Both of these methods depend on the eye's response to a strobed display. Figures 5.4.2-1a,b depict the sequence of steps necessary to form the character "H" using either row or column strobing techniques.

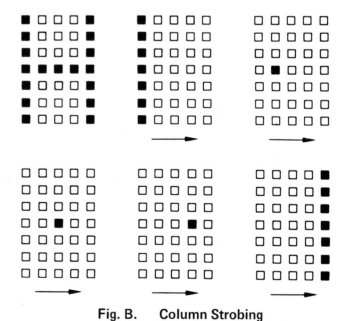

Fig. B.   Column Strobing

Figure 5.4.2-1a,b Character Generation Using Row (a) or Column (b) Strobe Methods.

In row strobing, data for one row of the desired character is applied to the column lines and the row is then energized. This process is repeated for each of seven rows. For row strobing of a multicharacter string, row data for each character in the string is decoded and stored in a data latch associated with each character. The row line is then enabled, displaying simultaneously one row of data for all of the characters in the string.

In column strobing, data for one column of the desired character is supplied to the row lines and the column is then energized. The process is repeated, varying the data, until all of the columns of the character have been selected. Multicharacter displays are treated as simply an extension of the single character case such that the number of subsets of data generated will be equal to 5N, where N is equal to the number of digits in the display.

If either of these sequences is repeated at a rate which insures that each of the appropriate matrix locations is re-energized a minimum of 100 times per second the eye will perceive a continuous image of the entire character. The apparent intensity of each of the display elements will be equal to the intensity of that element during the "ON" period multiplied by the ratio of "ON" time to REFRESH PERIOD. This ratio is referred to as the display DUTY FACTOR.

From a conceptual standpoint, column scanning is the easiest method to understand; however, it is more limited as to the number of digits which may be displayed using only one decoder. The block diagram in Figure 5.4.2-2 depicts the elements of a column strobing system. Circuit operation is as follows:

Coded character information from the keyboard is first stored in the input storage buffers, generally six bits per character. The timing circuitry then selects the first character word, from storage buffer number one, presenting this data to the seven line output ROM. The timing circuitry also inputs the column one select code to the ROM causing the ROM output states in turn to present column 1, character 1 data at its output terminals. These states turn on the row drivers for the appropriate diodes in the first column of the display. The first column is then enabled by its column driver for an appropriate "ON" period. The second column data is then selected and displayed, etc., until all five columns of the first character have been displayed.

The timing circuit then selects the data for the second character from buffer number two, and displays each of the columns. This procedure is repeated until all columns of all characters have been displayed. If each column is refreshed at 100 times per second, the eye will not perceive flicker in

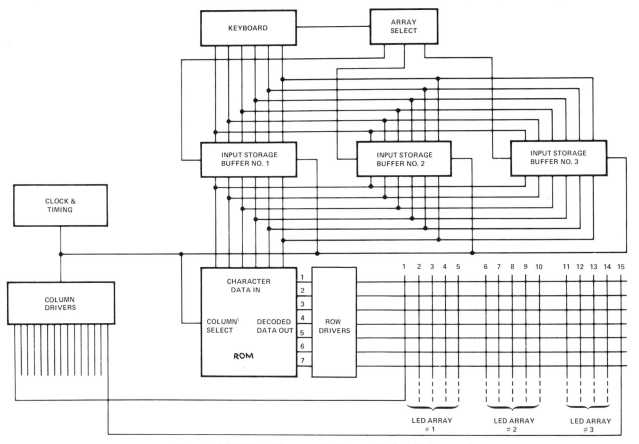

Figure 5.4.2-2 Block Diagram of a Column Strobed Alphanumeric Display.

5.45

the display. Herein lies the problem with column strobing. Since the maximum peak current per diode is limited to about 100 mA and the desired minimum average current per diode is around 1 mA, the maximum number of columns which could be displayed is 100 columns or 20 characters. Given the practicalities of timing and switching, this number is more realistically 16 characters.

The row strobing scheme eliminates the problem of high peak currents and consequent severe limitations on the number of digits which can be strobed in one display string utilizing only one decoder. However, there is a substantial increase in circuit complexity. The block diagram for a row strobing scheme is shown in Figure 5.4.2-3.

Coded alphanumeric information is sequentially entered and stored in the input storage buffers. Next, the timing circuitry enables the ROM and selects data word No. 1. An additional input code selects row I output data from the ROM. The first row of character information is stored in

the five bit output storage buffer No. 1. Each of the remaining words are selected, decoded into first row character information and loaded into the five bit output storage registers corresponding to each of the characters of the display. At this point in the sequence, each of the five output storage buffers has been loaded with the appropriate display information for the first row of each character. The timing circuitry now turns on the top row driver so current flows through and lights all of the appropriate LEDs in the top row of all of the characters. The complete cycle is repeated to decode and display row II, then row III, and so on through the seventh row. Repeating this scan at a rate above 100 Hz in most applications gives a flicker free alphanumeric character at each of the digit locations. The only practical limitation to the number of characters which can be operated from a single character generator in this mode is imposed by the maximum clock rate of the logic. In general, character string lengths well in excess of 100 digits should not pose a problem.

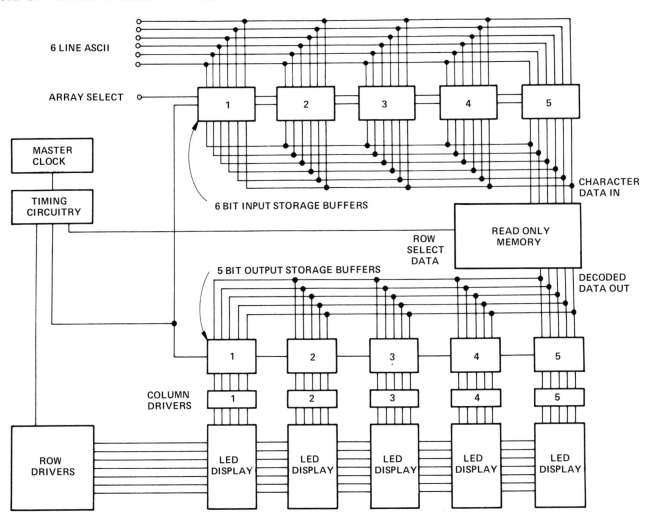

Figure 5.4.2-3 Block Diagram of a Row Strobed Alphanumeric Display.

5.4.3 Implementation of a 16 Character Row Scan Display

The circuit Figure 5.4.3-1 depicts a 16 character alphanumeric display utilizing the row strobing technique. One 7496 five bit shift register is used as the output data storage buffer for each character of the display. Data from the ROM is loaded into a parallel to serial shift register and then shifted into the data storage buffers on each positive transition of the clock. After five clock pulses a new input data word is selected from the ASCII data storage RAM (1½ Fairchild 93403). After all sixteen words (one word per character) have been decoded and shifted into the output data storage buffers, the shift register clock input is gated off and the row line is enabled to light the display. At the end of the display period, the row is turned off, the row counter is incremented by one count, and the second row of data is loaded into the output storage buffer. The row scanner is then enabled while the shift register clock is again gated off. This process is repeated for all rows of the display.

The high true output from the parallel to serial shift register is inverted before loading into the output data buffers. The shift register outputs will then be low for a dot "ON" condition. The output low state of the 7496 is rated to sink 16 mA and can, therefore, be used to directly drive the display columns at 16 mA peak. A 500 KHz input clock rate will give a refresh rate of about 93 Hz. The DUTY FACTOR expressed in Equation 5.4.3-1 is set by the ratio of the loading time to the display time multiplied by $1/m$ where m is equal to the number of rows to be addressed; seven in this case:

$$D.F. = (1 - \frac{2^4}{2^7})\ \frac{1}{m} = 12.5\% \qquad (5.4.3\text{-}1)$$

Average diode current is 2.0 mA.

5.4.4 Alphanumeric Displays with On Board Data Storage

The circuit in Figure 5.4.3-1 utilizes a shift register to store decoded row data during the display cycle. If the shift register and current limiting elements are included along with the display matrix in a single hybrid package, significant improvements are realized in the total cost of display implementation, packing density, and reliability. The HP HDSP-2000 display is designed to provide on-board storage of decoded column data plus constant current sinking row drivers for each of the 28 rows in the 4 character display. This approach allows the user to address each display package through just 11 active interconnections vs. the 176 interconnections and 36 components required to effect a similar function using conventional LED matrices.

Figure 5.4.4-1 is a block diagram of the internal circuitry of the display. The device consists of four LED matrices and two 14-bit serial-in-parallel-out shift registers. The LED matrix for each character is a 5x7 diode array organized with the anodes of each column tied in common and the cathodes of each row tied in common. The row cathode commons of each character are tied to the constant current sinking outputs of 7 successive stages of the shift register. The like columns of each of the 4 characters are tied together and brought to a single address pin (i.e., column 1 of all 4 characters is tied to pin 1, etc.). In this way, any diode in the four 5x7 matrices may be addressed by shifting data to the appropriate shift register location and applying a voltage to the appropriate column.

The two on-board shift registers act as a single serial-in-parallel-out (SIPO) register. The SIPO shift register has a constant current sinking output associated with each shift register stage. The output stage is a current mirror design with a nominal current gain of 10. The magnitude of the current to the reference diode is established from the output voltage of the V_B input buffer applied across the current reference resistors, R. The reference current flow is switched by a transistor tied to the output of the associated shift register stage. A logical 1 loaded into the shift register will turn the current source "ON" thereby sinking current from the row line. A voltage applied to the appropriate column input will then turn "ON" the desired diode.

Data is loaded serially into the shift register on the high to low transition of the clock line. The data output terminal is a TTL buffer interface to the 28th bit of the shift register (i.e., the 7th row of character 4 in each package). The Data Output is arranged to directly interconnect to the Data Input on a succeeding 4 digit display package. The Data, Clock and V_B inputs are all buffered to allow direct interface to any TTL or DTL logic family. The display is organized so that column strobing is utilized in the same manner as in a row strobed system; (i.e., the data for all of the like columns in the display string is loaded into shift register and then the column is energized to generate character information).

For a four character display, 28 bits representing the first column of each of the four characters are loaded serially into the on-board SIPO shift register and the first column is then energized for a period of time, T. This process is then repeated for columns 2 through 5.

If the time required to load the 28 bits into the SIPO shift register is t, then the duty factor is:

$$D.F. = \frac{T}{5(t + T)} \qquad (5.4.4\text{-}1)$$

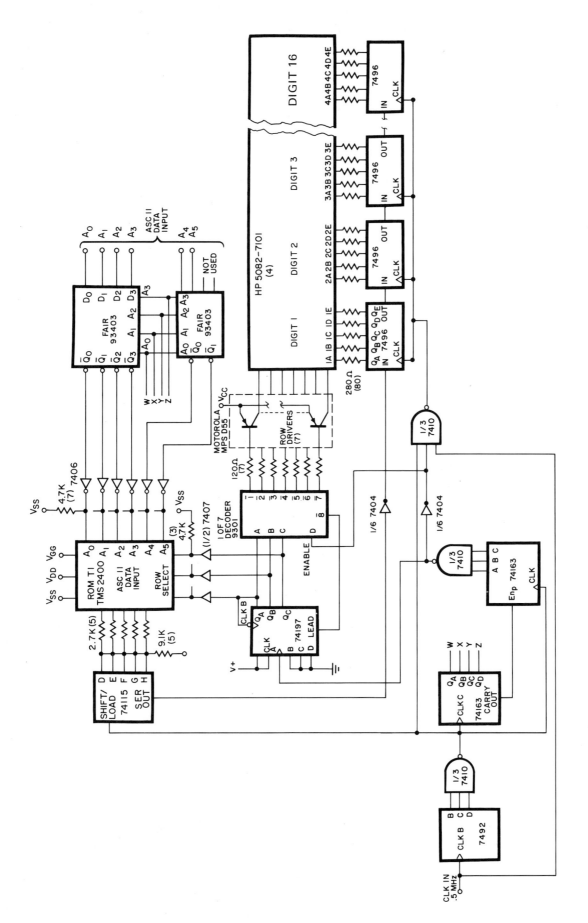

Figure 5.4.3-1   Practical Implementation of a Row Strobed Display.

5.48

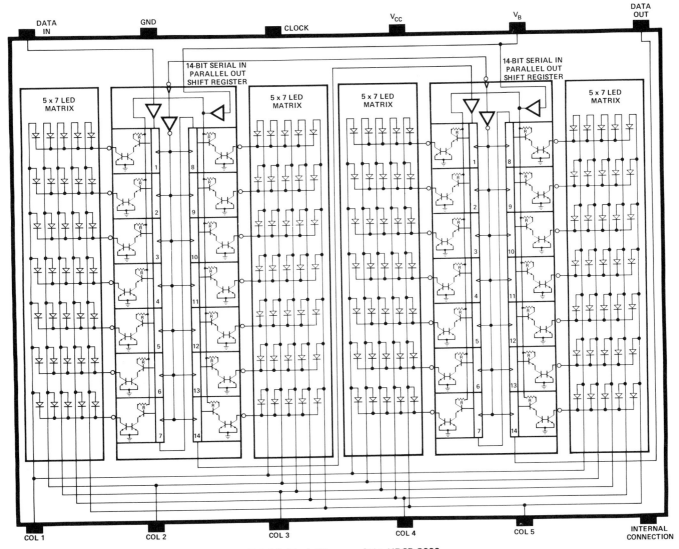

Figure 5.4.4-1 Block Diagram of the HDSP-2000.

the term $5(t+T)$ is then the refresh period. For a satisfactory display, the refresh period should be:

$$1/[5(t+T)] \geqslant 100 \text{ Hz} \qquad (5.4.4\text{-}2)$$

or conversely

$$5(t+T) \leqslant 10 \text{ ms} \qquad (5.4.4\text{-}3)$$

which gives

$$(t+T) \leqslant 2 \text{ ms} \qquad (5.4.4\text{-}4)$$

Two milliseconds then is the maximum time period which should be allowed for loading and display of each column location. For $t \ll T$, the duty factor will approach 20%. The number of digits which can be addressed in a single string is then dependent upon the minimum acceptable duty factor and the choice of clock rate. For instance, at 1 MHz clock rate, a 100 character string of 25 packages could be operated at a duty factor of

$$(5.4.4\text{-}5)$$

$$\text{D.F.} = \frac{(T+t) - (\# \text{ of bits to be loaded}) \cdot (1/1 \text{ MHz})}{5\,(t+T)}$$

$$= \frac{(2 \text{ ms}) - 700\,(1\ \mu\text{s})}{5 \times 2 \text{ ms}}$$

$$= 13\% \qquad (5.4.4\text{-}6)$$

For most applications, a duty factor of 10% or greater will provide more than satisfactory display intensity. In brightly illuminated ambient environments, a larger duty factor may be desirable whereas, in dim ambient situations, the duty factor may have to be reduced in order to provide a display with satisfactory contrast.

5.4.4.1 Drive Circuit Concept

A practical display system utilizing the HP HDSP-2000 display requires interfacing with a character generator and refresh memory. A block diagram of such a display system is depicted in Figure 5.4.4.1-1. In explanation, assume that this system is for a four character display. Therefore, the 1/N counter becomes a 1/4 counter where N is equal to the number of characters in the string.

The refresh memory is utilized to store the information to be displayed. Information can be coded in any one of several different standard data codes, such as ASCII or EBDIC, or the code and the display font can be customized through the use a custom coded ROM. The only requirement is the output data be generated as 5 subsets of 7 bits each. The character generator receives data from the refresh memory and outputs 7 display data bits corresponding to the character and the column select data input. This data is converted to serial format in the parallel to serial shift register for clocking into the display shift register. In the typical system, the right most character to be displayed is selected first and the data corresponding to the ON and OFF display elements in the first column is clocked into the first 7 shift register locations. In a similar manner, column 1 data for characters 3, 2, and 1 is selected by the 1/N counter, decoded and shifted into the display shift register. After 28 clock counts, data for each character is located in the shift register locations which are associated with the 7 rows of the appropriate LED matrix. The 1/N counter overflows, triggering the display time counter,

enabling the output of the 1/5 column select decoder and disabling the clock input to the display. The information now present in the shift registers will be displayed for a period, T, at the column 1 location. At the end of the display period, T, the divide by 5 counter which provides column select data for both the display and the character generator is incremented one count and column 2 data is then loaded and displayed in the same manner as column 1. This process is repeated for each of 5 columns which comprise the 5 subsets of data necessary to display the desired characters. After the fifth count, the 1/5 decoder automatically resets to "one" and the sequence is repeated. The only changes required to extend this interface to character strings of more than 4 digits are to increase the size of the refresh memory and to change the divide by four counter to a modulus equal to the number of digits in the desired string.

Since data is loaded for all of the like columns in the display string and these columns are then enabled simultaneously, only five column switch transistors are required regardless of the number of characters in the string. The column switch transistors should be selected to handle approximately 110mA per character in the display string. The collector emitter saturation voltage characteristics and column voltage supply should be chosen to provide a $3.0V \leqslant V_{COL} \leqslant V_{CC}$. To save on power supply costs and improve efficiency, this supply may be a fullwave rectified unregulated dc voltage as long as the PEAK value does not exceed the value of V_{CC} and the minimum value does not drop below 3.0 volts.

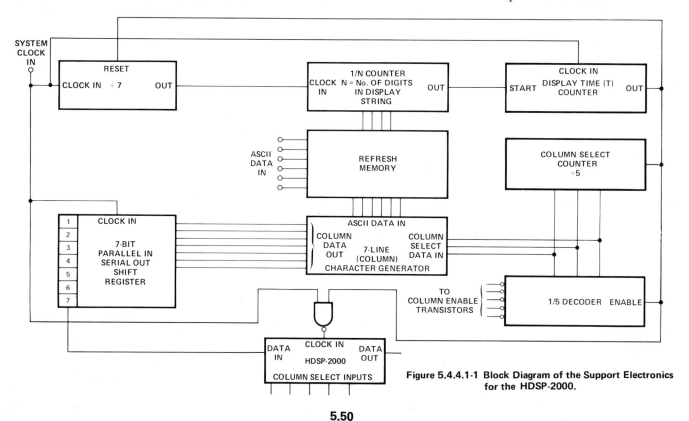

Figure 5.4.4.1-1 Block Diagram of the Support Electronics for the HDSP-2000.

5.50

Since large current transients can occur if a column line is enabled during data shifting operations, the most satisfactory operation will be achieved if the column current is switched off before clocking begins. I_{CC} will be reduced by about 10-15% if the clock is held in the logical 1 state during the display period, T.

5.4.4.2 Interface Circuits for HP HDSP-2000

There are many possible practical techniques for interfacing to the HP HDSP-2000 alphanumeric display. Two basic approaches will be treated here.

5.4.4.2.1 Instrumentation Interface Circuit

The circuit shown in Figure 5.4.4.2.1-1 is for a 16 character display and is designed to function primarily as a readout for general instrumentation systems. CMOS logic circuitry is utilized in this design; however, it should be a simple exercise to substitute TTL functions if CMOS is not desired. In this circuit, a CD 4022 and CD 4520 are combined to perform the functions of the divide by 7, divide by 16 (1/N) and display time counters as depicted in Figure 5.4.4.1-1. The timing diagram, Figure 5.4.4.2.1-2, demonstrates the relationship of the various critical outputs and inputs. The CD4022 actually acts here as a divide by 8 counter with the first count used to latch data into the parallel-in-serial-out (PISO) shift register and the other 7 counts shifting data out of the PISO and into the display shift register. The CD4520 is a dual 4 bit counter wired as an 8 bit binary ripple counter. The NAND gate, U_1, establishes the ratio of loading time to display time. In this case, loading will occur once in every 8×2^7 clock counts for a period of 8×2^4 clock counts. Duty factor is then from Equation 5.4.4-1

$$D.F. = \frac{(8 \times 2^7) - (8 \times 2^4)}{5\,(8 \times 2^7)} = 17.5\% \qquad (5.4.4.2.1\text{-}1)$$

and the refresh period is

$$5\,(8 \times 2^7)\,t, \qquad (5.4.4.2\text{-}1\text{-}2)$$

where t = clock period.

The four least significant bits of the CD4520 counter are used to continually address the CD4036 refresh memory. Data can be written into the desired memory address by strobing the WRITE ENABLE line when the appropriate memory address appears on the WRITE ADDRESS lines. This function can occur simultaneously with a read from memory.

Two counters, a CD4029 and a CD4022, are used for the column data generator and the column select decoder, respectively. Note that the Signetics 2516 character

generator requires column select inputs of binary coded 1 to 5 instead of 0 to 4. For this reason, the CD4029 is preset to a binary 1 by the same pulse which is used to reset the CD4022 column select decoder. To minimize I_{CC}, the V_B terminal is held low during data load operations, turning "OFF" the current mirror reference current. The column current switch is a PNP darlington transistor driven from a buffered NAND gate. The 1N4720 serves to reduce the column voltage by approximately 1 volt, thereby reducing on board power dissipation in the display devices. Due to maximum clock rate limitations of the CMOS logic, clock input should not exceed 1 MHz.

5.4.4.2.2 Microprocessor Interface for the HDSP-2000

The HDSP-2000 alphanumeric display is ideally suited for interface to a microprocessor. There are several different ways in which the hardware/software partitioning can be arranged in such a system. The choice of the technique will depend on how much of the microprocessor time the designer wishes to devote to supporting the display. The additional microprocessor software time required is traded off against additional hardware costs. Figures 5.4.4.2.2-1a,b,c illustrate three different partitioning techniques which can be utilized in interfacing the HP HDSP-2000 to a microprocessor. In 5.4.4.2.2-1a, the display and its interface hardware appear as an autonomous peripheral to the microprocessor. This system accepts data into a local RAM whenever the microprocessor updates the information to be displayed. A good example of this type of system is illustrated in Figure 5.4.4.2.1-1. This system, however, duplicates some of the functions available in the microprocessor system.

Figure 5.4.4.2.2-1b illustrates a technique in which the coded data RAM and character generator ROM have been removed from the display interface and included as a fraction of the microprocessor hardware. In this approach, the entire message is decoded and the data is sent to the display where it is stored and accessed by the display scanning logic. This approach is ideal where one microprocessor system may be utilized to service several displays.

The display interface may be even further simplified by developing a system in which the microprocessor supplies the display refresh information periodically in response to an interrupt request from the display. The HP HDSP-2000 due to the onboard storage of decoded data, is ideally suited to the implementation of this third type of interface. Depending on the number of digits in the display string, and the microprocessor clock rate, the display can be updated at a time burden to the microprocessor of less than 1% per digit. Figure 5.4.4.2.2-2 illustrates the practical implementation of the technique presented in Figure

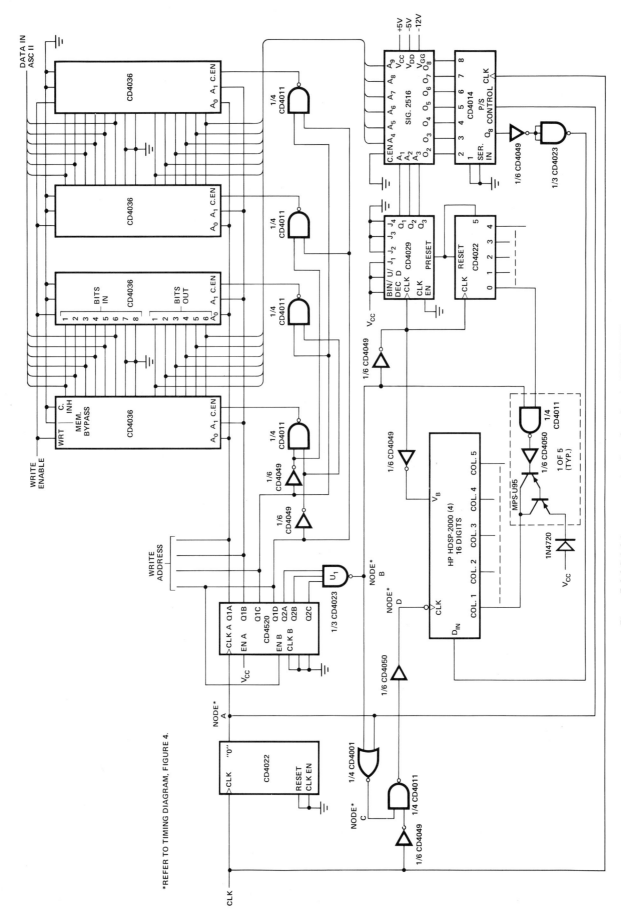

Figure 5.4.4.2.1-1 Instrumentation Interface for the HDSP-2000.

*REFER TO TIMING DIAGRAM, FIGURE 4.

5.4.4.2.2-1c, interfacing a 16 digit string of HP HDSP-2000 displays to an 8080A microprocessor. In this circuit, an astable timing element generates an interrupt request once every 2 msec. The interrupt subroutine outputs 16 bytes of data to the display via a parallel to serial converter. A local clock shifts this data into the display between output cycles of the microprocessor. Only 7 bits of each output word are shifted to the display. A column select word is sent to a separate address. Prior to data loading, this word is used to turn all column enable transistors off. After all of the data has been sent, this word gates the appropriate column data "ON". Decoded data for the display is stored in a memory stack in the microprocessor RAM. A memory pointer in the programmed subroutine is utilized to indicate the appropriate block of data to be sent to the display.

For a display refresh rate of 100 Hz (interrupt every 2 msec), the microprocessor time required to service the display as a percent of total time is expressed in the following equation:

Display strings of other lengths can be accomodated by changing the number of words which are written to the display from the RAM. This involves a simple one instruction software change.

In this situation, the microprocessor load time can be calculated from Equation 5.4.4.2.2-1 using

$$t_{LOAD} = (166 + 22N)\ \frac{1}{f} \qquad (5.4.4.2.2-2)$$

where:

  f = Microprocessor clock frequency

  N = Number of digits in the display string

$$\%\ t_{SERVICE} = \frac{t_{LOAD}}{t_{INTERRUPT}} \times 100\% \qquad (5.4.4.2.2-1)$$

$$= \frac{518 \times .5\ \mu s}{2\ ms} \times 100\%$$

$$= 12.55\%$$

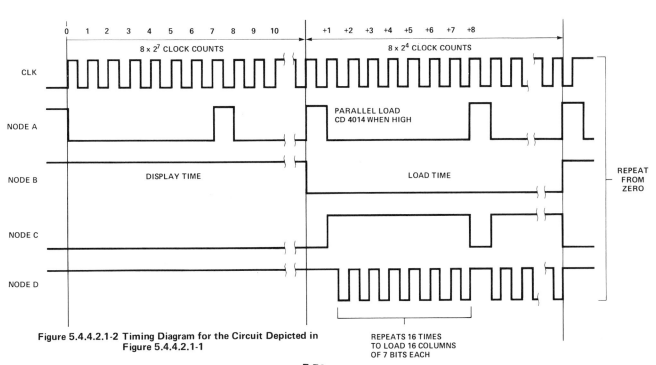

Figure 5.4.4.2.1-2 Timing Diagram for the Circuit Depicted in Figure 5.4.4.2.1-1

5.53

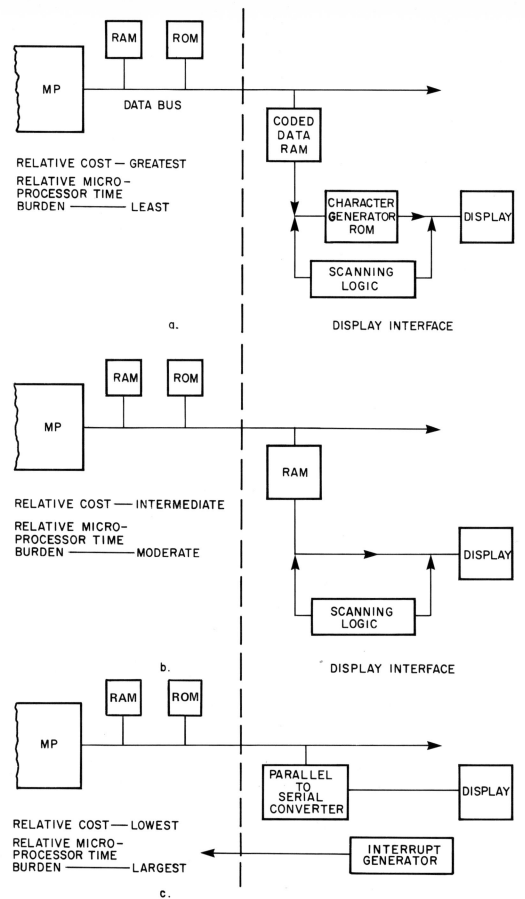

RELATIVE COST — GREATEST

RELATIVE MICRO-
PROCESSOR TIME
BURDEN ———————— LEAST

DISPLAY INTERFACE

a.

RELATIVE COST — INTERMEDIATE

RELATIVE MICRO-
PROCESSOR TIME
BURDEN ——————— MODERATE

DISPLAY INTERFACE

b.

RELATIVE COST — LOWEST

RELATIVE MICRO-
PROCESSOR TIME
BURDEN ——————— LARGEST

c.

Figure 5.4.4.2.2-1a, b, c Microprocessor Interface Partitioning
Techniques for Implementation of Display
Using the HDSP-2000.

5.54

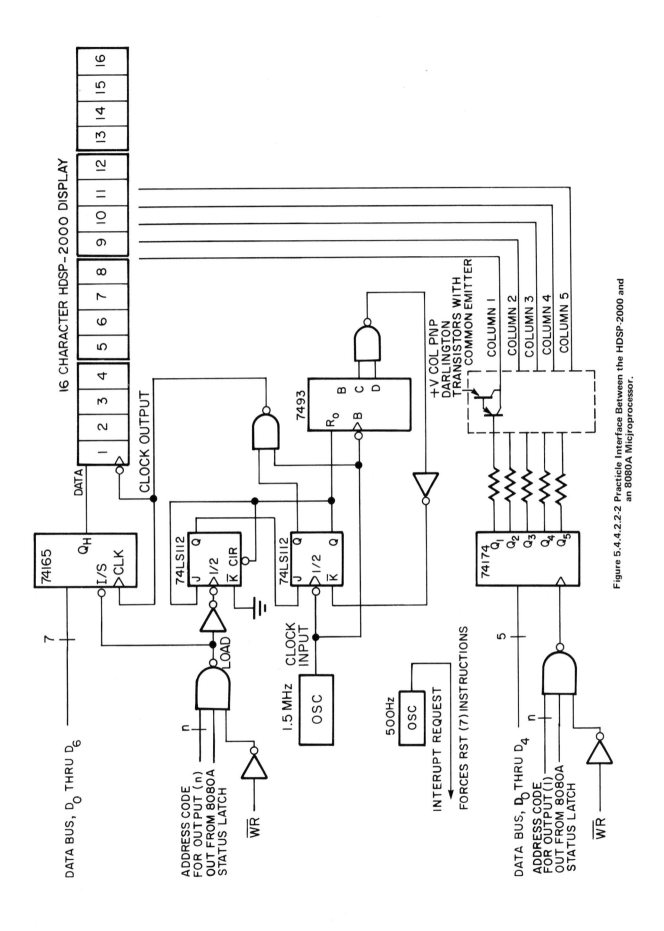

Figure 5.4.4.2.2-2 Practicle Interface Between the HDSP-2000 and an 8080A Micjroprocessor.

5.55

PROGRAM

ADDRESS	OP CODE	CLOCK CYCLES	COMMENTS
$(0038)_{16}$	PUSH PSW	(11)	
	PUSH HL	(11)	
	ORI	(7)	$A = (FF)_{16}$
	$(FF)_{16}$	------	
	OUT	(10)	TURNS OFF COLUMNS
	I	------	
	LHLD	(16)	HL = POINTER
	A_L	------	
	A_H	------	
	MOV A, M	(7)	A = (HL)
	OUT	(10)	DIGIT 16 = A
	n	------	
	INX HL	(5)	HL = HL + 1
	MOV A, M	(7)	A = (HL)
	OUT	(10)	DIGIT 15 = A
	n	------	
	INX HL	(5)	HL = HL + 1
	MOV A, M	(7)	A = (H2)
	OUT	(10)	DIGIT 14 = A
	n	------	
	⋮		
	INX HL	(5)	HL = HL + 1
	MOV A, M	(7)	A = (HL)
	OUT	(10)	DIGIT 1 = A
	n	------	
	INX HL	(5)	HL = HL + 1
	MOV A, M	(7)	A = (HL)
	OUT	(10)	TURNS A COLUMN ON
	I	------	
	MOV A, L	(5)	A = L
	CPI	(7)	COMPARE L TO ADDRESS
	$(64)_{16}$	------	OF LAST MEMORY LOCATION
	JNZ	(10)	JUMP IF A ≠ $(64)_{16}$
	(ADDRESS	------	
	OF LOOP)	------	
	MVI L	(7)	$L = (OF)_{16}$
	$(OF)_{16}$	------	
LOOP	INX HL	(5)	HL = HL + L
	SHLD	(16)	
	A_L	------	
	A_H	------	
	POP HL	(10)	
	POP PSW	(10)	
	EI	(4)	
	RET	(10)	

(b).

RAM

ADDRESS	CONTENTS (D_7-D_0)	COMMENTS
$A_H A_L$	POINTER L	
$A_H A_L + 1$	POINTER H	
$(XX10)_{16}$	X R_1 R_2 R_3 R_4 R_5 R_6 R_7	DIGIT 16, COL 1
⋮	⋮	⋮
$(XX1F)_{16}$	X R_1 R_2 R_3 R_4 R_5 R_6 R_7	DIGIT 1, COL 1
$(XX20)_{16}$	X X X 1 1 1 1 0	COL 1 ENABLE
$(XX21)_{16}$	X R_1 R_2 R_3 R_4 R_5 R_6 R_7	DIGIT 16, COL 2
$(XX30)_{16}$	X R_1 R_2 R_3 R_4 R_5 R_6 R_7	DIGIT 1, COL 2
$(XX31)_{16}$	X X X 1 1 1 0 1	COL 2 ENABLE
$(XX32)_{16}$	X R_1 R_2 R_3 R_4 R_5 R_6 R_7	DIGIT 16, COL 3
$(XX41)_{16}$	X R_1 R_2 R_3 R_4 R_5 R_6 R_7	DIGIT 1, COL 3
$(XX42)_{16}$	X X X 1 1 0 1 1	COL 3 ENABLE
$(XX43)_{16}$	X R_1 R_2 R_3 R_4 R_5 R_6 R_7	DIGIT 16, COL 4
$(XX52)_{16}$	X R_1 R_2 R_3 R_4 R_5 R_6 R_7	DIGIT 1, COL 4
$(XX53)_{16}$	X X X 1 0 1 1 1	COL 4 ENABLE
$(XX54)_{16}$	X R_1 R_2 R_3 R_4 R_5 R_6 R_7	DIGIT 16, COL 5
$(XX63)_{16}$	X R_1 R_2 R_3 R_4 R_5 R_6 R_7	DIGIT 1, COL 5
$(XX64)_{16}$	X X X 0 1 1 1 1	COL 5 ENABLE

X = DON'T CARE

(c.)

POINTER →
(POINTS AT NEXT WORD TO BE SENT TO HDSP-2000 DISPLAY)

Figure 5.4.4.2.2-2   Continued. Microcade and RAM Contents Used to Support Microprocessor Display Interface.

5.56

DEVICE DESCRIPTION	BASIC DEVICE NUMBER	I_{MAX} mA	MANUFACTURER
Quad Segment Driver, MOS to LED Anode	75491	50 (sink/source)	FAIR, MOT, TI
	75493	25 (source)	NS, TI
	7895/8895	19 (source)	NS
	501	40 (source)	ITT
	503	34 (source)	
	507	18 (source)	
	517	16 (source)	
	518	14.5 (source)	
	522	12 (source)	
	523	10 (source)	
	491	50 (source)	
HEX Digit Driver, MOS to LED Cathode	75492	250 (sink)	FAIR, MOT, NS, TI
	75494	150 (sink)	NS, TI
	8870	350 (sink)	
	8877	50 (sink)	
	8892	200 (sink)	
	500	250 (sink)	ITT
	502	200 (sink)	
	506	200 (sink)	
	510	160 (sink)	
	492	250 (sink)	
7-Digit Driver, MOS to LED Cathode	75497	150 (sink)	TI
	8866	50 (sink)	NS
	546	50 (sink)	ITT
	552	500 (sink)	
	554	500 (sink)	
	556	500 (sink)	
8-Digit Driver, MOS to LED Cathode	8863/8963	500 (sink)	NS
	8865	50 (sink)	
	8871	40 (sink)	
	514/525	40 (sink)	ITT
9-Digit Driver, MOS to LED Cathode	526	40 (sink)	ITT
	548	60 (sink)	
	558	40 (sink)	
	75498	150 (sink)	TI
	8855	50 (sink)	NS
	8874/8876/ 8879	50 (sink)	
	8973/8974/ 8976	100 (sink)	
12-Digit Driver, MOS to LED Cathode	8868	110 (sink)	NS
	8973	40 (sink)	
Segment Driver, MOS to LED	8861 (5-Seg)	50 (source)	NS
	8877 (8-Seg)	14 (source)	

TABLE 5.5-1 List of LED Display to Logic Interface Devices

DEVICE DESCRIPTION	BASIC DEVICE NUMBER	I_{OUT} mA SOURCE/SINK	FONT: 6 AND 9 WITH OR WITHOUT TAILS	MANUFACTURER
BCD to 7-Segment Decoder/Driver OUTPUTS: Active High, Interal Resistive Pull-up	7448	6.4 (sink) 2 (source)	W/O	FAIR, ITT, MOT, NS, SIG, TI
	74248	6.5 (sink) 2 (source)	W	TI
	74LS48	6 (sink) 2 (source)	W/O	NS, TI
	8T05	15 (sink) 4 (source)	W/O	SIG
	7856	7.5 (source)	W/O	NS
BCD to 7-Segment Decoder/Driver OUTPUTS: Active High, Current Source	8857	60 (source)	W/O	NS
	8858	50 (source)	W/O	NS
	14547	65 (source)	W/O	MOT
BCD to 7-Segment Decoder/Driver OUTPUTS: Active High, Open Collector	7449	10 (sink)	W/O	FAIR, MOT
	74LS49	8 (sink)	W/O	NS, TI
	74249	10 (sink)	W	TI
	74LS249	8 (sink)	W	TI
BCD to 7-Segment Decoder/Driver OUTPUTS: Active Low, Open Collector	7447	40 (sink)	W/O	FAIR, ITT, MOT, NS, SIG, TI
	74L47	20 (sink)	W/O	TI
	74LS47	12 (sink)	W/O	NS, TI
	74247	40 (sink)	W	TI
	74LS247	24 (sink)	W	TI
	9317B	40 (sink)	W/O	FAIR
	9317C	20 (sink)	W/O	FAIR
	8T04	40 (sink)	W/O	SIG
BCD to 7-Segment Decoder, CMOS; OUTPUTS: N-channel sink and NPN Bipolar Sources	74C48	3.6 (sink) 50 (source)	W/O	HAR, NS
BCD to 7-Segment Latch/Decoder/ Driver; CMOS with Bipolar Outputs	4511 (14511)	25 (source)	W/O	FAIR, HAR, MOT, RCA, SSS
BCD to 7-Segment Decoder/Driver, CMOS	14558	0.28 (sink) 0.35 (source)	W/O	MOT
BCD to 7-Segment Latch/Decoder/ Driver OUTPUTS: Constant Current	9368	22 (source)	6 (W) 9 (W/O)	FAIR
	9374	18 (sink)	6 (W) 9 (W/O)	FAIR
	8673/ 8674	18 (sink)	6 (W) 9 (W/O)	NS
BCD to 7-Segment Latch/Decoder/ Driver OUTPUTS: Active Low, Open Collector	9370	40 (sink)	6 (W) 9 (W/O)	FAIR

LIST OF MANUFACTURERS

FAIR = FAIRCHILD SEMICONDUCTOR

HAR = HARRIS SEMICONDUCTOR

ITT = ITT SEMICONDUCTOR

MOT = MOTOROLA SEMICONDUCTOR

NS = NATIONAL SEMICONDUCTOR

RCA = RCA SOLID STATE DIVISION

SIG = SIGNETICS

SSS = SOLID STATE SCIENTIFIC

TI = TEXAS INSTRUMENTS

TABLE 5.5.9 List of LED Display to Logic Interface Devices

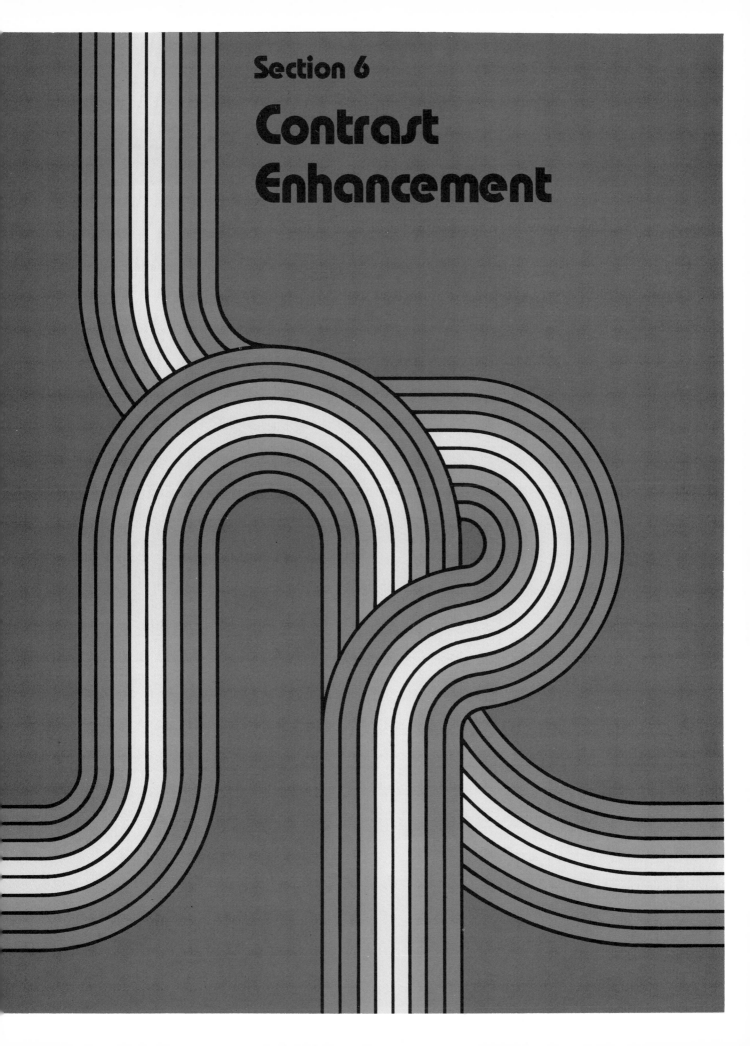

Section 6

Contrast Enhancement

6.0 CONTRAST ENHANCEMENT FOR LED DISPLAYS

The most important attribute of any equipment utilizing a digital readout is the ability to clearly display information to an observer. An observer must be able to quickly and accurately recognize the information being displayed by the instrument. The display, usually front panel mounted, must be visible without difficulty in the ambient light conditions where the instrument will be used.

Since most ambient light levels are sufficiently bright to impair the visibility of an LED display, it is necessary to employ certain techniques to develop a high viewing contrast between the display and its background. Since the quality of visibility is primarily subjective, it is not easily measured or treated by analytical means. Thus, human engineering plays a very important role in display applications. The best judge of the viewing esthetics of a display is the human eye, and the final display design must be pleasing to the eye when viewed in the end user ambient.

6.1 Contrast and Contrast Ratio

The objective of contrast enhancement is to maximize the contrast between display "ON" and display "OFF" conditions. This is accomplished by (1) reducing to a minimum the reflected ambient light from the face of the display and (2) allowing a maximum of the display's emitted light to reach the eye of an observer. The goal is to achieve a maximum contrast between "ON" segments and "OFF" segments as well as a minimum contrast between "OFF" segments and display package and background.

Both contrast and contrast ratio are used as the measure of the difference in the luminous sterance of a source with respect to the surrounding background. Contrast, C, for an LED display is defined as:

$$C = \frac{L_{vS} - L_{vB}}{L_{vS}} ; 0 \leqslant C \leqslant 1 \frac{L_{vS} - L_{vB}}{L_{vS}} \qquad (6.1\text{-}1)$$

where: L_{vS} = Source luminous sterance (cd/m^2)

L_{vB} = Background luminous sterance (cd/m^2)

Contrast varies from zero, where $L_{vS} = L_{vB}$, to a value of one, where the source intensity is at a level such that $L_{vS} \gg L_{vB}$.

$$CR = \frac{L_{vS}}{L_{vB}} ; 1 \leqslant CR \leqslant \infty \qquad (6.1\text{-}2)$$

Contrast ratio, CR, is the ratio of the source luminous sterance to the background luminous sterance.

Contrast ratio ranges in value from one, where $L_{vS} = L_{vB}$, to a value tending towards infinity, when $L_{vS} \gg L_{vB}$. Contrast and contrast ratio are related by the following two expressions:

$$C = 1 - \frac{1}{CR} \qquad (6.1\text{-}3)$$

$$CR = \frac{1}{1-C} \qquad (6.1\text{-}4)$$

As shown in the curve of Figure 6.1-1, the range of attainable contrast for LED displays is between a useable minimum of 0.90 (CR=10) to 0.98 (CR=50) with a typical filter providing a contrast of 0.95 (CR=20).

A spot photometer is used to measure luminous sterance. The display should be located behind the contrast filter, installed in its mounting assembly and packaged as the final product in order to obtain realistic values. Ambient lighting, which represents the end use environment, should be incident to the filter during the measurement. Care should be taken to prevent shading of the spot being measured by the photometer.

With stretched segment displays, it is difficult to achieve a high value of segment on/off contrast while effectively concealing the display package from view. For example, a display with a black package is easily concealed from view, however, the "OFF" segments will be visible. This is due to the difference in reflectivity between the "OFF" segments and the black package.

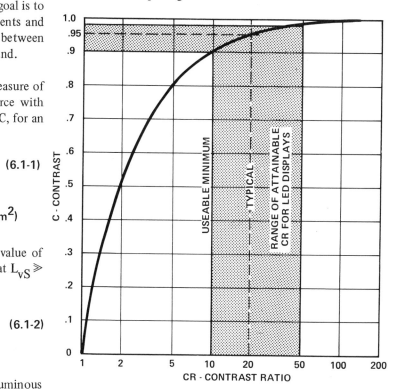

Figure 6.1-1 The Relationship Between Contrast and Contrast Ratio.

A reduction in the reflectivity difference between the "OFF" segments and the package of a stretched segment display may be obtained by adding a small amount of dye to color tint the segments, and the display package may be colored to match the "OFF" segment color. With the addition of an appropriate optical filter placed in front of the display, the "OFF" segments tend to be indistinguishable from the background. The trade-off is that a colored package is more visible than a black package. Because of this trade-off, a designer has to decide which is more important, concealing "OFF" segments or concealing the display package. Since the usual choice is to conceal "OFF", segments, Hewlett-Packard is using this colored package technique on its high-efficiency red, yellow and green stretched segment displays.

Contrast enhancement under artificial lighting conditions may be accomplished by use of selected wavelength optical filters. Under bright sunlight conditions contrast enhancement becomes more difficult and requires additional techniques such as the use of louvered filters combined with shading of the display. The effect of a wavelength optical filter is illustrated in Figure 6.1-2. The filtered portion of the display can be easily read while the "OFF" segments are not apparent. By comparison, reading the unfiltered portion of the display is difficult.

6.2 Eye Response, Peak Wavelength and Dominant Wavelength

The 1931 CIE (Commission Internationale De L'Eclairage) standard observer curve, also known as the photopic curve, is shown in Figure 6.2-1. This curve represents the eye response of a standard observer to various wavelengths of light. The vivid color ranges are also identified in Figure 6.2-1 to illustrate the sensitivity of the eye to the various colors. The photopic curve peaks at 555 nanometers (nm) in the yellowish-green region. This peak corresponds to 680 lumens of luminous flux (lm) per watt of radiated power (W).

Two wavelengths of the LED emission are important to a user of LED displays; Peak Wavelength and Dominant Wavelength. Peak wavelength (λ_p) is the wavelength at the peak of the radiated spectrum. The peak wavelength may be used to estimate the approximate amount of display emitted light that passes through an optical filter. For example, if an optical filter has a relative transmission of 40% at a given λ_p, then approximately 40% of the display emitted light at the peak wavelength will pass through the filter to the viewer while 60% will be absorbed. This gives a designer an initial estimate of the amount of loss of display emitted light he should expect.

Dominant wavelength (λ_d) is used to define the color of an

Figure 6.1-2 Effect of a Wavelength Optical Filter on an LED Display.

LED display. Specifically, the dominant wavelength is that wavelength of the color spectrum, which, when additively mixed with the light from the source CIE illuminant C, will be perceived by the eye as the same color as is produced by the radiated spectrum. CIE illuminant C is a $6500°$K color temperature source that produces light which simulates the daylight produced by an overcast sky. A graphical definition of λ_d and color purity is given on the CIE chromaticity diagram in Figure 6.2-2. The dominant wavelength is derived by first obtaining the x,y color coordinates from the radiated spectrum. These color coordinates are then plotted on the CIE chromaticity diagram. A line is drawn from the illuminant C point through the x,y color point intersecting the perimeter of the diagram. The point where the line intersects the perimeter is the dominant wavelength.

The color purity, or saturation, is defined as the ratio of the distance from the x,y color point to the illuminant C point, divided by the sum of this distance and the distance from the x,y point to the perimeter. The x,y color coordinates for LEDs plot very close to the perimeter of the chromaticity diagram. Therefore, the color purity approaches a value of 1, typical of the color saturation obtained from a monochromatic light source.

The dominant wavelengths and corresponding colors for LEDs are shown on the CIE chromaticity diagram in Figure 6.2-3. As defined by λ_d, the color of a standard red LED is red, a high-efficiency red LED is reddish-orange, a yellow LED is yellowish-orange and a green LED is actually greenish-yellow.

It is of value to know the actual colors of each LED when selecting a contrast filter, as the optimum filter will have the same color as the device. Both λ_p and λ_d are listed on LED data sheets.

6.3 Filter Transmittance

The contrast filter must meet certain basic requirements. It should be mechanically stable with temperature, have

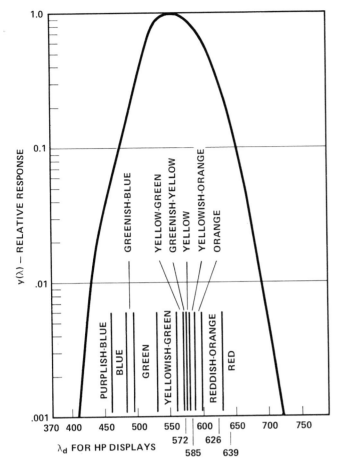

Figure 6.2-1 CIE Standard Observer Eye Response Curve (Photopic Curve), Including CIE Vivid Color Ranges.

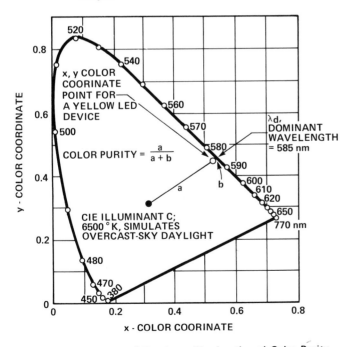

Figure 6.2-2 Definition of Dominant Wavelength and Color Purity, Shown on the CIE Chromaticity Diagram.

reasonable chemical resistance, be free from visual defects, have a homogeneous index of refraction and have sufficient transmission of the radiated spectrum of the LED. It is the last two requirements that are of primary optical importance.

The relative transmittance of an optical filter with respect to wavelength is defined to as: **(6.3-1)**

$$T(\lambda) = \frac{\text{Luminous Flux with Filter at Wavelength } \lambda}{\text{Luminous Flux without Filter at Wavelength } \lambda}$$

The relative transmittance is a function of the index of refraction of the filter material and the transmission through the material as determined by the coloring. Both are a function of wavelength. Specifically, the transmission through the filter is dependent upon the amount of incident light reflected at the filter/air interface and the amount of wavelength absorbtion within the filter material.

The index of refraction determines the amount of incident light that is reflected at the filter/air interface. The amount of reflected light is given by the following ratio:

$$R = \frac{(n_1 - n_2)^2}{(n_1 + n_2)^2} \qquad \text{(6.3-2)}$$

where: n_1 = Index of refraction of the filter material.

n_2 = Index of refraction for air = 1.0.

A plastic filter with an average index of refraction equal to 1.5, for the range of wavelengths encompassing the LEDs radiated spectrum, will reflect 4% of the normal incident light at each filter/air interface. Thus, 8% of the LED emitted light, passing through the filter to an observer, is lost due to reflection.

Inside the filter, light is lost due to absorbtion by the tinted material. The amount of absorbtion is a direct function of wavelength and is determined by the dye coloring and dye concentration. If the dye coloring is held at a constant density, the transmission through the filter material at any given wavelength, T_λ, is an exponetial function of the thickness of the material:

$$T_\lambda = e^{-ax} \qquad \text{(6.3-3)}$$

where: x = The quantity of unit thicknesses of filter material.

e = 2.71828

a = Absorbtion coefficient and is equal to −Int, where t is the transmission for a unit thickness.

As an example, the transmission through 1.0mm thickness of filter material is 0.875 at a wavelength of 655 nm. At x = 2.5, the thickness is 2.5mm and the transmission is 0.716:

$$-\ln(.875) = 0.1335$$

$$T_\lambda = e^{-(.1335)(2.5)} = .716$$

The relative transmission of a contrast filter at a particular wavelength, $T(\lambda)$, may be calculated with reasonable accuracy by using the following relationship:

$$T_{(\lambda)} = \left[\frac{2n}{n+1}\right] T_\lambda \qquad (6.3\text{-}4)$$

where: n = The index of refraction of the filter material.
T_λ = Transmission through the filter material at wavelength λ.

For a 1.0mm thick plastic filter with n = 1.5 and T_λ = .875 at a wavelength of 655 nm, the relative transmission is 0.808:

$$T(\lambda = 655 \text{ nm}) = \left[\frac{2(1.5)}{(1.5)^2 + 1}\right][.875] = .808$$

The same filter material at a thickness of 2.5 mm has a relative transmittance of 0.661, as shown in Figure 6.3-1.

6.3.1 Plastic Filters

Due to their low cost, ease in machining to size and resistance to breakage, plastic contrast filters are being used in a majority of display applications. The filter requirements for dim, moderate and bright ambients are different for each lighting condition. Therefore, it is advantageous to become familiar with the various relative transmittance characteristics that are available in plastic filters.

Most manufacturers of wavelength filters for use with LED displays provide relative transmittance curves for their products. Sample transmittance curves are presented in Figures 6.3.1-1, -2, -3 and -4. These curves represent approximate filter characteristics which may be used in various ambient light levels. The total transmittance curve shape and wavelength cut-off points have been chosen in direct relationship to the LED radiated spectrum. Each filter curve has been empirically determined and is similar to commercially available products. The higher the ambient light, the more optically dense the filter must be to absorb refected light from the face of the display. Because the display emitted light is also strongly absorbed, the display must be driven at a high average current to be readily visible. For dim ambient light, the filter may have a high value of transmittance as the ambient light will be at

levels much less than display emitted light. The display can now be driven at a low average current.

Dim ambients are in the range of 32 to 215 lx (3 to 20 footcandles), moderate ambients are in the range of 215 to 1076 lux (20 to 100 footcandles), and bright ambients are in the range of 1076 to 5382 lux (100 to 500 footcandles). Lux = (lm/m^2) and footcandle = (lm/ft^2).

Listed on each filter transmittance curve are empirically selected ranges of relative transmittance values at the peak wavelength which may give satisfactory filtering. For example, a filter to be used with a yellow display in moderate ambient lighting could have a transmittance value at the peak wavelength $[T(\lambda_p)]$ between 0.27 and 0.35. The filter wavelength cut-off should occur between 530 and 550 nm for best results.

When selecting a filter, the transmittance curve shape, attenuation at the peak wavelength and wavelength cut-off should be carefully considered in relationship to the LED radiated spectrum and ambient light level so as to obtain optimum contrast enhancement.

Three manufacturers of plastic wavelength filters are Panelgraphic Corporation (Chromafilter®), SGL Homalite and Rohm & Haas Company (Plexiglas®). The LED filters produced by these manufacturers are useable with all LED display and lamp devices. Table 6.9-2 lists some of the filter manufacturers and where to go for further information. Table 6.9-1 lists some specific wavelength filter products with recommended applications.

6.3.2 Optical Glass Fitlers

Optical glass filters are typically designed with constant density so it is the thickness of the glass that determines the optical density. This is just the opposite of plastic filters which are usually designed such that all material thicknesses have the same optical density.

The primary advantage of an optical glass contrast filter over a plastic filter is its superior performance. This is especially true for red LED filters. Figure 6.3.2-1 illustrates two thicknesses of a red optical glass filter for use with standard red LED displays. The relative transmittance is generally higher than that of a comparable plastic filter, and the slope of the relative transmittance curve is usually much steeper and follows more closely the shape of the radiated spectrum of the LED. This particular filter provides excellent contrast in a bright ambient. A reddish-orange optical glass filter which is suitable for use with a high-efficiency red display in a moderate ambient is shown in Figure 6.3.2-2. The relative transmittance of this filter follows almost exactly the leading edge of the shape of the LED's radiated spectrum.

A leading manufacturer of optical glass filters is the Schott Optical Glass, Inc. of Duryea, Pennsylvania and Munich, Germany.

6.4 Wavelength Filtering

The application of wavelength filters as described in the previous section is the most widely used method of contrast enhancement under artificial lighting conditions. However, they are not very effective in daylight due to the high level

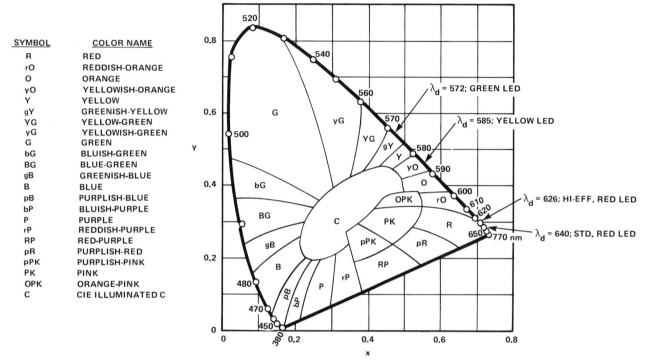

SYMBOL	COLOR NAME
R	RED
rO	REDDISH-ORANGE
O	ORANGE
yO	YELLOWISH-ORANGE
Y	YELLOW
gY	GREENISH-YELLOW
YG	YELLOW-GREEN
yG	YELLOWISH-GREEN
G	GREEN
bG	BLUISH-GREEN
BG	BLUE-GREEN
gB	GREENISH-BLUE
B	BLUE
pB	PURPLISH-BLUE
bP	BLUISH-PURPLE
P	PURPLE
rP	REDDISH-PURPLE
RP	RED-PURPLE
pR	PURPLISH-RED
pPK	PURPLISH-PINK
PK	PINK
OPK	ORANGE-PINK
C	CIE ILLUMINATED C

Figure 6.2-3 Dominant Wavelengths and Corresponding Colors for LEDs, Shown on the CIE Chromaticity Diagram.

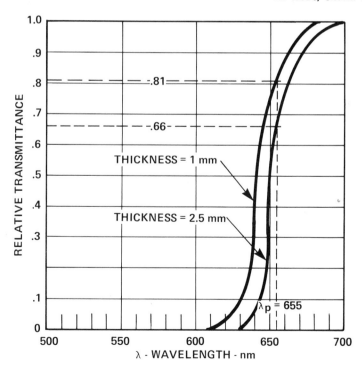

Figure 6.3-1 Variation in Relative Transmittance vs. Thickness for a Constant Density Filter Material.

ambient light. Filtering in daylight conditions is best achieved by using louvered filters (discussed in a later section).

The figures in sections 6.4.1 through 6.4.4 show the relationship between artificial lighting and the spectra of LED displays, both unfiltered and filtered. Figures 6.4.1-1 through 6.4.4-1 show the relationship between the various LED spectra and the spectra of daylight fluorescent and incandescent light. The photometric spectrum (shaded curve) is obtained by multiplying the LED radiated spectrum $[f(\lambda)]$ by the photopic curve $[y(\lambda)]$. Thus, photometric spectrum = $f(\lambda) \cdot y(\lambda)$. Figures 6.4.1-2 through 6.4.4-2 demonstrate the effect of a wavelength filter. The filtered photometric spectrum is what the eye perceives when viewing a display through a filter (shaded curve). Thus, filtered photometric spectrum = $f(\lambda) \cdot y(\lambda) \cdot T(\lambda)$. The ratio of the area under the filtered photometric spectrum to the area under the unfiltered photometric spectrum is the fraction of the visible light emitted by the display which is transmitted by the filter:

$$\text{Fraction of Available Light from Filtered Display} = \frac{\int f(\lambda) \cdot (\lambda) \cdot T(\lambda) \cdot d\lambda}{\int f(\lambda) \cdot y(\lambda) \cdot d\lambda}$$

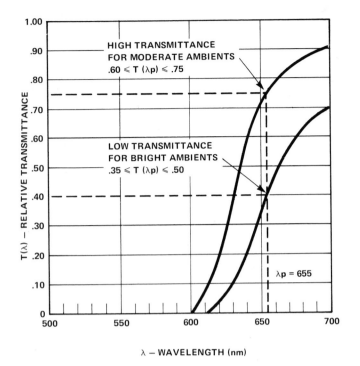

Figure 6.3.1-1   Typical Transmittance Curves for Filters to be Used with Standard Red Displays.

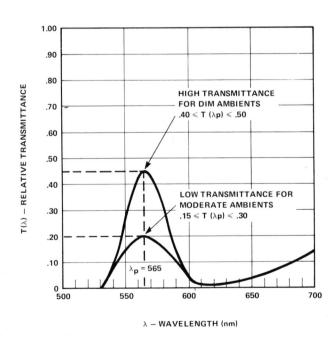

Figure 6.3.1-2   Typical Transmittance Curves for Filters to be Used with High-Efficiency Red Displays.

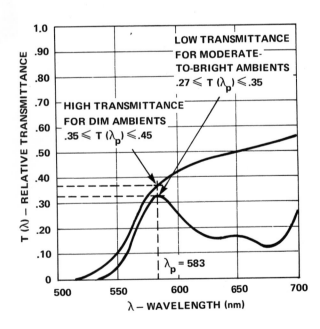

Figure 6.3.1-3   Typical Transmittance Curves for Filters to be Used with Yellow Displays.

Figure 6.3.1-4   Typical Transmittance Curves for Filters to be Used with Green Displays.

6.6

In addition to attenuating a portion of the light emitted by the display, a filter also shifts the dominant wavelength, thus causing a shift in the perceived color. For a given display spectrum, the color shift depends on the cut-off wavelength and shape of the filter transmittance characteristic. A choice among available filters must be made on the basis of which filter and LED combination is most pleasing to the eye. A designer must experiment with each filter as he cannot tell by transmittance curves alone. The filter spectra presented in Figures 6.3.1-1 through 6.3.1-4 are suggested starting points. Filters with similar characteristics are commercially available.

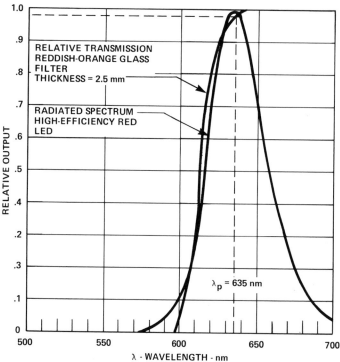

Figure 6.3.2-2 A Reddish-Orange Optical Glass Filter for Use with High-Efficiency Red Displays.

6.3.1-2 and 6.4.2-2). The resulting color is a rich reddish-orange.

6.4.3 Filtering Yellow Displays (λ_p = 583 nm)

The peak wavelength of a yellow LED display is in the region of the photopic curve where the eye is most sensitive (see Figure 6.4.3-1). Also, there is a high concentration of yellow in the spectrum of fluorescent light and a lesser amount of yellow in incandescent light. Therefore, filters that are more optically dense than red filters at the peak wavelength are required to filter yellow displays. The most effective filters are the dark yellowish-orange (or dark amber) filters as shown in Figure 6.3.1-3. The use of a low transmittance yellowish-orange filter, as shown in Figure 6.4.3-2 results in a similar color to that of a gas discharge display. Pure yellow filters provide very little contrast enhancement.

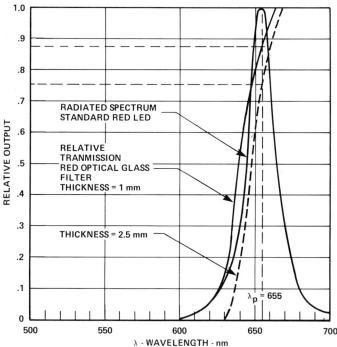

Figure 6.3.2-1 A Red Optical Glass Filter for use with Standard Red Displays.

6.4.1 Filtering Standard Red Displays (λ_p = 655 nm)

Filtering out reflected ambient light from red displays is easily accomplished with a long wavelength pass filter having a sharp cut-off in the 600 nm to 620 nm range (see Figures 6.3.1-1 and 6.4.1-2). Under bright fluorescent light, a red filter is very effective due to the low concentration of red in the fluorescent spectrum. The spectrum of incandescent light contains a large amount of red, and therefore, it is difficult to filter red displays effectively in bright incandescent light.

6.4.2 Filtering High-Efficiency Red Displays (λ_p = 635 nm)

The use of a long wavelength pass filter with a cut-off in the 570 nm to 590 nm range gives essentially the same results as is obtained when filtering red displays (see Figures

6.4.4 Filtering Green Displays (λ_p = 565 nm)

The peak wavelength of a green LED display is only 10 nm from the peak of the eye response curve (see Figure 6.4.4-1). Therefore, it is very difficult to effectively filter green displays. A long wavelength pass filter, such as is used for red and yellow displays, is no longer effective. An effective filter is obtained by combining the dye of a short wavelength pass filter with the dye of a long wavelength pass filter, thus forming a bandpass yellow-green filter which peaks at 565 nm as shown in Figure 6.3.1-4. Pure

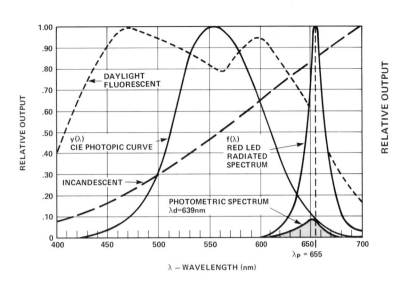

Figure 6.4.1-1 Relative Relationship Between the Standard Red LED Spectrum, Photopic Curve and Artificial Lighting.

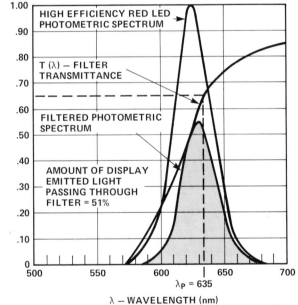

Figure 6.4.1-2 Effect of a Long Pass Wavelength Filter on a Standard Red Display.

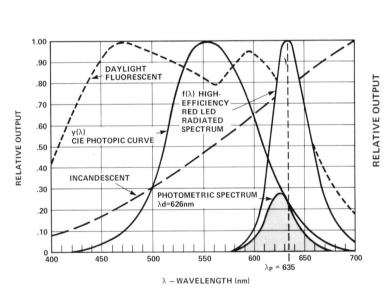

Figure 6.4.2-1 Relative Relationship Between the High-Efficiency Red LED Spectrum, Photopic Curve and Artificial Lighting.

Figure 6.4.2-2 Effect of a Long Pass Wavelength Filter on a High-Efficiency Red LED Display.

green filters peak at 520 nm and drop off rapidly in the 550 nm to 570 nm range and are not recommended. The best possible filters for green LED displays are those which are yellow-green bandpass, peaking at 565 nm and dropping off rapidly between 575 nm and 590 nm. As shown in Figure 6.4.4-2, this filter passes wavelengths 550 to 570 while sharply reducing the longer wavelengths in the yellow

region. To effectively filter green LED displays in fluorescent light would require the use of a filter with a low transmittance value at the peak wavelength. This is due to the high concentration of green in the fluorescent spectrum. It is easier to filter green displays in bright incandescent light due to the low concentration of green in the incandescent spectrum, see Figure 6.4.4-1.

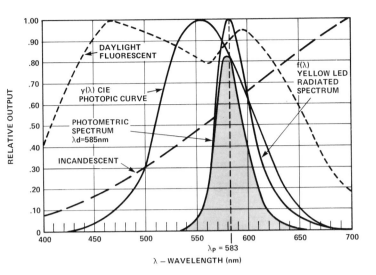

Figure 6.4.3-1   Relative Relationship Between the Yellow LED Spectrum, Photopic Curve and Artificial Lighting.

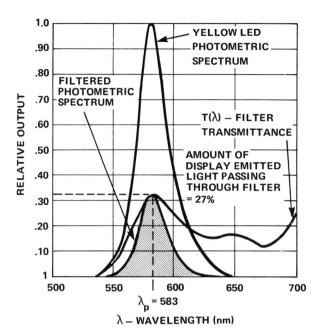

Figure 6.4.3-2   Effect of a Wavelength Filter on a Yellow LED Display.

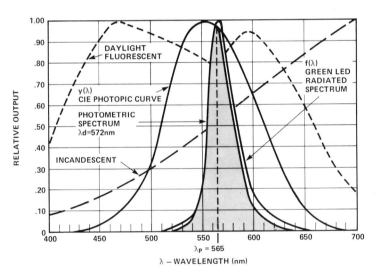

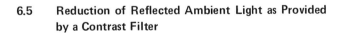

Figure 6.4.4-1   Relative Relationship Between the Green LED Spectrum, Photopic Curve and Artificial Lighting.

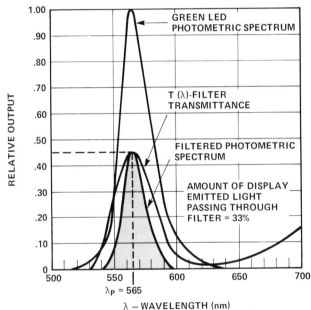

Figure 6.4.4-2   Effect of a Bandpass Wavelength Filter on a Green LED Display.

6.5   Reduction of Reflected Ambient Light as Provided by a Contrast Filter

Incident ambient light that is reflected back to an observer from the face of a filtered display travels twice through the contrast filter. The amount of light actually reflected from the face of the display is dependent upon the diffused reflectance, off the face of the display, as a function of wavelength, $R(\lambda)$.

Diffused reflectance is absorbtion dependent and is measured with the incident light beam normal to a surface. Those wavelengths which are not absorbed are reflected back as color. This is not the same as specular reflection. In specular reflection, the light is incident to a surface at some angle less than $90°$ where all wavelengths are essentially reflected equally. Figure 6.5-1 shows the diffused reflectance as a function of wavelength for stretched segment displays that have colored packages.

The amount of incident light reflected back to an observer from a filtered display as compared to an unfiltered display may be calculated from the ratio of the filtered reflected light to the unfiltered reflected light.

$$\frac{\text{Amount of Filtered}}{\text{Reflected Light}} = \frac{\int X(\lambda) \cdot R(\lambda) \cdot T^2(\lambda) \cdot Y(\lambda) \cdot d\lambda}{\int X(\lambda) \cdot R(\lambda) \cdot Y(\lambda) \cdot d\lambda}$$

where: $X(\lambda)$ = Spectral distribution of the incident ambient light.

$R(\lambda)$ = Relative diffused reflectance off the face of the display.

$T(\lambda)$ = Filter relative transmittance.

$Y(\lambda)$ = Photopic curve.

6.5.1 Effectiveness of a Wavelength Filter in an Ambient of Artificial Lighting

The contrast is very dependent upon the ambient lighting. If most of the spectral distribution of the ambient light is outside the radiated spectrum of the LED, it is very easy to reduce the reflected ambient light to a very low level while imposing only minimal attenuation on the emitted light from the display. Such is the case when using a red LED display in a fluorescent ambient or a green LED in an incandescent ambient. The opposite is also true. It is very difficult to effectively reduce the reflected incandescent light from the face of a red LED display, or to reduce

reflected fluorescent light from the face of a yellow or green LED display without significantly decreasing the emitted light from the LED.

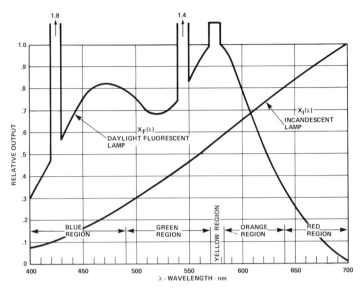

Figure 6.5.1-1   Spectral Distribution for a Daylight Fluorescent and an Incandescent Lamp used for Artificial Lighting.

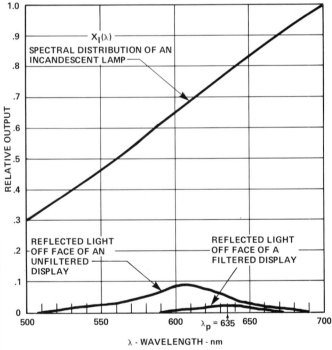

Figure 6.5.1-2   Reflected Incandescent Light Off an Unfiltered and Filtered High-Efficiency Red Display with a Colored Package, as Seen by an Observer. The Filtered Reflected Light is 17% of the Unfiltered Reflected Light.

Figure 6.5.1-1 reproduces the spectral distributions for fluorescent and incandescent lighting. Fluorescent lighting contains almost no red, yet contains a high level of yellow and long wavelength green. Incandescent, is just the

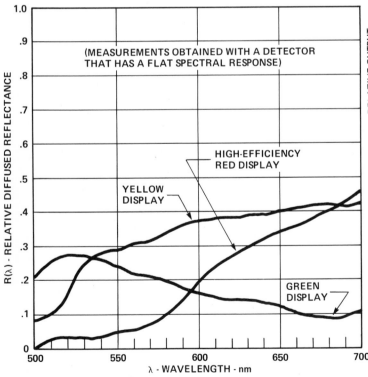

Figure 6.5-1 Diffused Reflectance for the Faces of Stretched Segment Displays that have Colored Packages.

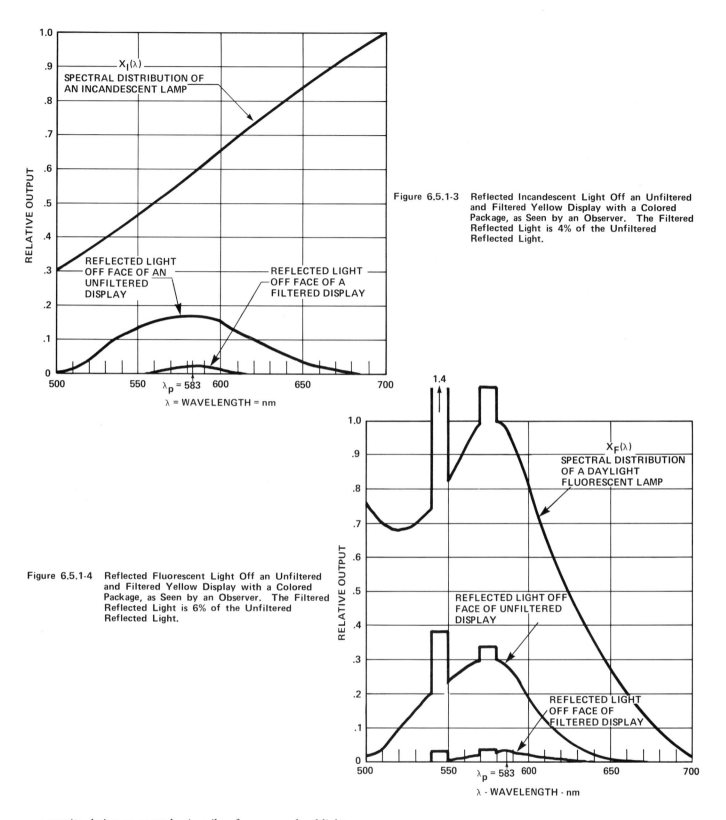

Figure 6.5.1-3 Reflected Incandescent Light Off an Unfiltered and Filtered Yellow Display with a Colored Package, as Seen by an Observer. The Filtered Reflected Light is 4% of the Unfiltered Reflected Light.

Figure 6.5.1-4 Reflected Fluorescent Light Off an Unfiltered and Filtered Yellow Display with a Colored Package, as Seen by an Observer. The Filtered Reflected Light is 6% of the Unfiltered Reflected Light.

opposite, being composed primarily of orange and red light. As can be seen from Figure 6.5.1-1 and the Figures in Section 6.4, the contrast filter may have to reduce the same wavelengths that are predominant in both the LED and ambient light in order to achieve adequate contrast. This is especially true for green LEDs under fluorescent lighting.

Figures 6.5.1-2 through 6.5.1-5 illustrate a comparative difference between the filtered and unfiltered diffused reflected light off the face of a stretched segment LED display as seen by the eye of an observer. The curve of the unfiltered diffused reflected light is obtained by the product of the lamp spectral distribution $[X(\lambda)]$, diffused reflectance off the face of the display [R(λ)] and the photopic curve [y(λ)]. Since the incident ambient passes through the filter twice, the curve of the filtered reflected light is the unfiltered curve multiplied by the square of the filter transmittance [$T^2(λ)$]. The filters used are the same as those used in Figures 6.4.2-2, 6.4.3-2 and 6.4.4-2. These curves illustrate that a wavelength filter substantially reduces the incident ambient light that is reflected off the face of a display. Even so, the peak of the reduced reflected ambient light is at or near the peak wavelength of the LED radiated spectrum.

6.5.2 Effectiveness of a Wavelength Filter in Daylight Ambients

The purpose of the wavelength filter is to reduce the luminous sterance of the background to a level that is very much less than the luminous sterance of the display's illuminated segments in order to achieve a high value of contrast. Referring to Equation 6.1-1 for contrast, the contrast goes to zero if the background luminance sterance equals the luminance sterance of the illuminated segments. This situation occurs when the ambient light level is sufficient to effectively mask the display's emitted light.

The spectral distribution of bright sunlight is nearly a flat curve across the complete color spectrum, as shown in Figure 6.5.2-1. Bright sunlight that is directly incident upon a display with a wavelength filter raises the background luminous sterance to a level that may actually exceed the luminous sterance produced by the emitted light from the illuminated segments. Therefore, the luminous sterance of the display segments, "ON" or "OFF", equals the luminous sterance of the background and the contrast goes to zero.

A similar situation occurs with overcast sky daylight that is directly incident onto the face of a wavelength filtered display. The situation is not as pronounced for standard red LED displays as for green LED displays, since the overcast sky filters out a considerable amount of the ambient red light, see Figure 6.5.2-1. It is obvious, then, that a wavelength filter by itself is not sufficient to achieve the necessary contrast in daylight ambient. Some shading or blocking out of the incident daylight is also required to achieve sufficient contrast between the display background and the illuminated segments.

It is almost impossible to achieve a minimum acceptable contrast if the incident daylight is parallel to the viewing axis (the incident daylight is perpendicular to the face of the display). If the incident daylight is not parallel to the viewing axis, other filter techniques may be employed to obtain acceptable contrast. These techniques include the use of louvered or cross hatch filters, incorporation of an eyebrow shade and recessing the display. Also, a circular polarizing filter may be employed with those displays that

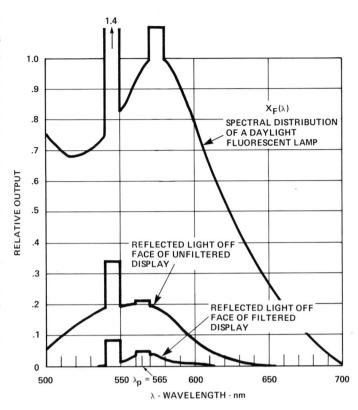

Figure 6.5.1-5 Reflected Fluorescent Light Off an Unfiltered and Filtered Green Display with a Colored Package, as Seen by an Observer. The Filtered Reflected Light is 10% of the Unfiltered Reflected Light.

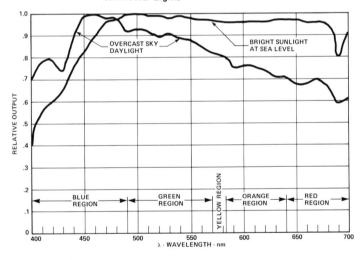

Figure 6.5.2-1 Spectral Distributions for Bright Sunlight and Overcast Sky Daylight.

have a face which is a specular reflecting surface. These techniques are discussed in more detail in later sections.

Indirect daylight that is incident to the face of a display is already reduced in level from direct daylight. Simple shading of a display, which is filtered by a louvered filter, will produce adequate contrast. It is the shading that reduces the indirect daylight to a low enough level to allow the filter to reduce the background luminous sterance to a level sufficient for adequate contrast.

6.6 Special Wavelength Filters and Filters in Combination

A designer is not limited to a single color wavelength filter to achieve the desired contrast and front panel appearance. Some unique wavelength filters and filter combinations have been successfully developed. One is the purple color filter for use with red LED displays, and another is the use of a neutral density filter in combination with a light amber filter to achieve a dark front panel for yellow LED displays.

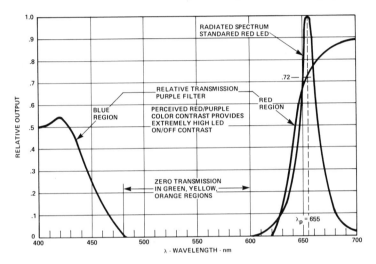

Figure 6.6.1-1 A Purple Color Wavelength Filter for Standard Red LED Displays.

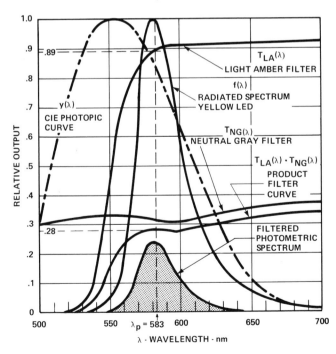

Figure 6.6.2-1 A Neutral Density Gray Filter in Combination with a Light Amber Filter for Use with Yellow Displays.

6.6.1 The Purple Contrast Filter for Red LED Displays

The wavelength filters that have been previously discussed have a distinct color that may be identified by a dominant wavelength. The contrast that they provide is essentially due to the high level luminous sterance of the display segments contrasted against the low level luminous sterance of a background of the same color. Another approach maintains the contrast ratio, but has as the background a different color than that of the LEDs in the display. This is easily accomplished by using a dark purple filter with standard red LED displays.

A most effective contrast filter is shown in Figure 6.6.1-1. The color, purple, is not defined by a dominant wavelength, see Figure 6.2-3. Purple is a mixture of red and blue light which is perceived by the eye as a distinct color from red. Psychologically, a purple contrast filter is more pleasing to many people than a red filter. The reason for this may be that when illuminated, the standard red display stands out so vividly against the purple background. Therefore, it is the color difference that enhances the contrast. This makes the purple contrast filter extremely effective in brightly lighted ambients.

6.6.2 Filters in Combination

A neutral density gray filter is often used in combination with other filters to provide a dark gray filter window as well as increased contrast in bright ambients. A typical example is given in Figure 6.6.2-1. The resulting filter is the product of the relative transmittance of the light amber, $T_{LA}(\lambda)$, and the relative transmittance of the neutral density gray, $T_{NG}(\lambda)$:

Product Filter = $[T_{LA}(\lambda) \cdot T_{NG}(\lambda)]$

The amount of light reaching the eye of a viewer is 24% of the unfiltered LED spectrum:

Fraction of Available Light Through a Combination Filter	$= \dfrac{\int f(\lambda) \cdot Y(\lambda) \cdot [T_{LA}(\lambda) \cdot T_{NG}(\lambda)] \cdot d\lambda}{\int f(\lambda) \cdot Y(\lambda) \cdot d\lambda}$

The advantage is a dark gray front panel window with very low luminous sterance (zero transmission below 525 nm) that retains its appearance in bright ambients. The trade-off is a considerable reduction in the luminous sterance of the display which reduces the contrast ratio. This is somewhat offset by the distinct color difference between the illuminated yellow segments of the display and the dark gray background.

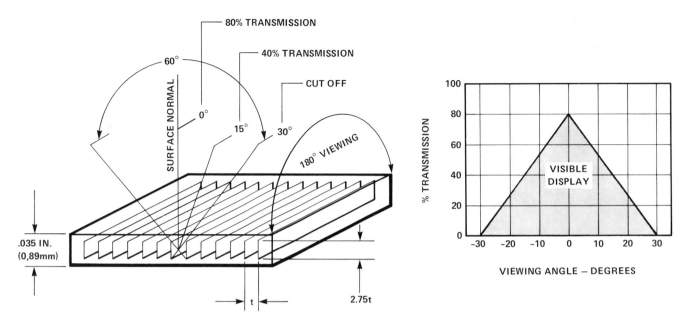

Figure 6.7-1 Construction Characteristics of a 0° Neutral Density Louvered Filter.

6.7 Louvered Filters

Louvered filters are very effective in reducing the amount of bright artificial light or daylight reflected from the face of a display, without a substantial reduction in display emitted light. The construction of a louvered filter is diagrammed in Figure 6.7-1. Inside a plastic sheet are thin parallel louvers which may be oriented at a specific angle with respect to the surface normal. The zero degree louvered filter has the louvers perpendicular to the filter surface.

The operation of a louvered filter is similar to a venetian blind as shown in Figure 6.7-2. Light from the LED display passes between the parallel louvers to the observer. Off-axis ambient light is blocked by the louvers and therefore is not able to reach the face of the display to be reflected back to the observer. This results in a very high contrast ratio with minimal loss of display emitted light at the on-axis viewing angle. The trade-off is a restricted viewing angle. For example, the zero degree louvered filter shown in Figure 6.7-1 has a horizontal viewing angle of 180°; however, the vertical viewing included angle is 60°. The louver aspect

AVAILABLE OPTIONS FOR LOUVERED FILTERS—
ANY COMBINATION IS POSSIBLE

ASPECT RATIO AND VIEWING ANGLE	LOUVER ANGLE	LOUVER COLOR
2.75:1 = 60° 2.00:1 = 90° 3.50:1 = 48°	0° 18° 30° 45°	OPAQUE BLACK TRANSLUCENT GRAY TRANSPARENT BLACK

EXAMPLE: 2.75:1 – 18° – TRANSPARENT BLACK

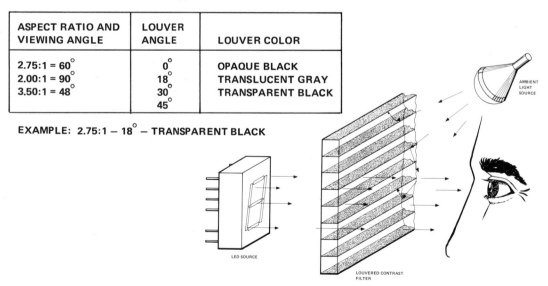

Figure 6.7-2 The Operation of a Louvered Filter.

6.14

ratio (louver depth/distance between louvers) determines viewing angle. A list of louver option possibilities is given in Figure 6.7-2.

Some applications require a louver orientation other than zero degrees. For example, an 18 degree louvered filter may be used on the sloping top surface of a point of sale terminal. A second, is the use of a 45 degree louvered filter on overhead instrumentation to block out ambient light from ceiling mounted lighting fixtures.

Louvered filters are effective filters for enhancing the viewing of LED displays installed in equipment operating under daylight ambient conditions. In bright sunlight, the most effective filter is the crosshatch louvered filter. This is essentially two zero degree neutral density louvered filters oriented at 90 degrees to each other, as illustrated in Figure 6.7-3. Red, yellow and green digits may be mounted side by side in the same display. Using only the crosshatch filter, all digits will be clearly visible and easily read in bright sunlight as long as the sunlight is not parallel to the viewing axis. The trade-off is restricted vertical and horizontal viewing. The effective viewing cone is an included angle of $40°$ degrees (for a filter aspect ratio of 2.75:1).

Neutral density louvered filters are effective by themselves in most bright ambient lighting conditions without the aid of a secondary wavelength filter. However, colored louvered filters may be used for additional wavelength filtering at the expense of display emitted light.

A most effective filter which provides exceptional contrast in bright sunlight is the $45°$ louvered filter. The difference from standard cross hatch is that the louvers are at a $45°$ angle with respect to the edge of the filter material as shown in Figure 6.7-4. The louvers are transparent black, with a cross light transmission of 12% to 15%. This small amount of cross transmission virtually eliminates the double image due to ghosting which was a problem in earlier designs. The horizontal and vertical included viewing angles are increased to about $60°$ (for a louver aspect ratio of 2.75:1).

A combination that provides significant contrast improvement for standard red LED displays being used in bright sunlight is the $45°$ cross hatch with a purple tint to the filter plastic. The purple tint provides a color contrast for the illuminated display segments, so they may be recognized when the sunlight rays are somewhat parallel the viewing axis.

The combination of a $45°$ cross hatch filter and a fresnel lens in tinted plastic form a special magnifying filter for LED displays. Both the cross hatch and fresnel grooves are made extra fine so that they are virtually invisible to the eye. The result is a superior lens/filter combination of

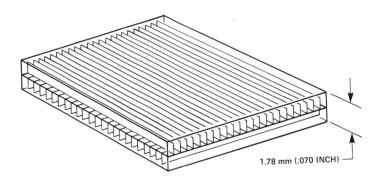

Figure 6.7-3 Conceptual Sketch of a Cross Hatch Filter, Two Louvered Filters Bonded Together in a 90° Orientation with Respect to Each Other.

minimum thickness that produces a magnified digit which is visible in bright daylight. Figure 6.7-5 illustrates a barrel lens magnifier in combination with a $45°$ cross hatch for use with unmagnified monolithic LED displays.

Louvered filters are available with either a "light matte" or "very light matte" anti-reflection surface which does not produce a fuzziness to the display appearance. Another option is a very hard scratch resistant surface which is incorporated directly into the plastic.

3M Company, Light Control Division, manufactures louvered filters for LED displays. Their product trade name is "Light Control Film", which is useable with all LED display and lamp products.

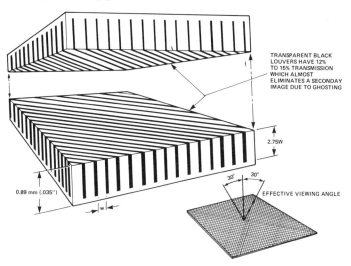

Figure 6.7-4 45° Cross Hatch Filter for Use with LED Displays in Bright Daylight Ambient.

6.8 Circular Polarizing Filters

Circular polarizing filters are effective when used with LED displays that have specular reflecting front surfaces. Specular reflecting surfaces reflect light without scattering. Displays that have polished glass or plastic facial surfaces belong to this category.

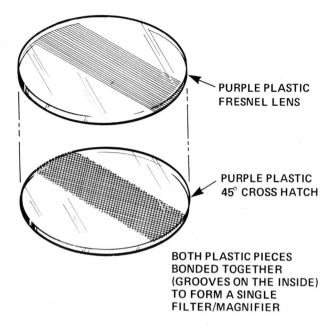

PURPLE PLASTIC
FRESNEL LENS

PURPLE PLASTIC
45° CROSS HATCH

BOTH PLASTIC PIECES
BONDED TOGETHER
(GROOVES ON THE INSIDE)
TO FORM A SINGLE
FILTER/MAGNIFIER

Figure 6.7-5 A Fresnel Lens Magnifier/45° Cross Hatch Purple Filter for Use with Unmagnified Standard Red Monolithic LED Displays.

The operation of a circular polarizer may be described as follows. As shown in Figure 6.8-1, the filter consists of a laminate of linear polarizer and a quarter wave plate. The quarter wave plate has its optical axis parallel to the flat surface of the polarizer and is oriented at 45° to the linear polarization axis. Non-polarized light is first linearly polarized by the linear polarizer. The linearly polarized light has x and y components with respect to the quarter wave plate. As the light passes through the quarter wave plate, the x and y components emerge 90° out of phase with each other. The polarized light now has x and y forming a helical pattern with respect to the optical path, and is termed circular polarized light. As this circular polarized light is reflected by the specular reflecting surface, the circular polarization is reversed. When the light passes back through the quarter wave plate it becomes linearly polarized at 90° to the linear polarizer. Thus reflected ambient light is blocked. The advantage of a circular polarizer is that reflected ambient light is reduced by more than 95%. However, the trade-off is that display emitted light passing through the circular polarizer is reduced by approximately 65% at the peak wavelength. This then necessitates an increased drive current for the display, more than that required for a wavelength filter.

Circular polarizers are normally colored to obtain additional selected wavelength filtering. One Caution: outdoor applications will require the use of an ultraviolet, uv, filter in front of the circular polarizer. Prolonged exposure to ultraviolet light will destroy the filter's polarizing properties.

Polaroid Corporation manufactures circular polarizing filters in the United States. In Europe, E. Käseman of West Germany produces high quality circuilar polarizers.

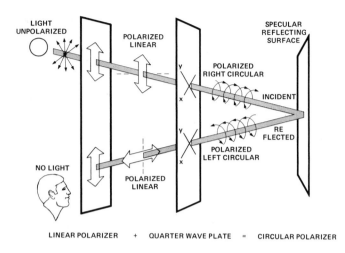

Figure 6.8-1 The Operation of a Circular Polarizer.

6.9 Anti-Reflection Filters, Mounting Bezels and Other Suggestions

Anti-reflection filters: A filtered display still may not be readable by an observer if glare is present on the filter surface. Glare can be reduced by the addition of an anti-reflection surface as part of the filter. Both sections of the display shown in Figure 6.9-1 are filtered. The left hand filter has an anti-reflection surface while the right hand filter does not.

An anti-reflection surface is a mat, or textured, finish or coating which diffuses incident light. The trade-off is that both incident ambient and display emitted light are diffused. It is therefore desirable to mount the filter as close to the display as possible to prevent the display image from appearing fuzzy.

Panelgraphic Chromafilters® come standard with an anti-reflection coating. SGL Homalite offers two grades of a molded anti-reflection surface. 3M Company and Polaroid also offer anti-reflection surface options. Optical coating companies will apply anti-reflection coating for specialized applications, though this is usually an expensive process.

Mounting bezels: It is wise to take into account the added appearance of a front panel that has the display set-off by a bezel. A bezel of black plastic, satin chrome or brushed aluminum, as examples, will accent the display and attract the eye of the observer. The best effect can be achieved by a custom bezel. Commercial black plastic bezels for digits up to 7.62mm (.3 inch) tall are available.

Other suggestions: When designing the mounting configuration of a display, consider recessing the display and filter 6.35mm (0.25 inch) to 12.7mm (0.5 inch) to add some shading effect. If a double sided printed circuit board is used, keep traces away from the normal viewing area or

Figure 6.9-1 The Effect of an Anti-Reflection Surface on an Optical Filter.

cover the top surface traces with a dark coating so they can not be seen. Mount the display panel in such a manner as to be easily removed if service should become necessary. If possible, mount current limiting resistors on a separate board to reduce the ambient temperature in the vicinity of the displays.

Filter Product	Type of LED Display	Ambient Lighting
Panelgraphic Chromafilter® with Anti-Reflection		
Ruby Red 60	Standard Red	Moderate
Dark Red 63		Bright
Purple 90		Bright
Scarlet Red 65	High-Efficiency Red	Moderate
Yellow 27	Yellow	Moderate to Bright
Green 48	Green	Moderate
Gray 10	**Yellow High-Efficiency Red**	**Sunlight**
SGL Homalite, Grade 100		
H100-1605	Standard Red	Moderate
H100-1804 (Purple)		
H100-1670	**High-Efficiency Red**	Moderate
H100-1726	Yellow	Dim
H100-1720		Moderate
H100-1440	Green	Dim
H100-1425		Moderate
H100-1266 (Gray)	**Yellow High-Efficiency Red**	**Sunlight**

Anti-Reflection
LR-72; 0.5 inch (12.70mm) Mounting Distance From Display
LR-92; Up to 3.0 inch (76.20mm) Mounting Distance From Display

Filter Product	Type of LED Display	Ambient Lighting
Rohm & Haas		
Plexiglas 2423	Standard Red	Moderate
Oroglas 2444		
3M-Company — Louvered Filters		
Red 655	Standard Red	Bright
Violet		
Red 625	High-Efficiency Red	Bright
Amber 590	Yellow	Bright
Green 565	Green	Bright
Neutral Density	All	Moderate to Bright
Schott Optical Glass		
RG-645	Standard Red	Bright
RG-630		Moderate
RG-610	High-Efficiency Red	Moderate

TABLE 6.9-1 Specific Wavelength Filter Products

Manufacturer	Product
Panelgraphic Corporation 10 Henderson Drive West Caldwell, New Jersey 07006 Phone: (201) 227-1500	Chromafilter® — Wave length filters with anti-reflective coating; Red, Yellow, Green
Thorn/Panelgraphic Great Cambridge Road Enfield, Middlesex ENGLAND Phone: 01-366-1291	
SGL Homalite 11 Brookside Drive Wilmington, Delaware 19804 Phone: (302) 652-3686	Wavelength filters; two optional anti-reflective surfaces; three plastic grades; Red, Yellow, Green
SGL Homalite Comtronic GMBH D8000 Munich 90 Theodolinden Str 4 GERMANY Phone: (089) 643 011	

3M - Company
Industrial Optics
3M Center, Bldg. 225-3
St. Paul, Minnesota 55101
  Phone: (612) 733-4403
	3M — Brand
	  Light control film; louvered filters

3M Europe S.A. 53/54 Avenue Des Arts 1040 Bruxelles, Belgium Phone: 12-39-00	
Rohm and Hass Independence Mall West Philadelphia, Penn. 19105 Phone: (215) 592-3000	Plexiglas; sheet and molding powder; wavelength filters
Polaroid Corporation Technical Polarizer Division 20 Ames Street Cambridge, MA 02139 (617) 577-3315	Circular polarizing

E. Käsemann GmbH D 8203 Oberaudorf WEST GERMANY Phone: (08033) 342	Circular Polarizing filters
Schott Optical Glass Duryea, Pennsylvania 13642 Phone: (717) 457-7485	Glass filters
Physikalische Optik Jenaer Glaswerk Schott Gen. Hattenberg Stra. 10 Mainz, W. GERMANY Phone: (06131) 6061	
Norbex Division Griffith Plastics Corporation 1027 California Drive Burlingame, California 94010 Phone: (415) 344-7691	DIGIBEZEL®; Plastic bezels for LED displays
Industrial Electronic Engineers, Inc. 7720-40 Lemona Avenue Van Nuys, California 91405 Phone: (213) 787-0311	Plastic bezels for .30 inch (7,62mm) tall LED displays
Rochester Digital Displays, Inc. 120 North Main Street Fairport, New York 14450 Phone: (716) 223-6855	Complete mounting kits for H.P. 5082-7300, -7700 and -7600 displays.

TABLE 6.9-2 List of Filter and Bezel Product Manufacturers

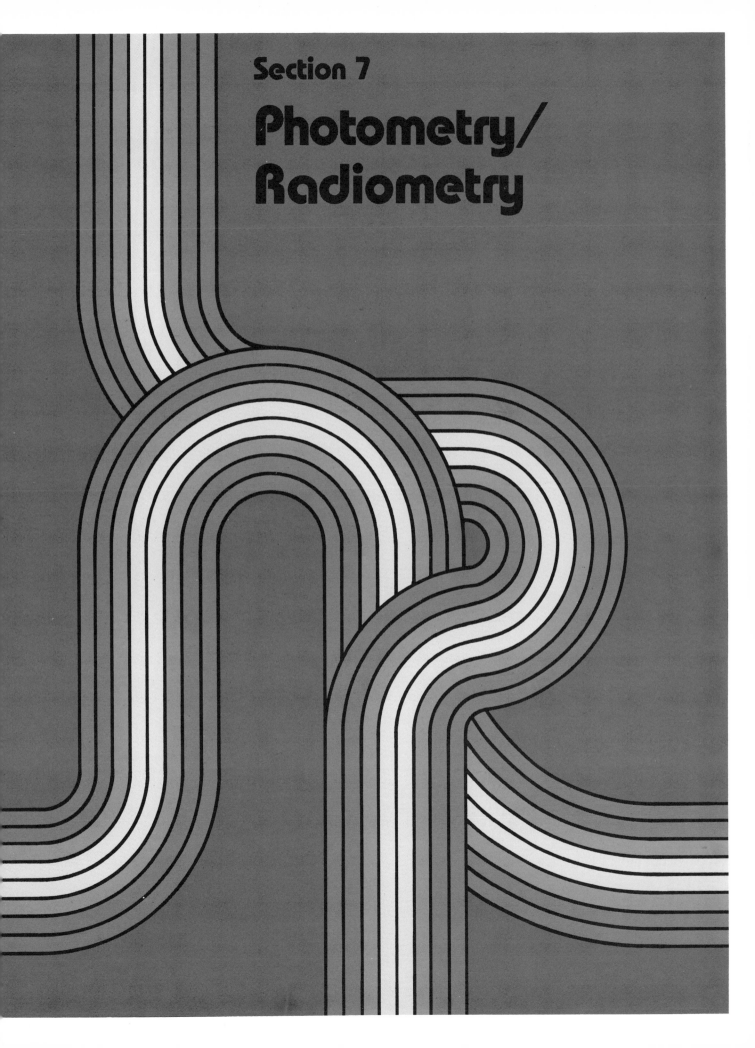

Section 7
Photometry/
Radiometry

7.0   PHOTOMETRY AND RADIOMETRY

7.1   Spectral Relationships

Photometry deals with flux (in lumens) at wavelengths that are visible, so the unit symbols have the subscript 'v' and the unit names have the prefix "luminous".

Radiometry deals with flux (in watts) at all wavelengths of radiant energy, so the unit symbols have the subscript 'e' and the unit names have the prefix "radiant".

Except for the difference in units of flux, radiometric and photometric units are identical in their geometrical concepts. Luminous flux is related to radiant flux by means of the "luminosity function", V_λ, also known as the "standard observer curve" or the "CIE curve". At the peak wavelength, 555 nm, of the luminosity function, the conversion factor is 680 lumens per watt. The luminosity function is given in Figure 7.1-1 on a log scale, so the accuracy is the same at all wavelengths. Values of V_λ taken from Figure 7.1-1 are accurate enough for all but the most exacting calculations.

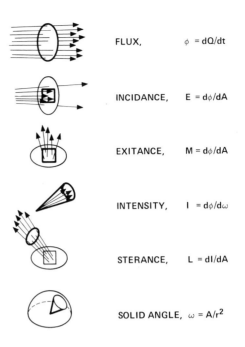

FLUX,	$\phi = dQ/dt$
INCIDANCE,	$E = d\phi/dA$
EXITANCE,	$M = d\phi/dA$
INTENSITY,	$I = d\phi/d\omega$
STERANCE,	$L = dI/dA$
SOLID ANGLE,	$\omega = A/r^2$

Figure 7.2-1 Generic Terms and Symbols and Their Geometrical Relationships.

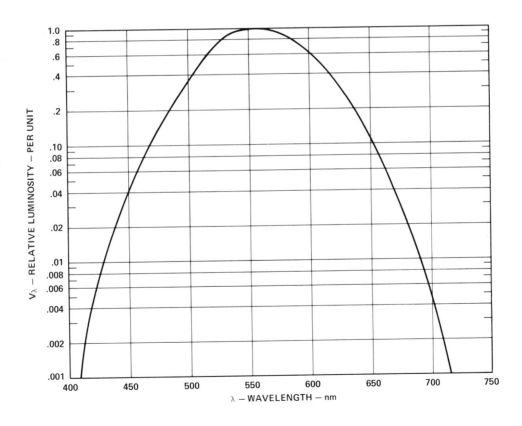

Figure 7.1-1 CIE Relative Luminosity Function. At Peak, 1 watt = 680 lumens.

For a spectrum of radiant flux, $(d\varphi_e/d\lambda)$, the total radiant flux is obtained by spectral integration:

$$\varphi_e(w) = \int_0^\infty (\frac{d\varphi_e}{d\lambda})\, d\lambda \qquad (7.1\text{-}1)$$

Luminous flux integration requires that each spectral element be weighted by the value of the luminosity function. It is therefore found as:

$$\varphi_v(\ell m) = 680\,(\frac{\ell m}{w}) \int_0^\infty (\frac{d\varphi_e}{d\lambda})\, V_\lambda\, d\lambda \qquad (7.1\text{-}2)$$

Conversion of photometric units to radiometric units is conveniently done by the luminous efficacy, η_v, which is just the ratio of the luminous flux to the radiant flux:

$$\eta_v(\frac{\ell m}{w}) = \frac{\varphi_v(\ell m)}{\varphi_e(w)} \qquad (7.1\text{-}3)$$

Luminous efficacy can be evaluated from any relative spectrum, i.e., one which is normalized, since the normalization factors cancel in the numerator and denominator of equation (7.1-3). Typical values of luminous efficacy for LEDs are:

	λ_{PEAK}(nm)	η_v(ℓm/w)
"STANDARD RED"	655	60
HIGH-EFFICIENCY RED	635	135
YELLOW	585	540
GREEN	565	640

7.2 Geometrical Relationships

Generically, there are only five units (and symbols) for radiant energy, as illustrated in Figure 7.2-1, and quantified in Table 7.2-1 and Table 7.2-2.

ϕ – FLUX, describes the rate at which energy is passing to, from, or through a surface or other geometrical entity.

E – INCIDANCE, describes the flux per unit area normally (perpendicularly) incident upon a surface.

M – EXITANCE, describes the flux per unit area leaving (diverging) from a source of finite area.

I – INTENSITY, describes the flux per unit solid angle radiating (diverging) from a source of <u>finite area</u>.

L – STERANCE, describes the intensity per unit area of a source.

ω – SOLID ANGLE; a solid angle, ω, with its apex at the center of a sphere of radius, r, subtends on the surface of that sphere an area, A, so that $\omega = A/r^2$ in steradians (sr).

Adding subscripts and prefixes quantifies these as radiometric or photometric units, e.g.:

I_e – RADIANT INTENSITY – watts per steradian, (W/sr)

I_v – LUMINOUS INTENSITY – lumens per steradian (lm/sr) or candelas (cd). cd = lm/sr

L_v – LUMINOUS STERANCE – candelas per square meter

As a practical matter, **flux** is used mainly in relating to the other terms. Few applications actually utilize all the flux available from a source. The same is true for **exitance**. The only situation in which "exitance" and "flux" have practical significance is with a receptor so tightly coupled to a source that virtually all the flux leaving the source enters the receptor, such as in "sandwich" type optoisolators.

Incidance has the same units as excitance but is of a vastly different nature. Exitance ignores the direction taken by the exiting flux. Incidance takes account of the flux component which is normal to a surface. If the flux direction at a surface is not normal, then the incidance is just the normal component of the angularly incident flux density. Incidance is most useful in describing photodetector properties.

Intensity is an extremely useful concept in both photometry and radiometry, and the candela unit of luminous intensity is the only universally utilized photometric unit. (Other photometric quantities are plagued by a profusion of English, metric, MKS, and CGS units of length.) Since flux passing through space is usually divergent, it is usually possible to define an equivalent point from which it diverges in terms of a solid angle, the flux therein, and hence the intensity of the equivalent point. As seen in Section 7.3, intensity is the most easily measured and most uniformly repeatable quantity. Therefore, although sterance is more fundamentally significant in many applications, the actual parameter to be measured for performance verification should be the intensity.

Sterance is most significant because of its constance in an optical (lens) system. This is discussed in much detail in the National Bureau of Standards publication NBS Technical Note 910-1 issued in March 1976. While magnification and image position varies, sterance does not.

Luminous sterance (or luminance) is the luminosity basis for visibility (i.e., distinguishing an object from its background). As such, luminous sterance is a useful measure for contrast and photographic exposure. However, luminous sterance is not the only basis for visibility -- color is another. For example, spots of brown gravy on a green tie are visible, even if the spots have the same luminous

GENERIC   RADIOMETRIC   PHOTOMETRIC

Term, Definition	Symbol	Defining Equation	New Term (Old Term)	Symbol	SI Units, Abbr.	New Term (Old Term)	Symbol	SI Units, Abbr.	Equivalent
FLUX, rate of flow of energy, Q. Q_e – radiant, Q_v – luminous, Q_q – photon [see note 1]	ϕ	$\dfrac{dQ}{dt}$	Radiant Flux (Radiant Power)	ϕ_e (P)	watts, W	Luminous Flux (Luminous Flux)	ϕ_v	lumens, lm	
INCIDANCE, flux per unit area on a reception surface	E	$\dfrac{d\phi}{dA}$	Radiant Incidance (Irradiance)	E_e (H)	watts per sq. meter, W/m²	Luminous Incidance (Illuminance or Illumination)	E_v (E)	lux, lx [see note 3]	lumens per sq. meter, lm/m²
EXITANCE, flux per unit area from an emitting surface	M	$\dfrac{d\phi}{dA}$	Radiant Exitance (Emittance)	M_e (W)	watts per sq. meter, W/m²	Luminous Exitance	M_v	lumens per sq. meter, lm/m²	[see note 2]
INTENSITY, flux per unit solid angle from a remote source	I	$\dfrac{d\phi}{d\omega}$	Radiant Intensity (Radiant Intensity)	I_e (J)	watts per steradian, W/sr	Luminous Intensity (Luminous Intensity)	I_v (I)	candelas, cd	lumens per steradian, lm/sr
STERANCE, flux per unit solid angle per unit area of emitting surface at angle θ with respect to surface normal.	L	$\dfrac{dI}{dA \cos\theta}$ $\dfrac{d^2\phi}{d\omega\, dA \cos\theta}$	Radiant Sterance (Radiance)	L_e (N)	watts per steradian per sq. meter, W/sr/m²	(Luminous Sterance) (Luminance)	L_v (B)	candelas per sq. meter, cd/m² [see note 3]	lumens per steradian per sq. meter, lm/sr/m² nit, nt

Note 1. Quantametric terms use the prefix word "photon" and their symbols have a subscript, "q".
Note 2. Lux and other units of luminous incidance DO NOT APPLY to luminous exitance.
Note 3. Other units in Table 7.2-2.

TABLE 7.2-1 Terms, Definitions, Symbols, and Units for Energy Measurement

7.3

In general, confusion is averted by use of self-explanatory units, such as: lumens per square meter for luminous incidance; and, candelas per square meter for luminous sterance. At times, for brevity, where the risk of confusion is negligible, other units used are:

Luminous Sterance		Lambert L	Footlambert fL	Apostilb asb	*stilb sb	cd/ft²	**cd/m²	nit
Unit, Abbr.	Equivalent							
Lambert, L	$1/\pi$ cd/cm² =	1	929	10,000	.3183	295.7	3183	3183
Footlambert, fL	$1/\pi$ cd/ft² =	.001076	1	10.76	.0003426	.3183	3.426	3.426
Apostilb, asb	$1/\pi$ cd/m² =	.0001	.0929	1	.00003183	.02957	.3183	.3183
*Stilb, sb	cd/cm² =	3.1416	2919	31,416	1	929	10,000	10,000
Candelas per sq. foot	cd/ft² =	.003382	3.1416	33.82	.001076	1	10.76	10.76
**Candelas/sq. meter	cd/m² =	.00031416	.2919	3.1416	.0001	.0929	1	1

Luminous Incidance		*Phot ph	*Footcandle fc	**,*Lux lx
Unit, Abbr.	Equivalent			
*Phot, ph	lm/cm²	1	929	10,000
*Footcandle, fc	lm/ft²	.001076	1	10.76
**,*Lux, lx	lm/m²	.0001	.0929	1

*CIE Unit
**Recommended SI Unit
Many of the other units are in common usage, but efforts are being made to standardize on the SI units.

USE OF TABLE: In any row, the quantities are all equal; for example, in the table for luminous sterance, cd/m² = .00031416 L = .2919 fL = 3.1416 asb = .0001 sb = .0929 cd/ft² = 1 nt; in the table for luminous incidance, lx = .0001 ph = .0929 fc.

TABLE 7.2-2 Conversion Factors for Luminous Sterance and Luminous Incidance

DEVICE DESCRIPTION	APPARENT EMITTING AREA (mm^2)	STERANCE INTENSITY , L_v/I_v	
		cd/m^2 per mcd	fL per mcd
LEDs			
T-1 3/4 SIZE DIFFUSED	3.0	330	97
UNDIFFUSED	0.5	2,000	580
T-1 SIZE DIFFUSED	1.2	830	240
UNDIFFUSED	0.3	3,300	970
SUBMINIATURE	0.4	2,500	730
HERMETIC	2.0	500	150
RECTANGULAR	18.0	56	16
DOT MATRIX DISPLAYS	0.09	11,000	3,200
LARGE 7-SEGMENT DISPLAYS			
10.9 mm (.43 in.) TALL	4.4	230	66
7.6 mm (.3 in.) TALL	1.7	590	170
MONOLITHIC 7-SEGMENT			
LARGE CHIP (5082-74XX)	0.18	5,600	1,600
SMALL CHIP (5082-743X)	0.24	4,200	1,200
CALCULATOR DIGITS			
8, 9 CLUSTER (5082-7440 SERIES)	0.17	5,900	1,700
12, 14 CLUSTER (-7442/4/5/7)	0.15	6,700	1,900
8, 9 CLUSTER (5082-7240 SERIES)	0.18	5,600	1,600
5,15 CLUSTER (-7265/75/85/95)	0.27	3,700	1,100
CHIPS			
1.35 x 1.50 mm (-7811/21)	0.041	24,000	7,100
1.42 x 2.24 mm (-7832/42)	0.085	12,000	3,400
1.91 x 2.72 mm (-7851/61/52/62)	0.11	9,100	2,700
2.34 x 3.25 mm (-7871/81)	0.13	7,700	2,200
0.38 x 0.38 mm (-7890/93)	0.063	16,000	4,600

$$\frac{L_v \; (\text{cd/m}^2)}{I_v \; (\text{mcd})} = \frac{L_v \; (\text{cd/m}^2)}{I_v \; (\text{mcd})} \left(\frac{I_v \; (\text{mcd})}{10^3 \, I_v \; (\text{cd})} \right) = \frac{10^{-3}}{A \; (\text{m}^2)} = \frac{10^3}{A \; (\text{mm}^2)} = \frac{10}{A \; (\text{cm}^2)}$$

$$\frac{L_v \; (\text{fL})}{I_v \; (\text{mcd})} = \frac{L_v \; (\text{fL})}{I_v \; (\text{mcd})} \left(\frac{\pi \, L_v \; (\text{cd/ft}^2)}{L_v \; (\text{fL})} \right) \left(\frac{I_v \; (\text{mcd})}{10^3 \, I_v \; (\text{cd})} \right) = \frac{\pi}{10^3} \cdot \frac{1}{A \; (\text{ft}^2)} \left(\frac{92903 \, A \; (\text{ft}^2)}{A (\text{mm}^2)} \right)$$

$$= \frac{\pi \cdot 92.9}{A \; (\text{mm}^2)} = \frac{291.9}{A \; (\text{mm}^2)}$$

$$1 \; \text{cd/m}^2 = 0.2919 \; \text{fL}$$

$$1 \; \text{fL} = 3.426 \; \text{cd/m}^2$$

TABLE 7.2-3 Intensity-to-Sterance Conversion $\left(\frac{L_v}{I_v} = \frac{1}{A} \right)$

sterance as the tie. On a brown tie of the same color as the gravy, a spot would be visible only if its luminous sterance is different from that of the tie by a factor of two or more.

While luminous sterance is the basis for visibility, luminous intensity is the popularly preferred performance parameter for LEDs and LED displays. The basic reasons are:

RELEVANCE — Most applications utilize the light radiated by all portions of the emitting area, but sterance is a measure of the intensity of only a small incremental portion. A large area device could have a higher intensity than a device of higher sterance but smaller area. Luminous intensity, being the product of sterance times area, ranks devices properly according to their effectiveness.

VIEWING ANGLE (RADIATION PATTERN) — Sterance of an LED remains nearly constant at all viewing angles but the projected light-emitting area does not. Luminous intensity, the product of sterance and area gives proper radiation pattern information.

EASE OF MEASUREMENT — Sterance measurement requires an optical system to image and field-stop the emitting surface; intensity measurement requires only a calibrated detector and a scale to find the LED-to-detector distance.

To understand the relationship between sterance and intensity, consider a uniform source whose area is one square meter and having an intensity of 1000 candelas. Its sterance is then 1000 cd/m^2. If this area were subdivided into 10^6 pieces, each having an area of one square millimeter, the intensity of each piece would be 10^{-3} candelas or one millicandela, but the sterance of each piece would still be 1000 cd/m^2.

Sterance can be an important consideration in product selection for a given application but is a poor parameter for specification. The figures in Table 7.2-3 give the sterance-to-intensity ratios for a number of product groups. They are simply the inverse of emitting area, with appropriate adjustment of units:

$$I_v \text{ (cd)} = L_v \text{ (cd/m}^2) \times A \text{ (m}^2) \qquad (7.2\text{-}1)$$

$$\frac{L_v}{I_v} \left(\frac{\text{cd/m}^2}{\text{mcd}}\right) = \frac{10^3}{A \text{ (mm}^2)} = \frac{10}{A \text{ (cm}^2)} \qquad (7.2\text{-}2)$$

7.3 Photometric and Radiometric Measurements

7.3.1 Spectral Effects

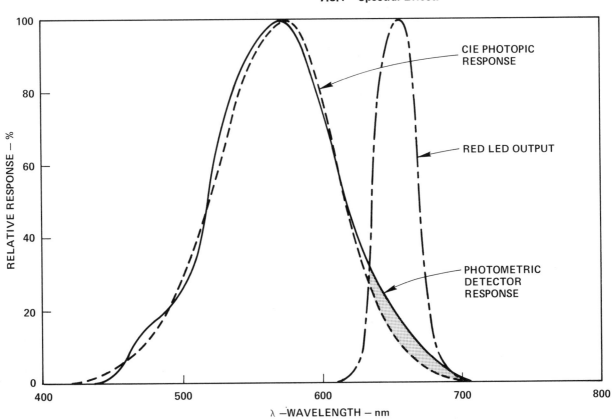

Figure 7.3.1-1 Spectral Response of a Photometer For Broad-Spectrum Radiation.

In radiometry, spectral effects are not much of a problem because radiometers can usually be made to have very nearly flat spectral response over the spectrum of interest. The flatest and broadest response is obtained from thermal-type radiometers (e.g. thermopile or thermocouple). Quantum-type (e.g. silicon diode) radiometers are much more sensitive than thermal-type, and have a higher speed of response; while the spectral flatness, is not as good, it is adequate for most LED measurements.

In photometric measurement of LEDs, it is important that the photometer be calibrated at the wavelength of the LED spectrum. This is necessary because most photometers do not have a spectral response that fits the luminosity function precisely at all wavelengths. To achieve the best overall fit, the response is adjusted (by colored filters) so that, over the spectrum, negative and positive deviations are balanced. This provides precise calibration for photometry of incandescent and other broad-spectrum sources, but for the narrow-spectrum LED sources, the entire spectrum might lie within a wavelength region where the photometer deviation is all positive or all negative, as seen in Figure 7.3.1-1.

In addition, for red LEDs, the spectral slope of the photometer must be very close to the $-.25$ dB per nm slope of the luminosity function. This is because LED spectra change with temperature and drive conditions, shifting as much as 10 nm. If a photometer has $-.20$ dB/nm (instead of the correct $-.25$ dB/nm), the incremental error would be 0.5 dB (+12%).

Suitable photometers are available from EG&G, Photo-Research Corp., Tektronix, and United Detector Technology.

7.3.2   Intensity Measurement

If the source to be measured has a plane of emission with respect to which intensity can be defined, then all that is necessary for intensity measurement is a photometer to read the incidance, E_v at a distance, d, from the source. The governing relationship is illustrated in Figure 7.3.2-1a. The flux, φ_v (lm) in the cone may be referred to the source as the product of the intensity, I_v (lm/sr) times the solid angle, ω (sr) of the cone, or it may be referred to the sensor as the product of the incidance, E_v (lm/m^2) times the area A (m^2). Since both relationships describe the same flux, they can be equated:

$$I_v \left(\frac{\ell m}{sr}\right) \times \omega \text{ (sr)} = E_v \left(\frac{\ell m}{m^2}\right) \times A \text{ (m}^2\text{)} \qquad (7.3.2\text{-}1)$$

If the distance, d, is large enough, then the value of ω can be expressed as:

$$\omega = A/d^2 \qquad (7.3.2\text{-}2)$$

Substitution in equation (7.3.2-1) gives the result:

$$I_v \left(\frac{\ell m}{sr}\right) = E_v \left(\frac{\ell m}{m^2}\right) \times d^2 \text{ (m}^2\text{)} \qquad (7.3.2\text{-}3)$$

The fact that the left side is in lumens per steradian while the right side is in lumens may seem incorrect, but note that the steradian is a dimensionless unit.

The relationship, $I_v = E_v\, d^2$, is valid only if d is large enough that the solid angle can be described as $\omega = A/d^2$. For the error in this assumption to be less than 1%, it is necessary that the distance, d, be at least ten times the diameter of the photodetector or ten times the diameter of the source, whichever is greater. This is called the TEN DIAMETERS RULE. Its purpose is to make the solid angle small enough that the flat photometer receptor can be regarded as a portion of a spherical surface.

To be absolutely correct, intensity should be measured with an incrementally small solid angle. The TEN DIAMETERS RULE defines a solid angle of 7.84 msr which is slightly larger than the 4.0 msr recommended by experienced observers.

When the distance, d, cannot be defined, intensity measurement can be made by placing the photometer at two different distances from the source, as in Figure 7.3.1-1 (b and c). With the "approximate" method, making θ small improves the accuracy, as far as equation (7.3.2-2) and (7.3.2-3) are concerned, but if θ is too small there may be other problems. One other problem is getting enough flux for an adequate signal-to-noise ratio. Another is the possible variation of the ratio φ/ω; undiffused LEDs, especially with very narrow radiation pattern, can cause the φ/ω ratio to vary substantially. The "precise" two-point method, reduces potential radiation pattern problems by keeping the solid angle nearly the same for both observations. Note in the expression for I_v by the "precise" method, that provision is made for $\varphi_2 \neq \varphi_1$; however, if optical bench positioning can be adjusted so that $\varphi_2 = \varphi_1$, the accuracy will be limited mainly by the accuracy obtained in measuring A_1 and A_2. These areas can be found as a ratio to the photometer area, A_0, as follows:

1.   Place the photometer at large distance ($>$1m) from a small source ($<$10mm) of high intensity. (Wavelength and spectral effects are unimportant here.) Note the photometric reading as E_0.

2.   Place A_1 over the photometer and note the reading as E_1. Then $A_1 = A_0(E_1/E_0)$.

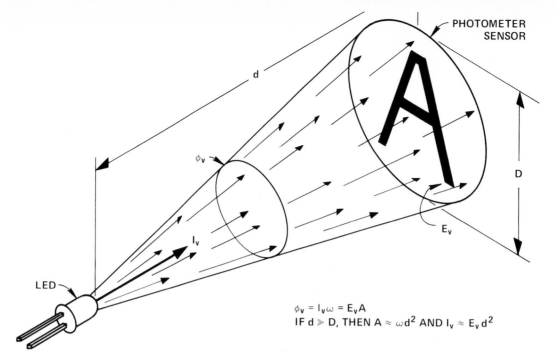

$$\phi_v = I_v \omega = E_v A$$
$$\text{IF } d \gg D, \text{ THEN } A \approx \omega d^2 \text{ AND } I_v \approx E_v d^2$$

**(a)   RELATIONSHIP BETWEEN LUMINOUS INTENSITY, I_v, OF AN LED
AND LUMINOUS INCIDANCE, E_v, AT THE PHOTOMETER**

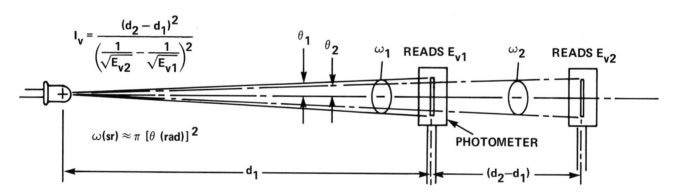

$$I_v = \frac{(d_2 - d_1)^2}{\left(\frac{1}{\sqrt{E_{v2}}} - \frac{1}{\sqrt{E_{v1}}}\right)^2}$$

$$\omega(\text{sr}) \approx \pi \, [\theta \, (\text{rad})]^2$$

(b)   APPROXIMATE TWO-POINT METHOD ASSUMES $\varphi_1/\omega_1 = \varphi_2/\omega_2$

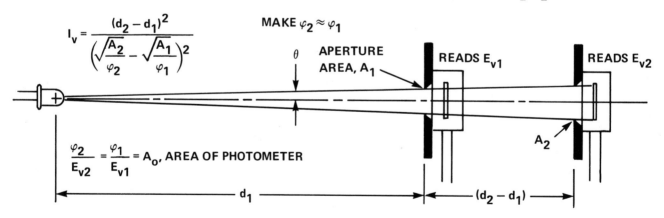

$$I_v = \frac{(d_2 - d_1)^2}{\left(\sqrt{\frac{A_2}{\varphi_2}} - \sqrt{\frac{A_1}{\varphi_1}}\right)^2}$$

MAKE $\varphi_2 \approx \varphi_1$

$$\frac{\varphi_2}{E_{v2}} = \frac{\varphi_1}{E_{v1}} = A_o, \text{ AREA OF PHOTOMETER}$$

**(c)   PRECISE TWO-POINT METHOD USES TWO APERTURES, A_1 AND A_2
SO THE SOLID ANGLE IS NEARLY THE SAME AT EACH POINT.
THIS REDUCES ERROR DUE TO RADIATION PATTERN PECULIARITIES.**

Figure 7.3.2-1   Intensity Measurement — Optical Arrangements.

3. Place A_2 over the photometer and note the reading as E_2. Thus $A_2 = A_0(E_2/E_0)$.

Several observations of A_1 and A_2 at different distances from the source will improve the accuracy.

7.3.3 Sterance Measurement

Sterance, being intensity per unit area, requires apparatus capable of resolving an incrementally small area. It is not actually necessary for this area to be precisely known because the measurement is usually done by comparing the unknown sterance to a sample of known sterance.

Figure 7.3.3-1 shows the arrangements that can be used for sterance measurement. Apparatus utilizing the beam splitter and aperture field stop is available commercially from Photo-Research Corp. and is called a brightness spot meter. Apparatus using the fiber-optic at the coaxial image plane is employed in the photometric microscope from Gamma Scientific Co. In both of these systems, a viewer can directly observe the precise portion of the LED for which the sterance is being measured. The area increment is the area of the fiber-optic core (or of the aperture field stop) divided by the square of the magnification of the objective lens. Spot diameters as small as .02 mm (.0008") can be resolved. The smaller the spot used, the larger will be the sterance variation observed across the surface of the source. This is the reason why sterance is a difficult specification to describe or use.

7.3.4 Flux Measurement

Total flux from a source is the integral of the flux radiated in all directions. If a source is placed at the center of a sphere, the total flux can be described as the integral of the incidence over the entire inside surface of the sphere. The completely general expression would require integration over both the azimuth angle (0 to 2π) and the polar angle (0 to π). However, most sources have radiation patterns that are very nearly a surface of revolution about the optical axis, so the integral is simplified as seen in Figure 7.3.4-1. That is, for a particular value of polar angle, θ, (also called the off-axis angle) the intensity is constant all the way around the polar axis. The differential flux can therefore be described as:

$$d\varphi = [I(\theta)] \times d\omega \qquad (7.3.4\text{-}1)$$

and the differential solid angle has the value:

$$d\omega = 2\pi \sin\theta \; d\theta \qquad (7.3.4\text{-}2)$$

Then the flux in a cone of half-angle θ is the integral:

$$\varphi(\theta) = \int_0^\theta I(\theta)\, 2\pi \sin\theta \; d\theta \qquad (7.3.4\text{-}3)$$

Total flux is that obtained for $\theta = \pi$:

$$\varphi = \int_0^\pi I(\theta)\, 2\pi \sin\theta \; d\theta \qquad (7.3.4\text{-}4)$$

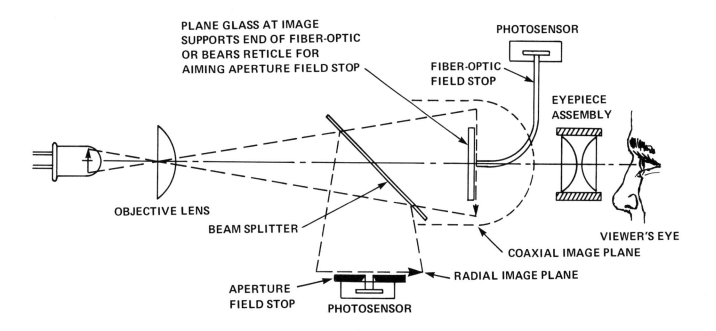

Figure 7.3.3-1 Sterance Measurement — Optical Arrangements.

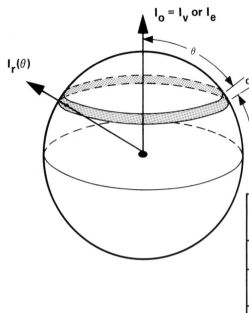

AXIAL INTENSITY

$$d\varphi = I(\theta) \times d\omega$$

$$= I_o I_r(\theta) \times d\omega$$

$$d\omega = 2\pi \sin\theta \, d\theta$$

$$\varphi(\theta) = I_o \int_o^\theta I_r(\theta) \, 2\pi \sin\theta$$

$$\varphi = I_o \int_o^\pi I_r(\theta) \, 2\pi \sin\theta \, d\theta$$

$$*\cos^n(\pi/2) = 0$$

$$\text{SO} \quad \varphi/I_o = \varphi(\pi/2)/I_o$$

RADIATION PATTERN	$I_r(\theta)$	$\theta_{1/2}$	$\varphi(\theta)/I_o$	φ/I_o
POINT SOURCE	1.00	NONE	$2\pi(1-\cos\theta)$	4π
LAMBERTIAN	$*\cos\theta$	$\dfrac{\pi}{3} = 60°$	$\pi\sin^2\theta$	π
COMMON APPROXIMATION FOR LEDs	$*\cos^n\theta$	$\cos^{-1}(.5^{1/n})$	$\dfrac{2\pi(1-\cos^{n+1}\theta)}{n+1}$	$\dfrac{2\pi}{n+1}$

Figure 7.3.4-1   Flux-to-Intensity Ratio; Analytic Integral of Radiation Pattern.

*Exponent n is derived from the LED radiation pattern and usually is a valid approximation out to $\theta_{1/2}$, where $I(\theta) = .5 \, I_o$. The value for n may be determined as follows: Assume $\cos^n(\theta_x) = y$, then $n = \text{Log}(y)/\text{Log}\cos\theta_x$.

$\varphi(\theta)$ and φ can be related to the axial intensity since the intensity at any angle, θ can be given as the product of the axial intensity, I_0, a constant, and the relative intensity, $I_r(\theta)$, for which $I_r(0) = 1.00$ as in customary radiation patterns. The axial intensity, of course, would be either I_v(cd) or I_e(w/sr).

$$\varphi(\theta) = I_o \int_o^\theta I_r(\theta) \, 2\pi \sin\theta \, d\theta \qquad (7.3.4-5)$$

Figure 7.3.4-1 gives the results of integrating radiation patterns with analytic functions. Also given is the value of $\theta_{1/2}$ which is the angle at which $I_r(\theta) = \frac{1}{2}$.

If the radiation pattern is not analytic but is a surface of revolution about the polar axis, then the integration is performed by converting equation 7.3.4-5 to a summation

$$\varphi(\theta) = I_o \, \Sigma \, I_r(\theta) \times [2\pi \sin\theta \, \Delta\theta] \qquad (7.3.4-6)$$

The term $[2\pi \sin\theta \, \Delta\theta]$ is called a zonal constant, C_Z, in which the incremental angle, $\Delta\theta$, is in radians. Of course, the magnitude of C_Z varies as $\sin\theta$, and depends on the size of the $\Delta\theta$ increment. A little experience will show that adequate accuracy (3%) is obtained by selecting $\Delta\theta$ equal to one fourth of the angle at which $I_r(\theta) = 0.5$ that is $\Delta\theta = (\theta_{1/2})/4$. Thus, if there are to be N equal parts over 180°, choose $N = 180°(4/\theta_{1/2}) = 720°/\theta_{1/2}$ rounding off to the next larger integer that yields convenient values for $\delta = 180°/N$.

Figure 7.3.4-2 shows how δ and N are used in zonal integration. The customary method appears acceptable in Figure 7.3.4-2, but if δ had been 5°, then it would have required using the values 2.5°, 7.5°, 12.5°, etc. for $I_r(\theta)$ and $C_Z(\theta)$. An accuracy check, using the analytic $\cos^n\theta$ radiation pattern, shows that with the customary method $\varphi(\theta)$ is slightly above the analytic (correct) value, while the modified method yields $\varphi(\theta)$ slightly below the correct value. Both methods have adequate accuracy.

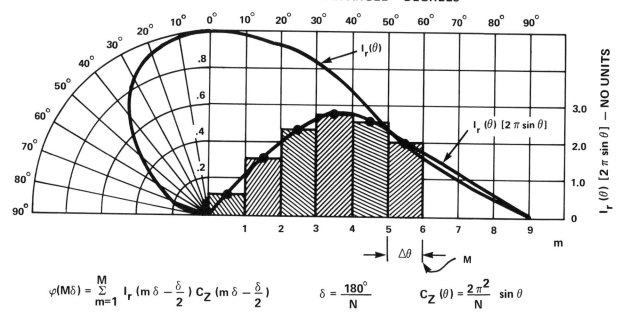

$$\varphi(M\delta) = \sum_{m=1}^{M} I_r\left(m\delta - \frac{\delta}{2}\right) C_Z\left(m\delta - \frac{\delta}{2}\right) \qquad \delta = \frac{180°}{N} \qquad C_Z(\theta) = \frac{2\pi^2}{N}\sin\theta$$

(a) CUSTOMARY METHOD – MAY REQUIRE AWKWARD VALUES FOR (mδ – δ/2)

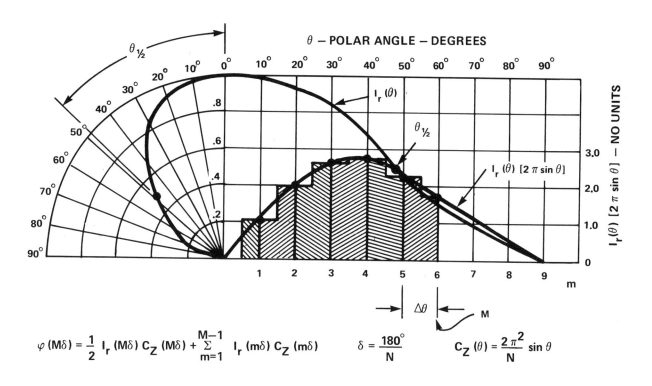

$$\varphi(M\delta) = \frac{1}{2} I_r(M\delta) C_Z(M\delta) + \sum_{m=1}^{M-1} I_r(m\delta) C_Z(m\delta) \qquad \delta = \frac{180°}{N} \qquad C_Z(\theta) = \frac{2\pi^2}{N}\sin\theta$$

(b) MODIFIED METHOD – CONVENIENT VALUES (mδ) FOR EVALUATING I_r(mδ) AND C_Z(mδ)

Figure 7.3.4-2   Zonal-Constant Flux Integration; 3% Accuracy with
N $\geqslant$ 720°/$\theta_{1/2}$.

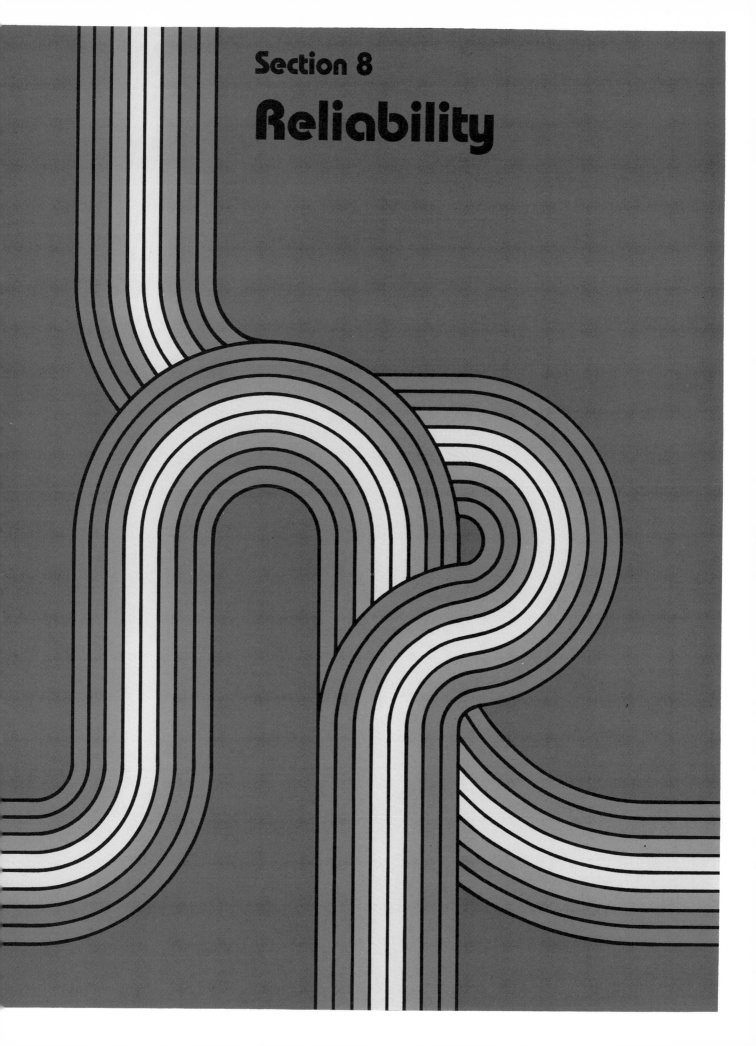

Section 8
Reliability

8.0 RELIABILITY OF OPTOELECTRONIC DEVICES

LED optoelectronic devices are solid state semiconductor p-n junctions. They are fabricated and packaged utilizing many of the technologies which are used in the manufacture of silicon and germanium semiconductor products. It is reasonable to expect, therefore, that LED products will exhibit reliability aspects similar to those observed in these other semiconductor technologies. An LED offers many reliability advantages with respect to other available light sources. Long life-time, wide operating temperature range, and reliable operation in adverse environmental conditions are only a few of the important attributes of LEDs when compared to other light emitting technologies. These attributes can, however, only be realized when a device is properly designed, fabricated, packaged, and operated. Optoelectronic device data sheets define the limits within which devices should be operated in order to achieve the expected reliability performance. It is important that the optoelectronic system designer be aware of how variations in these limits will effect the long term performance of a device.

The reliability of an optoelectronic device can be considered to be dependent upon two separate variable factors. These are the reliability characteristics of the LED semiconductor chip and the reliability characteristics inherent to the package into which that chip is assembled. The variables are not necessarily independent because, for instance, package failure which is not in itself catastrophic can trigger eventual failure or degradation of the semiconductor device.

8.1 Reliability Aspects of the LED Semiconductor Chip

From a theoretical standpoint, an LED like any p-n junction should have a nearly infinite lifetime. This operating life would be limited only by the infinitely slow natural diffusion of the dopants. In practice, an LED device when operated under reasonable stress conditions is normally expected to exhibit an operating lifetime of about 100K hours. In some cases, the end of operating life is defined as the point at which a device light output has been reduced to 50% of the initial value. This could occur slowly over an extended period or catastrophically as a result of a short or an open. The value of 50% has been chosen as a ratio at which an observer would be expected to be capable of detecting a visible change in the light output. There are no known failure modes in LED semiconductors which should result in occurrence of catastrophic failures after extended periods of operation (such as notching in a tungsten filament incandescent lamp which will result eventually in a catastrophic failure of the filament).

Conversely, the slow reduction of light output with time is a well known (though not well understood) characteristic

of an LED device. This degradation of light output exhibits a direct correlation to the current density in the junction. Figures 8.1-1 and 8.1-2 depict the normalized light output vs. time for direct gap and indirect gap plastic package lamps when operated at various junction current densities. As can be seen, increased current density resulted in a more rapid degradation and greater total degradation for stress times up to about 5000 hours. The maximum operating levels for the devices of Figure 8.1-1 represent a stress level of 200% of maximum device rating and 350% of maximum device rating in Figure 8.1-2. Figure 8.1-3 represents long term operating life data for hermetic package lamps operated at the maximum allowable device stress level for stress times up to 40,000 hours. This data represents the potential for LED devices to be successfully used for periods in excess of 100k hours.

Close inspection of Figures 8.1-1 to 8.1-3 will reveal that the rate of light output degradation decreases as a function of increasing stress time. Figure 8.1-4 depicts the range of the expected rate of degradation (in % per 100 hours) vs. stress time. This plot represents a composite of several different device types from many different production lots when operated at 20 mA dc. It is important to note that with increasing stress time, even those device lots which start out with a very high rate of degradation converge quickly to a very low degradation rate. This implies that devices showing a high initial degradation will not necessarily continue to degrade to failure.

The degradation of LED light output demonstrates some dependency on junction temperature. As can be seen in Figures 8.1-1 and 8.1-3, however, this dependency is so small as to be inconclusive.

The data sheet maximum dc operating level for an LED is generally established somewhat below the level at which unacceptable degradation would be encountered within 10k hours of operation. To achieve the lowest possible LED degradation consistent with useful operating levels, it is best to select an operating point at 40% to 50% of the maximum data sheet rating.

8.2 Reliability Aspects of LED Packaging

The package into which a semiconductor device is assembled should provide environmental protection for the device as well as providing electrical contacts and a mounting technique. For an LED, the package is often also required to provide an optical system to enhance coupling of the emitted light to the detector. An ideally designed and utilized package would perform these functions very reliably. There are, however, many variables which can affect this performance. To evaluate a package, tests such as solderability, temperature cycling, thermal shock, moisture resistance, mechanical shock, acceleration and terminal

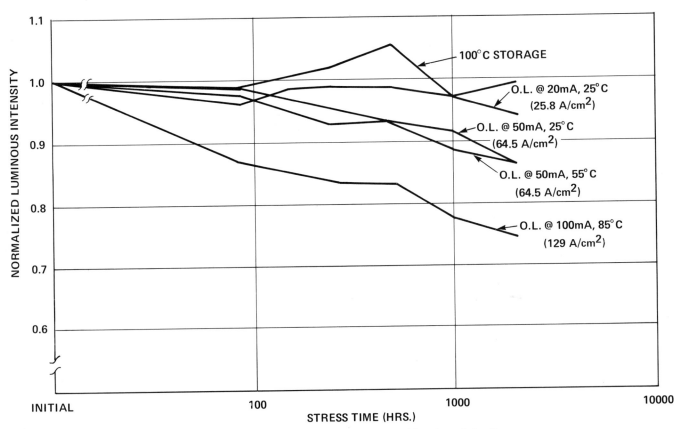

Figure 8.1-1 Direct Gap LED Lamp Normalized Luminous Intensity
vs. Operating Life and High Temperature Storage
T-1 3/4 Plastic Packages.

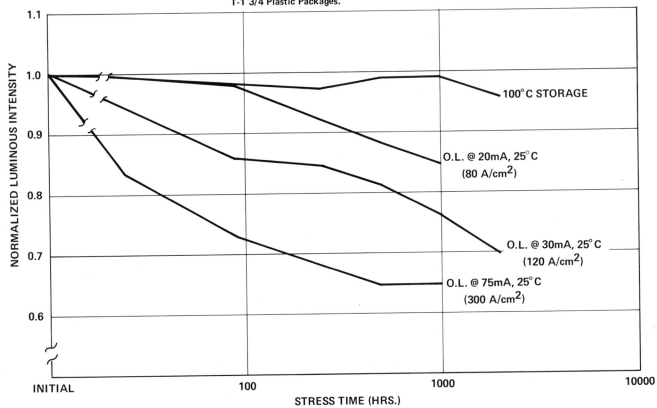

Figure 8.1-2 Indirect Gap LED Lamp Normalized Luminous
Intensity vs. Operating Life and High Temperature
Storage T-1 Plastic Package.

8.2

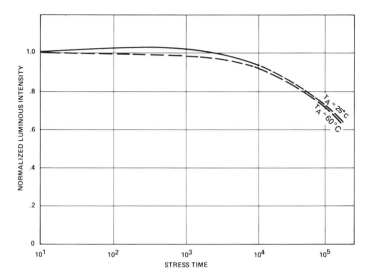

Figure 8.1-3 Direct Gap LED Lamp Normalized Luminous Intensity vs. Operating Life TO-18 Hermetic Package.

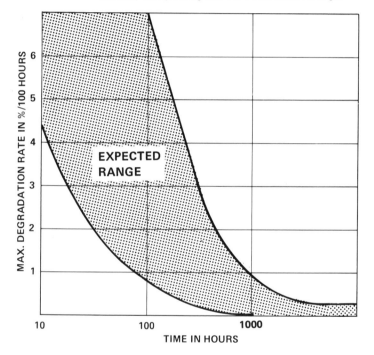

Figure 8.1-4 Expected Rate of Degradation vs. Time for T-1 3/4 Plastic Package.

strength are performed at various stress levels. These tests can be used to establish the worst case environmental stresses which should be applied to the package. Figure 8.2-1 is a copy of a typical plastic lamp reliability data sheet. Generally, such a data sheet will provide a reasonably good analysis of the performance a user can expect from a device. Deviation from this performance can, however, be caused by either variations in the manufacture of the package or by exceeding the defined environmental conditions in the end use product. For plastic packages, several environmental aspects are very important.

The package material is not impervious to the diffusion of water vapor, and long term exposure to high humidity will eventually subject the active elements to high humidity. The time required for moisture to saturate most plastic packages is in the order of 100 hours. High humidity can lead to failure from corrosion of the metallization or from increased surface leakage currents. The humdity associated with most normal applications does not harm plastic-packaged devices, but accelerated humidity testing should be considered a destructive test. Most units can be expected to withstand 10 days of non-operating environmental testing in high humidity and elevated temperature (e.g. 95% RH at 65°C).

Another package-related limitation of plastic products lies in the area of thermal fatigue life. Because the different materials (lead frame, plastic package, junction coating, die, bond wires, etc.) are in physical contact, and because coefficients of thermal expansion cannot be perfectly matched, temperature changes are accompanied by physical strain. Extended temperature cycling or thermal shock can lead to fatigue failure of the bond wires, or to die attach failures. It is not unusual to see 5% failures after 50 cycles temperature cycling -55°C to +125°C or thermal shock 0°C to +100°C.

Thermo-mechanical life could be a limiting factor for plastic products in markets where reliability is required through many temperature excursions, such as automotive applications. A sizeable effort of product improvement is currently underway to develop the materials that will realize the reliability objectives of these new markets.

There are theoretical limitations to the maximum temperature at which the device materials are stable and compatible. The gold wires bonded to aluminum metallization form intermetallic compounds at a rate which depends very strongly on temperature. At ordinary temperatures the reaction is so slow that it is not a significant contributor to failure rate. At +150°C bond life may be a few months. At +200°C, bond life may be a few days. Trace impurities and bonding conditions can significantly affect bond life. Stress from surroundings package materials can break a bond weakened by intermetallics.

The thermal expansion coefficients of epoxy package materials are relatively constant over normal service temperature, but at some high temperature the expansion coefficient increases significantly. This temperature is called the "glass transition temperature", or Tg. Tg of most of the presently utilized epoxy casting compounds is specified at +120°C, but it varies by several degrees, depending on batch history and assembly conditions. Also, the glass transition effect occurs over a range of several degrees, so that any exposure over a +115°C is potentially harmful. The epoxy and dye materials, being organic, are also subject to chemical degradation if overheated.

If proper care is not taken in assembly, plastic packages are potentially susceptible to the occurrence of thermal intermittents. The solid encapsulation around the bond wires keeps adjacent parts of a broken wire together, so that a defective bond, or a bond which failed for some reason, may not be a consistent failure. A unit with this phenomenon will often be good at room temperature and open when hot. A continuity test monitored over temperature ranging from room ambient to the maximum storage temperature is quite effective at detecting thermal intermittents. A high temperature function test is less effective, but still quite good.

Failure resulting from improper soldering operations have occurred frequently in LED applications. This is generally the result of soldering at too high a temperature and/or for too long a time period. As pointed out above, excess temperature above Tg, can cause substantial changes in the mechanical properties of the epoxies commonly available for LED products. Since most devices are relatively small, there is very little thermal capacitance available to distribute heat conducted into the package from the leads. Exposure of the epoxy to extreme excess temperature conditions can result in softening of the epoxy around the leads where mechanical stress could cause loosening of the die attach or lead bond posts or the formation of voids in the epoxy in the vicinity of the leads. Either condition may result in catastrophic failure of the device from broken wires or wire bond separation.

From a manufacturing standpoint, there has historically been a high degree of parametic variability in the epoxy materials used for LED encapsulation. Incoming inspection criteria such as gell time, hardness, clarty, and viscosity can be utilized by the LED manufacturer to improve product uniformity.

OED SOLID STATE VISIBLE LIGHT EMITTER

The following cumulative test results have been obtained from testing performed at OED Division, in accordance with the latest revisions of Military Semiconductor Specifications MIL-S-19500, MIL-STD-202 and MIL-STD-750. Because this device has a non-hermetic, cast epoxy enclosure intended for commercial and industrial markets, we do not recommend its use in applications requiring military high reliability performance.

TEST	MIL-STD-750 REFERENCE	TEST CONDITIONS	UNITS TESTED	FAILED
Physical Dimensions	2066	Device profile at 20X	50	0
Solderability	2026	SN 60, Pb 40, solder at 230°C	50	0
Temperature Cycling	1051.1	5 cycles from -65°C to +100°C, .5 hrs. at extremes, 5 min. transfer	50	0
Thermal Shock	1056.1	5 cycles from 0°C to +100°C, 3 sec. transfer	50	0
Moisture Resistance	1021.1	10 days, 90-98% RH, -10 to +65°C, non operating	50	0
Shock	2016.1	5 blows each X_1, Y_1, Y_2, 1500 G, 0.5 msec. pulse		0
Vibration Fatigue	2046	32 ±8 hours each X, Y, Z, 96 hr. total, 60 Hz, 20 G min.	50	0
Vibration Variable Frequency	2056	4, 4 minute cycles each X, Y, Z, at 20 G min., 100 to 200 Hz	50	0
Constant Acceleration	2006	1 minute each X_1, Y_1, Y_2, at 20,000 G.	50	0
Thermal Strength	2036.3	Condition A, 1 lb. for 10 sec.	50	0
Lead Fatigue	2036.3	Condition F, Method A, 8 oz., 10 sec.	50	0

Figure 8.2-1 Typical Mechanical Reliability Data Sheet for a T-1 3/4 Plastic Lamp.

Section 9

Mechanical

9.0 MECHANICAL HANDLING CONSIDERATIONS FOR LED DEVICES

The first stage in designing a circuit utilizing an optoelectronic device is selecting the proper device for the application. The second step is to establish the electrical operating conditions and design the circuit. The third step is to install the optoelectronic device into the physical assembly, be it a printed circuit board, front panel mounting or some other mounting arrangement. The mounting considerations are primarily mechanical in nature, requiring attention to such items as the similarity of LED packages, the bending of leads, silver plated lead frames, soldering and post solder cleaning, socket mounting and heat sinking if required. Reliable operation of the LED device is more positively assured when all of these mechanical considerations have been given careful attention.

9.1 Similarity in LED Packages

Most plastic encapsulated LED devices are assembled using the lead frame technology. The exceptions are some stretched segment display devices which are assembled on substrates. Independent of the lead frame or package design, a lead frame device has the LED die attached directly to one lead and wire bonded across to another lead, as shown in Figure 9.1-1.

The primary thermal path to the LED is the cathode lead. Any mechanical and thermal stress applied to the leads is transmitted directly to the LED, die attach and wire bonds. The plastic encapsulant forms the device package and is the only supporting element for the lead frame. Therefore, the integrity of the encapsulation must be maintained to insure reliable operation of the LED device throughout its expected operating life.

Devices that do not use an encapsulating epoxy are most likely assembled on a ceramic substrate. Thick film metallization is usually applied to the face of the substrate. The LED is die attached to one metallization pad and wire bonded across to another pad, as shown in Figure 9.1-2. The circuit metallization is the primary thermal path from the LED to the leads, and the substrate acts as a secondary thermal path to the external ambient. The substrate isolates the LEDs and wire bonds from mechanically applied stresses; the amount of isolation dependent upon the type of substrate and package configuration. Devices of this construction are mechanically more rugged than a plastic encapsulated device and are more tolerant of temperature extremes.

One other type of device is assembled on a TO header, such as the high-reliability LED lamp which is packaged on a TO-18 header. The devices will withstand considerable mechanical and temperature stress without any effect upon performance.

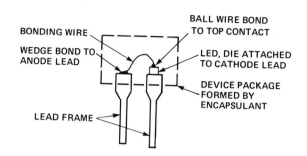

Figure 9.1-1 Basic Construction of a Plastic Encapsulated LED Device.

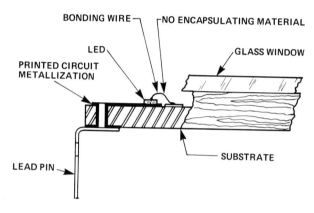

Figure 9.1-2 Basic Construction of an Unencapsulated, Substrate LED Device.

9.2 The Bending of Leads

In many LED lamp applications, it is necessary to bend the leads in order to mount the lamp at some angle other than 90° to the surface of a PC board. The leads of a lamp may be easily bent without introducing mechanical stresses inside the plastic package. The proper procedure for bending leads is illustrated in Figure 9.2-1. Bend the leads prior to soldering. Firmly grip the leads at the base of the lamp package with a pair of needle nose pliers. The pliers form a mechanical ground to absorb the stresses when bending. Bend the leads, one at a time, to the angle desired.

9.3 The Silver Plated Lead Frame

Since the price of gold has increased several times during the past few years, the cost of a gold plated lead frame has increased substantially, necessitating the search for an alternative. The impact of this cost increase has been industry wide. Many plating material alternatives were examined, and silver plating offered most of the desired properties of gold, while remaining price competitive when compared to other materials.

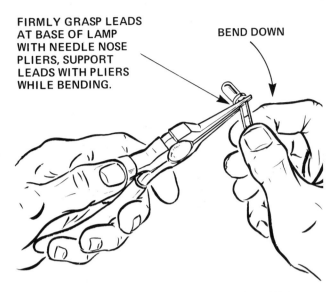

FIRMLY GRASP LEADS AT BASE OF LAMP WITH NEEDLE NOSE PLIERS, SUPPORT LEADS WITH PLIERS WHILE BENDING.

BEND DOWN

Figure 9.2-1 Correct Method to Bend the Leads of an LED Lamp.

By using silver plating, no additional manufacturing process steps are required. Silver has excellent electrical conductivity. LED die attach and wire bonding to a silver lead frame is accomplished with the same reliability as with a gold lead frame. Also, soldering to a silver lead frame provides a reliable electrical and mechanical solder joint. Soldering silver plated lead frame LED devices into a printed circuit board is no more complicated than soldering LED devices with gold plated lead frames.

9.3.1   The Silver Plating

The silver plating process is performed as follows: the lead frame base metal is cleaned and then plated with a copper strike, nominally 50 microinches (0.00127 mm) thick. Then a 150 microinch (0.00381 mm) thick plating of silver is added. A "brightener" is usually added to the silver plating bath to insure an optimum surface texture to the silver plating. The term "brightener" comes from the medium bright surface reflectance of the silver plate. Figure 9.3.1-1 illustrates the metallographic cross-section of the silver plating system as it would appear with a 1200X magnification.

150 μinch MINIMUM SILVER PLATING

50 to 80 μinch COPPER STRIKE

LEAD FRAME BASE METAL

Figure 9.3-1 Metallographic Cross-Section Through a Silver Plated Lead Frame, 1200X Magnification.

Since silver is porous with respect to oxygen, the copper strike acts as an oxygen barrier for the lead frame base metal. Thus, oxide compounds of the base metal are prevented from forming underneath the silver plating. Copper is miscible and readily diffuses into silver to form a solution that has a low eutectic point. This inter-diffusion between the copper strike and the silver overplate improves the solderability of the overall plating system. If basic soldering time and temperature limits are not exceeded, a lead frame base metal-copper-silver-solder metallurgical system will be obtained.

9.3.2   The Effect of Tarnish

Silver resists attack by most dry and moist atmospheres, such as carbon monoxide or high temperature steam. Halogen gases do attack silver, however, once the initial film layer is formed the process does not continue.

Silver reacts chemically with sulfur to form tarnish, silver sulfide (Ag_2S). The build-up of tarnish is the primary reason for poor solderability. However, the density of the tarnish and the kind of solder flux used actually determine the solderability. As the density of the tarnish increases, the more active the flux must be to penetrate and remove the tarnish layer.

9.3.3   Storage and Handling

The best technique for insuring good solderability of a silver plated lead frame device is to prevent the formation of tarnish. This is easily accomplished by preventing the leads from being exposed to sulfur and sulfur compounds. The two primary sources of sulfur are free air and most paper products, such as paper sacks and cardboard containers. The best defense against the formation of tarnish is to keep silver lead frame devices in protective packaging until just prior to the soldering operation. One way to accomplish this is to store the LED devices unwrapped in their original packaging. For example, Hewlett-Packard ships its seven segment display products in plastic tubes which are sealed air tight in polyethylene. It is best to leave the polyethylene intact during storage and open just prior to soldering.

Listed below are a few suggestions for storing silver lead frame devices.

1.   Store the devices in the original wrapping unopened until just prior to soldering.

2.   If only a portion of the devices from a single tube are to be used, tightly re-wrap the plastic tube containing the unused devices in the original or a new polyethylene sheet to keep out free air.

3. Loose devices may be stored in zip-lock or tightly sealed polyethylene bags.

4. For long term storage of parts, place one or two petroleum naphthalene mothballs inside the plastic package containing the devices. The evaporating naphthalene creates a vapor pressure inside the plastic package which keeps out free air.

5. Any silver lead frame device may be wrapped in "Silver Saver" paper for positive protection against the formation of tarnish. "Silver Saver" is manufactured by:

The Orchard Corporation
1154 Reco Avenue
St. Louis, Missouri 63126
314/822-3880

6. To reduce shelf storage time, it is worthwhile to use inventory control to insure that the devices first received will be the first devices to be used.

One caution: The adhesives used on pressure sensitive tapes such as cellophane, electrical and masking tape can soak through silver protecting papers and may leave an adhesive film on the leads. This film reduces solderability and should be removed with Freon T-P35, Freon T-E35, or equivalent, prior to soldering.

Petroleum naphthalene in the form of mothballs or chips may be obtained from the Frank Curran Company, 8101 South Lemont Road, Downers Grove, Illinois 60515, 312/969-2200.

9.4   Solders, Fluxes, and Surface Conditioners

Optoelectronic devices are usually soldered into a printed circuit board along with other components. It is necessary to achieve reliable soldering in order to obtain reliable circuit operation. A basic understanding of the types of solder, fluxing agents and surface conditioners that are typically used in the electronic industry will aid the process engineer in establishing a reliable soldering process.

9.4.1   Solders

The solder most widely used for wave soldering electronic components into printed circuit boards is Sn60 (60% tin and 40% lead) per Federal Standard QQ-S-571. Two alternatives are the eutectic composition Sn63 and the solder SN62 which contains 2% silver. Table 9.4.1-1 lists the composition, maximum acceptable contaminant levels and temperatures for these solders.

As the device leads pass through the solder wave of a flow solder process, the tin in the solder scavenges silver from the silver plating and forms one of two silver-tin intermetallics (Ag_6Sn or Ag_3Sn). This silver in the molten solder should not be considered a contaminant. As the silver content increases, the rate of scavenging decreases and the probability of obtaining the desired base metal-copper-silver-solder metallurgical system is improved. The result is that the silver content in solder, which reaches a maximum of 2½% in Sn60 at 230°C, aids in producing reliable solder joints on silver plated lead frames. Periodic replenishing of the solder as it becomes contaminated, with fresh solder, helps assure reliable soldering on a continual basis.

For hand solder operations, a high quality resin core wire solder is recommended. The solder may be obtained in wire diameters from .254mm (.010 inch) to 3.175mm (.125 inch) as common sizes. The core is typically an RMA flux in the amount of 2.2% or 3.3% by weight.

9.4.2   Fluxes

Solder flux classification per Federal Standard QQ-S-571, listed in order of increasing strength, are as follows:

Type R: Non-Activated Rosin Flux
Type RMA: Mildly Activated Rosin Flux
Type RA: Activated Rosin Flux
Type AC: Organic Acid Flux, Water Soluable

The Type R flux is a pure water white (WW) rosin without any additives. Rosin is a complex natural product obtained from the gum of live trees. It is mixed with a suitable solvent to form a homogeneous solution. Fluxes are graded as to the percentage by weight of rosin solids in the solution. The flux and its residue are non-corrosive and non-conductive.

The RMA flux is a homogeneous mixture of WW rosin in a blended alcohol vehicle into which a small amount of activating agent has been added. The flux and its residue are non-corrosive and non-conductive.

The RA flux is the same as RMA flux except that a greater amount of activating agents have been added. The flux and its residue are non-corrosive, and the residue is non-conductive only if all the solvent is volatilized and the residue remains dry.

An AC flux is considered full active and has a greater fluxing ability than a rosin flux. An AC flux may contain acids, organic or inorganic chlorides and therefore is corrosive. Due to their organic nature, the AC flux residues

Compo-sition	Tin	Lead	Silver	Anti-mony	Bis-muth max	Copper, max	Iron, max	Zinc, max	Alumi-num, max	Arse-nic, max	Cad-mium, max	Total of all others, max	Approximate melting range	
													Solidus	Liquidus
	%	%	%	%	%	%	%	%	%	%	%	%	°C	°C
Sn60	59.5 to 61.5	Remainder	----	0.20 to 0.50	0.25	0.08	0.02	0.005	0.005	0.03	----	0.08	183	191
SN62	61.5 to 62.5	Remainder	1.75 to 2.25	0.20 to 0.50	0.25	0.08	0.02	0.005	0.005	0.03	----	0.08	179	179
Sn63	62.5 to 63.5	Remainder	----	0.20 to 0.50	0.25	0.08	0.02	0.005	0.005	0.03	----	0.08	183	183

MAXIMUM CONTAMINANT LEVELS

TABLE 9.4.1-1 Composition, Maximum Contaminant Levels and Temperatures for Three Solders Commonly Used in Wave Soldering Operations.

9.4

decompose at soldering temperatures, thereby eliminating a large amount of the corrosive residue. All residues must be removed to prevent the formation of conductive ionic paths and corrosion caused by residual chloride salts.

Suggested applications of these flux types with respect to various tarnish levels are as follows:

Silver plated lead frames that are clean, contaminant and tarnish free may be soldered using a Type R flux such as Alpha 100.

Minor Tarnish: Since some minor tarnish or other contaminant may be present on the leads, a type RMA flux such as Alpha 611 or 611 Foam, Kester 197 or equivalent is recommended. Minor tarnish may be identified by reduced reflectance of the ordinarily medium bright surface of the silver plating. Type RMA fluxes which meet MIL-F-14256 are used in the construction of telephone communication, military and aerospace equipment.

Mild Tarnish: For a mild tarnish, a type RA flux such as Alpha 711-35, Alpha 809 foam, Kester 1544, Kester 1585 or equivalent should be used. A mild tarnish may be identified by a light yellow tint to the surface of the silver plating.

Moderate Tarnish: A type AC water soluable flux such as Alpha 830, Alpha 842, Kester 1429 or 1429 foam, Lonco 3355 or equivalent will give acceptable results on surface conditions up to a moderate tarnish. A moderate tarnish may be identified by a light yellow-tan color on the surface of the silver plating.

9.4.3 Surface Conditioners

If a more severe tarnish is present, such as a heavy tarnish identified by a dark tan to black color, a cleaner/surface conditioner must be used. Some possible cleaner/surface conditioners are Alpha 140, Alpha 174, Kester 5560 and Lonco TL-1. The immersion time for each cleaner/surface conditioner will be just a few seconds and each is used at room temperature. For example, Alpha 140 will remove severe tarnish almost upon contact; therefore, the immersion time need not exceed 2 seconds. These cleaner/surface conditioners are acidic formulations. Therefore, immediately, thoroughly wash all devices which have been cleaned with a cleaner/surface conditioner in cold water. A hot water wash will cause undue etching of the surface of the silver plating. A post rinse in deionized water is advisable.

CAUTION: These cleaner/surface conditioners may etch exposed glass surfaces and may have a detrimental effect upon the glass filled encapsulating epoxies used in optoelectronic devices. Complete immersion of an optoelectronic device into a surface conditioner solution is NOT recommended. For best results, immerse only the tarnished leads.

Three major suppliers of soldering supplies are:

Alpha Metals, Inc.
56 G Water Street
Jersey City, New Jersey 07304
302/434-6778

Kester Solder Company
4201 G Wrightwood Avenue
Chicago, Illinois 60639
312/235-1600

London Chemical Company (Lonco)
240 G Foster
Bensenville, Illinois 60106
312/287-9477

9.5 The Soldering Process

Before the actual soldering begins, the printed circuit boards and components to be soldered should be free of dirt, oil, grease, finger prints and other contaminants. Fluorinated cleaners such as Genesolve DI-15 or DE-15 may be used to preclean both the printed circuit boards and LED devices. Operators may wear cotton gloves to prevent finger prints when loading components into the printed circuit boards.

If the silver lead frames have acquired an unacceptable layer of tarnish, remove this tarnish layer with a cleaner/surface conditioner just prior to soldering. Since a cleaner/surface conditioner does slightly etch the surface of the silver plating, the silver leads are now more susceptible to tarnish formation. Therefore, use a cleaner/surface conditioner only on those silver lead frame devices which will be soldered within a four hour time period.

9.5.1 Wave Soldering

The temperature of a solder wave should be at least $38°C$ ($100°F$) above the melting temperature (solidus) of the solder. For Sn60 solder, this $221°C$ ($430°F$) minimum. Most wave solder operations maintain a solder wave temperature between $230°C$ ($446°F$) and $260°C$ ($500°F$). At $230°C$, Sn60 solder dissolves silver at the rate of 60 microinches per second, and at $260°C$ it dissolves silver at the rate of 80 microinches per second. Therefore, with an initial silver plating thickness of 150 microinches, a dwell time of less than 2 seconds in a $230°C$ solder wave will provide the desired lead base metal-copper-silver-solder

Figure 9.5.1-1   The Desired Metallurgical System After a Solder Operation.

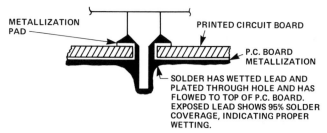

FIGURE A. IDEALLY SOLDERED LEAD

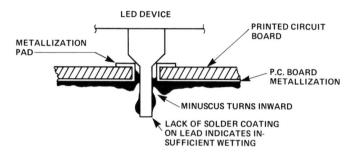

FIGURE B. UNDESIRABLY SOLDERED LEAD

Figure 9.5.4-1   Comparison Between an Ideally Soldered Lead and an Undesirably Soldered Lead.

metallurgical system, as illustrated in Figure 9.5.1-1. The copper strike is intact. At 4 seconds dwell time in a 230°C solder wave, all of the silver has been dissolved and the solder must adhere to the copper strike; at 5 seconds, it is possible to have completely dissolved the copper and now the solder must adhere to the lead base metal. For plastic LED devices, the 230°C solder temperature is preferable to 260°C.

9.5.2 Hand Soldering

Contrary to popular belief, hand soldering can be more injurious to an LED device than wave soldering, the reason being that it is difficult to control dwell time and temperature. To effectively hand solder, it is best that the operator realize that "the hotter and longer the better" is not true. It is best that a 15 watt iron (25 watt maximum) be used. Maintain the iron in contact with the lead for only that time which is sufficient to allow the solder to flow evenly around the joint. Heat sinking the leads may be worthwhile, though this should not be necessary if sufficient care is observed. Always keep the tip of the iron clean and well tinned.

9.5.3 Cutting the Leads

The best rule to follow is to cut the leads of an LED device after soldering, not before. This accomplishes two things. First, the additional lead length provides heat sinking during the soldering operation. Secondly, the soldered joint provides a mechanical ground which prevents the mechanical stresses due to cutting from being transmitted into the device package. If it is necessary to cut the leads prior to soldering, support the device by the leads with a pair of long nose pliers during the cutting operation as is illustrated in Figure 9.2-1.

9.5.4 Printed Circuit Board

Printed circuit boards, either single sided, double sided or multi-layer, may be manufactured with plated through holes with a metal trace pad surrounding the hole on both sides of the printed circuit board. The plated through hole is desirable to provide a sufficient surface for the solder to wet, and thereby be pulled up by capillary attraction along

the lead through the hole to the top of the printed circuit board. This provides the best possible solder connection between the printed circuit board and the leads of the LED device. Figure 9.5.4-1A illustrates an ideally soldered lead. The amount of solder which has flowed to the top of the printed circuit board is not critical. A sound electrical and mechanical joint is formed. Figure 9.5.4-1B illustrates a soldered lead which is undesirable.

9.6 Post Solder Cleaning

Unlike most semiconductor devices, LED devices for visual applications are limited in the kinds of materials that may be used for encapsulation or hermetic sealing. In the case of encapsulating materials, the optical properties required limit the selection to only a few highly specialized epoxies. Exposure to a solvent used in a cleaning operation must not alter these optical properties in any way. For this and other reasons, only certain cleaning solvents may be used to post solder clean LED devices.

9.6.1 Type of Cleaners

It is important to remove both the rosin and ionic residues after soldering to insure reliable operation of the complete circuit. This is best accomplished by using an azeotrope of fluorocarbon and alcohol. The fluorocarbon is used to dissolve the residual rosin and the ionic contaminants are removed by the alcohol.

The type of fluorocarbon is very important. Tests have demonstrated that the only fluorocarbon that is compatible

CLEANING AGENT	COMPOSITION	CLEANING OPERATION
WATER		60°C (140°F) Wash to remove an AC flux residue.
ETHANOL, ISOPROPANOL	Alcohol	General cleaning after hand soldering.
FREON TF GENESOLV D ARKLONE P	100% Fluorocarbon (F113)	General cleaning agent for removing grease, oils, etc.
FREON TE ARKLONE A	≈4% Ethanol	Improved general cleaning agent.
GENESOLV DE-15 DI-15 BLACO-TRON DE-15 DI-15	≈15% Ethanol ≈15% Isopropanol ≈15% Ethanol ≈15% Isopropanol	Vapor cleaning at boiling — up to 2 minutes in the vapors — best for post solder cleaning.
ARKLONE K	≈25% Isopropanol	Vapor cleaning at boiling — up to 2 minutes in vapors. Best for post solder cleaning.
FREON T-E 35 T-P 35	≈35% Ethanol ≈35% Isopropanol	Room temperature post solder cleaning.

TABLE 9.6.1-1 Table of Suggested Post Solder Cleaning Agents that are Compatible with LED Devices.

with plastic LED devices is trichloro-tri-fluoroethane (F113), sold under tradenames as Freon, Genesolv D and Arklone. Some suggested cleaning products are listed in Table 9.6.1–1.

Cleaning solvent mixtures based on the fluorocarbon tetrachloro-di-fluoroethane (F112) are not recommended for cleaning plastic LED parts. Also, such cleaning agents from the keystone family (acetone, methyl ethyl ketone, etc.) and from the chlorinated hydrocarbon family (methylene chloride, trichloroethylene, carbon tetrachloride, etc.) are not recommended for cleaning LED parts. All of these various solvents attack or dissolve the encapsulating epoxies used to form the packages of plastic LED devices.

9.6.2. Bulk Cleaning Processes

Post cleaning of soldered assemblies when a type RMA or Type RA flux has been used may be accomplished via a vapor cleaning process in a degreasing tank, using an azeotrope of fluorocarbon and 15% to 25% alcohol as the cleaning agent. A recommended method is a 15 second suspension in vapors, a 15 to 30 second spray wash in liquid cleaner, and finally a one minute suspension in the vapors. When a water soluable Type AC flux such as Alpha 830 or Kester 1429/1429F is used, the following post cleaning process is suggested: thoroughly wash with water, neutralize using Alpha 2441 or Kester 5761 foaming, then thoroughly wash with water and air dry.

Four major suppliers of solvents are:

Allied Chemical Corp. (Genesolv)®
Speciality Chemicals Division
P.O. Box 1087R
Morristown, New Jersey 07960
201/455-5083

E.I. DuPonte de Nemours & Company
Freon Products Division
Wilmington, Delaware 19898
302/774-8341

Baron-Blakeslee (Blaco-Tron)®
1620 S. Laramie Avenue
Chicago, Illinois 60650
312/656-7300

Imperial Chemical Industries, Ltd. (Arklone)
Imperial Chemical House, Millbank
London SW1P3JF, England

9.6.3 Special Cleaning Instructions for Monolithic PC Board Displays

The monolithic printed circuit board display devices use a lens made from a special plastic chosen for its superior optical properties. This plastic does not lend itself to vapor cleaning processes. Also, the lens does not environmentally protect the LED chips. Therefore, the following special cleaning procedures are suggested.

For cleaning after a solder operation, the following process is recommended: wash display in clean liquid Freon T-P 35 or Freon T-E 35 solvent for a time period up to 2 minutes maximum. Air dry for a sufficient length of time to allow solvent to evaporate from beneath display lens. Maintain solvent temperature below 30°C (86°F). Methanol, isopropanol, or ethanol may be used for hand cleaning at room temperature. Water may be used for hand cleaning if it is not permitted to collect under display lens.

Solvent vapor cleaning at elevated temperatures is not recommended as such processes will damage display lens. Ketones, esters, aromatic and chlorinated hydrocarbon solvents will also damage display lens. Alcohol base active rosin flux mixtures should be prevented from coming in contact with display lens.

9.7   Socket Mounting

LED devices may be socket mounted in the same fashion as other semiconductor devices. The selection of a suitable socket is generally based on a quality vs. cost trade-off, with cost usually the primary factor.

When selecting a socket to mount silver lead frame devices, it is the performance of the socket that becomes the important factor. The socket contacts must have sufficient force and contact area to form an air tight mechanical seal at the lead/contact interface to prevent the formation of tarnish at this interface. Tarnish (silver sulfide) is a semiconductor, thus would introduce an undesirable resistance if allowed to form in the contact area. Figure 9.7-1 illustrates the desired condition. A good quality

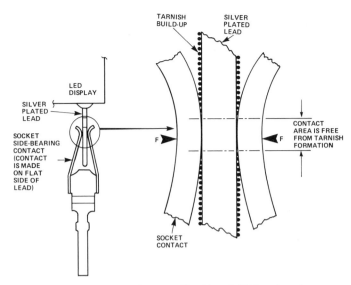

Figure 9.7-1   Connector Contacts That Have Sufficient Insertion Force and Contact Area to Form an Air Tight Mechanical Seal with the Silver Plated Lead to Prevent Tarnish Build Up in the Contact Area.

socket will require considerable insertion force and provide good wiping action upon insertion to scrape away any initial tarnish. The socket should be capable of securely holding the device without relative movement of the device lead with respect to the socket contact in the presence of mechanical shock or vibration. Sockets with side-bearing contacts that grip the flat sides of a lead are preferred for the mounting of DIP LED devices that have silver plated leads. Sockets that have edge-bearing contacts may be used to mount devices that have square silver plated leads, such as plastic LED lamps, providing the insertion force is sufficient to provide the required air tight mechanical seal along the edge of the lead.

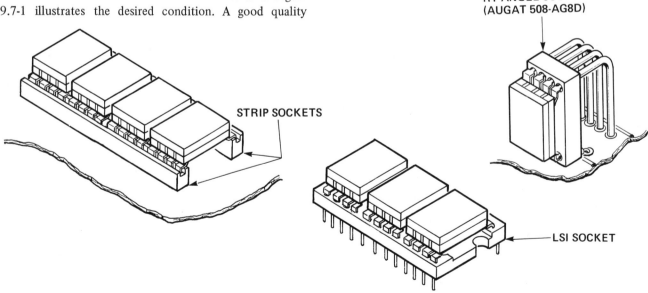

Figure 9.7.1-1   Optional Socket Mounting Configurations for the HP 5082-7300 Series of OBIC Displays.

9.8

MANUFACTURER	PART NUMBER	PINS	APPLICATIONS
AUGAT	508-AG8D	8	(1) HP 5082-7300; RT Angle Mtg
33 Perry Avenue	514-AG21D	14	90°
Attleboro, Mass. 02703	514-AG25D		45° (1) Stretched 7-Segment
617/222-2203	514-AG26D		60°
	314-AG5D-2R	14	(1) Stretched 7-Segment
	324-AG6D	24	(3)
	336-AG6D	36	(4) HP 5082-7300
	340-AG6D	40	(5)
	314-AG39D Low Profile	14	(1) Stretched 7-Segment
	324-AG39D Low Profile	24	(3) HP 5082-7300
	340-AG39D Low Profile	40	(5)
	325-AG1D	25	Strip Socket
BURNDY	HBRB2S-1	2	(1) LED Lamp; T-1 or T-1 3/4
Richards Avenue	DILB14P-108	14	(1) Stretched 7-Segment
Norwalk, Conn. 06856	DILB24P-108	24	(3) HP 5082-7300
203/838-4444	DILB40P-108	40	(5) HP 5082-7300
CAMBION	703-3777-04-12	14	(1) Stretched 7-Segment
445 Concord Avenue	703-5151-01-04-16		
Cambridge, Mass. 02138	703-5153-07-04-16	24	(3) HP 5082-7300
617/491-5400			
CIRCUIT ASSEMBLY	CA-14S-10SD	14	(1) Stretched 7-Segment
3169 Redhill Avenue	CA-24S-10SD	24	(3) HP 5082-7300
Costa Mesa, Calif. 92626	CA-40S-10SD	40	(5)
714/540-5490	CA-25STL-10SD	25	Strip Socket
	CA-S36 SP100	36	Stripline Plug (for pc board Monolithic displays)
ROBINSON-NUGENT	IC-143-S1	14	(1) Stretched 7-Segment
800 E. 8th Street	IC-163-S1	16	(2)
New Albany, Indiana 47150	IC-246-S1	24	(3) HP 5082-7300 Displays
812/945-0211	IC-406-S2	40	(5)
	ICN-143-S3 Low Profile	14	(1) Stretched 7-Segment
	ICN-406-S1	40	(1) HP 5082-7300 Display
	SB-25	25	Strip Socket
JERMYN	A1237	14	(1) Stretched 7-Segment
712 Montgomery Street	A1252AM Low Profile		
San Francisco, CA 94111	A23-2023 Low Profile	24	(3) HP 5082-7300
415/362-7431	A23-2030	40	(5)
AMP	583640 Series	12	(1)
449 Eisenhower Blvd.		24	(2) HP HDSP-2000
Harrisburg, Penn. 17105	Low Profile	36	(3)
717/564-0100	583773 Series	3 to 22	Strip Connectors

TABLE 9.7-1 A List of Sockets for Use With LED Displays

9.7.1 Special Socket Assemblies for LED Displays

LED display devices may be mounted in DIP sockets rather than being directly soldered into a PC board. Stretched seven segment displays may be mounted in standard 14 pin dip sockets. OBIC displays may require the use of strip sockets, LSI sockets or special sockets as illustrated in Figure 9.7.1-1. A small alphanumeric OBIC display may be mounted in standard DIP sockets that have been machined down in length to accept the 12 pin devices so as to form an end stacked display string.

Front panel mounting kits are available for LED displays. Rochester Digital Displays, Inc., 120 North Main Street, Fairport, New York 14450, 716/223-6855, offer two such kits. One kit accepts the HP 5082-7300 series display and the other accepts stretched seven segment displays.

PC board monolithic display products may be mounted using any one of several techniques. The most straight forward is the use of standard PC board edge connectors. A more cost effective approach can be implemented through the use of standard (or custom) stamped or etched metal mounting clips such as those available from Burndy, Richards Avenue, Norwalk, Connecticut 06850, 203/838-4444, (series LED-B) or J.A.V. Manufacturing, Inc., 125 Wilbur Place, Bohemia, New York 11716 (series 022-002). Circuit Assembly Corporation, 3169 Red Hill Avenue, Costa Mesa, California 92626, 714/540-5490 and Burndy each manufacture low cost connectors especially designed for mounting the display board at a given angle as illustrated in Figure 9.7.1-2. A third approach would be to use a series of etched clips which are first soldered to the PC mother board. The display board is then pressed into place, with each clip being inserted into one of the plated through holes at the edge of display board.

Front panel mounting hardware for use with T-1 3/4 plastic LED lamps is manufactured by the Eldema Division of the Genisco Technology Corporation, 18435 Susana Road, Compton, California 90221, 213/537-4750. Two of their devices are illustrated in Figure 9.7.1-3.

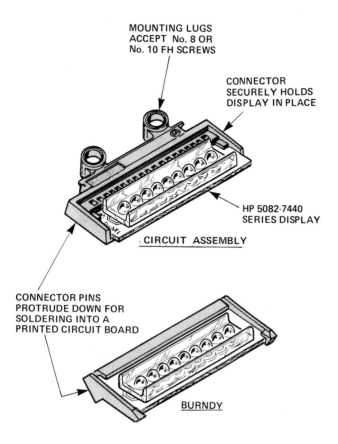

Figure 9.7.1-2 Mounting Connectors for Monolithic LED PC Board Display Devices.

9.8 Heat Sinking

Most LED devices may be operated in elevated ambient temperatures without heat sinking by utilizing input power derating. Devices which usually fall into this category are those devices that do not contain on-board integrated circuits. Heat sinking may be required for those on-board integrated circuit (OBIC) devices that have considerable power dissipation occuring within the integrated circuit.

The specific devices which may require heat sinking are the OBIC LED numeric and alphanumeric displays. An OBIC alphanumeric display, for example, combines a significant amount of integrated circuit logic and LED display capability in a very small package. As such, on board power dissipation is relatively high and the thermal design of the mounting assembly becomes an important consideration. The display is designed to operate over a wide temperature range with full power dissipation. This capability may be utilized only if the mounting assembly has a thermal resistance to ambient of 35°C/W per device, or less.

The primary thermal path for power dissipation is through the device leads. The thermal resistance junction-to-lead is typically 15 to 25°C/W, depending upon the package configuration. For a plastic device, the maximum LED

junction temperature is 110°C. A hermetic device will tolerate a much higher IC junction temperature, but the maximum LED junction temperature should not exceed 125°C. Reliable operation is obtained when the junction temperatures are maintained below these maximum values.

The thermal resistance to ambient of 35°C/W for the display mounting structure assumes that the mounting surface for the display is an isothermal plane. If only one OBIC display is operated on this isothermal plane at a power level of 1.7 watts maximum, as an example, the temperature rise above ambient would be 42.5°C:

$$T_{RISE} = (35^{\circ}C/W)(1.7W) = 42.5^{\circ}C$$

If a second display is placed on this same thermal plane, with no increase in thermal dissipation capability, the temperature rise would double to 85°C, reaching catastrophic levels very quickly. Hence, for each OBIC device added to the display string, there must be an appropriate increment of heat dissipating capability added to the mounting plane. For short display strings (4 to 6 OBIC devices), the 35°C/W per device may be achieved by utilzing a printed circuit board design which maximizes the amount of metal surface area remaining on the board. For longer display strings, or if sufficient metal surface area cannot be achieved, an external heat sink must be employed to obtain the 35°C/W per device case-to-ambient thermal resistance.

In practice, heat sink design involves the optimization of techniques used to dissipate heat through the device leads. The heat transfer is from the device leads to the PC board metallization to the heat sink and is dissipated in the surrounding ambient. A heat sink of approximately 52 square centimeters (8 square inches) per OBIC device will

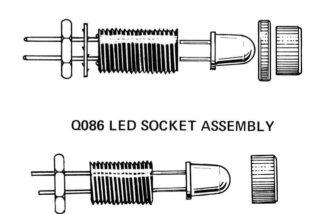

Q086 LED SOCKET ASSEMBLY

Q084 LED MOUNTING ASSEMBLY

Figure 9.7.1-3 Front Panel Mounting Hardware for T-1 3/4 LED Lamps.

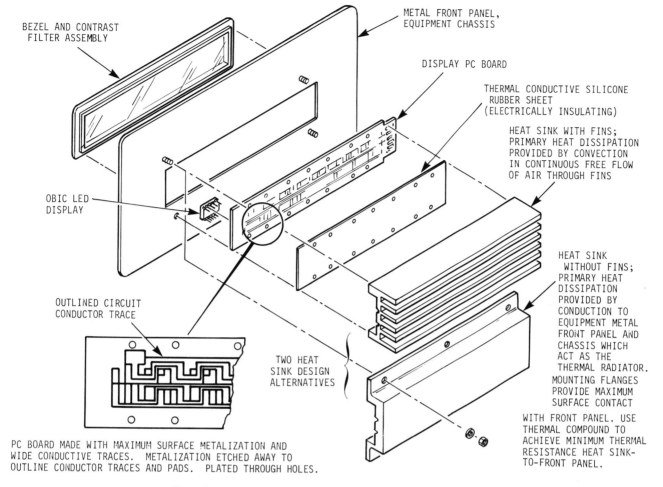

BEZEL AND CONTRAST FILTER ASSEMBLY

METAL FRONT PANEL, EQUIPMENT CHASSIS

DISPLAY PC BOARD

THERMAL CONDUCTIVE SILICONE RUBBER SHEET (ELECTRICALLY INSULATING)

HEAT SINK WITH FINS; PRIMARY HEAT DISSIPATION PROVIDED BY CONVECTION IN CONTINUOUS FREE FLOW OF AIR THROUGH FINS

OBIC LED DISPLAY

HEAT SINK WITHOUT FINS; PRIMARY HEAT DISSIPATION PROVIDED BY CONDUCTION TO EQUIPMENT METAL FRONT PANEL AND CHASSIS WHICH ACT AS THE THERMAL RADIATOR. MOUNTING FLANGES PROVIDE MAXIMUM SURFACE CONTACT

OUTLINED CIRCUIT CONDUCTOR TRACE

TWO HEAT SINK DESIGN ALTERNATIVES

WITH FRONT PANEL. USE THERMAL COMPOUND TO ACHIEVE MINIMUM THERMAL RESISTANCE HEAT SINK-TO-FRONT PANEL.

PC BOARD MADE WITH MAXIMUM SURFACE METALIZATION AND WIDE CONDUCTIVE TRACES. METALIZATION ETCHED AWAY TO OUTLINE CONDUCTOR TRACES AND PADS. PLATED THROUGH HOLES.

Figure 9.8.1-1   Heat Sink and Panel Mounting Configuration for an OBIC Alphanumeric LED Display.

typically permit operation at a power dissipation of 1 watt/device in an ambient temperature of 70°C.

As an alternative to soldering, OBIC displays may be mounted in DIP or stripline sockets which are themselves soldered into a PC board with maximum surface area metallization. These sockets will allow enough space between the PC board and the OBIC displays to permit a heat sink bar to be inserted to conduct heat to an external sink. These sockets add a thermal resistance of about 2°C/W between the device leads and the PC board.

9.8.1   OBIC Display Assembly with On-Board Heat Sink

Heat sinking an OBIC display is most effectively accomplished by constructing a direct thermal path between the device leads and the heat sink. A direct thermal path may be achieved by implementing the design concept illustrated in Figure 9.8.1-1. The printed circuit board is designed with a maximum amount of metallization remaining on the board. Thin lines are etched in the

metallization to outline the circuit conductors. The PC board metallization is used as part of the thermal path between the device leads and the heat sink.

The heat sink is designed to fit flush against the PC board, with grooves cut into it to provide clearance for the device leads. A sheet of thin thermally conductive silicone rubber is inserted between the heat sink and the printed circuit board. This silicone rubber completes the thermal path and at the same time, electrically insulates the heat sink from the PC board metallization.

Heat transfer from the heat sink to the ambient air may be accomplished in one of two ways. One method is to mechanically fasten the heat sink to the metal chassis or front panel. This extends the display heat sink by utilizing the complete chassis as a thermal radiator. If this is not possible, then fins may be made as part of the display heat sink to increase its surface area. Heat is now transferred to the local ambient air and carried off by convection. This process is assisted by insuring a continuous free flow of air passing through the fins of the heat sink.

Thermal conductive silicone rubber is sold under the trade name "Sil-Pad", by the Bergquist Company, Inc., 4350 West 78th Street, Minneapolis, Minnesota 55435. The sheet is a nominal 0.305mm (.012 inch) thick, has a thermal resistance of approximately .08°C/W per square centimeter and a Shore A hardness of 76. Anticipated useful life at a temperature of 140°C is 300 thousand hours, in a non condensing atmosphere.

The heat sink may be made from a length of aluminum extrusion or purchased from any one of a number of heat sink manufacturers. In either case, the heat sink will be a custom part and should be designed so as to (1) have maximum surface area contact with the PC board, (2) have as much surface contact area with the supporting chassis or (3) have the maximum possible surface area if the local ambient air must carry off the heat by convection.

9.8.2 A Display Assembly with Heat Pipe

An emerging technology for the efficiency transfer of heat from one point to another is the heat pipe. When it is not possible to incorporate an on-board heat sink, these devices are a cost effective means of efficiently transferring the heat from the display PC board to a remote heat sink or to the equipment chassis. The heat pipe by itself does not function as the heat sink. The input to output thermal resistance of a 305mm (12 inch) long heat pipe is typically .10°C/W when transporting 10 watts. The contact thermal resistance at either end of the heat pipe can be maintained to about .062°C/W per square centimeter (.4°C/W per square inch).

9.8.3 List of Manufacturers

The following list of manufacturers are only suggestions, as those not listed may well be equally qualified:

Heat Sinks:

Aham
968 W. Foothill Blvd.
Azusa, California 91702

Aham
2 Gill Street, Bldg. 5
Woburn, Massachusetts 01810

International Electronic Reserach Corp. (IERC)
135 W. Magnolia Blvd.
Burbank, California 91502

Industrial Heat Sink Corporation
5338 Alhambra Avenue
Los Angeles, California 90032

Heat Pipes:

Noren Products, Inc.
3511 Haven Avenue
Menlo Park, California 94025

Tecknit
129 Dermody Street
Cranford, New Jersey 07016

Jermyn
712 Montgomery Street
San Francisco, California 94111

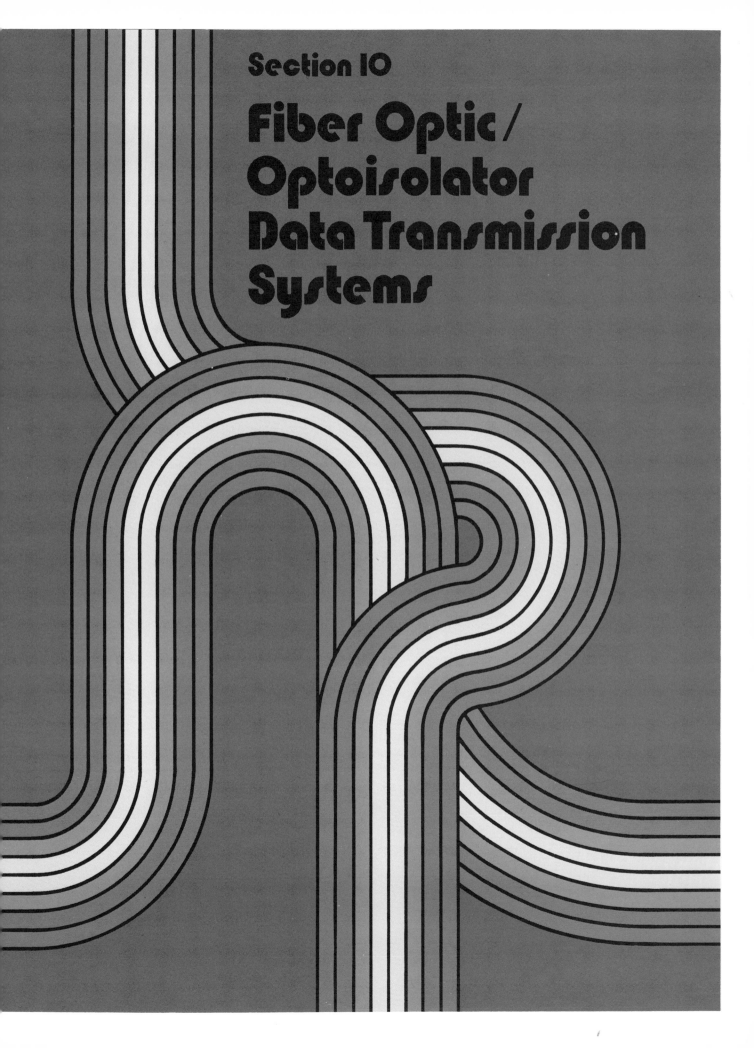

Section 10

Fiber Optic/ Optoisolator Data Transmission Systems

10.0   FIBER OPTICS AND OPTOCOUPLERS FOR DIGITAL DATA TRANSMISSION

It is frequently necessary to transmit digital information between modules which are physically separated and which may have significant voltage differences between their common points. Physical separation raises the problem of bandwidth limitations and the voltage differences may be large enough to cause errors in data transmission or to present a hazard to personnel or equipment.

This chapter deals with the use of optocouplers and fiber optics in overcoming these problems. Guidelines are given for making the proper choice of transmission medium (wire or fiber optic). Separate subsections describe the hardware for wire and fiber optic data transmission links. The final subsection presents hardware/software arrangements to be used for interfacing either wire or fiber optic links with other terminal equipment.

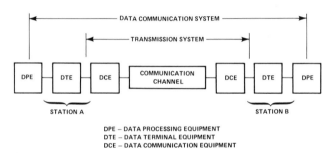

DPE – DATA PROCESSING EQUIPMENT
DTE – DATA TERMINAL EQUIPMENT
DCE – DATA COMMUNICATION EQUIPMENT

Figure 10.0-1.   Block Diagram of a Data Communication System.

The block diagram of a communication system is shown in Figure 10.0-1. Any of the blocks may actually represent more than one element. For example, the Coummunication Channel could be a bi-directional or a multiple-path link joining also the DCE elements of additional Stations not shown. Within each Station, there may be multiple DCE, DTE, and DPE elements. The DPE is the "user" of the communication system, either providing or requiring data to be transmitted. The DTE performs electrical interfacing between the DPE and the DCE; in addition to providing signal ports which are electrically and mechanically compatible with the DCE requirements, the DTE may also perform part of the coding, formatting, prioritization, and timing of the data. The DCE performs interfacing between electrical ports of the DTE and the electrical, mechanical and/or optical requirements of the Communication Channel. The Channel itself could be as simple as a pair of wires, but could also employ fiber optic, microwave, optocoupler, or telephone line linkage. The Transmission System, which includes the DCE's, should be "transparent" to the DTE's, that is, except possibly for transit-delay differences any Transmission System joining the DTE's should provide functional equivalence.

Protocol requirements of a communication system are represented in Figure 10.0-2 by the dashed lines which describe implied paths. The actual paths, of course, are via the Communication Channel, but to make the transmitted data appear in the right form at the correct location and with proper timing or priority requires the exchange of information at all of the levels shown. Physical Protocol involves coding and timing; Link Control governs preparation for transmission. These lower levels can be provided in the DCE, as described in Subsection 10.5. The higher levels of protocol are provided in the DTE and/or the DPE, depending on system architecture.

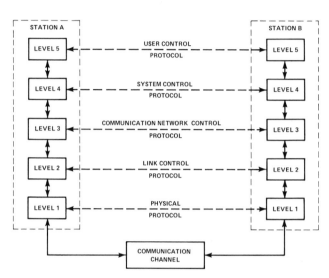

Figure 10.0-2.   Protocol Requirements for a Data Communication System.

10.1   Fundamental System Arrangements

There are three basic commonly utilized data transmission system arrangements: simplex, duplex, and multiplex. Simplex transmission describes transmission in one direction only, from a single transmitter (or driver) to one or more receivers, as shown in Figure 10.1-1. The use of several receivers is sometimes called a multi-drop system. In duplex transmission there are two stations, each having "transmit" as well as "receive" capability, so that data can move in either direction. Simultaneous transmission of data in both directions is called "full-duplex;" alternate transmission in one direction or the other is called "half-duplex." Depending on the coding scheme used, either full- or half-duplex transmission can share the transmission medium, but in full-duplex transmission of digital data, separate paths are usually required.

"Multiplex" transmission is similar to "half-duplex" with a shared transmission medium, except that more than two stations are involved. In bus multiplexing, the stations are arranged along a bus transmission medium. The star multiplex arrangement is shown symbolically in Figure 10.1-1; in practice, there may be separate transmission mediums from one master station to each of several slave stations. Loop multiplexing is symbolized by the dotted lines.

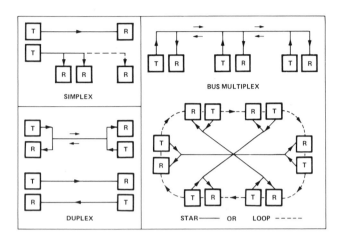

Figure 10.1-1. Basic System Arrangements for Data Transmission.

10.2   Fiber Optic vs. Wire/Optocoupler Considerations

Guidelines for selecting fiber optics or optocoupler/wire transmission mediums involve the consideration of several performance criteria as well as cost. On the basis of a simple comparison of the cost per meter of wire cable and fiber optic cable, the wire cable appears to be more attractive, but the peripheral costs for both mediums must be considered. These peripheral costs (repeaters, armor, shielding, space, power, labor of installation, etc.) can move the economic advantage either way, depending on system performance requirements.

10.2.1   Ground Loop Effects

There are two effects that ground loops can exhibit: interference with reception of the transmitted signal and interference with other components of the system. In Figure 10.2.1-1, it is clear that a fiber optic system provides a larger physical separation between the nodes of the common mode voltage, e_{CM}, and is therefore superior to optocouplers in dealing with either of these types of interference. However, a fiber optic system may be more costly and may not be needed to solve a ground loop problem.

If the problem is only one of interference with the signal transmission, there are several optocoupler circuit tricks for solving the ground loop problem. The most effective of these is the use of polarity-reversing drive and an exclusive-or flip-flop (see page 3.54) at the receiver. This is beneficial whether the optocoupler is employed as a line driver or a line receiver.

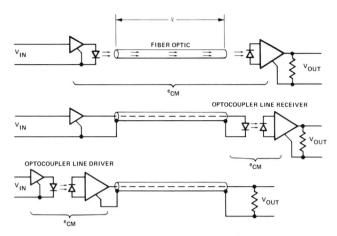

Figure 10.2.1-1.   Fiber Optic and Wire/Optocoupler Data Link Diagrams Showing Common-Mode Node Separation.

On the other hand, if the problem is one of capacitively coupled ground loop current causing noise elsewhere in the system, then the small input-to-output capacitance of an optocoupler ($<$ 1 pF per channel) may provide adequate opening of the ground loop. Subsection 10.3.2 has more on ground loop effects.

10.2.2   Cross Talk and EMI Effects

Electro-Magnetic Interference (EMI) may enter into, or be radiated by, wire cable but has no effect on fiber optics. Cross talk is simply a particular kind of EMI resulting when adjacent wires carry signals which may mutually interfere. The effects may be reduced (by use of twisted paris, shielding, and impedance balancing) to the extent that performance is adequate; however the cost of doing so should be considered when making the wire-cable/fiber optic comparison.

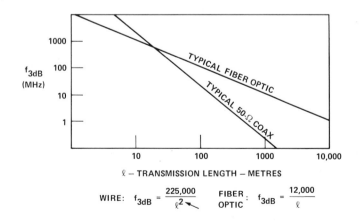

Figure 10.2.3-1.   Bandwidth (Data Rate) Comparison Between Fiber Optic Cable and Wire Cable.

10.2.3   Bandwidth Comparison

In general, the 3 dB bandwidth of wire cable varies inversely as the square of length, while in fiber optic cable, it varies as only the first power of length. Figure 10.2.3-1 shows some typical values of 3 dB bandwidth as a function of length. It is clear that over long transmission distances, a wire system would require more repeater stations if the system bandwidth requirement is much over 3 MHz.

10.2.4   Terminal Requirements

In a system using wire cable with optocouplers, the optical coupling is essentially the same at any transmission length. Although the electrical signal strength may vary with distance, this effect is easily compensated, so terminal circuitry for optocouplers is relatively simple. In a fiber optic system, the optical coupling is influenced by choice of connectors, fiber optics, and transmisison length, requiring the optical receiver to have a large dynamic range. This makes fiber optic receiver terminal equipment more complicated and costly than the simple buffering that can be used with optocouplers.

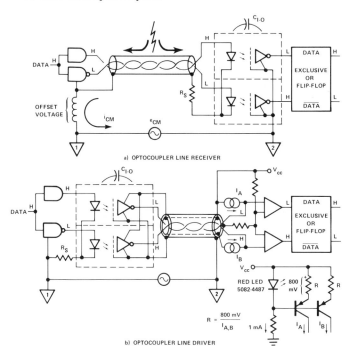

Figure 10.3.1-1.   Optocouplers Directly Connected to Transmission Line as Drivers or Receivers.

10.2.5   Mechanical/Optical Considerations

In most respects, fiber optic cable is superior to wire on the basis of weight, strength, and durability. The main concern with fiber optics is the minimum bend radius. Optical fibers are therefore made very small to allow bending to a small radius without breaking. In addition to the fiber itself being very small, in a proper cable, the inner buffer jacket, tensile support material, and cover jacket produce a mechanical system which is far lighter in weight and less costly than a wire cable armored well enough to withstand the same crush and impact abuse. On the other hand, making cable-to-cable and cable-to-terminal connections is at present much simpler with wire than with fiber optics.

In optocouplers, there are no optical concerns; the signal ports are all electrical. With fiber optics, there are concerns for optical losses in connectors, splices, and transmission. These are discussed in Subchapter 10.4.

10.3   Optocoupler/Wire-Cable Data Link Realizations

With appropriate external circuitry, optocouplers can provide data links having a high degree of isolation, insulation adequate for many applications, and channel capacity up to 15 M baud. The actual data rate achievable for a given channel capacity depends on the efficiency of the coding scheme (number of code symbols per data bit). This is discussed in Subchapter 10.5.

10.3.1   Simplex Data Transmission

With data moving in only one direction, the external circuitry to be used with optocouplers is very straightforward. In many cases, optocouplers can operate directly on the transmission line, as either drivers or receivers. This is shown in Figure 10.3.1-1 for a simplex system. Note that whether the optocouplers are employed as line drivers or as line receivers, the only element allowing ground loop current, i_{CM}, to flow is the input-to-output capacitance, C_{I-O}. Since C_{I-O} is typically only about 0.7 pF for each optocoupler, ground loop current is quite low, even when a pair of optocouplers is used for a single data channel.

A pair of optocouplers in each data channel is recommended for dealing with interference. Where interference is no problem, a single optocoupler working against ground may be adequate. Where there is a known or suspected interference problem, the split-phase drive offers infinite rejection of e_{CM} (the common-mode intereference) providing the split-phase outputs are applied to an exclusive-or flip-flop (see Section 3.6.2). This split-phase drive and reception is beneficial even when optocouplers with internal electric shielding are used.

Split-phase drive has two other advantages in addition to common mode interference rejection. In the split-phase dircuits of Figure 10.3.1-1, both optocouplers work against the shield common, but the line-to-line signal is balanced. Consequently, the threshold point is the same for low-to-high and high-to-low data transitions. This reduces the delay distortion and consequent inter-symbol interference that is almost always present when single-ended (unbalanced) drive is used. The other advantage is that in split-phase drive, either one side or the other is "on," so that, with respect to the shield, the impedances are constant. This simplifies the problem of dealing with line reflections by load matching with a line-to-line resistor at the receiving end. Such a resistor imposes some drive capability requirement on the line driver when optocoupler line receivers are used because the voltage drop across the line-to-line resistance must exceed the LED turn-on voltage. When directly-coupled line receivers are used, the voltage drop on the line-to-line resistance needs only (as a minimum) to exceed the few millivolts of offset voltage in the differential line receivers, so a fairly low line-to-line resistance can be connected as a load match.

Notice in Figure 10.3.1-1b that current-source pullups are recommended rather than simple resistor pullups. There are two reasons for this. First, the rate of rise of either line voltage is limited by the current which can be sunk by the optocoupler line drivers. With current-source pullup, this rate of rise remains constant up to the 1.0V compliance of the current source, whereas, with resistive pullup, the rate of rise diminishes as the voltage rises, making the time delay to threshold 39% longer than for current-source pullup. Second, the high impedance of the current sources allows more freedom in the choice of a line-to-line terminating resistor. An adequate current-source design is shown in Figure 10.3.1-1b.

In situations requiring an optically coupled line driver with very high current capability (e.g., RS-422) or with single-ended polarity-reversing output (e.g., RS-423), the simplest and most effective solution is the use of a buffer at the driver, as in Figure 10.3.1-2a. The active pullup at the output of the buffer permits higher data rate transmission.

A buffered receiver, as in Figure 10.3.1-2b, also offers performance advantages over a directly connected opto-coupler receiver. The high input impedance of the buffer amplifier allows it to draw from the signal line only a small fraction of the current required by the input diodes of most optocouplers. It is, of course, true that some optocouplers can be operated with an input current as small as 0.5 mA, but at such a low input current, the data rate is limited to less than 50 kb/s. The use of an input buffer amplifier thus allows high speed operation along with small line loading. This combination establishes two important benefits. One benefit is that proper low-reflection termination of both differential-mode and common-mode impedance can be made. The other benefit is that low-reflection "taps" can be connected anywhere along the signal line.

The only disadvantage in using buffered optocoupler line drivers and receivers is that it always requires provision of conductors for the remote V_{cc} and GND. Some types of drivers require a negative, as well as a positive, V_{cc}, but

these types are usually for single-ended (unbalanced) operation, which is not recommended where interference is troublesome.

The realization for the buffered line driver in Figure 10.3.1-2a requires little comment other than to point out that a single split-phase line driver, such as half of a Fairchild 9634 (see Figure 10.3.2-3) could be used, providing the receiver is configured as in Figure 10.3.1-2b.

Realization of a buffered receiver as in Figure 10.3.1-2b does require a little attention. Notice that the R_1/R_2 voltage divider sets the switching threshold. The threshold should be set at the average of the high and low line-to-ground (shield) voltage, V_H and V_L. This is done to balance the propagation delays for H→L and L→H transitions. If these delays are unbalanced, intersymbol interference may become too high. It would be simpler to perform delay balancing by cross-connecting the + and − inputs of the buffer amplifiers, but this has a high risk of causing erroneous transition in the exclusive-or flip-flop. In the presence of a large common mode transient, the buffer outputs are not specifically defined. By referring them to the R_1/R_2 divider, their outputs are never undefined — a common mode transient will make both outputs high or low, depending on its polarity, and the exclusive-or flip-flop does not change when this occurs.

The terminating resistors R_{DM}/R_{CM} are optional. They are needed only if reflections are a problem. If they are used, the line driver must be capable of sourcing enough current to pull the line voltage higher than the threshold established by R_1/R_2.

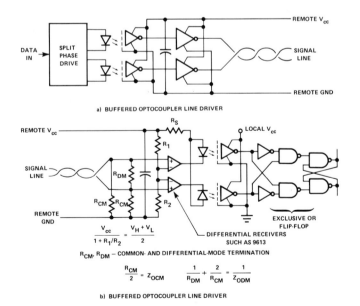

a) BUFFERED OPTOCOUPLER LINE DRIVER

$$\frac{V_{cc}}{1 + R_1/R_2} = \frac{V_H + V_L}{2}$$

R_{CM}, R_{DM} — COMMON- AND DIFFERENTIAL-MODE TERMINATION

$$\frac{R_{CM}}{2} = Z_{OCM} \qquad \frac{1}{R_{DM}} + \frac{2}{R_{CM}} = \frac{1}{Z_{ODM}}$$

b) BUFFERED OPTOCOUPLER LINE DRIVER

Figure 10.3.1-2. Optocouplers with Buffered Connections as Transmission Line Drivers and Receivers.

10.3.2 Duplex Data Transmission

Duplex transmission involves movement of data in both directions between two stations. In half-duplex, the direction of data movement alternates, often along the same time-shared path. In full-duplex, there is simultaneous bi-directional transmission. Use of separate carrier frequencies, directional couplers, and other special techniques allow simultaneous bi-directional use of the path, but it is more common with wire cable transmission to use separate paths for the two directions. When the presence of common mode interference suggests considering the use of optocouplers, it is usually not satisfactory to consider their use only as line receivers. As seen in Figure 10.3.2-1a, the ground loop current, i_{CM}, can be coupled by the interwire capacitance, C_{W-W}. Interwire capacitance may be more than 50 pF/m so that even a very short length can lead to severe ground loop current. Only at very low frequencies of the common mode interference could the benefit of using optocouplers be realized.

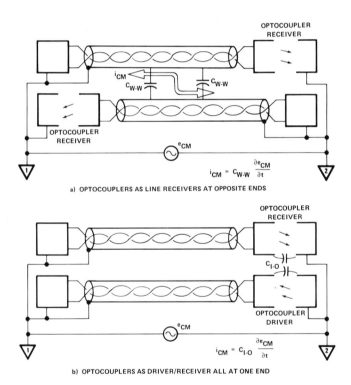

Figure 10.3.2-1. **Full-Duplex Arrangement Showing the Need for Optocoupler Line Driver as well as Optocoupler Line Receiver to Reduce to Reduce Ground Loop.**

Much greater reduction of common mode interference is obtained by having the optical coupling placed all at one end, or at BOTH ends. When this is done, the only capacitance coupling the ground loop current is the input-to-output capacitance, C_{I-O}, of the optocouplers, as shown in Figure 10.3.2-1b.

Circuits for full-duplex realization are the same as those in Figures 10.3.1-1 and 10.3.1-2 because each line in the full-duplex system may be regarded as a simplex line. It is also possible to use a buffered driver with non-buffered receiver and vice versa, providing attention is given to tracing out polarity and adequacy of drive current.

Half-duplex with a time-shared transmission line is partly shown in Figure 10.3.2-2 for a non-buffered connection and in Figure 10.3.2-3 for a buffered connection. Here the buffered and non-buffered circuits are usually not interchangeable. The half-duplex system is only partly shown because at the other end of the line the driver and receiver circuits are the same. The only difference is that the other end of the line does not require the I_A and I_B current source in the non-buffered system. In the buffered system, the I_A and I_B sources are not used because the 9634 driver has the capability of driving (sourcing current to) a 50Ω line. If the transmission line is a low-loss type, it may be necessary to apply the R_{CM}/R_{DM} termination at both ends to prevent false differential response from reflections.

In Figure 10.3.2-2, the open-collector outputs of the optocouplers must have a current-sinking capability equal to or greater than I_A. These outputs are not both on at the same time, so a simple current limiting resistor can serve both input diodes. With the Enable input low, both optocouplers are off so both sides of the Signal Line are high, which makes the Line Status output high. Because a common-

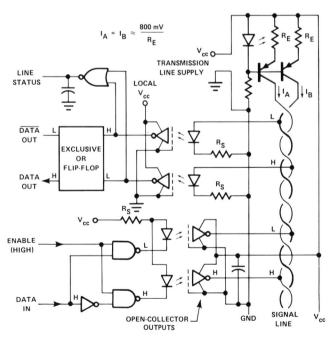

Figure 10.3.2-2. **Half-Duplex or Multiplex Station Having Drive, Receive, and Standby Functions, with Optocouplers Connected Directly to the Line.**

mode transient could momentarily cause the same condition, a filter capacitor is needed to prevent a false indication that the line is not in use. During data transmission, either one side or the other of the Signal Line is always low, providing the line driver has $t_{PHL} < t_{PLH}$, which is generally true; but should $t_{PLH} < t_{PHL}$, there would be only a momentary "low-low" at the Line Status gate, so the capacitor will keep its output low until a genuine "line available" condition occurs.

In Figure 10.3.2-3, the 9634 driver has a low impedance output, whether it is at "low" or "high" logic state. Consequently, the Signal Line does not require either resistive or current-source pullups. The 9634 also has split-phase outputs, operating from a single-ended input, so only one optocoupler is needed for carrying data. The second optocoupler operates the Strobe input (pin 7); when the Enable input is low, pin 7 is high and the tri-state outputs of the 9634 are "open" — that is, neither side of the Signal Line is pulled high or low. However, the R_{CM} resistors will make both sides of the Signal Line go low, the length of time required depending on their values, and on the line length and characteristics. The Line Status output will then go low, indicating that the line is not in use. Here, again, a filter capacitor is used to prevent a momentary low-low on the line (or a large negative common mode transient) from giving a false indication. Whenever the line is not in use, it is desirable to have all the LED input diodes of the optocouplers turned off to avoid unnecessary degradation; that is the reason for the apparently redundant gating of the Data In signal.

10.3.3   Multiplex Data Transmission

A multiplex data transmission system is like half-duplex, except that more than two stations time-share their transmission. The circuit realization for multiplex would have at each station the same circuits as those shown in Figure 10.3.2-2 or in Figure 10.3.2-3. The transmission line arrangement deserves a bit of thought, however, especially if the circuits of Figure 10.3.2-2 are to be used, since the drivers are current-sinks only, rather than being alternately low-impedance source/sink like the 9634 in Figure 10.3.2-3.

The bus and star arrangements in Figure 10.3.3-1 are drawn with the resistance of the transmission line represented by the discrete resistors, R_W. The open-collector outputs of the drivers are represented by the Schottky-clamped transistors. If the distance from I_A in the bus arrangement to the most remote terminal is very far, the voltage drop on the line (across the $R_{W's}$) may prevent the voltage at the nearer station(s) from going low enough to operate the receiver correctly. This situation is much less likely to occur if the Star configuration is used (Figure 10.3.3-1b). The Star arrangement is superior also with the high speed buffered driver/receiver circuits in Figure 10.3.2-3 because in the Star arrangement, the transmission delay is the same from any station to any other station in the system.

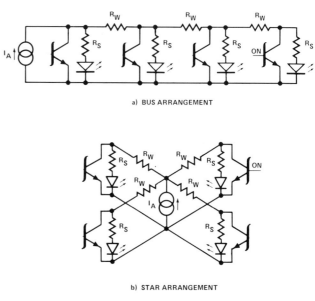

a) BUS ARRANGEMENT

b) STAR ARRANGEMENT

Figure 10.3.3-1.   Bus and Star Arrangements for Parallel (Shunt) Multiplex Interconnections.

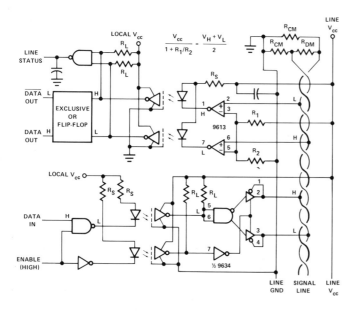

Figure 10.3.2-3.   Half-Duplex or Multiplex Station Having Drive, Receive, and Standby Functions, with Buffers Between Optocouplers and Signal Line.

Another arrangement by which several stations can communicate in multiplex fashion is the current loop. A basic current loop is symbolized at the top of Figure 10.3.3-2. When none of the stations in the loop are transmitting, the loop is closed all around. When a given station transmits, the switching transistor is turned off and the resulting drop in loop current, I_L, is sensed at all other stations in the loop. In the Mark condition (see Figure 10.3.3-2), it is desirable to have the terminal voltage at each station as low as possible to allow a large number of stations for a given V_{cc}. When any one station transmits (transistor off), the switching transistor will have nearly the full value of V_{cc} applied, depending on how many stations are in the loop and what terminal voltages they have during a Space condition produced by another station.

Current looping is simple and inexpensive and is an ideal application for optocouplers. The main limitation on data rate lies in the choice of the optocoupler. With optocouplers having separate photodiode and amplifier, data rates up to 20k b/s are possible .

Without optocouplers, current loop realization has the difficulty of the voltage buildup around the loop. With optocouplers, that is no problem and the realization can be very simple indeed, as seen in Figure 10.3.3-2. With this simple realization, the number of stations is limited by the voltage tolerance of the optocoupler output. The number of possible stations can be expanded by adding a buffer transistor that can tolerate the voltage that appears when it is switched off.

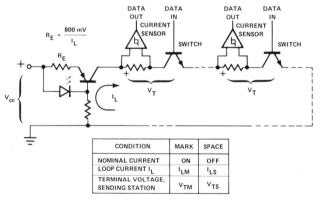

CONDITION	MARK	SPACE
NOMINAL CURRENT	ON	OFF
LOOP CURRENT I_L	I_{LM}	I_{LS}
TERMINAL VOLTAGE, SENDING STATION	V_{TM}	V_{TS}

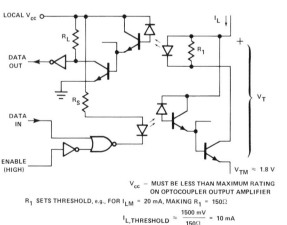

V_{cc} — MUST BE LESS THAN MAXIMUM RATING ON OPTOCOUPLER OUTPUT AMPLIFIER

R_1 SETS THRESHOLD, e.g., FOR I_{LM} = 20 mA, MAKING R_1 = 150Ω

$$I_{L,THRESHOLD} \approx \frac{1500\ mV}{150\Omega} = 10\ mA$$

MAXIMUM NO. STATIONS = V_{cc} (MAX) $\div V_{TM}$

Figure 10.3.3-2.   Current Loop Multiplex Connection with Unbuffered Optocouplers in Loop with Current Limited at 20 mA.

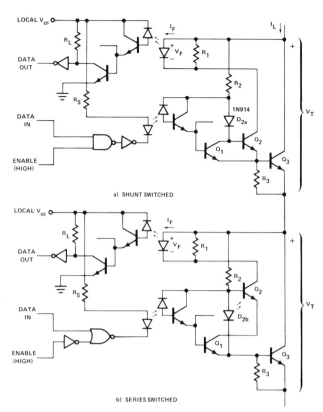

a) SHUNT SWITCHED

b) SERIES SWITCHED

Figure 10.3.3-3.   Current Loop with Optocoupler Buffered by a Switching Transistor, Q2, for High V_{TS} and a Regulator Transistor, Q3, for High I_{LM}.

Figure 10.3.3-3 shows the circuit for adding a switching transistor, Q_2, to take the higher voltage. It can be either shunt switched or series switched. In shunt switching, only the base current of Q_2 is switched by the optocoupler so it can be operated with a smaller input current than is required for series switching where the optocoupler output must switch the emitter current of Q_2. The shunt switch requires the silicon diode D_{2a} to allow the collector of Q_1 to go low enough to cut off Q_2. In series switching the LED, D_{2b}, keeps the base voltage of Q_2 (and hence the collector voltage of Q_1) from rising too high when Q_1 is off; a Zener diode could also be used.

In either series or shunt switching, for the Space condition, some current does flow, via R_2. For high speed switching and low V_{TM}, R_2 should be relatively small in value, but making R_2 smaller raises I_{LS}. To deal with this, R_1 is made small enough that under worst-case conditions the voltage across R_1 will not be high enough to allow significant current in that input diode – making it less than 1.2 volts is adequate. The worst worse-case current in R_2 is V_{cc}/R_2, and leads to the the inequality:

$$R_1 \ (V_{cc}/R_2) < 1.2V \qquad\qquad (10.3.3\text{-}1)$$

The source driving the loop may or may not have built-in

current limiting, but even if it has, the level at which the loop current is clamped may be much higher than I_F, the level of current desired for the optocoupler input diode. For this reason, the shunt regulator, R_3, Q_3 is added. This does add an additional V_{be} to the terminal voltage for the Mark condition, but it also adds safety. When Q_2 comes on, loop current flows in R_3, and if the voltage drop exceeds V_{be}, then Q_3 is turned on to bypass the excess loop current around the input diode.

In selecting resistor values, here are the steps:

1. Determine what I_F and I_{LS} should be.

2. Select $R_2 \approx V_{cc}/I_{LS}$.

3. Select $R_1 \leqslant 1.2V/I_{LS}$.

4. Select $R_3 \leqslant V_{be}/(I_F + V_F/R_1)$.

An estimate can be made of what the terminal voltage will be in the Mark condition:

SHUNT SWITCHED:

$$[V_{be} + V_{SAT,Q_2} + V_F] < V_{TM}$$

$$< [2V_{be} + V_{D2a} + I_L R_2/(\beta + 1)]$$

SERIES SWITCHED:

$$[V_{be} + V_{SAT,Q_1} + V_{SAT,Q_2} + V_F] < V_{TM}$$

$$< [2V_{be} + V_{SAT,Q_1} + I_L R_2/(\beta + 1)]$$

Insertion of typical values makes it clear that, unless β is very large, the series switched circuit will usually provide a lower terminal voltage, but does require the loop-driving optocoupler to operate at a higher input current than would be needed for shunt switching. The terminal voltage for either circuit may depend on the loop current in the Space condition as well as the Mark condition because the choice of R_2 depends on I_{LS}.

10.4   Fiber Optic System Realization

In the design of a fiber optic data transmission system, there are a number of variables that enter the tradeoff considerations. Some of these are in the design or selection of the fiber optic cable and connectors. Others involve the transmitter and receiver. Still others relate to the system arrangement. All of these are discussed in this section, while the details of coding schemes and their effect on data rate are discussed in Section 10.5.

10.4.1   Properties of Fiber Optic Systems

Optical fibers are all made with a core index of refraction, n_1, which is greater than the cladding index of refraction, n_2. The difference in refractive index causes optical signals to propagate within the core because of Total Internal Reflection (TIR) taking place at the interior of the cladding. The range of angles of incidence at which TIR takes place is limited, and leads to definition of an acceptance cone. As seen in Figure 10.4.1-1, this acceptance cone is described by its half-angle, called the Acceptance Angle, θ_A. The sine of the acceptance angle is called the Numerical Aperture, N.A. The N.A. is perhaps the most important property of an optical fiber because it describes the efficiency with which optical signals can be coupled into a fiber and relates also to the bandwidth limitation. N.A. can be derived from the core and cladding indices of refraction:

$$\text{N.A.} = \sqrt{n_1^2 - n_2^2} = \sqrt{n_1 + n_2}\sqrt{\Delta n} \qquad (10.4.1\text{-}1)$$

but a more practical value of N.A. is obtained by a measurement. With an optical signal source of very large N.A. (e.g., an LED) coupled to one end of the fiber, the radiation pattern of the opposite end is observed. The Acceptance Angle, θ_A, is then defined as that angle in the radiation pattern at which the intensity is 10% of the axial intensity, and the measured N.A. is the sine of θ_A.

For reasons which are beyond the scope of this discussion, N.A. is not a constant, but is somewhat smaller for longer lengths of the same fiber, so a reported N.A. is precisely valid only for the particular length of fiber used in the measurement. Also the rate at which N.A. varies with length is not a constant, but does become nearly zero in a length called stabilization length. Stabilization length varies from a few metres for lossy fibers to several hundred metres in very low loss fibers.

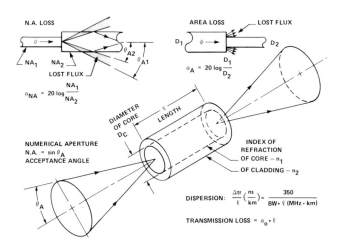

Figure 10.4.1-1.   Important Properties of Optical Fibers.

Other properties mentioned in Figure 10.4.1-1 also change with length. The Transmission Loss Constant, α_O, and the Dispersion, $\Delta t_r/\ell$ change slightly over lengths shorter than the stabilization length. These changes are due to the mode structure.

Optical rays in the fiber propagate at various angles up to the angle allowed by θ_A. Those rays propagating at small angles are called low-order modes, and those propagating at larger angles are called high order modes. In a particular length of fiber, the higher order modes travel a longer path length than the low order modes. Optical losses, resulting from scattering and absorption, are a function of path length, so the higher order modes are more lossy. Also, in travelling a longer path length, the higher order modes have a longer travel time. Consequently, for an optical signal in which all modes start out at the same instant, the lower-order mode fraction will reach the other end of the fiber earlier than the higher-order fraction. The Modal Dispersion is a main cause of rise-time dispersion. Since the dispersion increases with length, it is characterized in Figure 10.4.1-1 as the increase of rise time per unit of length in (ns/km). Dispersion may also be described in terms of a bandwidth-times-length product in (MHz • km).

The relationship between N.A. and dispersion is only a loose one, but generally fibers with high N.A. are more lossy and have higher dispersion. This relationship is important in the design tradeoff involving large bandwidth (tens or hundreds of megahertz) over a kilometre or more.

N.A. is important also in coupling fibers (with splices or connectors) where the sending fiber has a larger N.A. than the receiving fiber. As seen in Figure 10.4.1-1, N.A. loss varies as the square of the N.A. ratio.

Core diameter also is important in coupling. It is clear, in Figure 10.4-1, that the area loss will vary as the square of the core diameter ratio. The tradeoff parameter against which core diameter is weighed is the mechanical strength of the fiber optic cable. Paradoxically, a smaller optical fiber makes a stronger cable — that is, one which tolerates more crush, impact, and flexure abuse than a cable made with larger fiber. This is because the minimum bend radius rises greatly as the diameter of the optical fiber increases.

The core-cladding DIFFERENTIAL index of refraction is important in establishing N.A. The ABSOLUTE value of n_1, the core index of refraction, is responsible for fresnel (reflection) loss when fibers are coupled without an index matching material. In a splice, where index-matching material can be applied, fresnel loss is less than 0.05 dB; in a connector, without index matching, fresnel loss is typically 0.35 dB, corresponding to $n_1 = 1.5$. The absolute value of n_1 also establishes the travel (or transit) time for optical

signals. The propagation delay time constant is given by:

$$\frac{t_p}{\ell}\left(\frac{ns}{m}\right) = \frac{n_1}{c} \qquad (10.4.1\text{-}2)$$

where c = velocity of light in space = 0.3 m/ns

Since all of the fiber optic properties, except core diameter, depend on index of refraction, and since index of refraction varies with wavelength, it should come as no surprise that these properties also vary with wavelength. In most applications, the variation is not large enough to be of concern, but there are two exceptions.

As seen in Equation (10.4.1-2), the travel time can vary with index of refraction, so if the spectrum of the source extends over a wide enough range of wavelengths, different parts of the spectrum will have different travel times, the difference depending on the core material. This causes Material Dispersion which adds linearly to the Modal Dispersion described earlier.

The other important wavelength-dependent property is the transmission loss. The Transmission Loss Constant, α_O, varies with wavelength, mainly due to the wavelength-dependent absorption and scattering coefficients. An example of how much α_O varies with wavelength is shown in Figure 10.4.1-2 for the glass-clad silica fiber used in the Hewlett-Packard cable.

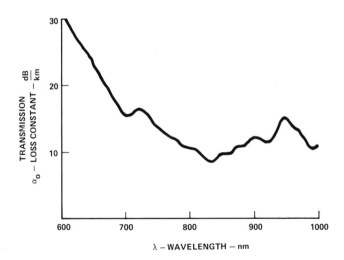

Figure 10.4.1-2. Wavelength Dependence of Fiber Optic Transmission-Loss Constant, α_O.

The most important design equation for a fiber optic system is called the Flux Budget. The terms in this equation appear as illustrated in Figure 10.4.1-3. The flux available and flux required are defined with respect to the accessible optical ports. The total flux produced by the LED, for example, is much higher than the flux available at the optical port of the transmitter because there is at least some N.A. loss at the LED which has N.A. = 1 while N.A. of the coupling stub is lower. An exception here would be a laser diode which has N.A. $\ll$ 1. The term α_{TC} (Transmitter-to-Cable coupling loss) results primarily from N.A. loss and Area loss while α_{CR} (Cable-to-Receiver coupling loss) has neither N.A. nor Area loss if the receiver port has a larger N.A. and size than the cable but α_{CR} does have 0.3 dB of fresnel loss. The term α_{CC} (Cable-to-Cable coupling loss) also has fresnel loss, but the more significant loss factors are from gap loss or misalignment; since the fibers on both sides of the joint are presumed to have the same core diameter and N.A., α_{CC} has no area or N.A. losses. The values in Figure 10.4.1-3 are examples drawn from the Hewlett-Packard system, and are selected for minimum ϕ_T, maximum ϕ_R, and maximum losses to yield the worst-case MINIMUM value for ϕ_M, the flux margin. ϕ_M must be more than zero dB if the system is to function as expected.

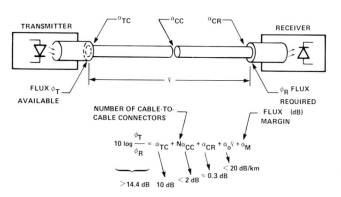

Figure 10.4.1-3. Loss-Factor Definitions and Flux Budget Equation.

To obtain the worst-case MAXIMUM value of ϕ_M, the same equation is used, but with maximum ϕ_T, minimum ϕ_R and minimum loss values. The value of ϕ_M from this computation describes the dynamic range requirement for the receiver.

In most systems, this dynamic range requirement is so large that DC coupled receiver designs cannot be utilized. The flux must be modulated in some way and the receiver must have ac-coupled response along with gain control so the detected and amplified signal can have excursion amplitudes that are adequate for the logic slicing levels without overdriving. Overdriving is undesirable because it causes excessive propagation delay, and thus impairs the signaling rate. Extreme overdrive causes cut-off or saturation, resulting in loss of some data bits.

At the lower end of the flux margin range, the performance is limited by noise in the receiver. The flux required for proper performance of the receiver is established by the Bit Error Rate (BER) and data rate (bandwidth) required,

relative to the noise present. By allowing higher error rate or lower data rate, or by reducing the unfiltered noise, the flux requirement, ϕ_R, can be reduced. Of the three alternatives, noise reduction, if it is possible, is the most attractive because ϕ_R varies jointly as the first power of the effective input noise current (i_n) or noise voltage (e_n). Relaxing the data rate (BW) requirement is the next choice because ϕ_R varies as the square root (one-half power) of bandwidth due to the bandwidth-limitation of noise. Allowing a higher bit error rate (BER) is the LEAST attractive tradeoff because the error rate varies as the complementary error function (Figure 10.4.1-4) requiring an order-of-magnitude increase in error rate to obtain a 5% reduction in ϕ_R.

Usually, though not always, the errors that occur in a fiber optic link at lower flux levels are random. This means that the probability of error, P_e, can be calculated from the complementary error function shown in Figure 10.4.1-4. For a given data rate, P_e and BER are numerically the same. For example, if the BER is 10^{-9}, then the probability that any given bit is erroneous is 10^{-9}. In a measurement, this would signify that in an interval of 10^9 bits, the probability is even (0.5) that one error will be found. Therefore, to establish BER $< 10^{-9}$ would require observation for a long enough time to be certain, to some level of confidence, that $P_e < 0.5$ in an interval of 10^9 bits. Such a measurement can be very time-consuming, but the use of the complementary error function, erfc(x) (Figure 10.4.1-4) allows observation to be made at a lower flux level, where the BER (P_e) is high enough that it can be measured in a reasonable period of time. The MEASURED flux margin of the system is the amount of attenuation (in dB) which was inserted to make

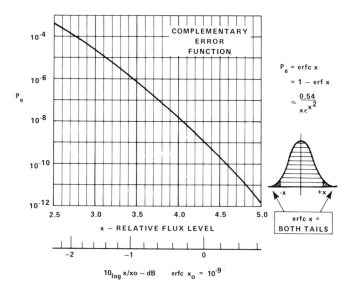

Figure 10.4.1-4. Probability of Error, P_e, the Complementary Error Function, with Argument in dB for 0dB at $P_e = 10^{-9}$.

P_e high enough to be conveniently measured. The OPERATING flux margin is lower by the ratio (in dB) of the relative flux level at the desired P_e to the relative flux level at the P_e observed in the measurement.

To make sure that errors are indeed occurring randomly, it is advisable to measure P_e at several flux levels. If errors are random, the results should fit the erfc(x) curve in Figure 10.4.1-4.

10.4.2 Description of the Hewlett-Packard Fiber Optic System

In the design of the Hewlett-Packard system, the optimization suggested in Subsection 10.4.1 has already been worked out for a symbol rate of 10 M baud (10 M b/s NRZ) at a bit error rate (BER) less than 10^{-9}. There are also some convenience features such as the single +5V supply requirements; the easily-mounted, low-profile, shielded all-metal packages; and pre-aligned fiber optic connectors. Perhaps the greatest convenience feature of the HFBR-1001 Transmitter and HFBR-2001 Receiver is the flux modulation scheme and the resultant benefits in the design of the Receiver.

Figure 10.4.2-1 shows the block diagram of the Transmitter. There are two modes of operation. With Mode Select "high," the refresh multivibrator is disabled with the Q output high and $\overline{Q}$ low. This makes both gates active so that whenever Data Input is high, both I_A and I_B are on and when Data Input goes low, both I_A and I_B are turned off. Thus, with Mode Select high, the output flux, ϕ_{OUT}, is simply a digital replica of the Data Input signal; if Data Input is unmodulated, ϕ_{OUT} is unmodulated.

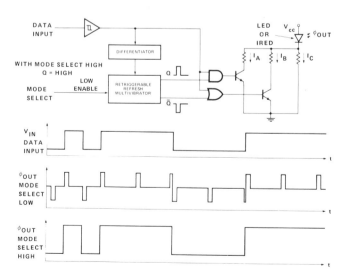

Figure 10.4.2-1. Hewlett-Packard Fiber Optic Transmitter Block Diagram and Input/Output Waveforms.

With Mode Select low, the refresh multivibrator is enabled. During most of the 5000-ns period, Q is low and $\overline{Q}$ is high so that only I_B is on. At each transition of Data Input, a new train of refresh pulses is initiated. If Data Input is high, the pulse gates I_A on while I_B remains on, causing ϕ_{OUT} to be a train of positive flux excursions from mid level to high level. With Data Input low, I_A is held off and each pulse gates I_B off causing ϕ_{OUT} to be a train of negative flux excursions from mid level to low level. A Data Input transition is permitted to occur at any time — if it occurs during a refresh pulse, the multivibrator is retriggered so the first pulse is always of full duration (50 to 60 ns). When applied to the Hewlett-Packard Receiver, this pulse bipolar flux waveform makes the overall system transparent to any data format. That is, the duration of either high or low level at the Data Input can be completely arbitrary — from dc to 100 ns — and the Data Output of the Receiver will follow.

Figure 10.4.2-2 shows the block diagram of the Receiver. It is a four-chip hybrid circuit assembled in the same type of package as the Transmitter. Two of the chips are the filter capacitors for the dc restorer and ALC; a third chip is the PIN photodiode, and the fourth chip has all of the remaining circuits shown in the figure.

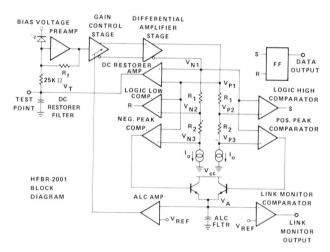

Figure 10.4.2-2. Hewlett-Packard Fiber Optic Receiver Block Diagram.

The Receiver has four major functional blocks:

1. The amplifier, with current-to-voltage preamplifier and differential output.

2. The dc restorer amplifier.

3. The logic comparators with an R-S flip-flop for Data Output.

4. The peak detector comparators driving the ALC and Link Monitor amplifiers.

With no input flux, the voltage at the Test Point (output of the dc restorer) will be nearly 1300 mV. When flux enters the Receiver, the positive-going output voltage, V_{P1} rises and the negative going output, V_{N1} falls. This differential, applied to the dc restorer amplifier will bring the Test Point voltage down until it is low enough to pull away the average photocurrent from the preamplifier input, making $(V_{P1} - V_{N1}) \approx 0$ whenever the input flux is at the average level. With $(V_{P1} - V_{N1})$ near zero, all of the comparator input voltages are negative. The dc restorer amplifier has a very narrow bandwidth and, hence, does not respond to rapidly changing input signals. When input flux rises faster than the dc restorer output voltage can follow, $(V_{P1} - V_{N1})$ rises, and if the excursion amplitude is greater than $I_o R_1$, it will make the input voltage positive at the Logic High comparator, whose output will Set the R-S latch. With the R-S latch Set, Data Output remains high until there is a negative input flux excursion of sufficient amplitude to operate the Logic Low comparator and Reset the R-S latch to make Data Output low.

As mentioned earlier, it is undesirable to allow receivers to be overdriven. In the Hewlett-Packard Receiver, this is controlled by the ALC circuit. If the amplifier output voltage $(V_{P1} - V_{N1})$ excursion is so high that it exceeds not only the logic threshold $(I_o R_1)$ but goes beyond the peak detector threshold $[I_o (R_1 + R_2)]$, either the Positive or Negative Peak Detector will cause current to be switched into the ALC filter. When the ALC filter voltage has risen to V_{REF}, the ALC amplifier becomes active and limits the gain of the main amplifier until peak output voltage excursions are barely enough to exceed the threshold, $I_o (R_1 + R_2)$. Since the ALC Amplifier and Link Monitor Amplifier both operate against the same reference voltage (V_{REF}), a high at the output of the Link Monitor signifies that input flux excursions are high enough to activate the ALC, which means the main amplifier output voltage excursions are more than high enough to reliably operate the logic comparators.

The Link Monitor output is an indicator of the amplitude of the voltage excursions at the output of the main amplifier — it is not an indicator that these excursions are noise free. In most cases, it is true that when Link Monitor is high, the flux level is adequate for $BER < 10^{-9}$. Due to variations in component values, it is sometimes possible to have the Link Monitor high even though the input flux level is so low that $BER > 10^{-9}$.

For assurance that BER is adequately low, the system performance should be checked. This is done by using BER-measuring or bit-error-counting equipment. The screw-on type connector in the Hewlett-Packard system can be used as a variable attenuator at the Transmitter. With the flux reduced, a BER measurement is made at some high BER

(e.g., at 10^{-6} or 10^{-4}). Then, with a sensitive radiometer substituted for the Receiver at the end of the fiber optic cable, a measurement is made of the flux ratio resulting from before and after the connector at the Transmitter is screwed on again. This flux ratio is a measure of the flux margin for the higher BER used in the measurement. To obtain the flux margin for some lower BER, a ratio can be drawn from Figure 10.4.1-4 (e.g., if, at $BER = 10^{-6}$ a flux margin of 10 dB were observed, then, reading from $P_e = 10^{-6}$ in Figure 10.4.1-4 a value of -0.96 dB is found. This means that at $BER = 10^{-9}$, the flux margin of the system is $10\,dB - 0.96\,dB = 9.04\,dB$). Such measurement techniques are described in greater detail in the product data sheets and accompanying application notes.

10.4.3 Simplex Transmission

The complexity involved in achieving simplex data transmission via fiber optics depends mainly on the degree of sophistication built into the transmitter and receiver. At low data rates and with a large flux margin, a dc system with a fixed logic slicing level can be successfully operated. In general, however, flux modulation is required at the transmitter along with ac coupled response and gain control in the receiver.

Optically, simplex transmission requires only application of the flux budget to make sure the flux margin is within the range which the receiver can tolerate. The operating range is limited at the upper end by the flux level at which the receiver is overloaded, and at the lower end by the flux level at which the BER is at the maximum tolerated by the overall system. The use of error-correcting and/or error-detecting codes can extend the lower limit somewhat, although at a tradeoff of system data rate.

Simplex transmission with several receivers operating from the signal produced by a single transmitter (multi-drop) is possible but requires transmitter and receiver characteristics that provide a flux margin high enough to allow the insertion loss of the optical coupler. This insertion loss depends on the number of taps provided by the coupler:

$$\text{INSERTION LOSS (dB)} = 10 \log N + 10 \log \left[\underbrace{\frac{\phi_{IN}}{\phi_{IN} - \phi_{LOST}}}_{\text{4 to 6 dB}} \right]$$

where N = number of taps

10.4.4 Duplex Transmission

Half-duplex transmission can be accomplished with fiber

optics by using an optical coupler, but the insertion loss of the coupler imposes severe flux budget requirements.

Another kind of half-duplex arrangement requires an optical fiber with a core diameter large enough to allow side-by-side coupling of both the LED source and silicon detector chips to the single fiber. With such an arrangement at each end, signals can be transmitted, alternately, in either direction. The insertion loss of such shared optical ports is 6dB, or more, depending on the relative sizes of the emitting and detecting chips.

Full-duplex transmission can also be done with optical directional couplers, but a much easier approach is to use two separate fibers, just as is done with a wire cable.

10.4.5   Multiplex Transmission

For several stations to interchange information via a single optical medium is also possible. This is usually done with the optical fibers from each station joined in a "star" coupler. This makes the loss from any one transmitter to any other receiver very nearly the same. When the stations are "bussed" (that is, distributed along a single fiber), optical couplers can still be used for multiplexing, but the insertion loss is not the same from any transmitter to any other receiver, so dynamic range limits the number of stations.

An alternative to optically-arranged multiplexing is the electrically-arranged fiber optic multiplexing shown in Figures 10.4.5-1 and 10.4.5-2. Each fiber optic link is treated as a simplex system. The Master/Slave configuration of Figure 10.4.5-1 requires more transmit/receive (T/R) pairs than the loop configuration in Figure 10.4.5-2, but has the advantage that the transit time is nearly the same among all stations. In this way, the Master/Slave configuration resembles the multiplex star arrangement described in Section 10.3.3.

While the loop configuration of Figure 10.4.5-2 is more economical of T/R pairs, the repeated re-transmission of data raises three effects to be considered. One effect is

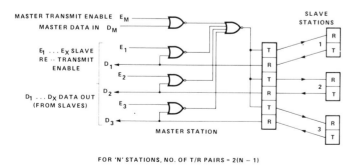

Figure 10.4.5-1.   Master/Slave Multiplex Arrangement.

transit time around the loop; in fiber optics, the propagation velocity causes a delay of about 5 ns per metre of fiber. The second effect is distortion of the signal due to unequal propagation delay, of rising and falling edges, in the transmitters and receivers; if this is a problem, it can be averted by inverting before each re-transmission. The third effect is error accumulation; the overall probability of error is approximately the linear summation of the individual probabilities of error in each of the links.

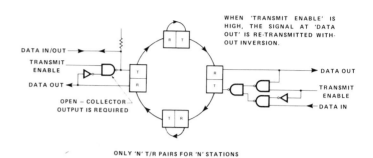

Figure 10.4.5-2.   Repeater Stations for Looped Multiplex Arrangement.

10.5   Data Formatting

In order for information to be useful when it is received, the data must be arranged in a format that is recognizable by the terminal equipment at the receiving station. This section describes some popular coding formats, circuits for encoding, decoding, and techniques for multiplexing several data channels through a single communication channel, then demultiplexing at the receiving end.

10.5.1   Channel Capacity

There are various ways of describing channel capacity. A useful term is the "baud" rate. This describes the number of code symbols per second which the channel can carry. A code symbol is a logic level: 1 or 0, high or low state, mark or space condition. In some systems, more than just two levels are used, but these are beyond the scope of this discussion. The choice of a coding format affects the rate at which data can be transmitted. Some formats require more than one code symbol per data bit. Baud rate is the reciprocal of the time duration of the shortest code symbol.

The most efficient use of channel bandwidth is with the NRZ (Non Return to Zero) format. It is called NRZ because when consecutive 1's are transmitted, the signal remains in the 1 state; the signal returns to the 0 state only when the data bit is a 0. As seen in Figure 10.5.1-1, this requires only a single code symbol per data bit, so the baud rate and data rate are equal. In the RZ (Return to Zero) code, the signal remains in the 0 state when the data bits are 0, but when a 1 is to be transmitted, the signal goes to the high state for part of the bit interval, but must return to the low state for the remainder of the bit interval. If the time in each state is evenly divided, then for RZ code the data rate is half the baud rate. If the duration in the high state is not 50% of the bit interval (i.e., has a duty factor, D.F. $\neq 0.5$), then the data rate is the baud rate multiplied by D.F.$_\tau$ or by $(1 - \text{D.F.})$, whichever is least.

The other codes in Figure 10.5.1-1 also require two code symbols per data bit, such that the data rate is half the baud rate (assuming the durations at high and low state are balanced). Manchester and Bi-Phase codes have two important properties that NRZ and RZ codes lack. One is that the average time spent at low state is equal to the average time in the high state. The other is that they are self-clocking, the importance of which is described later. The RZ code has the advantage that it might require a lower average power from the transmitter. Manchester and Bi-Phase can be transmitted via an a-c coupled communication channel while NRZ and RZ require a "transparent" channel, i.e., one that can hold a high or low logic level for an indefinite period of time.

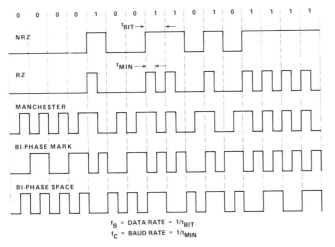

f_B = DATA RATE = $1/t_{BIT}$
f_C = BAUD RATE = $1/t_{MIN}$

Figure 10.5.1-1. **Coded-Data Waveforms and Definitions of Baud Rate (Symbols per Second) and Data Rate (Bits per Second).**

10.5.2   Clocking

The necessity for clocking is clearly seen with reference to the NRZ format. In a long sequence of 1's or 0's, it is only by means of a clock signal that the number of 1's or 0's in the sequence can be discerned. A clock signal is usually a square wave (D.F. = 0.5) but it need not be. The reference edge of the clock (either rising or falling) must be properly timed relative to the edges of the data bit interval. This is true whether the clock is generated in the DTE (Data Terminal Equipment) or the DCE (Data Communication Equipment) at either sending or receiving end.

In a bandwidth-limited system having rise-times or fall-times which are long, relative to the bit interval, it may be desirable to arrange the clock phasing to have the reference edge occur near the end of the bit interval. On the other hand, if bandwidth is no problem, but jitter is the major impairment of signal quality, then the clock should be phased with its reference edge as far as possible from the edges of the bit interval, i.e., near the center. If both jitter and bandwidth limitations exist, the reference edge of the clock should be halfway between the middle and the end of the bit interval.

When data are transmitted in the NRZ format, the clock signal is generated at the sending end and transmitted on a separate channel. This is usually done when there are several channels of data to be simultaneously transmitted at a rate so high that they cannot be conviently multiplexed into a single communication channel. A single clock channel then provides the timing reference for all of the data channels. The disadvantage of having separate data and clock channels lies in the possiblity of phase shift between them. Such relative phase shift is most likely to arise from transit time variations when transmission is over long distances, but can also occur in the short-distance between DTE and DCE due to thermal variations in propagation delays in the circuits.

To deal with variation in either transit time or propagation delay, a self-clocking data format is used. The use of self-clocking code usually requires the receiving DCE to be designed for a particular data rate or range of data rates. Without such a-priori information designed into the system, the receiving DCE may require a lengthy period of "training" before data can be transmitted. During such a training interval, the receiving DCE sets up the phasing of a phase-locked local oscillator. Then when data are transmitted, the PLL (Phase-Locked Loop) is applied to demodulate the self-clocked data, yielding recovered data (NRZ) and a recovered clock which is then applied to maintain the PLL in proper phase.

10.5.3   Self-Clocking Codes

The two most popular forms of self-clocking codes are the Manchester and Bi-Phase codes. As transmitted, they both have a combination of symbol times with durations of either a full- or half-bit interval. Thus, in addition to a-priori information on data (or clock) rate designed into the receiving DCE, there must be a-priori agreement between the sending and receiving DCE's as to whether the coding is Manchester or Bi-Phase.

While their transmitted waveforms (see Figure 10.5.1-1) appear similar, Bi-Phase and Manchester codes are quite different and have different properties. In Bi-Phase Mark, each bit interval begins with a transition (either H to L or L to H); then if the datum being encoded is a 1, a second transition occurs during the bit interval, usually halfway through; if the datum is a 0, there is no second transition during the bit interval – the next transition begins the next bit interval. Bi-Phase Space is essentially the same, except that the second transition within the bit interval signifies 0, rather than 1. Manchester always has a transition at or near mid-interval, falling for a 1 and rising for a 0. Thus, for Bi-Phase, the reference (clock) edge is always at the beginning of the bit interval and for Manchester it is at mid-interval.

Here are the main differences in their properties:

With Bi-Phase, demodulation yields the same NRZ data, whether the transmitted waveform is inverted or not (hence the name, Bi-Phase). With Manchester, inversion of the modulated data will cause the demodulator's NRZ output to be inverted (1's become 0's and vice versa). If the modulated data consist of alternate H's and L's, each having a duration of half a bit interval, it is clear that in Bi-Phase this signifies a string of connective 0's, whereas in Manchester such half-bit H's and L's could be either consecutive 0's or consecutive 1's. Thus, if a burst of interference causes a momentary loss of phase, a Bi-Phase demodulator of the kind shown in Figure 10.5.3-2 would immediately resume demodulation of the half-bit H-L sequence as consecutive 1's, although possibly working from the wrong edge; even this would be cleared up when the next 0 came along, although the 0 bit would vanish. In the same situation, if a Manchester demodulator got onto the wrong edge, it would switch from decoding the sequence as consecutive 1's to decoding it as consecutive 0's; it also would be rephased at the next change from a coded 1 to a coded 0. The point is this: the one-shot type of demodulator in Figure 10.5.3-2 is retrained to Bi-Phase by a single coded 0 while retraining to Manchester requires either 1-0 or 0-1 sequence. Either way, one bit is lost, but with Manchester code, a sequence of 1's or 0's could be inverted.

It would appear that a modulator to generate Manchester code would require only an exclusive-or gate, with the data at one input and clock at the other. This will work, but there are glitches which occur when the data change from 1 to 0 or from 0 to 1. If the glitch duration is much less than a bit interval, a simple low-pass filter eradicates it.

The modulator in Figure 10.5.3-1 is inherently free of glitches and can be used for either Bi-Phase or Manchester encryption by selection of the appropriate input at the exclusive-or gate for DATA. The two exclusive-or gates that clock the flip-flops generate split-phase clocking. If sync is not needed and split-phase clocking is available, these exclusive-or gates can be omitted. (They can be omitted even if sync is needed, but the sync gating is more complicated, requiring SYNC and $\overline{\text{SYNC}}$ applied to NAND and NOR gating at the flip-flop clock inputs.)

Q_2 always changes state at the rise of the clock and generates the reference edges for the coded output. These reference edges occur at the beginning of each interval for Bi-Phase and at mid-interval for Manchester. The sync pulse, S, must be timed to arrive while the clock, K, is low, otherwise a glitch will occur and might cause incorrect clocking. Q_1 is clocked when K falls and may or may not change state, depending on what logic level is present at D_1. The fall of the clock should therefore occur at some time near the center of the datum element.

In generating Bi-Phase, the sequence is:

1. At clocking of Q_2, start the bit interval by inverting M.

2. At clocking of Q_1 (mid-interval), either

    a. re-invert M if D = 1 by making $D_1 = \overline{Q}_1$

    b. NOT re-invert M if D = 0 by making $D_1 = Q_1$.

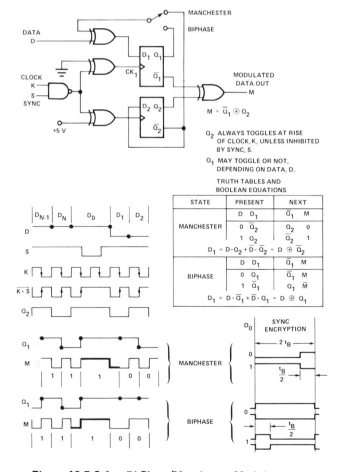

Figure 10.5.3-1. **Bi-Phase/Manchester Modulator with Sync Encryption.**

In generating Manchester, the sequence is:

1. At clocking of Q_1, start the bit interval with M = D by either

   a. making $D_1 = Q_2$ if D = 1

   b. making $D_1 = \overline{Q}_2$ if D = 0

2. At clocking of Q_2 (mid-interval), invert M.

The Bi-Phase-coded bit intervals are in phase with the un-coded (NRZ) data elements while the Manchester-coded bit intervals lag by approximately half a bit interval. This is because the encryption of data takes place at mid-interval for Bi-Phase and at the beginning of the bit interval for Manchester. This is no problem because in the demodulator (Figure 10.5.3-2), the opposite occurs: the NRZ output from demodulating Bi-Phase comes a half-bit-interval later than the NRZ from Manchester.

Encryption of the sync pulse requires removal of a rising and a falling clock edge, so that there is a period in the coded waveform during which there are no transitions, then followed by a reference edge (rise of the clock). The generation of a sync pulse is usually required only in multiplexing. A method of generating a properly timed sync pulse is described in Section 10.5.4 on Multiplexing.

Demodulation of encoded waveforms requires some kind of clocking at the receiving end. A common method of generating such clocking is with a phase-locked loop (PLL). The PLL is locked in phase with the clock signal derived from the demodulation, and there is a certain amount of settling time associated with this phasing. Thus a demodulator clocked by a PLL would require a definite period of "training" during which the transmitted encoded signal contains only reference edges. For Bi-Phase, this would be a sequence of consecutive 0's to be encoded; for Manchester, the sequence would be 1, 0, 1, 0, 1, 0, etc. In either case, the encoded waveform for training is a series of alternately high and low levels with each having a full bit interval duration.

The demodulator in Figure 10.5.3-2 does not require a long training; it requires only one full bit interval with no edges and followed by a reference edge (in Bi-Phase, a single encoded 0; in Manchester, a 1,0 or 0,1). The reason for the rapid training is that a pair of monostable multivibrators (one-shots) are used, rather than a PLL for demodulation. The second one-shot is used only to recover a sync pulse, so if sync is not required, it can be omitted. Demodulation of Bi-Phase requires both the A and B flip-flops, but for Manchester only the first one-shot (A) is required. The sync

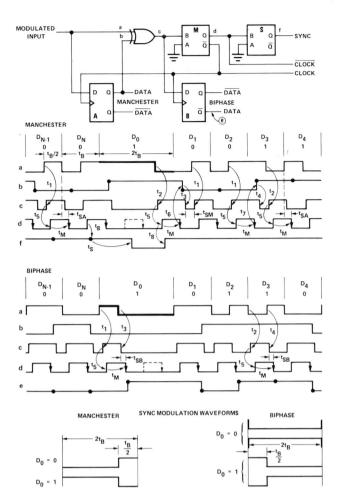

Figure 10.5.3-2. Bi-Phase/Manchester Demodulator with Outputs of Sync, Clock, and Data.

bit, D_0, can also be one of the $D_0 \ldots D_N$ data bits. Even if the system does not require sync, its inclusion is a way of providing assurance that, in the event of a loss of proper phase, the demodulator becomes retrained at the sync bit, regardless of what the data are — even if they are all 1's.

As the reference edge (beginning of Bi-Phase bit interval or middle of Manchester bit interval) of the input waveform, **a**, occurs it causes a rise in the output of the exclusive-or gate (waveform **c**) which triggers the main timer, M. The rise of Q_M triggers the sync timer, S. The duration, t_M, of the pulse from M must be such that the falling edge (waveform **d**) occurs before the next reference edge, but after the beginning of the next bit interval, plus a delay time, t_{SA}, the setup time required by flip-flop A. The duration of the pulse from S must be long enough that during most clock cycles, S is not "timed out" before the next rise of Q_M retriggers; it must be short enough that it DOES "time out" during the sync interval where there is a pulse "missing" from the train of pulses at **d**. This will allow Q_S to fall, thus producing a sync pulse.

At the fall of Q_M (waveform **d**), $\overline{Q}_M$ rises to clock both A and B. At that time, the D input of A will have the same level as **a**, and clocking A then stores that level — the level of **a** just before the reference edge occurs — so that when **a** changes (as it always must at the reference edge), there will be a rise at **c** to start the demodulation cycle again.

The timing of t_M can be quite critical at high data rates. The analysis of the restriction is taken from the waveforms, noting that there are several kinds of delay involved:

	REF.	DESCRIPTION
Exclusive-Or Delay	t_1	t_{PLH}, Opposite Input L
	t_2	t_{PLH}, Opposite Input H
	t_3	t_{PHL}, Opposite Input L
	t_4	t_{PHL}, Opposite Input H
Flip-Flop Prop. Delay	t_5	t_{PHL} of Q_M
	t_6	t_{PLH} of Q_A
	t_7	t_{PLH} of Q_A
	t_8	t_{PLH} of Q_S
Setup Times for Flip-Flops and One-Shots	t_{SA}	For Manchester
	t_{SB}	For Bi-Phase
	t_{SM}	Main Demod. One-Shot
	t_{SS}	Sync Demod. One-Shot

Applying the timing restriction as to when $\overline{Q_M}$ must rise, the following expressions are obtained:

FOR CONSECUTIVE 0's (MANCHESTER), ALSO CONSECUTIVE 1's (BI-PHASE):

$$t_B/2 + t_{SA} - t_1 - t_5 < t_M < t_B - t_1 - t_5$$

FOR CONSECUTIVE 1's (MANCHESTER), ALSO CONSECUTIVE 1's (BI-PHASE):

$$t_B/2 + t_{SA} - t_2 - t_5 < t_M < t_B - t_2 - t_5$$

FOR 1,0 SEQUENCE (MANCHESTER), ALSO CONSECUTIVE 0's (BI-PHASE):

$$0 < t_M < t_B + t_1 - t_2 - t_5 - t_6 - t_3 - t_{SM}$$

FOR 0,1 SEQUENCE (MANCHESTER), ALSO CONSECUTIVE 0's (BI-PHASE):

$$0 < t_M < t_B + t_2 - t_1 - t_5 - t_7 - t_4 - t_{SM}$$

In addition, for Bi-Phase demodulation, t_{SB}, the setup time for the B flip-flop must be considered:

$$t_B/2 + t_{SB} + t_3 - t_1 - t_5 < t_M$$

$$t_B/2 + t_{SB} + t_4 - t_2 - t_5 < t_M$$

These timing restrictions make it possible to have t_M very nearly equal to $t_B/2$. Since the falling edge of waveform **d** is just before the DATA transition, if $t_M = t_B/2$, the falling edge of the CLOCK would be very near the center of the DATA element, as required by the EIA standard, RS-334. Of course, the best choice for t_M should allow for jitter and make it halfway between the minimum upper and maximum lower found in the expressions above. Since some delays add and others subtract, it makes sense to use typical values to select t_M, then test to see if worst-case delays may cause a problem. By using logic elements from the 74S . . . . series, it is easily possible to operate at 5 Mb/s (t_B = 200 ns). To go faster would probably require the use of ECL logic elements.

10.5.4   Multiplexing

As used here, the term "multiplex" does not mean the interchange of data among several stations in a bus or star connection. Here it means the use of a single communication channel to transmit data from several data channels and, at the receiving end, to deliver data to outputs arrayed in the same order as the inputs.

Multiplexing several channels is useful in applications where the data rate in each channel is much less than the communication channel rate capability. Some possible examples are in the interchange of data and address information among isolated elements of a microprocessor system. Multiplexing is useful also in transferring information according to the standards given in IEEE-488, also known as HP-IB, in which bused data are transmitted in 8-bit bytes and control information is transmitted on an 8-line bus.

Figure 10.5.4-1 shows a block diagram of a multiplex/demultiplex system. Data arrives byte-serial with all n bits in parallel. The Multiplexer serializes the n bits in NRZ format, with one of the bits held for two bit intervals to allow insertion of the sync bit by the Modulator. The Modulator converts the data from NRZ format to some form of code, such as Bi-Phase or Manchester. The modulated output is carried by the Transmission System to the receiving end.

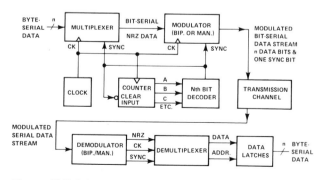

Figure 10.5.4-1.   Block Diagram of a Complete Multiplex, Transmission, and Demultiplex System.

The Transmission System could be a fiber optic link or a wire cable with appropriate line drivers, receivers, and opto-couplers as described earlier in Section 10.3 and 10.4.

At the receiving end, the Demodulator accepts modulated data and delivers three outputs: NRZ data (bit-serial), a clock signal, and a sync pulse. These three outputs operate the Demultiplexer and the Data Latches. The outputs of the Data Latches have the data arrayed with the bits appearing in the same position they were presented to the input of the Multiplexer.

There are many ways in which all of this can be accomplished, and the design approach is governed, to some extent, by the number of bits per byte, the data rate requirement, and channel capacity. The channel capacity and serial data rate are related as in Figure 10.5.1-1. For modulated data, the maximum ratio of data rate to baud rate is:

$$f_B/f_C = 1/2 \qquad \begin{aligned} f_B &= \text{data rate (bit/second)} \\ f_C &= \text{baud rate (symbol/second)} \end{aligned} \qquad (10.5.4\text{-}1)$$

Since multiplexing requires one extra bit for sync, the ratio of byte rate to bit rate is:

$$(10.5.4\text{-}2)$$
$$f_D/f_B = 1/(n+1) \qquad \begin{aligned} f_D &= \text{data rate (bytes/second)} \\ n &= \text{byte size (bits/byte)} \end{aligned}$$

Combining these expressions yields the multiplexed scan rate (scans/sec., words/sec., bytes/sec.):

$$f_D = f_C/2\,(n+1) \qquad (10.5.4\text{-}3)$$

For example, with a 10 M baud channel, with 16 bits per byte, the data rate is $10^7/2\,(16+1) = 294{,}118$ bytes per second, and this is the maximum data rate (bit rate) in each of the multiplexed channels.

Assuming that 16-bit bytes are to be transmitted, the number of 10 M baud channels required is computed from Equation (10.5.4-3):

NO. OF CHANNELS	BITS PER CHANNEL	DATA RATE k byte/sec
1	16	294
2	8	556
4	4	1,000
16	1	5,000

NOTE: When there is only one bit per channel, no sync pulse is required, making $f_D = f_C/2$.

Figures 10.5.4-2 and 10.5.4-3 show the circuits required in a system for multiplex transmission of any number of bits up to 16. If less than 8 bits are required, it would make sense to leave off the extra 8-bit shift register in the multiplexer and the extra 8-bit latch in the demultiplexer. In the demultiplexer, omitting the extra latches allows other gates to be omitted also. This modification will be discussed later.

In Figure 10.5.4-2, the waveforms for the multiplexers are shown. The n bits to be multiplexed are numbered $D_0 \ldots D_N$ corresponding to states of the counter. Thus, if 7 bits are to be multiplexed, the Nth bit decoder should be designed to decode state 6 of the counter (0 1 1 0 binary) with a low logic level occurring when the counter arrives at state 6 (for n bits, decode state number n−1). This low does not immediately clear the counter because it is synchronously cleared by the next rising edge of the clock. The arrival of a logic low from the Nth Bit Decoder begins the synchronizing sequence. At the falling edge of the clock during the Nth bit, Q_A is clocked low which applies a logic low to the Shift/Load inputs of the shift registers, if the Data Entry Enable is at logic high. However, new data are not loaded into the shift registers until the clock rises at the

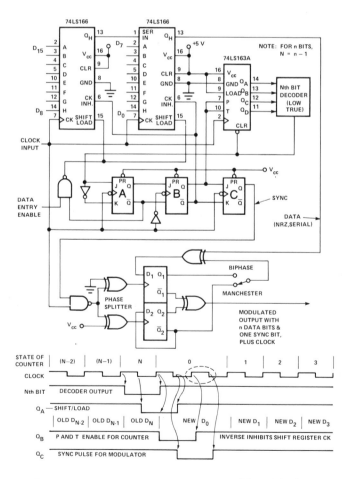

Figure 10.5.4-2.   Schematic Diagram of 0-to-16-Bit Multiplexer and Modulator.

end of the Nth bit. At that time, if Data Entry Enable is high, whatever data are present at the $D_0 \ldots D_N$ terminals are loaded; also at that time, Q_B is clocked low, to inhibit advancement of the shift register ($Q_B = H$) and the counter ($Q_B = L$) so the shift register and counter will remain in state 0 until the next rise of the clock after Q_B is clocked high. This causes state 0 to persist for two clock periods. During the interval produced by Q_B, the sync signal is produced by Q_C. The sync pulse must be timed so as to gate out the positive clock pulse in the second half of state 0 — the one indicated by dotted lines. It is not necessary to preset or clear the J-K flip-flops because they are all clocked to specific states. With all actions controlled by either rising or falling clock edges, there are no race problems in the multiplexing.

In the modulator, there is a small possibility of a race when the system is operated at a very high data rate. It is necessary that the correct level be established at D_1 when Q_1 is clocked. The data level is clocked by the rising edge of the clock, then delayed by the propagation delay of the shift register and of the exclusive-or gate. Q_1 is not clocked until the falling edge of the clock, delayed by the propagation delay of the sync gate and of the exclusive-or phase splitter. These delays, while not perfectly balanced, allow nearly half a bit interval for the data level to be established at D_1. More critical is the delay of feedback from $\overline{Q_2}$ to D_1 when the modulator is operated in the Manchester mode. For high data rates (over 5 M b/s), the modulator logic elements should be from the 74S.... series.

Figure 10.5.4-3 shows a demultiplexer circuit using a counter to generate the address for data latches which receive and store the data bits as they come along. This arrangement has the advantage that it requires very little wiring to manipulate the data bits into their proper slots. It has the disadvantage that not all n bits change simultaneously. This could be remedied by adding an array of D-latches clocked by the rising of the sync pulse (delayed to avoid a race); however, it would be simpler to omit the counter and addressable latches and use a serial-in-parallel-out (SIPO) shift register, asynchronously cleared by the sync pulse and having its parallel outputs gated into an array of D-latches by the rise of the sync pulse. This would require some attention to proper amounts of relative delay in the clock and NRZ data signals.

Timing for the demultiplexer of Figure 10.5.4-3 is shown in Figure 10.5.4-4. The relative timing of R-NRZ (recovered data) and CLOCK from the demodulator are taken from the waveforms in Figure 10.5.3-2, where waveform **d** is that of CLOCK while the data timing is taken from waveform **b** for Manchester and **e** for Bi-Phase.

In applying these timing and data signals to the counter and latches, it is necessary to make the edges occur at the proper relative time. The 74LS259 latches are pulse-type (i.e., not edge triggered). When the enable signal, G, drops, whatever is present at the D input will appear at whichever output is being addressed at that time. The addressed output will follow whatever is at D as long as $\overline{G}$ remains low. Therefore, both the Delayed Data and the Address State should be timed to overlap the low state of $\overline{G}$. The falling edge of CLOCK clocks the data out of the demodulator and also causes the rising edge of the clock input of the counter, so there is little risk of a race problem at the beginning of each bit interval. The rising edge of $\overline{CLOCK}$ generates the falling edge of $\overline{G}$, so if the address and data are delayed too much, or if $\overline{CLOCK}$ has a duty factor much over 50% (high most of the bit interval), there could be a race at the drop of $\overline{G}$.

If less than 16 bits are multiplexed, the counter will be momentarily clocked into the (N + 1) state. This is of no concern because $\overline{G}$ remains high, and the sync pulse occurs

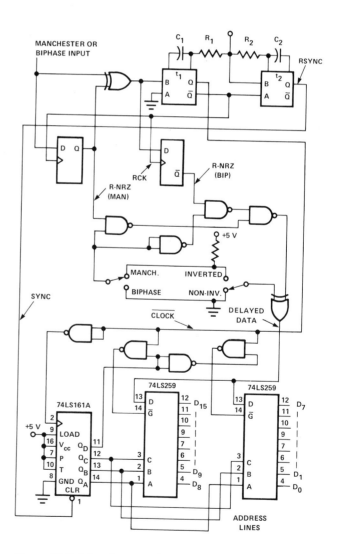

Figure 10.5.4-3. Schematic Diagram of a Demodulator with 0-to-16-Bit Demultiplexer.

before $\overline{G}$ drops. If the full 16 bits are used, the (N + 1) state will be the 0 state, but this does not eliminate the need for an asynchronous clear on the counter. The asynchronous clear is needed because there is a pulse missing from $\overline{CLOCK}$ during the sync interval.

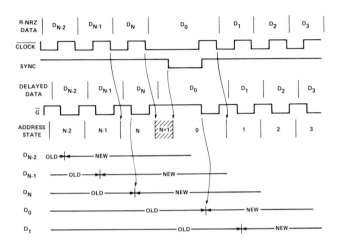

Figure 10.5.4-4.   Timing Diagram for Addressable-Latch Demultiplexer.

10.5.5   Interfacing With Established Standards

There are a number of standards which have been established for information interchange. They describe signal quality for both data and timing elements, and function assignments for the various terminals in a standard multi-pin connector.

RS-232-C is for a 25-pin connector. It is for data rates up to 20 kb/s over distances up to 18 metres (60 feet) and uses polarity-reversing signals at voltages from 3 V minimum to 25 V maximum. Of the 25 pins, 23 have assigned functions, but not all 23 are required in every application. For example, the minimum required for a "transmit-only" application could be four pins: Ground, Transmitted Data, Clear to Send, and Data Set Ready. A more general application of RS-232-C is in two-way communication in either full- or half-duplex mode. This requires a minimum of ten pins as shown in Figure 10.5.5-1. The pin numbers, the circuit designator letters, and their functions are listed at the bottom along with a chart relating the RS-232-C characteristics to the levels used in the DCE (Data Communicaton Equipment) circuits.

Interfacing the ten pins and their functions with a Hewlett-Packard fiber optic link is given here as an example of how to apply standards for information exchange. Some of the functional requirements are imposed by the RS-232-C standard, others by the characteristics of the fiber optic link.

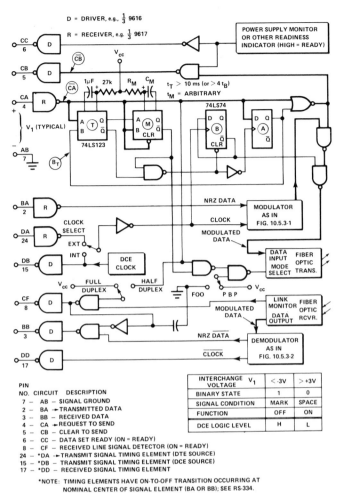

Figure 10.5.5-1.   RS-232-C Interface Control Circuit for Use with a Fiber Optic Link.

When the terminal is not in use for transmitting data, the fiber optic transmitter should have its Data Input at a logic low and its Mode Select input high; this operates the LED in the Transmitter at its minimum current of about 8 mA to allow the LED to degrade less rapidly than it would if left fully on all the time. Then, when a Request-to-Send signal appears at CA, the Clear-to-Send response, CB, should not be given until the Transmitter, along with the Receiver at the other end, are ready. This may require as much as 10 milliseconds. There are two timers to deal with this requirement; the first, T, is the Training timer and the second, M, is the Maintenance timer. After a long idle period, when CA rises, it is prevented by Q_A from causing a high at CB until Q_T falls. Because this Training period is such a long time, it is desirable to have a "keep alive" function that maintains the system ready for use for a period of time extending as long as desired into the idle period when CA is off ($\overline{CA}$ = H). This Maintenance period of time is provided by M; if a Request-to-Send signal ($\overline{CA}$ = L) reappears before M is "timed out," it allows an immediate Clear-to-Send response ($\overline{CB}$ = L).

During the Training period (Q_T = H) and also during the Maintenance period (Q_M = H), it is necessary to have the fiber optic transmitter provided with an f/2 signal at Data Input while Mode Select is either low or high, depending on whether data transmission is to be done in Pulse Bi-Polar (PBP) or Full On-Off (FOO) mode. The f/2 signal simulates either the "1,0" required for "training" a Manchester demodulator or the "0" required for Bi-Phase. Of course this f/2 signal is replaced by data when $\overline{CB}$ = L, $(\overline{CA} + Q_A)$ = L, so there is something happening at all times to keep the system "trained" until after the Request-to-Send signal has remained off $(\overline{CA}$ = H) for a period of time longer then the maintenance period provided by M. If the Maintenance period expires before $\overline{CA}$ = L, the fall of Q_M clocks Q_A with D = H. This makes Q_A = H, so the B input on the training timer, T, is activated (B_T = H) and steady idle conditions are applied to the Transmitter inputs.

Timing elements for the data may come from either an internal source (DCE CLOCK) or be externally generated. Either way, they must conform to RS-232-C specifications requiring an "ON-to-OFF" transition at the center of the data element. The Modulator (Figure 10.5.3-1) is designed for H-to-L transition at the datum center, which is the inverse of RS-232-C clock specification, hence the inverter following the Clock Select switch. The NRZ data need not be inverted because the RS-232-C definitions of 1 and 0 match those used in the design of the Modulator.

At the Demodulator output, $\overline{DATA}$ is selected (rather than DATA) because it goes through a NAND gate before reaching the RS-232-C interface point. $\overline{CLOCK}$ is selected to meet the RS-232-C specification as described in the preceding paragraph. The Link Monitor output of the fiber optic receiver simplifies generation of the CF — Received Line Signal Detector function. Without this convenience, a peak-to-peak detector, or other circuit for checking received signal quality would be required. It is used also to gate in the Received Data when the system is in Full-Duplex mode. In Half-Duplex, the Received Data is gated by $\overline{CA}$ = H, which occurs when the Request-to-Send (CA) is "OFF."

Obviously, the 20 kb/s data rate of RS-232-C is far below the 10-M baud channel capacity of the Fiber Optic link. Several ten-pin RS-232-C stations can be used with a single Transmit/Receive pair. To do this, it is only necessary to make a few minor changes:

1. Provide fan-in to allow triggering of Q_T from the CA terminal in any station.

2. Fan-out of CB, CC, CF, DD to all stations.

3. A multiplexer with the Modulator to accept data from each of the BA terminals.

4. A demultiplexer with the Demodulator to deliver data to proper BB terminals. (All stations would be in the same mode, so full- half-duplex gating of DATA is not necessary and DATA would go to the BB drivers directly from the $D_0 \ldots D_N$ outputs of the demultiplexer, Figure 10.5.4-3.)

The Modulator would have to work from a single clock source so it would be best to use the DCE CLOCK to clock transmitted data (BA) from all stations (via DB). Signal ground (AB) must, of course, be common to all.

Optocoupler driver and receiver circuits (Section 10.3) can be used in place of the fiber optic transmitter and receiver shown in Figure 10.5.5-1. In making such a substitution, the selection of FOO/PBP is unnecessary since optocouplers are always operated in the FOO mode. Peak-to-peak detectors operating directly from the optocoupler-receiver outputs and applied to threshold comparators would be used to generate the CF (Received Line Signal Detector) function. With optocouplers, the 10-ms training time limit does not apply.

RS-449 interface circuitry is much like RS-232-C, with a few important differences. The electrical specifications for transmission lines used with RS-449 are given in the standards RS-422 for balanced lines and RS-423 for single-ended. The RS-423 drivers and receivers allow data rates up to 20 kb/s but over longer distance — the RS-232-C recommendation is for distances less than 15 metres (50 feet) while the RS-423 guidelines are 60 metres (200 feet). The RS-422 balanced drivers are capable of data rates up to 2,000 kb/s NRZ (2 M baud) over the same distance and clock signals up to 4 MHz. A larger, 37-pin connector is prescribed to accomodate the balanced (two-wire) interchange points and also to allow additional functions not defined in RS-232-C such as local and remote loop tests. Not all the functions require the high-speed balanced circuit, so the pin assignments are single for the low speed (unbalanced) circuits and double for the high speed circuits.

The C.C.I.T.T. Recommendation V.24 is similar to RS-449 in the circuit functions described. For unbalanced circuits, Recommendation V.10 prescribes a distance/data-rate tradeoff extending from a maximum of 1,000 metres at 1 kb/s down to 10 metres at 100 kb/s. For higher-speed balanced circuits, Recommendatiaon V.11 prescribes two distance/data-rate tradeoffs according to whether or not the lines are terminated; for unterminated lines, the range is from 1,000 metres at 10 kb/s to 10 metres at 1 Mb/s and for terminated lines the range is from 1,000 metres at 100 kb/s to 10 metres at 10 Mb/s.

The EIA RS-232-C and RS-449, as well as the C.C.I.T.T. Recommendation V.24, deal only with bit-serial transmission of data. Instrumentation and control systems are often better suited to bus arrangements in which data are transmitted in bytes — usually at 8 bits per byte. Examples of these are the HP-IB and CAMAC systems. HP-IB, described by IEEE-488, uses an 8-bit data bus and an 8-bit control line for interchange of data over distances up to 20 metres at 1 Mb/s. For CAMAC, the distance/data-rate tradeoffs are prescribed by C.C.I.T.T. Recommendations, such as V.10 or V.11 and in IEEE-583. In some CAMAC systems, data are transmitted bit-serial, but the control and data information is organized into 8-bit bytes. CAMAC systems can also be arranged along parallel "dataways," transmitting the 8-bit bytes on buses.

Section II
Sunlight Viewable Displays

11.0 SUNLIGHT VIEWABLE LED DISPLAYS

The rapid growth of sophisticated electronic systems for use in avionic, automotive, shipboard, military, and other equipment has created the need for an electronic information display that is viewable in bright sunlight. Display technologies other than light emitting diodes have been used in many applications where readability in bright sunlight is a requirement. The reason for the exclusion is that LED displays have been traditionally cited as having one major disadvantage: "the devices wash out in bright sunlight." This criticism was due to two fundamental reasons: 1) low light output characteristics, and 2) the lack of appropriate attention to contrast enhancement techniques.

This criticism is not necessarily true when applied to the LED display devices of latest design. By combining the newest LED product design technology and the most recent techniques for contrast enhancement, the light emitting diode industry is now providing the display system designer with devices that are useable in direct 107,000 lumens per square metre (10,000 foot candle) sunlight.

11.1 Contrast Enhancement Techniques Needed to Achieve Readability in Bright Sunlight

Achieving readability in direct sunlight is possible when the display device produces adequate light output and when the proper contrast enhancement techniques are incorporated into the design of the display system.

The following parameters contribute to the readability of an LED display in bright sunlight:

a) LED Color

b) Luminance Contrast

c) Chrominance Contrast

d) Front Surface Reflections

Each parameter is considered separately and the combined solutions are then brought together in the overall display system design.

Luminance contrast and chrominance contrast are two parameters that may be combined into a visual discrimination index. This concept was introduced at the 29th Agard Avionics Panel Technical Meeting on Electronic Displays in Paris, France by Jean-Pierre Galves and Jean Brun. Mr. Andre Martin then elaborated on the concept in his article "The CRT/Observer Interface," Electro-Optical Systems Design, June 1977, pp. 35-41. The advanced contrast enhancement techniques presented herein make use of this discrimination index concept which is based on the 1960 CIE—Uniform Chromaticity Scale Diagram.

11.1.1 The LED Color

The importance of the LED color is evaluated from two aspects: LED chip technology and the color discrimination ability of the eye. First, the gallium phosphide substrate LED technology has a significantly higher quantum efficiency than does the gallium arsenide substrate technology. Therefore, the high-efficiency red, yellow, and green devices of the GaP substrate technology are possible colors for use in sunlight visible displays.

The second consideration is the ability of the eye to discern a given color. The ability of the eye to discern a color can be evaluated by examining the chrominance distance between dominant wavelengths that are one nanometre apart throughout the visual spectrum. The 1960 CIE—UCS chromaticity diagram is illustrated in Figure 11.1.1-1. Since the color purity of an LED is better than 0.99, the u,v color coordinates for LEDs essentially plot at the corrdinates for their dominant wavelengths, λ_d, located on the perimeter of the chromaticity diagram.

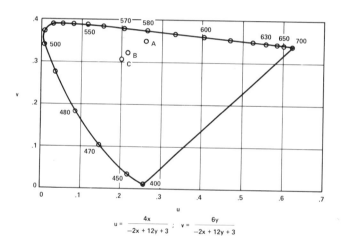

$$u = \frac{4x}{-2x + 12y + 3} \; ; \; v = \frac{6y}{-2x + 12y + 3}$$

Figure 11.1.1-1. The 1960 CIE — UCS Chromaticity Chromaticity Diagram.

Therefore, this analysis will directly predict the relative ability of the eye to discern the color of the radiated spectrum for a given LED.

The 1960 CIE—UCS chromaticity coordinates u,v are determined by measuring the normalized radiated spectrum of the LED and performing the following calculations.

1. Only two of the three tristimulus values, X and Y, need to be used to determine the chromaticity coordinates for an LED. Since the dominant wavelengths are greater than 560 nm, deleting the Z tri-

stimulus value from the calculation introduces an error of only 0.25% or less.

$$X = \Sigma f(\lambda) \, \overline{x}_\lambda$$

(11.1.1-1)

$$Y = \Sigma f(\lambda) \, \overline{y}_\lambda$$

where $\overline{x}_\lambda$ and $\overline{y}_\lambda$ are the 1931 CIE color matching functions

$f(\lambda)$ is the normalized radiated spectrum

Performing the summations at 5 nm intervals is usually sufficient to determine color coordinates within the accuracy of ½ nanometre.

2. The 1960 chromaticity coordinates for LEDs may be calculated directly from X and Y or from the 1931 chromaticity coordinates x, y.

(11.1.1-2)

$$u = \frac{4X}{X + 15Y} = \frac{4x}{-2x + 12y + 3}$$

$$v = \frac{6Y}{X + 15Y} = \frac{6y}{-2x + 12y + 3}$$

where $x = \dfrac{X}{X + Y}$; $y = \dfrac{Y}{X + Y}$

The chromatic distance, EC, between two color coordinate locations on the 1960 CIE–UCS Chromaticity Diagram is determined from the following equation:

$$EC = \sqrt{\Delta u^2 + \Delta v^2}$$

(11.1.1-3)

Calculating the chromatic distance for λ_d values one nanometre apart and plotting the results normalized to 1.0 at 595 nm produces the graph shown in Figure 11.1.1-2. This graph may be considered to be the first derivative of the perimeter of the 1960 CIE–UCS Chromaticity diagram. It shows how the normalized chromatic distance varies across the visual spectrum. This graph illustrates the ability of the eye to discern a color difference, or more specifically, the ability of the eye to distinguish a given color. The ability of the eye to distinguish a color peaks at 480 nm in the blue region, has limited capability in the green region near 520 nm, peaks again at 595nm in the orange region, and then diminishes in the red region.

The positions of the dominant wavelengths for the four basic LED colors are located on the curve of Figure 11.1.1-2. These positions indicate the relative ability of the

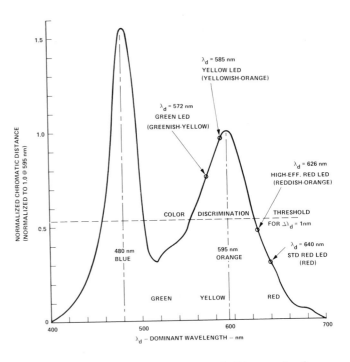

Figure 11.1.1-2. Normalized Chromatic Distance for 1 nm Differences in Dominant Wavelength on the Perimeter of the 1960 CIE – UCS Chromaticity Diagram.

eye to distinguish each color, especially in a bright ambient such as direct sunlight. The yellowish-orange color of the yellow LED is a color that is highly discernible to the eye. For this reason, the yellow LED displays are used extensively in direct sunlight applications. Though the ability of the eye to distinguish the greenish-yellow color of the green LED is better than that of the two red LEDs, the green LED is not used in sunlight applications for two basic reasons. First, the light loss to achieve a color shift towards green reduces luminous contrast. Secondly, the quantum efficiency is not sufficient to produce the needed light output at reasonable power dissipation levels. Both of these limitations may be removed with future development. Though the ability to distinguish the reddish-orange color of the high-efficiency red LED is much less than that of the yellow LED, the superior quantum efficiency of the high-efficiency red LED offsets this deficiency by producing very high light output at reasonable power levels. Thus, the high-efficiency red LED is an attractive device for use in sunlight applications. Based on color discrimination, the standard red LED is the least attractive for use in sunlight applications.

11.1.2 Luminance Contrast and the Luminance Index

Contrast may be defined as the ability to exhibit noticeable differences in color, brightness, and other characteristics in a side by side comparison. The objective in applying contrast enhancement is to obtain improved readability in a given ambient lighting condition. The concept is to implement the following three conditions: 1) achieve as large a contrast as possible between the illuminated elements and the surround; 2) maintain minimal contrast between the nonilluminated elements and the surround; and 3) provide as high a value of ON/OFF contrast for the light emitting elements as possible.

In the past, efforts have been concentrated on obtaining the best possible luminance contrast as the primary method of obtaining readability. However, both luminance contrast and chrominance contrast need to be employed in order to achieve readability in bright sunlight. Figures of merit can be established for both luminance and chrominance contrasts in the form of index values. These two index values may then be combined into a single discrimination index within a photocolorimetric space.

Conceptually, luminance contrast is the observed brightness of the illuminated light emitting element compared to the brightness of the surround. The brightness of the illuminated element is obtained from both electroluminescence and reflected light while the brightness of the surround is due only to reflected light. The definition of luminous contrast for a light emitting element with respect to the surround may be mathematically expressed as:

$$(11.1.2\text{-}1)$$

$$C = \frac{L_v S_{OFF} + L_v S_{ON}}{L_v B} \; ; \; 1 \leqslant C < \infty$$

where

$L_v S_{ON}$ = luminous sterance of the illuminated light emitting element

$L_v S_{OFF}$ = luminous sterance of the non-illuminated light emitting element reflecting ambient light

$L_v B$ = luminous sterance of the surrounding background reflecting ambient light

Mathematically:

$$L_v S_{ON} = \int f(\lambda) Y(\lambda) T(\lambda) d\lambda$$

$$L_v S_{OFF} = \int X(\lambda) Y(\lambda) T^2(\lambda) R_S(\lambda) d\lambda$$

$$L_v B = \int X(\lambda) Y(\lambda) T^2(\lambda) R_B(\lambda) d\lambda$$

where

$f(\lambda)$ = radiated spectrum of the illuminated light emitting element

$Y(\lambda)$ = the 1931 CIE photopic curve

$T(\lambda)$ = relative transmission characteristic of the filter

$X(\lambda)$ = radiated spectrum of the ambient light

$R_S(\lambda)$ = relative reflection characteristic of the light emitting element

$R_B(\lambda)$ = relative reflection characteristic of the surrounding background

There are two values for Equation (11.1.2-1) that may be used to form an ON/OFF contrast ratio. The first is the ON—contrast value, C_{ON}, when the light emitting element is illuminated by electroluminescence and $L_v S_{ON} > 0$. The second is the OFF—contrast value, C_{OFF}, when the light emitting element is not illuminated so that $L_v S_{ON} = 0$ and C_{OFF} is due only to reflected ambient light.

To obtain the best possible readability, it is desirable to have two conditions of contrast occur simultaneously. First, the value of $L_v S_{ON}$ should be as high as possible. Secondly, the value of $L_v S_{OFF}$ should equal $L_v B$. With these two conditions achieved, the two contrast values will be

C_{ON} = large number

C_{OFF} = 1.0

and the ON/OFF contrast ratio, $C_{ON} : C_{OFF}$, will equal the value of C_{ON}.

In some display packaging technologies, it is possible to design the device such that the luminous sterance of the nonilluminated light emitting element, due to reflected light, is approximately equal to the luminous sterance of the package. Such is the case with the stretched segment sunlight viewable LED displays. Therefore, in the following discussions, the value of $L_v S_{OFF}$ is assumed equal to $L_v B$ by design so that Equation (11.1.2-1) may be reduced to:

$$C = \frac{L_v B + L_v S}{L_v B} \qquad (11.1.2\text{-}2)$$

where $L_v S \equiv L_v S_{ON}$

Achieving an ideal luminous contrast so that the value of Equation (11.1.2-2) is a large number is usually not possible. The limitation comes from the lack of sufficient display emitted light to overcome the optically dense filter that is required to make $L_v B$ a small value in bright sunlight. Therefore, a figure of merit as a measure of achievable luminous contrast needs to be defined. This figure of merit takes the form of a luminous index.

Psychophysical studies have shown that the visual perception of luminance is a logarithmic function of the light source luminous sterance, $L_v S$. From this observation, the concept of a luminous difference, EL, may be defined as:

$$EL = \log_{10} C \qquad \text{(11.1.2-3)}$$

A generally accepted value for a minimal discernible contrast, based on empirical data, is 1.05. This corresponds to a threshold luminous difference of 0.021.

$$EL_{TH} = \log_{10}(1.05) = 0.021$$

Though a threshold luminous difference has been established, it is necessary to define a luminous difference that is based on a readily discernible level of contrast. This is called a unitary luminance difference, ELU, and is derived from a one half tone step on the 1960 CIE Uniform Luminance Scale. A one half tone step has a luminance contrast of $\sqrt{2}$. Inserting this $\sqrt{2}$ value for contrast into Equation (11.1.2-3) yields 0.15 as the unitary luminance difference.

$$ELU = \log_{10}\sqrt{2} = 0.15$$

The value of ELU turns out to be approximately seven times the value of EL_{TH}.

The luminous index, IDL, is defined to be the ratio of luminous difference obtained from an achievable luminous contrast to a unity luminous difference.

$$IDL = \frac{EL}{ELU} = \frac{\log_{10} C}{0.15} \qquad \text{(11.1.2-4)}$$

11.1.3 Chrominance Contrast and the Chrominance Index

Chrominance contrast is a normal part of everday life. For example, an observer can easily distinguish the difference between two adjacent surfaces of the same area, shape, texture, and luminance if they are of two different colors. Even so, the concept of Chrominance Contrast has only recently been applied to light emitting displays in order to achieve readability in bright sunlight. As will be illustrated later, it is chrominance contrast that is the dominant factor in achieving readability in sunlight, not luminance contrast.

Color perception tests have indicated that the smallest discernible color difference, EC_{TH}, as defined by Equation

(11.1.1-3), is a chromatic distance of 0.00384 on the 1960 CIE–UCS Chromaticity Diagram. An appreciation of this small chromatic difference is obtained by examining Figure 11.1.1-2. This minimal discernible color difference corresponds to a normalized threshold distance of 0.525 on the perimeter of the 1960 CIE–UCS Chromaticity Diagram. For dominant wavelengths of one nanometre difference, the ability of the eye to discern the color difference is limited in the deep blue, green, and red regions. Conversely, the ability of the eye to discern a color is very good in the blue, yellow, and orange regions.

A unitary color difference for comfortable discrimination between two colors, ECU, may be defined as that color difference which gives the same visual impression of contrast that is observed with a unitary luminance difference. Since a unitary luminance difference is seven times the minimal discernible luminous contrast, the same physiological result should be obtained for the unitary chrominance difference being set at seven times the chrominance threshold:

$$ECU = 7\, EC_{TH} = 0.027$$

The chrominance index, IDC, is defined as the ratio of the chromatic difference between the illuminated light emitting elements and the surrounding background to a unitary chromatic difference.

$$IDC = \frac{EC}{ECU} = \frac{\sqrt{\Delta u^2 + \Delta v^2}}{0.027} \qquad \text{(11.1.3-1)}$$

11.1.4 The Discrimination Index

The total contrast required for readability in sunlight may be described as the geometric combination of luminance and chrominance contrast as described within a photocolorimetric space is defined by the 1960 CIE–UCS Chromaticity Diagram as illustrated in Figure 11.1.4-1. The luminance and chrominance values of the illuminated display elements and the display background/nonilluminated element combination are plotted as two distinct points within this photocolorimetric space. The goemetric distance between these two points is called the discrimination index, ID.

$$ID = \sqrt{IDL^2 + IDC^2} \qquad \text{(11.1.4-1)}$$

The discrimination index is then a figure of merit used to evaluate the achieved total contrast between the illuminated display elements and the display background.

For minimum readability, the value of the discrimination index is calculated using the unitary luminance and chrominance differences.

$$ID_{MIN} = \sqrt{IDL^2_{MIN} + IDC^2_{MIN}} = \sqrt{2}$$

where

$$IDL_{MIN} = \frac{\log_{10}ELU}{0.15} = 1.0$$

$$IDC_{MIN} = \frac{ECU}{.027} = 1.0$$

The readability of a light emitting display increases proportionately as the achieved value of ID increases above ID_{MIN}.

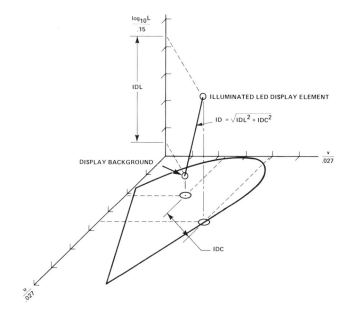

Figure 11.1.4-1. The Photocolorimetric Space, based on the 1960 CIE — UCS System, Used to Define the Discrimination Index, ID.

11.1.5   Front Surface Reflections

The glare due to nonspecular reflectance off the front surface of a filter window directly back to the eyes of an observer can significantly reduce the readability of the display. The extreme example of this is the reflected glare from the sun which is so bright that it will mask the displayed message from view. Therefore, techniques must be employed to reduce front surface reflections to as low a level as possible.

One simple technique is to cant the filter window forward as illustrated in Figure 11.1.5-1. The reflected glare is directed downward away from the eyes of the observer.

Plastic filter materials use a textured front surface to reduce front surface reflections. The texture may be either cast,

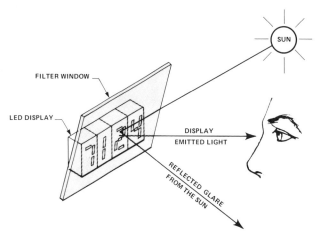

Figure 11.1.5-1. A Filter Window Canted Forward with Respect to the Plane of the Face of the Display Directs Reflections Away from the Eyes of an Observer.

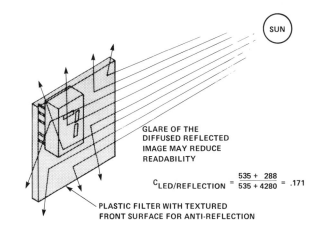

$$C_{LED/REFLECTION} = \frac{535 + 288}{535 + 4280} = .171$$

Figure 11.1.5-2. The Textured Antireflection Surface of a Plastic Filter May Reduce Readability in Direct Sunlight.

molded, or etched into the surface or applied in the form of a coating. This textured surface has the effect of reducing the amount of display emitted light that is transmitted through the filter to an observer by an additional 5% to 12%.

The textured surface disperses the incident light rays so that a reflected image is not discernible. This method does work very well in moderately bright ambients. However, in direct sunlight, the diffused reflected sunlight causes a bright glare to cover the face of the filter which may reduce readability to an unacceptable level, as shown in Figure 11.1.5-2. Approximately 4% of the direct sunlight is reflected back to the observer to form the bright glare. The resulting contrast ratio between the illuminated LED elements and the luminous sterance of the diffused reflected image is very much less than one, indicating that insufficient contrast is present for readability.

A superior technique for reducing front surface reflections to improve readability in bright sunlight is available with optically coated glass filters. Glass filters may be surfaced with quarter wave optical coatings using the same technology that is applied to camera lenses. The quarter wave optical coating is applied to both the front and back surfaces of the filter window to reduce the reflection loss that occurs at these two glass-to-air interfaces.

The quarter wave optical coating reduces reflected light by quarter wave cancellation. The effect of reducing reflection results in improved readability and, as an added benefit, improved optical transmission.

Mathematically, for a nonabsorbing system, the internal transmittance, T, and the surface reflectance, R, add to equal one.

$$T + R = 1 \qquad (11.1.5\text{-}1)$$

The amount of loss due to the reflection at the glass-to-air interfaces may be calculated as follows:

$$R = \left(\frac{n_1 - n_2}{n_1 + n_2}\right)^2 \qquad (11.1.5\text{-}2)$$

where n_1 = 1.6, a typical index of refraction for glass

n_2 = 1.0, index of refraction for air

$$R = \left(\frac{1.6 - 1.0}{1.6 + 1.0}\right)^2 = 0.053$$

The amount of light loss at the two glass-to-air interfaces is:

Reflection Loss = 2(0.053)100 = 10.6%

The amount of transmitted light for a nonabsorbing glass filter is typically 89.4%. If a neutral gray tinting is added to reduce the transmittance of the substrate material to 25%, the total light transmission is 22.4%.

A quarter wave coating has the effect of reducing the apparent index of refraction of the glass filter to a value that closely approximates the index of refraction for air. For example, Figure 11.1.5-3 presents the front surface reflectance characteristics of a piece of glass that has both surfaces otpically coated. The typical reflectance across the visual spectrum is 0.25%. The apparent index of refraction for this optically coated glass is calculated by rewriting Equation (11.1.5-2).

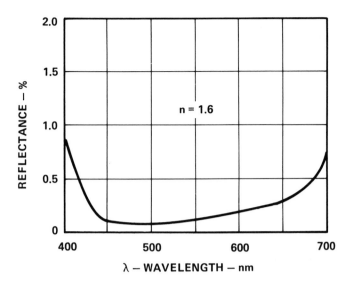

Figure 11.1.5-3. **Front Surface Reflectance of Glass with Double Sided ¼ Wave Optical Coating.**

$$n_2 = \frac{1 + \sqrt{R}}{1 - \sqrt{R}} \qquad (11.1.5\text{-}3)$$

$$n_2 = \frac{1 + \sqrt{.0025}}{1 - \sqrt{.0025}} = 1.105$$

The amount of LED display emitted light that is transmitted by a nonabsorbing glass filter is:

T = [1 − 2(.0025)] 100 = 99.5%

The total light that is transmitted by a 25% neutral density substrate is:

$$T_{ND} = (.995)(25\%) = 24.9\%$$

This is an improvement by a factor of 1.11X over an uncoated glass substrate.

As illustrated in Figure 11.1.5-4, the nonspecular reflected glare from the sun does not restrict readability. The luminous sterance of the reflected glare is only 267 cd/m^2 which is half the value of the luminous sterance of the display background.

Present technology has not been able to solve the problem of reducing the brilliance of a specular reflected image of the sun off the face of a filter window. A specular reflected image of the sun will mask the display emitted light from view. All of the above discussions are valid for the nonspecular reflecting condition.

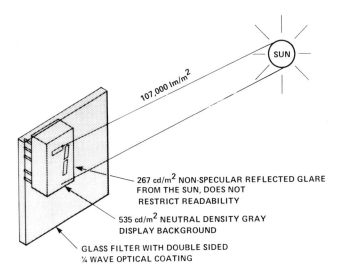

107,000 lm/m²

267 cd/m² NON-SPECULAR REFLECTED GLARE FROM THE SUN, DOES NOT RESTRICT READABILITY

535 cd/m² NEUTRAL DENSITY GRAY DISPLAY BACKGROUND

GLASS FILTER WITH DOUBLE SIDED ¼ WAVE OPTICAL COATING

Figure 11.1.5-4. The Improved Readability Obtained from a Glass Filter that has Both Surfaces Optically Coated.

11.1.6   A List of Neutral Density Gray Filters For Use With Sunlight Viewable LED Displays

FILTER PRODUCT	MANUFACTURER
Chromafilter® Gray 10 (Plastic)	Panelgraphic Corporation West Caldwell, NJ 07006 (201) 277-1500
H100-1266 Gray (Plastic)	SGL Homalite Wilmington, DE 19804 (302) 652-3686
Light Control Film N0210 50% N.D. Gray N0220 22% N.D. Gray (Louvered Filters – Plastic)	3M Company St. Paul, MN 55101 (612) 733-0128
Optically Coated Glass HEA® Double Sided Antireflection Coating Trade Name is "Sungard"	Optical Coating Laboratory, Inc. 2789 Giffen Avenue Santa Rosa, CA 95401 (707) 545-6440
Optically Coated Glass Circularly Polarized HNCP10	Polaroid Corporation Technical Polarizer Division 20 Ames Street Cambridge, MA 02139 (617) 577-3315

11.2   The Display System Concept to Achieve Viewability in Sunlight

The sunlight viewable LED displays are designed to make use of chrominance contrast in combination with luminous contrast to enhance readability in direct sunlight. To achieve the chrominance contrast, a neutral gray display package is placed behind a 23% to 27% neutral density gray filter. The off segments reflecting ambient light become indistinguishable by blending into the gray surround provided by the filter/package combination. The illuminated segments provide a distinctly visible chrominance contrast, color difference, with respect to this same gray surround. It is the implementation of chrominance contrast, coupled with the high sterance of the illuminated segments to achieve luminance contrast, that makes an LED display readable in direct sunlight.

11.2.1   The Display Devices

The sunlight viewable display devices are assembled with large junction gallium phosphide, GaP, LED chips that produce the needed light output when driven with peak currents up to 120mA. The high-efficiency red and yellow LEDs are used to take advantage of the efficient light output characteristics of a GaP substrate LED. The package configuration uses a neutral gray body and untinted segments to allow the display system designer to achieve readability by obtaining an optimum combination of both luminous and chrominance contrasts.

The high-efficiency red displays, reddish-orange in color, provide the most light output for a given input current and can easily be made visible in bright sunlight. The yellow displays, yellowish-orange in color, take advantage of that region in the color spectrum where the ability of the eye to distinguish color is most sensitive. Though both colors are widely used for sunlight applications, the yellow displays are preferred for use in such applications as avionics and automotive instrument panels since red is reserved as a warning color.

11.2.2   A Practical Example

A practical example is worth while to illustrate the use of the discrimination index with sunlight viewable displays. A 7.6mm yellow LED display is pictured behind a 25% neutral density gray filter in Figure 11.2.2-1. The display gray package and untinted segments have nearly the same reflective characteristics, as shown in Figure 11.2.2-2, so that $L_v S_{OFF} = L_v B$ and Equation (11.1.2-2) may be used to calculate the contrast. The incident sunlight is at the 107,000 lumens per square metre level. The display is being driven at data sheet limits and produces a light output of 3 millicandelas per segment. The resulting luminous contrast as seen by an observer is C = 1.54. The chromaticity coordinates for the yellow LED emitted light and the gray background are plotted on the 1960 CIE–UCS

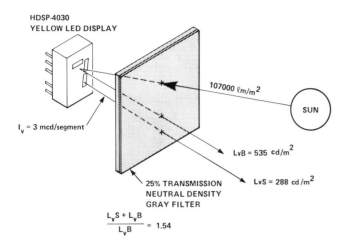

Figure 11.2.2-1. An Achievable Luminous Contrast for a Yellow LED/Neutral Gray Filter in an Ambient of Direct Sunlight.

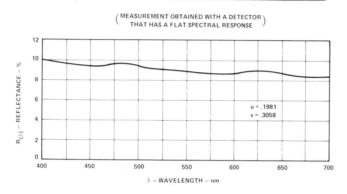

Figure 11.2.2-2. Reflectance Characteristic for the Face of a Gray Body Stretched Segment Sunlight Viewable LED Display.

Chromaticity Diagram in Figure 11.2.2-3. The chromatic difference is EC = .1342.

From the above information, the discrimination index may be determined by calculating the luminous and chrominance indices.

$$IDL = \frac{\log_{10} 1.54}{0.15} = 1.25$$

$$IDC = \frac{.1342}{0.027} = 4.97$$

$$ID = \sqrt{(1.25)^2 + (4.97)^2} = 5.12$$

This value of 5.12 is 3.62 times the ID_{MIN} value of $\sqrt{2}$.

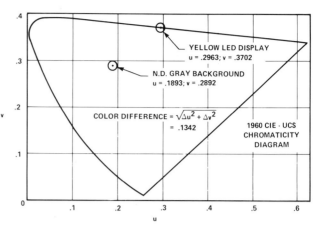

Figure 11.2.2-3. The Chromaticity Coordinate Locations for a Yellow LED Display and a Neutral Density Gray Background.

If the light output of the LED display is reduced from the 3 mcd value to 1.5 mcd per segment, the luminous contrast is reduced to C = 1.27. The luminance index is reduced to a value of 0.69. The discrimination index is reduced to a value of 5.02, which is only a 2% decrease. From this example, it can be seen that the dominating factor in achieving readability in bright sunlight is the effect of chrominance contrast. A suggested lower limit value for ID for reasonable readability in bright sunlight is ID = 5.00.

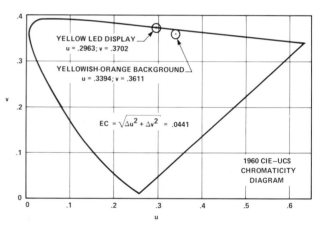

Figure 11.2.2-4. The Chromaticity Coordinate Location for a Yellowish-Orange Background.

The above example was illustrated with a gray background established by the neutral density gray filter/gray package combination. It is useful to compare this same display when placed behind a semi-bandpass yellowish-orange filter. The color coordinates of the resulting background are plotted in Figure 11.2.2-4. The light transmission through the filter is 27%. The luminous contrast remains at the 1.54 value. The chromatic distance between the coordinates of the display emitted light and the yellowish-orange background is only 0.0441. The resulting discrimination index is 2.05:

$$IDL = \frac{\log_{10} 1.54}{0.15} = 1.25$$

$$IDC = \frac{0.0441}{0.027} = 1.63$$

$$ID = \sqrt{(1.25)^2 + (1.63)^2} = 2.05$$

This discrimination index value is only 40% of the value that is achieved with a neutral density gray filter. This example illustrates the loss of visual discrimination when chrominance contrast is not included as part of the overall contrast enhancement design.

11.3 A Special Filter for Yellow Alphanumeric LED Displays

The need for presentation of alphanumeric information across the man/machine interface is expanding with the advent of new microprocessor controlled systems. Electronic alphanumeric displays are being incorporated into such sophisticated systems as electronic bank teller stations and the latest generation of avionics for use in the 1980 vintage commercial jet aircraft. The worst case ambient lighting condition, in which a user of these systems must be able to read the displayed information, is bright sunlight.

The difficulty in using an electronic display in one of these systems is that most of the available light emitting alphanumeric displays were not designed for use in bright sunlight. The light output of these displays is insufficient to meet the requirements for readability in 107,000lm/m^2 sunlight. However, a contrast enhancement filter technique is available that does provide for readability of a yellow LED alphanumeric display, such as the HDSP-2001, in sunlight ambients up to 80,000 lm/m^2.

The concept of this filter is to combine the benefits of a circular polarizer sandwiched between two pieces of optically coated glass. The filter described herein was developed by Optical Coating Laboratory, Santa Rosa, California. The construction of the filter is illustrated in Figure 11.3-1, and the relative transmission characteristics are shown in Figure 11.3-2.

The circular polarizer is used to reduce the background reflection to a low level. The specular reflections from the glass face of the display package, the top side surfaces of the LEDs, and the on-board ICs are reduced to a very low level, as indicated by the light extinguishing characteristics of the polarizer, A(λ), as shown in Figure 11.3-2.

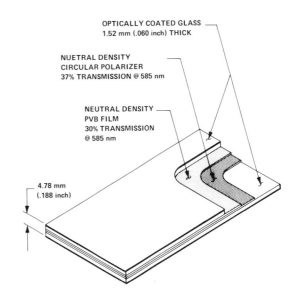

Figure 11.3-1. An Optically Coated Glass/Circular Polarizing Filter for use with Yellow LED Alphanumeric Displays in Moderate Sunlight Ambients.

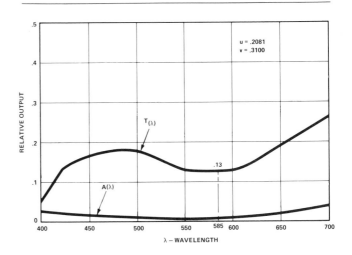

Figure 11.3-2. The Relative Transmission, $T_{(\lambda)}$, and Relative Background Light Extinguishing, $A_{(\lambda)}$, Characteristics of an Optically Coated Glass/Circular Polarizing Filter.

The light output of each LED element in an alphanumeric display, such as the HDSP-2001, is typically 600μcd. When placed behind this optically coated glass/polarizing filter in a sunlight ambient of 80,000 lm/m^2, a contrast ratio of 2.32 is achieved. The IDL value is 2.4.3. The chromatic distance between an illuminated yellow LED element and the neutral background is 0.1067. The IDC value is 3.954 and the resulting discrimination index is ID = 4.64.

11.9

11.4 The Importance of Low Thermal Resistance Design

In achieving readability in bright sunlight, it is necessary to incorporate into the display system not only those factors that contribute to contrast enhancement but also those factors that contribute to light output. Since the light output of an LED device is an exponential function of the junction temperature rise above ambient, it is important to keep this temperature rise to a minimum in order to obtain as much light output as possible in relation to the electrical power input. The product of the electrical power dissipation within the device and thermal resistance-to-ambient determines the temperature rise of the LED junction. Therefore, it is important to incorporate as low a thermal resistance-to-ambient as possible into the design of the display system to insure the maximum possible light output.

11.4.1 The Effect of LED Junction Temperature On Light Output

The thermal resistance design is simplified when it is considered on a per LED chip basis, such as on a per segment basis for the sunlight viewable displays.

The thermal resistance-to-ambient, $R_{\theta JA}$, is the sum of two terms:

$$R_{\theta JA} \ (^\circ C/W/LED) = R_{\theta J\text{-}PIN} + R_{\theta PC\text{-}A} \qquad (11.4.1\text{-}1)$$

where $R_{\theta J\text{-}PIN}$ = the device thermal resistance LED junction-to-pin, $^\circ$C/W, on a per LED basis

$R_{\theta PC\text{-}A}$ = the thermal resistance PC board-to-ambient, $^\circ$C/W, on a per LED basis

The thermal resistance device pin-to-PC board is usually very small, negligible for soldered devices and on the order of 2 to 5°C/W for socketed devices; therefore, it is included as part of the $R_{\theta PC\text{-}A}$ value.

The LED junction temperature rise above ambient, ΔT_J, is a direct function of the thermal resistance:

$$\Delta T_J \ (^\circ C) = P_{LED} \ R_{\theta JA} \qquad (11.4.1\text{-}2)$$

where P_{LED} = the electrical power in Watts dissipated within the LED junction

The display system designer has control over the value of $R_{\theta JA}$ by virtue of the fact that the design of the display PC board will determine $R_{\theta PC\text{-}A}$. It is prudent to design for as low a value of display PC board-to-ambient thermal resis-

tance as possible in order to achieve the most light output for the established input current drive condition. One method of obtaining a low $R_{\theta PC\text{-}A}$ is to design a double sided ground plane PC board with plated through holes, wide circuit traces, and maximum metallization. $R_{\theta PC\text{-}A}$ values on the order of 50°C/W/LED are achievable.

The benefit of reducing ΔT_J can be illustrated by listing the improvement in light output that corresponds to a decrease in the LED junction temperature. The improvement in light output, ΔI_v (percent), for a given decrease in ΔT_J is determined from Equation (5.2.4-2) rewritten in the following form:

$$\Delta I_v \ (\%) = (e^{K \Delta T_J} - 1) \ 100 \qquad (11.4.3\text{-}1)$$

where K = a constant in 1/$^\circ$C for each LED color

ΔT_J = LED junction temperature decrease, $^\circ$C, due to a reduction in $R_{\theta PC\text{-}A}$

As an example, Table 11.4.1-1 gives the light output improvement that is achievable from a corresponding reduction in the junction temperature of a yellow LED sunlight viewable display.

ΔT_J (DECREASE)	ΔI_v (INCREASE)
-1°C	1.1%
-5°C	5.8%
-10°C	11.9%
-20°C	25.1%

Table 11.4.1-1. The Light Output Improvement for a Corresponding Decrease in LED Junction Temperature for a Yellow Sunlight Viewable Display; k = −.0112/$^\circ$C

11.4.2 Thermal Resistance and the Allowed Operating Ambient Temperature

One generic parameter that limits the amount of input power that may be dissipated within an LED junction is the junction temperature, T_J.

$$T_J \ (^\circ C) = T_A + \Delta T_J \qquad (11.4.2\text{-}1)$$

where T_A = operating ambient temperature, $^\circ$C, within the immediate vicinity of the LED device

ΔT_J = junction temperature rise above ambient as defined by Equations (11.4.1-2) and (11.4.1-1)

The maximum allowed LED junction temperature for plastic encapsulated LED devices is typically +100°C, and +125°C for nonplastic devices.

An examination of Equation (11.4.2-1) points out that the value of ΔT_J is the limiting factor on the ambient temperature in which the device may be operated at maximum power input. Since the value of ΔT_J is determined by $R_{\theta PC\text{-}A}$, it is this thermal resistance-to-ambient through the display PC board that determines the allowed operating ambient temperature for a given input power.

It is desireable to operate a sunlight viewable LED display at or near the maximum allowed drive conditions to obtain the most light output regardless of T_A. To do this, a sufficiently low value of $R_{\theta PC\text{-}A}$ is designed into the display PC board so that T_J does not exceed the maximum value when the device is operated in the highest expected ambient temperature. An example will illustrate the concept.

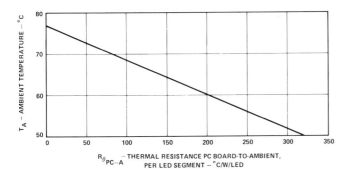

Figure 11.4.2-1. Operating Ambient Temperature vs. the Maximum Allowed Thermal Resistance-to-Ambient Through the Display PC Board for Device Operation Without Derating; P_{LED} = 83 mW/LED.

The device thermal resistance of a plastic sunlight viewable display is $R_{\theta J\text{-}PIN}$ = 282°C/W/LED segment. The maximum allowed power dissipation is 83 mW per LED segment. Figure 11.4.2-1 gives the maximum allowed thermal resistance-to-ambient through the display PC board that will permit operation in a specified ambient temperature at maximum input power, without derating. Though the data sheet calls for a power derating when the device is operated in ambient temperatures above 50°C, the display may be operated without a power derating at 70°C when the PC board thermal resistance-to-ambient has been designed to be 80°C/W/LED or less.

11.4.3   LED Junction Temperature vs. Pulsed Operation

A factor that has an effect on light output is the strobing refresh rate used in pulsed operation. The refresh rate controls the difference between the peak LED junction temperature, $T_{J\ PEAK}$, and the time average LED junction temperature, $T_{J\ AVG}$. Since the light output from an LED is determined by the peak junction temperature, it is beneficial to strobe at a fast refresh rate, where the difference between $T_{J\ PEAK}$ and $T_{J\ AVG}$ is minimal, to obtain the best possible light output.

When strobing at a slow refresh rate, for example 100 Hz, $T_{J\ PEAK}$ is many degrees above $T_{J\ AVG}$ due to the thermal time constant of the device. As the refresh rate is increased, $T_{J\ PEAK}$ approaches the $T_{J\ AVG}$ value and the light output increases accordingly, as shown in Figure 11.4.3-1. At a very fast refresh rate, 10 KHz for example, $T_{J\ PEAK}$ and $T_{J\ AVG}$ are equal and the light output is governed by $T_{J\ AVG}$.

Since the difference between $T_{J\ PEAK}$ and $T_{J\ AVG}$ becomes small for refresh rates in excess of 1 KHz, it is advisable to strobe at a 1 KHz rate or faster. The light output may then be calculated based on the $T_{J\ AVG}$ value.

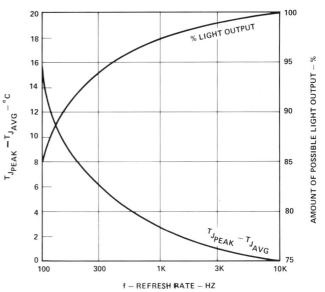

Figure 11.4.3-1. The Difference Between the Peak and Average LED Junction Temperatures and its Effect on Light Output as a Function of Strobing Refresh Rate for a Yellow Sunlight Viewable Display.

Section 12
Backlighting

12.0   BACKLIGHTING

Many modern electronic systems use some kind of annunciation to transmit information across the man/machine interface. This annunciation may be composed of indicator lights, lighted switches or illuminated legends to indicate the status of the system under control. The annunciator devices may be individual indicators placed at specific locations throughout the system or be grouped into a single comprehensive panel that accomodates all of the annunciator functions.

The primary difference between an annunciator and an information display is that the annunciator presents a fixed message while an information display presents a variable message. An electronic information display is typically composed of many light emitting elements that are activated in proper sequence to form the characters of the message being displayed. The annunciator device is typically composed of a single element that is activated when the indicated status of the system is true.

Since an electronic annunciator device is either illuminated (ON) when the status is true and not illuminated (OFF) when the status is false, it is usually transilluminated by a dedicated light source. The light source may be remote to the annunciator proper or may be an integral part of the unit's design.

Many factors go into the selection of the light source and relate directly to the design requirements of the annunciator device, such as:

1.   The required brightness, color and contrast necessary to achieve recognition of the ON condition in the end use ambient lighting.

2.   The kind of annunciator device, size, shape and legend format to be illuminated.

3.   Whether the annunciator is to have a self contained light source or be transilluminated by backlighting or edge lighting through a panel.

4.   Electrical power requirements and temperature limitations.

5.   Mechanical packaging options.

6.   Long term reliability.

7.   Cost vs. performance objectives.

12.1   Annunciator Brightness

Simply stated, the illuminated annunciator needs to be bright enough for an observer to recognize the ON condition and at the same time be inconspicuous in the OFF condition. Technically, the brightness of an area that is self-luminous is defined as that attribute of visual sensation which determines if an area is emitting a large or small amount of light. The concept of luminous sterance, introduced in Chapter 7, may be used as a determination of brightness.

12.1.1   The Concept of Luminous Sterance

To effectively evaluate the brightness requirements, the luminous sterance of the illuminated annunciator needs to be evaluated at the widest off-axis viewing angle where the apparent brightness of the illuminated annunciator is at its dimmest value. The brightness increases as the observer moves to the on-axis position and the off-axis viewing angle goes to zero.

The observed luminous sterance at an off-axis angle, $L_v(\theta)$, may be defined as the ratio of the observed luminous intensity to the apparent size of the light emitting area, and may be expressed mathematically by rewriting Equation (7.2-1):

$$L_v(\theta) \text{ cd/m}^2 = \frac{I_v(\theta)}{A(\theta)} \tag{12.1.1-1}$$

where: $I_v(\theta)$ = the observed luminous intensity at the off-axis viewing angle, θ; candelas

$A(\theta)$ = the apparent size of the light emitting area at the off-axis angle, θ; square metres

The apparent size of a light emitting area that is a flat surface varies as the cosine of the off-axis viewing angle.

$$A(\theta) = A \cos\theta \tag{12.1.1-2}$$

where: A = the actual area size when viewed from the on-axis position, metres

θ = the off-axis viewing angle

The observed luminous intensity may be considered to vary as the cosine of the off-axis angle raised to some power.

$$I_v(\theta) = I_{v_0} \cos^n\theta \tag{12.1.1-3}$$

where: I_{v_0} = the on-axis intensity

n = the exponent that is derived from the intensity radiation pattern (See Figure 7.3.4-1)

Placing Equations (12.1.1-2) and (12.1.1-3) into Equation (12.1.1-1) yields:

$$L_v(\theta) = \frac{I_{v_o} \cos^n \theta}{A \cos \theta} \qquad (12.1.1\text{-}4)$$

This equation will calculate the observed luminous sterance for a flat surface light emitting annunciator when the radiation pattern of the intensity is known.

To gain an appreciation of Equation (12.1.1-4), it is useful to examine two separate radiation patterns for the same flat light emitting surface area. The first is a Lambertian radiation pattern and the second is a non-Lambertian radiation pattern.

12.1.2 Luminous Sterance From A Lambertian Light Emitting Surface

A uniformly diffusing light emitting flat surface has a radiation pattern such that the luminous sterance is a constant in any direction. Termed Lambertian, this ideal light emitting surface appears equally bright regardless of the off-axis angle from which it is observed, as illustrated in Figure 12.1.2-1. As shown in Figure 12.1.2-2, the radiation pattern is such that the intensity varies as the cosine of the off-axis angle in the same fashion as does the apparent size of the light emitting surface. Therefore, the value of n is 1.0 and the value for Equation (12.1.1-4) is a constant ratio, I_{v_o}/A, and the value of $L_v(\theta)$ is a constant.

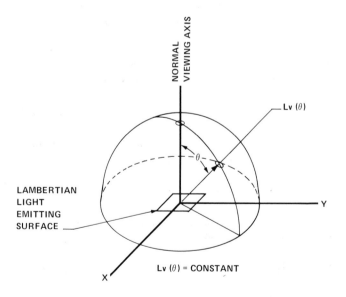

Figure 12.1.2-1. Luminous Sterance Variation with Off-Axis Viewing Angle, Lv (θ), for a Lambertian Light Emitting Surface.

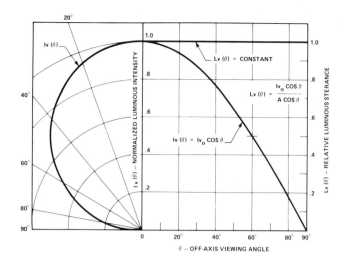

Figure 12.1.2-2. The Light Output Radiation Pattern and Relative Luminous Sterance for a Lambertian Light Emitting Surface. Note: 50% Intensity Occurs at 60°.

12.1.3 Luminous Sterance From A Non-Lambertian Light Emitting Surface

A light emitting surface that is not Lambertian will have a radiation pattern such that the luminous sterance decreases as the off-axis viewing angle increases, as illustrated in Figure 12.1.3-1. Most light diffusing films do not diffuse light uniformly and typically have an on-axis light output gain that is greater than 1.0. As the off-axis viewing angle increases, the light output gain decreases and the result is a radiation pattern that is less than Lambertian.

The radiation pattern for a typical light diffusing film is shown in Figure 12.1.3-2. The value for the exponent n that describes the off-axis intensity variation is 4.8. Inserting this value for n into Equation (12.1.1-4) results in values for $L_v(\theta)$ that vary as $\text{cosine}^{3.8} \theta$. $L_v(\theta)$ is not a constant, and an appraisal of Figure 12.1.3-2 shows that $L_v(\theta)$ essentially varies as light output radiation pattern.

12.1.4 Definition of Minimum Luminous Sterance

Noting that the luminous sterance for a flat non-Lambertian light emitting surface typically varies as the radiation pattern, it is reasonable to select the angle where the luminous intensity is one half of the on-axis value, $\theta_{1/2}$, as the defined maximum off-axis viewing angle. The included viewing angle is twice this value, $2\theta_{1/2}$. Designing for sufficient luminous sterance at the $\theta_{1/2}$ angle insures the recognizability of the ON-state annunciator over the defined viewing angle. The minimum luminous sterance is then defined as:

$$L_v \text{ MIN} \equiv L_v(\theta_{1/2})$$

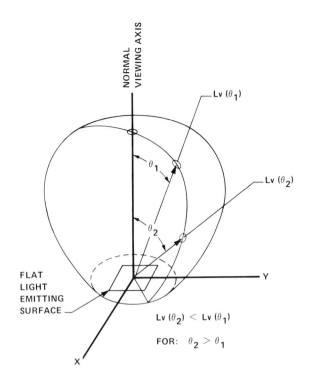

Figure 12.1.3-1. Luminous Sterance Variation with Off-Axis Viewing Angle, Lv (θ), for a Non-Lambertian Light Emitting Surface.

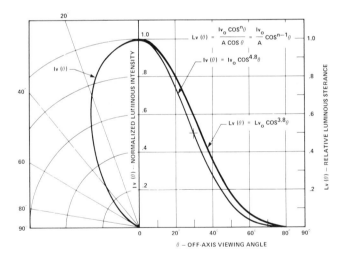

$$Lv\,(\theta) = \frac{Iv_o \cos^n\theta}{A \cos\theta} = \frac{Iv_o}{A} \cos^{n-1}\theta$$

$$Iv\,(\theta) = Iv_o \cos^{4.8}\theta$$

$$Lv\,(\theta) = Lv_o \cos^{3.8}\theta$$

Figure 12.1.3-2. The Light Output Radiation Pattern and Relative Luminous Sterance Variation for a Non-Lambertian Light Emitting Surface. Note: 50% Intensity Occurs at 30°.

$$L_v\,MIN = \frac{0.5\,I_{v_o}}{A\,\cos\theta_{\frac{1}{2}}} \qquad (12.1.4\text{-}1)$$

where $L_v\,MIN$ = Minimum luminous sterance in candelas per square metre

I_{v_o} = Axial intensity in candelas

A = Area in square metres

$\theta_{\frac{1}{2}}$ = The off-axis angle where the intensity is half the on-axis value

12.1.5   The Effect of Color On Observed Brightness

The theory of luminous sterance presented thus far makes no reference to the effect of color differences. The theory holds true for a given radiated spectrum that remains constant for each condition. However, duplicate annunciators that are illuminated by different colored light sources may not appear to have equal brightness even though the luminous sterance values are the same. Although the CIE photometric system using the 1931 photopic function $\overline{y}_\lambda$ predicts observed brightness-to-luminous sterance ratios of 1.0, there can be substantial differences between the measured luminous sterance values and the brightness as seen by an observer when different radiated spectra of saturated colors are involved.

The luminous sterance values of annunciators that are illuminated by different radiated spectra may be color adjusted so that all the annunciators will be observed to have equal brightness. The color correction is a function of such factors as the angular size of each annunciator subtended at the eye, the separation between annunciators, the ambient lighting, the color, saturation and luminance of the area around the annunciators, the dominant wavelengths of the radiated spectra, and the optical filtering. Of these, the dominant wavelengths of the radiated spectra and the choice of optical filtering are at the disposal of the designer.

LED COLOR AND DOMINANT WAVELENGTH	UNFILTERED	WITH A 25% NEUTRAL DENSITY GRAY FILTER
GREEN, λ_d = 572 nm	2.00	1.45
YELLOW, λ_d = 585 nm	1.84	1.34
HIGH-EFFICIENCY RED, λ_d = 626 nm	1.00	1.00
STANDARD RED, λ_d = 640 nm	0.48	0.73

Table 12.1.5-1. Relative Luminous Sterance for Equal Brightness as Seen by an Observer. Values are for a Backlighted Untinted Legend with a Gray Surround.

Two sets of luminous color correction factors are presented in Table 12.1.5-1. Both sets have been established for an annunciator consisting of a backlighted untinted legend with a gray surround. Both are normalized to a factor of 1.0 for a high-efficiency red LED. Each factor establishes the relative luminous sterance necessary to achieve equal brightness as seen by an observer. For example, the required luminous sterance of an unfiltered legend illuminated by a green LED is twice that of an unfiltered legend illuminated by a high-efficiency red LED.

For equal observed brightness of two similar legends that have the same area, A:

$$L_v \text{ (GREEN)} = 2 L_v \text{ (HI–EFF RED)}$$

From Equation (12.1.1-1):

$$\frac{I_v \text{ (GREEN)}}{A} = \frac{2 I_v \text{ (HI–EFF RED)}}{A}$$

12.1.6   Luminous Sterance Levels For Comfortable Viewing

The luminous sterance that is sufficient to permit comfortable viewing of an annunciator varies exponentially with the increase in the level of ambient light. As a guide, the suggested light output, in candelas per square metre, for an annunciator that is composed of a backlighted untinted legend with a gray surround is given versus a range of ambient light levels in Figures 12.1.6-1 and 12.1.6-2. The solid lines indicate the luminous sterance that is attainable from a typical LED backlighting device under nominal drive conditions. The dashed lines extend the curves to a lighting level that is experienced in daylight shade.

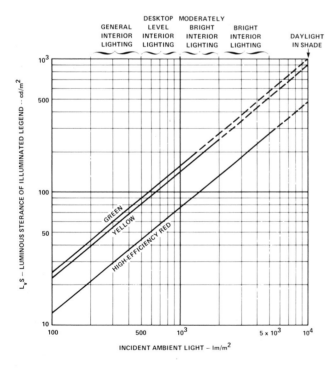

Figure 12.1.6-1. Light Output of an Unfiltered Annunciator with Untinted Legend and Gray Surround for Comfortable Viewing vs. Level of Incident Ambient Light.

When an annunciator is placed behind an optical filter, the required luminous sterance for comfortable viewing increases. However, when the optical filter is a 25% neutral density gray, the following is observed. First, the increase in the required luminous sterance is only 2.6 times the required unfiltered value. This is much less than the 4X value that would be theoretically derived due to the filter attenuation. The reason is that the luminous sterance of the background around the legend is reduced by a factor of 94% and that the chrominance contrast with respect to the gray background provided by the filter enhances viewability. Secondly, the differences in luminous sterance values for different LED colors are reduced. This is another effect of the increase in viewability due to the chrominance contrast provided by the neutral density gray filter.

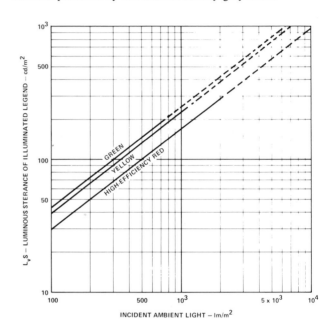

Figure 12.1.6-2. Light Output of an Annunciator with Untinted Legend and Gray Surround for Comfortable Viewing When the Annunciator is to be Placed Behind a 25% Neutral Density Gray Filter vs. the Level of Ambient Light.

The luminous sterance of the filtered legend as seen by an observer is the sterance value from Figure 12.1.6-2 multiplied by the relative transmission of the filter, 0.25. For example, the filtered luminous sterance of a backlighted legend, illuminated to a light output level of $170 \, \text{cd/m}^2$, would be seen by an observer as $43 \, \text{cd/m}^2$.

$$(170 \, \text{cd/m}^2)(0.25) = 43 \, \text{cd/m}^2$$

The luminous sterance provided by a self luminous surface used as a light source remains constant irrespective of the amount of opaque area in the legend that restricts light transmission. The reason is that both the transmitted light and light transmitting area are reduced in the same direct proportion by the opaque region. For example, a 3.8 mm x 19.1 mm self luminous surface may have a luminous intensity of 21 mcd at a given drive condition. The luminous sterance is 289 lm/m^2. A legend that has a 2.2 mm diameter transparent circle on an opaque background is attached to the face of the light emitting surface. The measured luminous intensity is reduced by 94.2% to 1.22 mcd and the light transmitting area has been reduced by 94.7%. The luminous sterance of the 2.2 mm diameter circle is 287 cd/m^2, a reduction of only 0.5% due to reflection losses off the backside of the legend.

12.2   Legends

An annunciator configuration that is extensively used is the backlighted legend. The backlighted legend is typically used to give such information as status; "ON," "OFF," mode of operation; "PHONO," "FM," "TAPE," equipment selection; "NAV/COM1," "ADF," or switch function identification; "FINE," 'COARSE." In all of these cases, a light source is placed behind the legend and the light passes through the legend to the observer. In the discussion that follows, the legend is considered to be directly attached to the face of the LED light source package to form an annunciator module.

12.2.1   Legend Font

There are two basic legend font configurations as illustrated in Figure 12.2.1-1: the transparent legend with opaque surround and the inverse, the opaque legend with transparent surround. Both font configurations have advantages depending upon the design objectives.

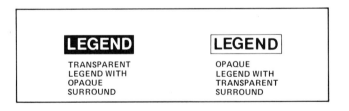

Figure 12.2.1-1.   The Two Basic Legend Formats.

The transparent legend with opaque surround is typically used in those applications where the lettering or symbol must be read by the observer. Color coding may not be important and the same colored light source may be used to illuminate other annunciators that display a variety of messages. Only the light that passes through the transparent characters reaches the eyes of an observer. In general, the characters and symbols are rather small and if not precisely made can be misinterpreted. For example, an "h" could be mistaken for an "n" and an "a" for an "o." Therefore, to insure readability, it is best to use boldfaced type and

symbols and upper case lettering whenever possible, as emphasized in Figure 12.2.1-2.

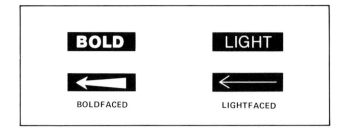

Figure 12.2.1-2.   Boldfaced Characters and Symbols Make Legends that are Easy to Read.

A suggested typeface for good readability is Futura Demi Bold. For instance, a six character legend in 12 point Futura Demi Bold can be incorporated within a 19 mm (0.75 inch) linear spacing and is easily readable at a distance of 1.2 metres.

The opaque legend with a transparent surround is an attractive device for use as a status indicator. In this case, color coding may be important. An obvious example is an annunciator used to indicate an overheated condition. The word "HOT" would be surrounded by a large bright field of red light that will quickly command the attention of an observer, who upon closer examination would read the status as overheating.

As a rule of thumb, the character height for a particular viewing distance may be determined from the following formula:

$$(12.2.1-1).$$

Character Height (mm) = Viewing Distance (m) x 1.454

Equation (12.2.1-1) is based on the definition of normal visual activity, 20/20 vision, where a character subtends an angular height of five minutes of arc as viewed by the eye at the viewing distance. The value $1.454 = 10^3$ times the tangent of a five minute angle. For example, a letter with a character height of 3.81 mm (0.15 inch) can be viewed comfortably at a distance of 2.62 metres (8.6 feet).

The most vivid visual difference between an ON and OFF annunciator is obtained by the use of an untinted transparent substrate for the legend. If a tinted substrate is used, the apparent color of the OFF legend due to the reflected ambient light will be similar to the color of the LED light. Confusion as to the ON/OFF state of the annunciator may result if a tinted substrate legend is used in moderate to bright ambients.

The use of an untinted transparent substrate eliminates this

ON/OFF confusion. The color of the OFF annunciator is neutral and does not resemble the LED light. When the annunciator is on, the transparent portion of the legend is illuminated with the vivid color of the LED light which will catch the attention of an observer.

12.2.2 Fabricating A Legend

A legend that is cost effective is one that may be attached directly to a self luminous surface. Such a legend is typically made by silkscreening the font onto a transparent substrate. The legend is then bonded to the face of the LED light source.

12.2.2.1 The Transparent Substrate

A material that performs well as a substrate is a transparent polycarbonate sheet, 0.127 mm (0.005 inch) thick, such as Lexan 8070-112 by General Electric, Plastics Division, Pittsfield, Massachusetts. Silkscreening paints adhere very well to polycarbonate, whereas considerably less adhesion is experienced with other materials such as polyesters. Also, polycarbonate will withstand the elevated temperatures that are necessary to cure the paint and adhesive.

12.2.2.2 The Silkscreen Paint

Epoxy paints offer two distinctive advantages. Though somewhat difficult to handle in a silkscreening process, an epoxy paint will air dry to a soft state which permits the legend to be sheared to size without the occurence of chipped or peeling paint. When cured, the epoxy paint has a high abrasion and chemical resistance. The adhesion of the cured epoxy to the polycarbonate is sufficient to withstand a tape pull test.

One such epoxy is Wornow M22278-X with B-3 catalyst made by the Hysol Division of the Dexter Corporation, City of Industry, California. The paint thickness should not exceed 0.025 mm (0.001 inch). A layer of paint thicker than this limit may cause the legend substrate to curl due to shrinkage during the curing cycle. Even so, paint coverage must be sufficient to insure opacity. Black pigmentation may be added to the paint to help insure opacity.

A measure of opacity is optical density, D.

$$D(\lambda) = \log_{10} \frac{1}{T(\lambda)} \qquad (12.2.2.2\text{-}1)$$

where $T(\lambda)$ = the relative transmission

A simple method to determine the optical density of the painted area of a legend is to measure the relative transmission at the peak wavelength for the LED. The optical density is then determined from Equation (12.2.2-1).

Optical densities in the range of 3.6 to 4.0 are achievable.

12.2.2.3 Legend Artwork For Alignment

Placing the silkscreened legend onto the face of the LED light source requires the use of some kind of alignment technique. One method is to have the silkscreened legend contain alignment guiding as part of the artwork. Two alignment format methods are pictured in Figures 12.2.2.3-1 and 12.2.2.3-2. The artwork of Figure 12.2.2.3-1 contains alignment marks and notched corners in the face of the legend proper. The concept is to align the legend such that the corners of the package of the LED light source just show through the notches. In Figure 12.2.2.3-2, the concept is to eliminate the need for alignment marks by having the artwork oversized by 0.127 mm (0.005 inch) on each side. When the legend is placed onto the face of the LED package, it is only necessary to have some overhang present along each edge. The slight misalignment that may occur is not noticeable. The legend is cut to size after it is attached to the LED light source, but prior to the cure cycle.

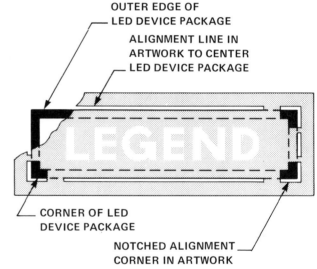

Figure 12.2.2.3-1. Legend Artwork with Alignment Marks.

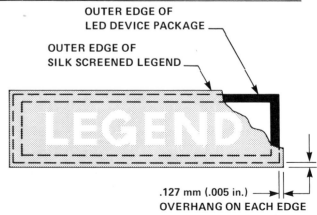

Figure 12.2.2.3-2. Legend Artwork with Overhang to Allow for Alignment.

12.2.3 Attaching the Legend to the Face of the LED Light Source

The silkscreened legend, not cut to size, is attached to the face of the package of the LED light source via a pressure sensitive optically transparent adhesive. The adhesive, M69 supplied by Connecticut Hard Rubber Company, New Haven, Connecticut, may be obtained in the form of a double back tape slit to the proper width. Each layer of adhesive is 0.051 mm (0.002 inch) thick on a 0.051 mm (0.002 inch) carrier. The adhesive forms a permanent bond after being subjected to an oven cure cycle.

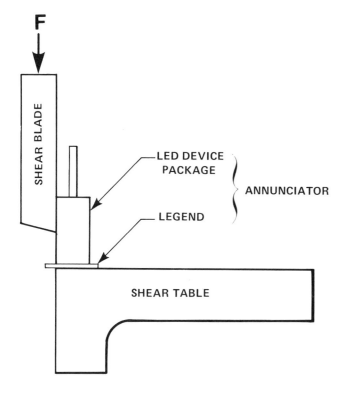

Figure 12.2.3-1. Shearing Legend to Size Prior to the Cure Cycle.

The legend is aligned with respect to the face of the LED package and then pressed into place. The legend is cut to size using a shear with the side of the LED device package being used as a guide, see Figure 12.2.3-1. After the legend is cut to size, the assembled annunciator is put through an elevated temperature cure cycle of 115°C for four hours to cure both the epoxy paint and the adhesive.

Care must be taken to insure that all of the legend area is securely attached to the entire face of the LED device package before cure. If not, a void will be created between the legend and the face of the LED device package. The difficulty is that the face of the LED device package may not be perfectly flat as shown in Figure 12.2.3-2. If a void is allowed to occur between the legend and the face of the

LED device package, light loss through the void will cause uneven illumination of the legend.

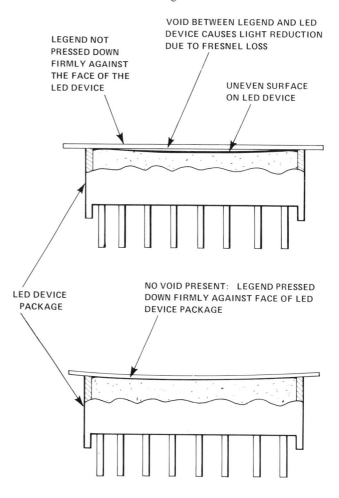

Figure 12.2.3-2. Securely Attached Legend Prevents Reduction of Light Due to Fresnel Loss.

12.2.4 Cross Talk Across the Face of a Multi-Function Annunciator

A multi-function annunciator may be assembled using a single substrate multi-function legend with each legend area backlighted by a separate LED light source. The LED light sources may be of different colors and one or more may be illuminated at any given time. The difficulty with this configuration is the partial illumination of an OFF legend due to light leakage through the legend substrate, called cross talk.

The light leakage path is through the legend's transparent substrate and the transparent adhesive as illustrated in Figure 12.2.4-1. Cross talk may cause confusion as to the ON/OFF state of an OFF legend.

There are a number of things that can be done to reduce the cross talk to a minimum. One method is to use a diffused transparent substrate with the legend silkscreened

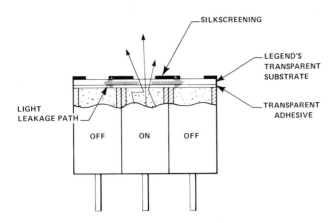

Figure 12.2.4-1. **Light Leakage Path Through the Legend Substrate and Adhesive of a Multi-Function Annunciator.**

on the backside surface, as shown in Figure 12.2.4-2. This essentially eliminates most of the cross talk through the substrate leaving the transparent adhesive as the primary light leakage path. A diffused substrate is necessary to illuminate front surface reflections that will reduce ON annunciator recognition and legend readability. One difficulty with this approach is that once the adhesive is applied and the legend is placed on the face of the LED device, it can not be adjusted for any misalignment. Any attempt to peel back the legend or move the legend causes the adhesive to smear the uncured epoxy paint.

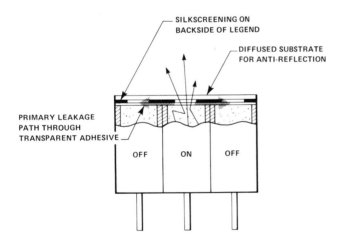

Figure 12.2.4-2. **Reduced Cross Talk in a Multi-Function Annunciator with the Legend Silkscreened on the Backside Surface of a Diffused Substrate.**

Other techniques that can be used to reduce the cross talk in a multi-function annunciator using a single legend substrate are presented in Figure 12.2.4-3. One simple effective technique is to paint the top surface edges of the LED light source devices black. The adhesive tends to dissolve and absorb some of the black paint reducing its light trans-

mission properties. The scribe line reduces the cross sectional area of the legend substrate, thus reducing the light pipe effect of the substrate. Both the selective placement of adhesive and the backside scribe line are effective if the LED devices are separated by some small distance. If the LED devices are not separated, the adhesive will fill the scribe line nullifying its effect.

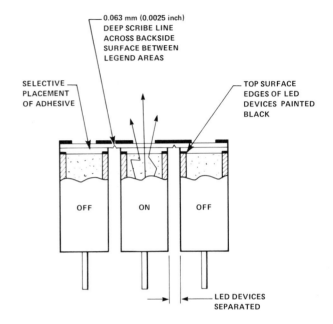

Figure 12.2.4-3. **Various Techniques that can be used to Reduce the Cross Talk in a Multi-Function Annunciator.**

12.3 **Front Panel Design for use with LED Backlighted Legends**

The design of a front panel may include one or more of three types of illuminated legend annunciators:

Individual annunciator modules mounted in cut outs within the panel

Backlighted transparent legend windows assembled as an integral part of the face of the panel

Edge lighted legends screened on a transparent plastic subpanel, see Section 12.6

The simplest method of incorporating an illuminated legend annunciator is to install an annunciator module, with the legend attached, into a cutout in the front panel. The annunciator module may be soldered to a small printed circuit board which is mounted on the backside surface of the panel. A bezel made from plastic or machined from stock adds a finishing touch as shown in Figure 12.3-1.

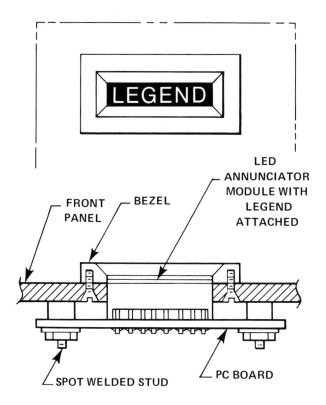

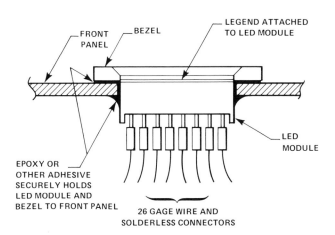

Figure 12.3-1. A Simple Front Panel Installation for an LED Annunciator with Legend Attached.

An inexpensive way to install the annunciator is to epoxy both the LED module, with legend attached, and the bezel directly to the front panel as illustrated in Figure 12.3-2. Electrical connection to the LED module may be made by use of small solderless connectors.

Figure 12.3-2. An Inexpensive Way to Install an LED Legend Annunciator to a Front Panel.

When many illuminated legends are to be placed in close proximity with respect to each other, or one or more illuminated legends are to be used with LED displays, it is economical to silkscreen the legends onto a common front panel window and backlight each legend area with a

separate LED light source. If the displays and illuminated legends are to be the same color, then the legends may be silkscreened onto the same filter window used for the LED displays.

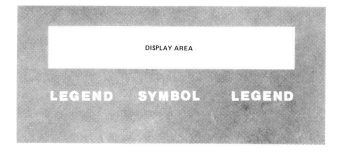

Figure 12.3-3. A Neutral Density Gray Filter Window Silkscreened to Accomodate Different Colored Legends and a Multi-Digit Display.

Usually the legends will be of different colors. If different colored legends are to be used in conjunction with displays, the simplest approach is to use a neutral density gray filter window with the legends silkscreened on the outside face of the filter. The silkscreening may be allowed to cover most of the filter to form an opaque field surrounding the legend characters/symbols and the display area as shown in Figure 12.3-3. The LED displays and LED modules for backlighting the legend areas are soldered to the same PC board and mounted directly behind the filter window as shown in Figure 12.3-4.

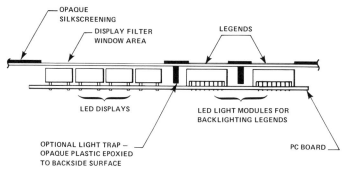

Figure 12.3-4. LED Displays and Light Modules for Backlighting Legends Mounted on the Same PC Board and Placed in Close Proximity to the Same Neutral Density Gray Filter Window.

In some applications, it is desirable to have a "dead front" appearance to the front panel. A reasonable approximation to a "dead front" appearance can be achieved by using a diffused material, such as the Diffused Light Control Film by 3M or Chromafuse® by Panelgraphic. It is not possible to recognize detail behind a diffused window so a dead front appearance is obtained. The silkscreening is done on the backside surface so that all detail is hidden when the displays or light source modules are not on. Care must be taken to insure that the displays and light source modules are in direct contact with, or at least in close proximity to, the backside surface of the diffuser to insure sharpness and minimal light loss.

More complex annunciator panels can be made by laminating various colored filters to an untinted transparent substrate or by silkscreening transparent colors onto an un-tinted substrate. In either case, care must be taken to use the correct color to match the radiated spectrum of the LED light source. The relative transmission of the tinting applied to the untinted substrate becomes an optical filter which adds light transmission loss and a possible color shift. The applied colors should have relative transmission characteristics similar to those presented in Figures 6.3.1-1 through 6.3.1-4 in order to obtain the best results when a color shift is not desired.

LED	COLOR SHIFT OF LED LIGHT	FILTER COLOR	LIGHT TRANSMISSION
HIGH EFFICIENCY RED	RED	RED	41%
	REDDISH–ORANGE	REDDISH–ORANGE	65%
YELLOW	YELLOWISH–ORANGE	AMBER	31%
	YELLOW	YELLOW–GREEN	37%
GREEN	YELLOW–GREEN	YELLOW–GREEN	50%

Table 12.3-5. Five Distinct Colors Achievable from Three LEDs and Four Filters.

It is possible to trade light transmission loss for a color shift. Table 12.3-5 lists five distinct colors that can be obtained from the use of three LEDs and four filters. The drive currents of each color LED would have to be adjusted separately in order to achieve apparent equal brightness.

12.4   The LED as a Light Source

The light emitting diode has been used extensively as the light source for lamps and displays for many years. The description of this technology that has been presented in earlier chapters has not evaluated the capability of the LED as a light source from the standpoint of achieving a desired level of luminous sterance. In making the determination that in fact an LED is a viable light source for annunciator illumination, it is useful to derive the size of the surface area that can be successfully illuminated at the desired sterance level by a single LED chip. Once this has been determined, the quantity of LED chips that are necessary to illuminate an annunciator of a given size is known. Also, it is important to compare electrical/optical parameters,

reliability and expected operating life with other light sources such as the incandescent bulb.

12.4.1   LED Flux Output and the Size of an Illuminated Area for a Given Luminous Sterance

As shown in Figures 12.1.6-1 and 12.1.6-2, the desired light output of an annunciator for comfortable viewing in a moderately bright ambient is in the range of $200 \, cd/m^2$. The relationship between this sterance level, the flux available from a single LED chip and the light output efficiency of the packaging configuration determine the size of the illuminated area.

The light output from the top surface of an LED chip is Lambertian. As demonstrated in Chapter 7, the luminous flux, ϕ_v, and axial intensity, I_{v_0}, for a Lambertian emitter are related as:

$$\phi_v = I_{v_0} \int_o^\pi 2\pi \sin\theta \cos\theta \, d\theta = \pi I_{v_0}$$

Thus: $I_{v_0} = \dfrac{\phi_v}{\pi}$ (12.4.1-1)

Inserting this value for intensity into Equation (12.1.1-1) yields an expression that defines the size of an illuminated area in terms of available flux and the required luminous sterance:

$$L_v = \frac{I_v}{A} = \frac{\phi_v}{\pi}\left(\frac{1}{A}\right)$$

$$A = \frac{\phi_v}{\pi L_v}$$ (12.4.1-2)

The ability of an LED chip to produce luminous flux from an electrical current is called quantum efficiency, η, and is expressed as lumens/amp. A mathematical definition of quantum efficiency is given by Equation (2.1.5-1). The net quantum efficiency of an encapsulated LED device is a function of chip quantum efficiency and package cavity efficiency:

$$\eta_{NET} = \eta_{CHIP} \cdot \eta_{CAVITY}$$ (12.4.1-3)

The total flux output from an encapsulated LED chip is the product of the net quantum efficiency and the forward current:

$$\phi_v = \eta_{NET} \cdot I_F$$ (12.4.1-4)

Placing this expression for total flux output into the equation for surface area yields the surface area expressed in terms of the net quantum efficiency and LED drive current.

$$A = \frac{\eta_{NET} \, I_F}{\pi L_v} \qquad \text{(12.4.1-5)}$$

Using some relatively typical values, it is possible to use Equation (12.4.1-5) to predict the size of an area that may be illuminated by a single high efficiency red LED chip.

Assume the following values:

$\eta_{CHIP} = 0.76$

$\eta_{CAVITY} = 0.8$

$I_F = 20$ milliamps

For the required luminous sterance of $200 \, cd/m^2$, an area of $19.4 \, mm^2$ can be illuminated.

$$A = \frac{(0.76)(0.8)(0.020A)}{(3.1416)(200 \, cd/m^2)} = 19.4 \times 10^{-6} \, m^2$$

From the above calculation, a quantity of four LED chips will successfully illuminate an encapsulated high efficiency red device with a light emitting surface that has an area four times this size, 3.81 mm (0.150 inch) x 19.05 mm (0.750 inch), to a light output level of $200 \, cd/m^2$.

12.4.2 The LED Light Source vs. The Incandescent Bulb

Until recently, it has been customary to use an incandescent bulb as the light source for backlighting an annunciator. For example, the type 683 and 715 T-1 incandescent lamps have been commonly used as the primary light source for the instrumentation and edge lighted panels used in aircraft cockpits. The reasons for choosing the incandescent bulb include:

High flux output — the primary reason.

Broadband light source — may use white or filter to any desired color.

Most readily available light source for the longest period of time — use of traditional and well understood design procedures.

A very wide variety of types and styles of lamps available.

However, there are some fundamental difficulties that are associated with the use of an incandescent bulb.

High heat generation — more than 98% of the energy generated is in the infrared region.

Susceptability to damage from mechanical shock and vibration.

Relatively short expected operating life to burn out — life expectancy is a function of selected color temperature to obtain the desired radiated spectrum and flux output.

Critical placement required to obtain maximum flux due to filament location.

Inability to maintain a constant radiated spectrum when voltage is reduced for dimming purposes— observed color shifts towards the red region with a decrease in color temperature.

Increased susceptability to failure when strobed — due to temperature cycling of the filament and increased stresses caused by a periodic repetition of high input currents.

Must optically filter to obtain any color other than white — usually with a considerable loss of radiated flux.

An LED device offers the annunciator designer an attractive alternative to the incandescent lamp:

All of the radiated flux is concentrated with the visual spectrum.

Low heat generation — light output generation is due to a semiconductor phenomenon, not due to incandescence from resistive heating.

Mechanically rugged — will withstand high levels of shock and vibration.

Extremely long expected operating life to 50% degradation of light output — actual useful life may extend well beyond this point.

Critical placement to obtain maximum flux output is not required — Lambertian radiation pattern.

The radiated spectrum (the color of the device) is independent of drive current — it is a small function of temperature.

LIGHT OUTPUT SPECIFICATION $T_A = 25°C$ (MILLICANDELAS)	TOTAL FLUX OUTPUT $T_A = 25°C$ (MILLILUMENS)	VOLTAGE DROP ACROSS DEVICE (VOLTS)	RATED CURRENT FOR LIGHT OUTPUT (MILLIAMPS)	POWER DISSIPATED WITHIN DEVICE (MILLIWATTS)	DEVICE TEMPERATURE RISE ABOVE T_A (°C)	COLOR	% OF FLUX AVAILABLE FOR LEGEND ILLUMINATION	% USABLE FLUX AFTER FILTERING TO OBTAIN RED COLOR	LUMINOUS STERANCE OF LEGEND (CANDELAS/METER²)	DIRECT LEGEND ATTACHMENT TO LIGHT SOURCE	EXPECTED USEFUL LIFE TO BURN OUT (HOURS)
HIGH EFFICIENCY RED LED LIGHT BAR MODULE											
15	47	1.9	80	152	22	REDDISH–ORANGE	100%	NOT REQUIRED	206	YES	IN EXCESS 100k
TYPE 683 T-1 INCANDESCENT LAMP											
50	628	5.0	60	300	92	WHITE	50%	10%	68	NO	25k
TYPE 715 T-1 INCANDESCENT LAMP											
150	1885	5.0	115	575	122	WHITE	50%	10%	207	NO	40k

Figure 12.4.2-1.   A Parameter Comparison Between an LED Light Bar Module and Two Incandescent Lamps Used to Backlight a 3.81 mm x 19.05 mm Legend with Red Light.

An LED is strobable — most efficient light output occurs under strobed operating conditions — an LED also performs efficiently under DC drive conditions.

Vivid color — the color purity of an LED is nearly equal to one.

There are a few limitations when choosing an LED device as an annunciator light source.

White is not available — the narrow band spectrum of the LED produces a specific color that falls within the range of colors from greenish-yellow to red.

Total flux output is somewhat less than that of many incandescent bulbs — the optical densities of "dead front" panels must be much less than that used for incandescents.

The light output is an exponential function of temperature — careful low thermal resistance design helps to insure light output at elevated temperatures.

A comparison between an LED device designed specifically for legend illumination and two commonly used incandescent lamps, types 683 and 715, is given in Table 12.4.2-1. All three are evaluated as to the possibility of being used to backlight a six character legend with red light. The overall legend size is 3.81 mm (0.150 inch) x 19.05 mm (0.750 inch).

An examination of the data listed in Table 12.4.2-1 brings forth the following observations. First, the cost of making the annunciator with the LED device is much less than using either of the incandescents. Secondly, the actual power dissipated within the LED device produces a lower package temperature rise. Third, the legend illumination using the LED is comparable to that achieved using the 715 lamp and is better than the 683 lamp. Finally, the expected operating life for the LED Light Bar Module includes the effects of mechanical shock, vibration and discontinuous electrical operation, whereas, the expected operating life for the incandescent lamps is based on a minimal amount of mechanical shock, vibration and turning on and off.

12.5   LED Light Bar Modules

LED devices that are intended for use as light sources have in the past been available primarily in T-1 and T-1 3/4 package sizes which are similar to the subminiature incandescent lamp, see Figure 2.2.2-1. Many applications pressed for an even smaller LED and the subminiature LED lamp pictured in Figure 2.2.3-1 was developed. The first lamp device designed with backlighting as one of the major uses was the rectangular lamp shown in Figure 2.2.4-1. Recently, the LED industry has introduced Light Bar Modules that have been designed specifically to illuminate legends using the backlighting technique.

Display lighting modules using incandescent lamps are assembled in a fashion similar to that illustrated in Figure 12.5-1. The design and use of such a module necessitates careful consideration as to the means of heat dissipation and access for lamp replacement.

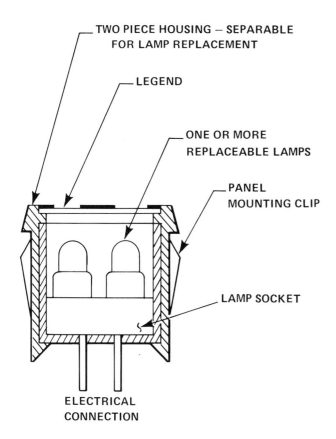

TWO PIECE HOUSING – SEPARABLE FOR LAMP REPLACEMENT

LEGEND

ONE OR MORE REPLACEABLE LAMPS

PANEL MOUNTING CLIP

LAMP SOCKET

ELECTRICAL CONNECTION

Figure 12.5-1.   The Basic Assembly Concept of a Legend Annunciator Using Incandescent Lamps as a Light Source.

An alternative is the LED Light Bar Module shown in Figure 12.5-2. The Light Bar Module is a plastic encapsulated device that capitalizes on the highly developed assembly technology used in the stretched seven segment displays, see Section 5.2.1. The light from each LED is optically scattered throughout the glass filled epoxy to form a large evenly illuminated light emitting surface, to which a legend may be directly attached as described in Section 12.2.3. The lead frame brings the anode and cathode of each LED chip out on separate pins. This universal pinout arrangement permits the LEDs to be wired in any one of three possible configurations; parallel, series, or series/parallel. The cathode pins are the primary thermal path for heat dissipation. Being an encapsulated device, the Light Bar Module exhibits a high tolerance to mechanical shock and vibration and a mean time between failure that is equivalent to other encapsulated LED devices.

In utilizing a Light Bar Module, a designer should consider it optically and electrically in the same manner as an LED display. These devices utilize GaP high efficiency red, yellow, and green LED chips. All of the LED chips in a single module may be driven simultaneously at 30 mA DC or strobed with peak currents up to 120 mA. To achieve legend viewability in a particular ambient lighting, the

contrast enhancement techniques presented in Chapters 6 and 11 may be applied.

Common practice has been to rely on the incandescent lamp as the primary light source for edge lighted panels. The four major difficulties associated with the use of incandescent lamps are:

1.   Inefficient coupling of the light into the panel.

2.   Optical filtering to achieve a color other than white.

3.   Heat dissipation.

4.   Panel design to accomodate periodic lamp replacement.

Light Bar Modules are available in a variety of area formats, packaged in either a standard single-in-line, SIP, or dual-in-line, DIP, pin configuration. Two of the area formats available are shown in Figure 12.5-3.

12.6   The Use of LED Light Bar Modules in Edge Lighted Panels

A common use of the edge lighting concept is the edge lighted panel used in the construction of control panels. The objective is to illuminate a legend from a remote light source by transmitting the light through the panel in a path that is parallel to the surface of the panel. The basic panel is made of a transparent plastic such as Plexiglas, Lucite, or Lexan. All of the components related to the edge lighting function are contained within the panel thickness, usually 4.32 mm (0.170 inch) to 9.53 mm (0.375 inch). The outer surface of the panel is painted black or some other suitable opaque color as a finish coat to prevent stray light leakage.

The concept of an edge lighted panel is illustrated in Figure 12.6-1. Shown is a typical one piece panel with the legend area to be illuminated unpainted. Light from the remote light source travels through the transparent panel, in similar fashion to a light pipe, and is reflected outwards through the legend.

The design and construction of an edge lighted panel is plagued by a significant problem. The problem is achieving even illumination of all the legends across the face of the panel. The difficulty is basically inefficient light insertion into the transparent panel from the various light sources imbeded within it. Only a small percentage of the flux generated by each light source is actually available for legend illumination.

Figure 12.6-2 shows a typical placement of an incandescent bulb in an edge lighted panel. This placement orients the

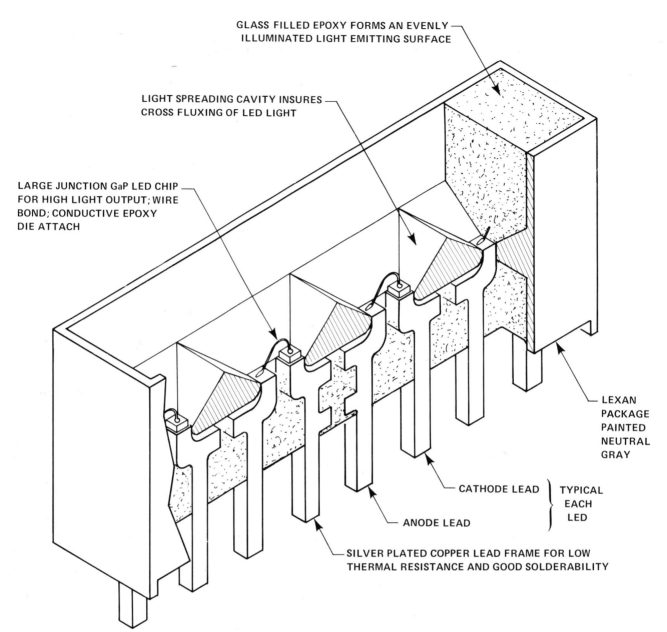

GLASS FILLED EPOXY FORMS AN EVENLY ILLUMINATED LIGHT EMITTING SURFACE

LIGHT SPREADING CAVITY INSURES CROSS FLUXING OF LED LIGHT

LARGE JUNCTION GaP LED CHIP FOR HIGH LIGHT OUTPUT; WIRE BOND; CONDUCTIVE EPOXY DIE ATTACH

LEXAN PACKAGE PAINTED NEUTRAL GRAY

CATHODE LEAD }
ANODE LEAD } TYPICAL EACH LED

SILVER PLATED COPPER LEAD FRAME FOR LOW THERMAL RESISTANCE AND GOOD SOLDERABILITY

Figure 12.5-2. The Assembly Details of a Plastic Encapsulated LED Light Bar Module. A Legend may be Placed Directly onto the Light Emitting Surface Using a Transparent Adhesive.

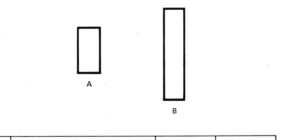

DEVICE	SIZE OF ILLUMINATED AREA	QTY. OF LEDs	PACKAGE
A	3.81 x 8.89 mm (.150 x .375 inch)	2	SIP
B	3.81 x 19.05 mm (.150 x .750 inch)	4	SIP

Figure 12.5-3. LED Light Bar Module Illuminated Area Formats.

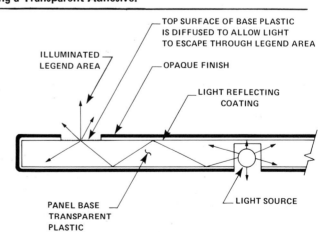

TOP SURFACE OF BASE PLASTIC IS DIFFUSED TO ALLOW LIGHT TO ESCAPE THROUGH LEGEND AREA

ILLUMINATED LEGEND AREA

OPAQUE FINISH

LIGHT REFLECTING COATING

PANEL BASE TRANSPARENT PLASTIC

LIGHT SOURCE

Figure 12.6-1. The Concept of an Edge Lighted Panel.

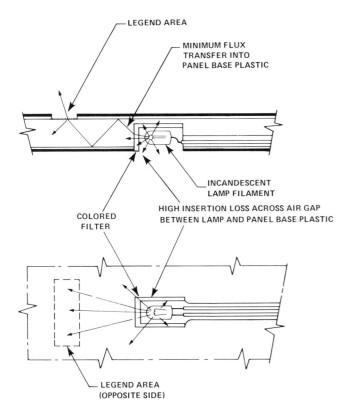

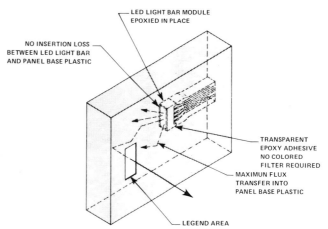

Figure 12.6-3. Typical Installation of an LED Light Bar in an in an Edge Lighted Panel.

Figure 12.6-2. Typical Placement of an Incandescent Lamp in an Edge Lighted Panel.

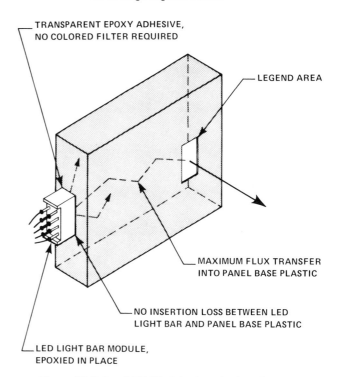

Figure 12.6-4. LED Module Attached to Outer Edge of Panel.

filament so as to achieve the maximum possible flux insertion into the base plastic in the direction of the desired light path. Even so, only a small portion of the flux generated by the filament actually enters the base plastic due to very high insertion losses. If a color like yellow is desired, the white light from the lamp must be otpically filtered to achieve the desired color shift. This may be accomplished by a filter piece inserted between the base plastic and the lamp, tinting the legend, or using a yellow lamp. All of these approaches to color shifting result in light loss.

To improve light insertion into the base plastic, the incandescent lamp could be epoxied in place. This would eliminate the Fresnel losses but would provide an undesirable thermal path between the lamp and the base plastic as well as make periodic lamp replacement rather difficult.

The use of an LED Light Bar Module eliminates all of these problems. The significant advantage is that all of the flux generated by the Light Bar Module is directly coupled into the base plastic in the desired direction of travel, as shown in Figures 12.6-3 and 12.6-4. Since the LED Light Bar does not require periodic preplacement, it may be epoxied in place, eliminating any Fresnel loss between the Module and the base plastic. This presents no thermal problems since heat generation is much less than incandescents. A consideration cost savings is realized since the panel design is much simpler.

The process of achieving even illumination of the various legends in an edge lighted process is called "light balancing." Light balancing of an incandescent edge lighted process is a tedious and costly procedure. Each panel must be light balanced individually by painting with black paint to dim bright legends, a procedure called "black patching," or removal of base plastic to reduce the light path to a bright legend. To enhance a dim legend, reflective paint may have to be descretely added, selective machining may be necessary to improve reflection, or a light pipe may have to be inserted between the legend and the lamp, buried within the base plastic. Light balancing by voltage adjustment is normally not practical.

Light balancing with LED Light Bar Modules is rather simple. An LED is a current operated device, as opposed to an incandescent which is a voltage operated device. Therefore, light balancing is easily accomplished by adjusting the current limiting to each Light Bar Module. This is done by inserting the proper current limiting resistors between each LED module and the common voltage source. Since the LED Light Bar Modules are categorized for light output, the amount of required light balancing is minimized when a panel is assembled with LED devices from one single light output category. No modification to the panel proper should be necessary.

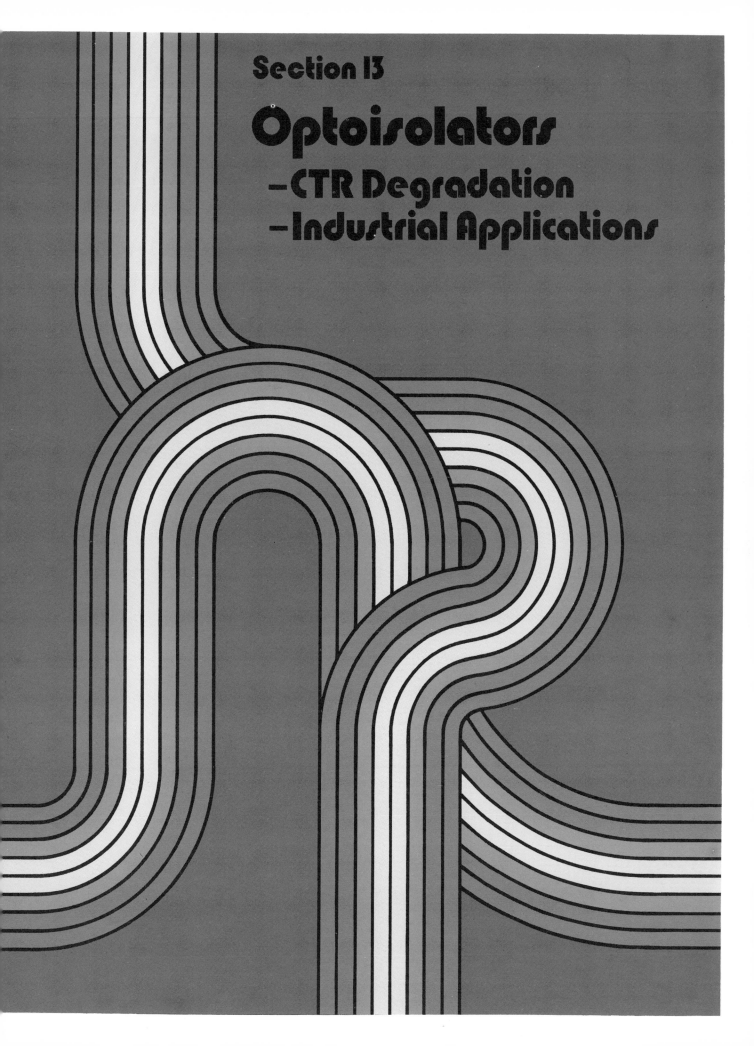

Section 13

Optoisolators
–CTR Degradation
–Industrial Applications

13.0   Optocouplers in Industrial Control Systems

The use of the computer, microprocessors, hard wired process controllers, and other forms of sequential state logic machines in industrial control systems and sophisticated appliances generates special input-output (I/O) interface demands. The new and sophisticated logic electronic system replaces, in some instances, older manual, pneumatic or electromechanical systems. What these older logic systems lacked in speed of response and sophistication was partially compensated by an almost total immunity to electrical stress caused by common mode signals, ground loop currents, or voltage transients which are encountered in the normal operating environment of these systems. Almost all forms of digital electronic logic can be disrupted, or even destroyed, if sufficient noise levels can reach the logic circuits.

Hence, a need exists for an interface unit which can provide an isolated signal path from the electrically noisy and destructive domain of power control switches, actuators, motors, and sensors to the logic electronics. Common interface units which provide the needed isolation are the electromechanical relay, the transformer, and the optically coupled isolator. While each technique has its merits and deficiencies, the optical isolator offers an optimum combination of speed of response with dc − 10 MHz frequency response; high isolation, see Section 3.2.1 − $C_{I-O} < 1$ pf; small package size, six or eight pin DIP; and low cost. These devices are easily interfaced to digital logic and complement the speed and processing efficiency achievable in sequential state machines. Figure 13.0-1 is a block diagram of an optocoupler application in the signal path from the control environment to the logic environment.

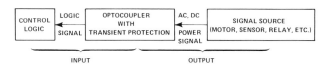

Figure 13.0-1.   Optocoupler Application − Control System to Logic System.

The design considerations for input systems, that is to the logic system as illustrated in Figure 13.0-1, versus design considerations for output systems from the logic electronics, illustrated in Figure 13.0-2, are significantly

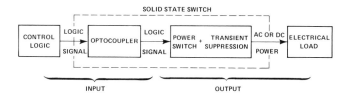

Figure 13.0-2.   Optocoupler Application − Logic System to Control System.

different. Output systems require the digital logic signal to be interfaced to a power control device, such as a power transistor for dc power switching, SCR/TRIAC for ac power switching, an electromechanical relay, or a solid state relay (SSR). In addition to switching power loads, appropriate measures must be taken to protect the power switch in order to prevent reliability or failure problems. Different possible techniques include voltage clamping, snubber (RC) networks, varistors, Transzorbs ®, Zener diodes, zero voltage crossing turn on network, zero current turn off, use of high power factor or real impedance loads, etc. These considerations are fundamental to power switch **design and will not be treated in this section. Section 3.4.6 treats design techniques for transmitting a digital** control signal from a logic source to a power control system.

Input signals to a logic system can be subdivided into two generalized categories: first, those signals which represent the transmission of preconditioned digital logic data and secondly, those signals which represent unconditioned sensor and control signals. Sensor signals could be generated by monitors for flow rate, pressure, temperature, etc. Control signals could be from contact closures on a relay, or float switches, limit or proximity switches, etc.

The first category of signals is interfaced using the techniques described in Section 3.4.6 (digital interface), Section 3.5.7.1 (analog to digital interfaces), Sections 3.6.1 through 3.6.4 and Section 10 (transmission line systems). The second category of signals, those involving primarily the unconditioned inputs, will be treated in the remainder of this section.

13.1   Application Methods for Optocouplers in Industrial Control Systems

The application of optocouplers to the industrial control system environment fulfills a fundamental need: isolation. The possible methods with which to use an optocoupler are simple in principle and practice. To provide the isolation and insulation that is needed between the electrical sensor data source and the command logic, an optocoupler is used with appropriate additional elements to attenuate large signal levels, filter transients, convert ac to dc, and threshold detect. The following optocoupler application areas will be considered in this order: ac voltage sensing, dc voltage sensing, current sensing, threshold control.

13.1.1   AC Voltage Sensing − Wet and Dry

AC voltage sensing is a very common application in the industrial control environment. Since ac electrical power is readily available for use, monitoring for the presence of ac voltage is a must. There are two approaches to use when sensing an ac voltage. These are wet sensing and dry sensing.

Wet sensing is used mainly with electromechanical relay (EMR) contacts, and is defined by the current level required to produce the amount of energy needed to create a plasma (ion-electron gas) around the metal contacts which will reduce or eliminate corrosion or contamination of the contact surface. The amount of energy needed in this application depends upon the metal surface to be cleaned, the environment of the contact (generally air), and many other factors. The amount of energy for "wetting" the contacts in order to clean them varies widely between different types of relays. The relay manufacturer's specifications should be consulted to determine the appropriate wetting current. Wet sensing is utilized when the load current being switched is not sufficient to create a cleansing plasma; otherwise, larger load current switching provides enough energy to clean contacts. If a low amount of energy is used, hydrocarbons will be broken down from the surrounding air, but they will not be cleaned away. Too large of a wetting current is generally not a problem since relays may be designed to handle larger load current conditions. Also, maximum power dissipation for the relay contacts must be obeyed to insure proper relay operation.

Solid state relays, in general, conduct an off-state current (5–10 mA) for their integral snubber (RC) network for transients, or for their zero crossing network for low current turn on feature, or due to the semiconductor device leakage current. This off-state current is not a wet sense current but must be taken into account at low load current values (< 100 mA). Otherwise, the off-state current could approach the load current value and create problems with relay release or small motor power heating.

Dry sensing is simply voltage sensing while conducting negligible current into the sensing branch. The actual load current may be large enough to clean the contacts and no wetting current is needed. Another possible dry sense condition can occur when the load voltage and current is small enough not to produce any significant contaminates on the relay contacts, removing the need for any sizeable sensing current.

13.1.2   AC Voltage Rectification

One of the major concerns of ac voltage sensing is which way to convert the ac voltage to dc voltage. This conversion can be done automatically by the light emitting diode in the optocoupler. The choice of conversion can be either half-wave rectification or full-wave bridge rectification. Full-wave rectification is not possible in most applications because, generally, there is no center tap on relay coils, motor windings, etc., to provide the needed symmetrical voltages. Half-wave rectification is certainly simpler and uses less components, but this technique provides only a pulse of energy, or information, every other half period of the ac voltage (@ 60 Hz, T/2 = 8.33 ms). This may require

the detection circuit, after the optocoupler, to be more tightly designed for tolerance variations to insure that missing information is detected quickly. Figure 13.1.2-1a shows a typical half-wave rectification circuit.

With full-wave bridge rectification, the pulses of energy or information are present every half period of the ac voltage. This provides quicker sensing of the presence of information as well as being an easier waveform to filter to dc for threshold detection. An illustration of full-wave bridge rectification is given in Figure 13.1.2-1b.

13.1.3   Detection Techniques

Threshold detection techniques can be applied prior to the input or after the output of the optocoupler. There are advantages and disadvantages to either method. If detection takes place at the output of the optocoupler, then the rectified ac can be adequately filtered with an RC circuit as shown in Figure 13.1.2-1a. The time constant of the RC circuit should be at least five times as long as the period of the ac voltage. Filtering on the output is more effective and allows the use of less expensive components. After filtering,

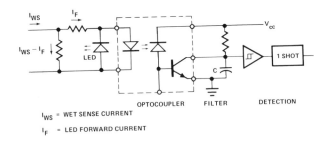

I_{WS} = WET SENSE CURRENT

I_F = LED FORWARD CURRENT

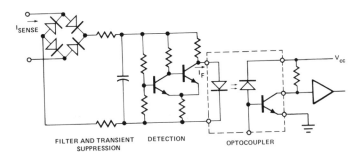

Figure 13.1.2-1.

a) **Half-Wave Rectification with Filtering and Detection After Optocoupler.**

b) **Full-Wave Bridge Rectification with Filtering and Detection Before Optocoupler.**

a threshold level can be established via a logic gate, or a Schmitt trigger, or both. The advantage of a Schmitt trigger circuit is that it provides noise immunity for the subsequent logic circuits. The main disadvantage of detecting at the output is that the current transfer ratio, CTR, of an optocoupler is a nonlinear function of input current, as well as the CTR variance from optocoupler to optocoupler.

The main advantage to having detection prior to the input of the optocoupler is that of basically eliminating the CTR variation influence. Also, filtering the rectified ac voltage at the input has the advantage of helping to provide additional transient filtering for the LED of the optocoupler.

Input power signals to the optocoupler are brought to the device via transmission lines that are adjacent to other power signals, and as a result, capacitive coupling can introduce high common mode voltage, or transients, to the optocoupler. With detection either prior to or after the optocoupler, the optocoupler provides high common mode transient protection (> 500 V/μs), as well as high voltage insulation capability (3,000 Vdc), see Sections 3.2.1 and 3.2.2 respectively. The common mode transient rejection capability is the key factor which provides clean signal transmission from power system environment to logic environment. Insulation voltage capability of the optocoupler insures protection from destructive voltage magnitudes entering the logic control circuitry. Differences in ground potential at different system/equipment locations which create system malfunctions or failure are easily eliminated with the optocoupler.

13.1.4   DC Voltage Sensing

DC voltage sensing is identical to ac voltage sensing techniques with the exception of the need to convert ac voltage to dc voltage. DC wet or dry sensing is determined in the same fashion as is ac wet or dry sensing. Simple illustrations of dc voltage sensing are given in Figure 13.1.4-1.

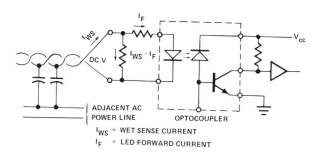

Figure 13.1.4-1.   DC Voltage Sensing with Common Mode Coupling from Adjacent AC Power Line.

13.1.5   Current Sensing

Current sensing is not a common, or practical, method in which to sense a load on-off condition when using optically coupled isolators. Only in the case of very small load current of the magnitude of the typical forward current rating for the LED emitter of an optocoupler can true current sensing occur, as shown in Figure 13.1.5-1. This low load current is rare in the industrial control environment.

Otherwise, current sensing is actually a variation of voltage sensing. The problem with current sensing for large currents is that of developing sufficient voltage across the series resistor, of Figure 13.1.5-2, in order to turn on the LED emitter ($V_F = 1.75$V max). This voltage is subtracted from the needed voltage at the load and can present a problem. In particular, a motor load which has insufficient voltage, at rated current, will conduct more current in order to maintain an operating power level. However, the motor can over heat quickly and fail with excessive load current demand. The advantage of the optocoupler isolation in current sensing applications is that it allows the input to float at any voltage level up to the maximum input-output voltage. Possible configurations for current sensing are given in Figures 13.1.5-1, 13.1.5-2, and 13.1.5-3.

13.1.6   Threshold Detection

Threshold control is an important application for data monitoring or direct control of industrial control systems. Threshold detection provides a means by which to accurately measure a quantity needed for control purposes. An example of threshold control would be to monitor line voltage to an ac motor to protect the motor from over

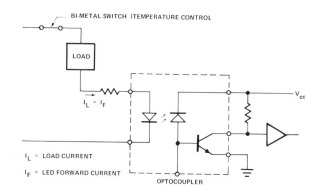

Figure 13.1.5-1.   True Current Sensing, $I_L = I_F$.

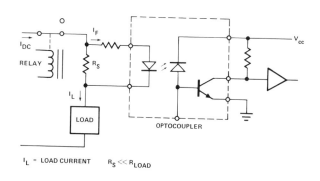

Figure 13.1.5-2.   DC Current Sense Floating Input.

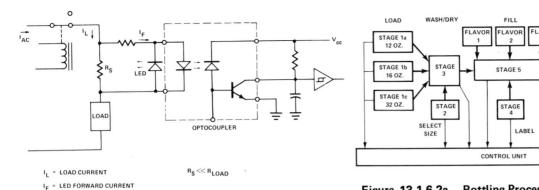

I_L = LOAD CURRENT $R_S \ll R_{LOAD}$

I_F = LED FORWARD CURRENT

Figure 13.1.5-3.   AC Current Sense with Half-Wave Rectification.

heating if the line voltage should drop to an unsafe operating voltage. Figure 13.1.6-1 demonstrates this particular example. Another application for threshold control within the industrial control environment would be to protect process sequence control. Again, if the line voltage at a certain point in the process should fall to an unsafe level because of a "brown-out" condition or accidental short on the line, the threshold detection would shut down the operation to prevent incorrect sequences occurring. This early shut down could provide for a proper start sequence when power returns to normal. An illustration of threshold control protection is given in Figure 13.1.6-2.

Threshold detection not only applies to power circuit monitors, but to sensor devices as well. Sensors, such as limit or proximity switches, temperature controlled bi-metal switches, or flow indicators will transmit their digital signals over lines which have troublesome electrical noise coupled onto them. Threshold level detection will provide the needed signal to noise discrimination for the subsequent logic system. If a sensor should present an open circuit in one of its states, coupled noise into the signal line could

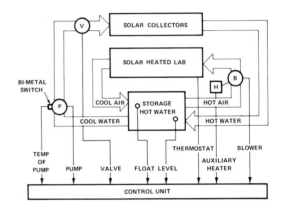

Figure 13.1.6-2a.   Bottling Process Application Where Optocouplers with Threshold Detection can Monitor Each Stage to Determine Whether Equipment has Functioned Properly to Insure Employee Safety, as well as Prevent Waste of Material or Damage to Equipment.

Figure 13.1.6-2b.   Solar Heating System Where Optocouplers with Threshold Detection Interface Data from System Components to Control Unit While Eliminating 115 V AC Electrical Noise.

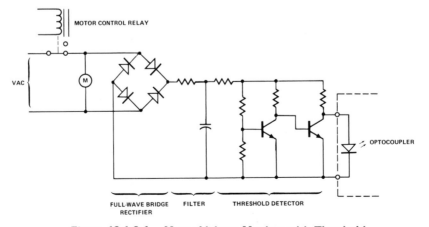

Figure 13.1.6-1.   Motor Voltage Monitor with Threshold Detection.

easily give erroneous results. Hence, a threshold level circuit is a must in an electrically noisy environment.

It should be noted that commercially available operational amplifier threshold level detectors are difficult to apply before the isolation is performed between the control system and logic system. Unless extra power supplies are employed, which is an undesirable situation, the signal to the operational amplifier is also the supply voltage. A very difficult and unpredictable situation to design for reliable operation. However, after isolation has been performed, threshold detection is easy, but the variations of the optocoupler will now have to be considered. In Section 13.3, some design examples are given to guide the designer.

13.2   Protection Considerations for Optocouplers in Industrial Control Systems

Optocoupler inputs which are connected to the power control circuits will see transients that can damage the LED emitter by current overstress, by excessive power dissipation, or by exceeding an optocoupler insulation voltage specification. Capacitive coupling between power lines will introduce a "leakage" input current which must be taken into account. This coupled current is difficult to determine because it is dependent directly on the specific power line locations. Also, if electromagnetic relay contacts are monitored by an optocoupler, suitable bounce filtering must be done to prevent false information from being transmitted. Over voltage and over current protection at the input of an optocoupler can be provided to protect the device. All of these protection considerations will be elaborated in the following sections.

13.2.1   Line Transients

Line transients need to be attenuated at the input of an optocoupler mainly to protect the light emitting diode from being overstressed or destroyed. Transients are always present on power signals and it is sometimes extremely difficult to predict their occurence or to remove them from the line. Transients can be in the form of a sharp spike of voltage, pulse of voltage, or a short burst of damped, oscillating voltage, as observed in Figure 13.2.1-1a-c.

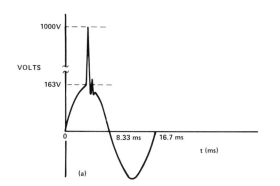

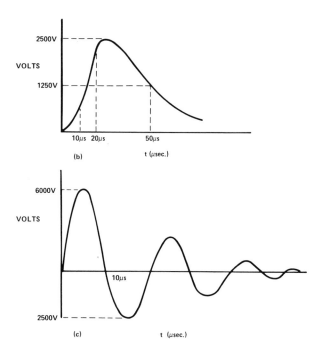

Figure 13.2.1-1.   Typical Transient Waveforms.
  a) **Worst Case Timing Voltage Spike**
  b) **Transient Energy Pulse**
  c) **Underdamped Oscillatory Transient**

Generally, transient generation comes from an electromagnetic relay or solid state relay which switches a motor or other low power factor load. The sudden change in load current generates the high transient voltage waveform via the relationship, $V = L(di/dt)$, where L equals inductance of load, i equals load current, t is time, and V is the voltage transient.

13.2.2   Transient Suppression Techniques

Three similar methods of attenuating transients are shown in Figure 13.2.2-1a-d. The resistive network of Figure 13.2.2-1a provides attenuation for the monitored signal from the power system, but also provides some attenuation to transients. However, this resistive attenuator does not provide the best filtering possible because the filtering impedance needs to be small in comparison to the line transient impedance. A better transient filter utilizing three components would be the resistive-capacitive combination for Figure 13.2.2-1b. The capacitor bypasses any transients to the input of the optocoupler since the capacitor and resistors form a low pass filter (low impedance voltage divider) to the high frequency components of the voltage transient. A third possible combination of resistive and capacitive components which can be used with ac voltage only is that given in Figure 13.2.2-1c. This arrangement provides for less power dissipation in the attenuation filter elements. A problem that may occur with this con-

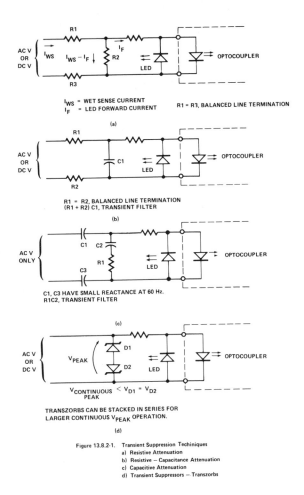

Figure 13.2.2-1. Transient Suppression Techniques.
  a) Resistive Attenuation
  b) Resistive — Capacitance Attenuation
  c) Capacitive Attenuation
  d) Transient Suppressors — Transzorbs ®

figuration is that if the capacitors should fail in a shorting manner, the optocoupler would be destroyed, and if the transient voltage was too large, it could be passed directly to the logic circuits inflicting more damage. This network will filter transients but not as well as the configuration of Figure 13.2.2-1b.

Commercial transient suppressors are available which can aid in limiting transient influences to a circuit. Such devices are Zener diodes, metal oxide varistors (MOV ®), Transzorbs ®, and the like. If the waveform of the transient is known, then the selection of a suppression device is made easier because the transient energy is approximately known, as well as the speed needed to respond to the transient. Zener diodes are slower in response to a transient than a varistor or the extremely fast Transzorbs ®. When the transient waveform is unknown, then the choice of a suppressor can best be judged by the amount of energy that is contained in an inductive load ($\frac{1}{2}LI^2$), or a capacitive

load ($\frac{1}{2}CV^2$). I, or V, is the peak value possible through/across the inductive (L)/capacitive (C) load. Selection of a **Zener diode, MOV ® device, or Transzorb ®** must be made with the continuous peak operating voltage less than the reverse standoff voltage, see Figure 13.2.2-1d.

13.2.3   Capacitive Coupling

Capacitive coupling from adjacent electrical power lines to input lines for an optocoupler creates an additional input "leakage" current. It is difficult to account for this current since it is largely dependent upon the physical proximity of neighboring wires which, generally, is not known. Instances have been observed where 10 VRMS of common mode ac voltage has been coupled into the open input lines to an optocoupler. This additional voltage/current can upset a threshold level design. A larger threshold voltage or current would have to be determined to detect the desired signal level. For example, Figure 13.2.3-1 illustrates the situation where a twisted pair signal transmission line connects between a supply voltage, via a remotely located switch, to an LED emitter of an optocoupler. This transmission line parallels power circuit lines which capacitively couple current into the switch line. When the switch is closed, the coupled current is added to the signal current and is not a problem. However, when the switch is opened, along with a transmission line which is a leaky line at 100kΩ impedance, 100μA of coupled ac current would develop 10 V ac on the line if the LED was not connected. Since the LED is connected and presents a lower impedance than the line, the coupled current will flow through the LED. In turn, the LED will partially turn on which may cause a false output from the optocoupler.

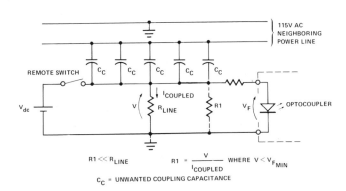

Figure 13.2.3-1.   Capacitively Coupled Current from Neighboring Power Line.

An elementary method by which to eliminate the influence of capacitive coupling is to parallel the line with resistor R1. Coupled current will shunt through R1 instead of the LED. The value of R1 is determined by maintaining the open line voltage sufficiently below the minimum forward voltage, V_F, of the LED. Use of optocouplers which have a

13.6

higher forward voltage will assist in this problem. Beware that R1 is not too low in value in order to prevent excessive power from being dissipated in R1 when the signal is present. Threshold circuits can be applied to this problem instead of R1, but they will be more complex.

13.2.4   Bounce Filtering

Besides transient filtering on the input to the optocoupler, bounce filtering will be needed when electromagnetic relays are used. When relay contacts swing closed, they have momentum which is not transferred completely to the stationary contact and, hence, the moving contacts recoil, or bounce. The amount of time in which the bouncing effect occurs depends mainly on the type of construction of the electromechanical relay. Bounce time can vary as long as approximately 10 ms. These interruptions of the input signal can be integrated to prevent false information from being sensed. A simple method for dc voltage bounce filtering can be to place a capacitor across the input in order to charge the capacitor to the signal voltage and maintain that charge long enough to last until the bouncing effect has ceased. The previous transient filter network of Figure 13.2.2-1b will allow the input capacitor to double in function to serve as a bounce filter for the electromechanical relay contacts. Contact bounce filtering with ac voltage is not as direct a method as in the dc voltage case. With interrupted ac voltages, the only simple means of contact bounce filtering is via the ripple filter network. The

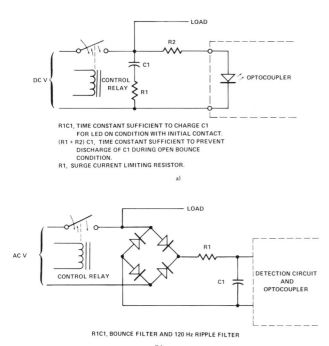

R1C1, TIME CONSTANT SUFFICIENT TO CHARGE C1
FOR LED ON CONDITION WITH INITIAL CONTACT.
(R1 + R2) C1, TIME CONSTANT SUFFICIENT TO PREVENT
DISCHARGE OF C1 DURING OPEN BOUNCE
CONDITION.
R1, SURGE CURRENT LIMITING RESISTOR.

a)

R1C1, BOUNCE FILTER AND 120 Hz RIPPLE FILTER

(b)

Figure 13.2.4-1.   Bounce Filtering Techniques.
a) DC Bounce Filtering
b) AC Bounce Filtering

ripple filter network will charge to near the final dc voltage level at which it operates and maintains that charge until the bouncing ceases. Examples of bounce filtering are given in Figure 13.2.4-1.

13.2.5   CTR Degradation

The expected lifetime of an optocoupler can be enhanced by proper consideration of protection or design techniques. As was mentioned earlier, transient voltage or current must be filtered from the circuit or they can destroy the optocoupler via excessive power dissipation in the LED emitter. However, if large steady state or peak forward current is present in the LED emitter, the emitter could be under an excessive current stress condition. The larger the forward current, the higher the current stress in the device which results in a greater rate of reduction of light output from the LED emitter. In turn, this directly reduces the current transfer ratio, CTR, of the optocoupler. This CTR degradation over time is explained in detail in Sections 3.4 through 3.4.4 with calculation procedures in Section 3.4.5. A basic method to regulate forward current through the LED emitter is given in Figure 13.3.2-1. This design allows excess emitter forward current to be shunted away via a parallel path provided by the Q1 transistor. A complete explanation of the design of this circuit is given in Section 13.3.2.

13.2.6   Insulation

Maximum allowable input to output voltage insulation must be observed in order to prevent damage occurring to the optocoupler, or to logic devices isolated by the optocoupler. Hewlett-Packard optocouplers are designed to withstand a specified minimum V_{I-O} of 3,000V dc.

However, it is important to recognize that operation at high input to output voltage will electrically stress the device more than at lower V_{I-O}. Corona can possibly occur within the insulation material resulting in the generation of ions which can degrade the quality of the insulation material. This is a cumulative effect over time. Hence, it is advantageous to operate the optocoupler at as low a V_{I-O} as possible in order to extend the optocoupler lifetime.

13.2.7   Power Dissipation

Major power dissipation of the control signal to the input of the optocoupler occurs external to the optocoupler. The input attenuation/transient circuit will dissipate the input power. External power dissipation keeps the optocoupler at a lower temperature for better electrical performance. Of course, design considerations must insure that input signal power does not exceed the maximum input power dissipation for the LED emitter. Also, output power dissipation in the amplifier stage of the optocoupler must be kept below the rated maximum level. A problem which is often overlooked is the power dissipation that will occur on a printed circuit board with a collection of input channels. For example, if an input board has 32 channels (@ 230V ac) and each channel must dissipate approximately 2.3 watts, then the total dissipation is 73.6 watts. This dissipated power can present a problem in keeping the ambient temperature within the equipment at a reasonable temperature level.

13.2.8   Overvoltage Protection

Overvoltage protection may need to be provided to an optocoupler to insure against accidental application of voltages larger than the optocoupler input is designed to withstand. Power transients are one source of an applied overvoltage condition. Transient suppression techniques were discussed in Section 13.2.2. However, another possible application of overvoltage is through an accidental short of input lines to higher voltage lines. Protection can be provided for dc overvoltage or reverse polarity, by using a Zener diode as shown in Figure 13.2.8-1a. The maximum dc input voltage will clamp at the Zener voltage. If the polarity should be reversed, the Zener would conduct as an ordinary diode. However, care must be taken to prevent excessive power dissipation from occurring in the Zener diode.

AC overvoltage protection can be easily applied for half-wave or full-wave bridge rectification as shown in Figure 13.2.8-1b-c. With the half-wave rectification circuits, the Zener diode protection gives additional protection to the LED emitter. The full-wave bridge rectification circuit of Figure 13.2.8-1c, which uses Zener diodes instead of regular diodes, provides a convenient method with which to clamp the input ac voltage to a maximum value depending upon the Zener diode voltage. As illustrated in Figure 13.2.8-1c, a particular ac cycle load current, I_L, is shown. A voltage equation can be written relating the forward, V_F, and reverse bias, V_R, Zener diode voltages and the load voltage, V_L: $V_R = V_L + V_F$. As the ac voltage magnitude, V_{AC}, is increased, the load voltage V_L is also increased. When $V_L + V_F$ equals V_Z of the Zener diode ($V_R = V_Z$ condition), the Zener diode conducts and clamps the input ac voltage to $V_{AC} = V_Z + V_F$. Of the two back biased Zener diodes in the bridge circuit, the one which has the lower V_Z will begin to Zener first. The clamping action is symmetrical for the opposite portion of the cycle. Again, caution must be exercised to prevent excessive power dissipation within the Zener diodes.

13.3   Design Examples

Two design examples are given in the following sections.

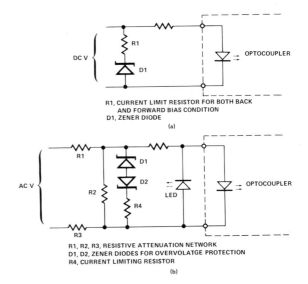

R1, CURRENT LIMIT RESISTOR FOR BOTH BACK
AND FORWARD BIAS CONDITION
D1, ZENER DIODE

(a)

R1, R2, R3, RESISTIVE ATTENUATION NETWORK
D1, D2, ZENER DIODES FOR OVERVOLATGE PROTECTION
R4, CURRENT LIMITING RESISTOR

(b)

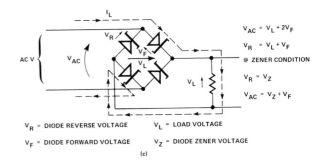

$$V_{AC} = V_L + 2V_F$$
$$V_R = V_L + V_F$$
@ ZENER CONDITION
$$V_R = V_Z$$
$$V_{AC} = V_Z + V_F$$

V_R = DIODE REVERSE VOLTAGE V_L = LOAD VOLTAGE

V_F = DIODE FORWARD VOLTAGE V_Z = DIODE ZENER VOLTAGE

(c)

Figure 13.2.8-1.   Overvoltage Protection.
   a) Zener Diode Protection — DC V.
   b) Zener Diode Protection — AC V.
   c) Zener Diode Protection and Full-Wave Bridge Rectifier — AC V.

Design example 1 is a voltage sensing application, while design example 2 is a current sensing application.

13.3.1   Design Example 1

A 115V ac, 60 Hz line monitor for 50% threshold line drop with **hysteresis**, transient protection, and a TTL load is shown in Figure 13.3.1-1.

Design of this 115V ac line monitor begins with an analysis of the threshold detection circuit with hysteresis. Analysis is direct in manner with few initial assumptions.

The operation of this circuit can be seen from Figures 13.3.1-2 and 13.3.1-3. Initially, Q1 is off and Q2 is on. As the voltage V' is increased, I_{IN} increases, and when the threshold voltage for Q1 is reached, Q1 turns on, turning off Q2. At this point, I_{IN} decreases abruptly to a new, lower value. Further increase in V' will continue to increase I_{IN} again. With the series input resistance, R5, the circuit will exhibit hysteresis about this voltage threshold, V'_{TH}.

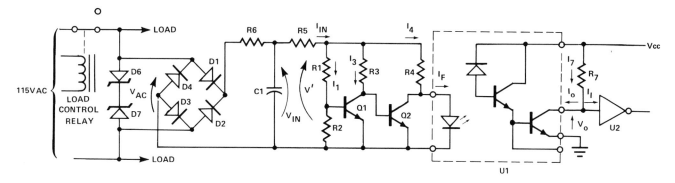

Figure 13.3.1-1. Line Monitor Circuit.

Figure 13.3.1-3 illustrates **the hysteresis effect** with this threshold circuit in the following manner. As V_{IN} increases from zero volts toward V_{TH+}, the voltage V′ also increases following the relationship defined by the path OB until V′ equals V'_{TH}, point B. At this point, V′ discontinuously increases from point B to D, due to the corresponding decrease in I_{IN} of Figure 13.3.1-2 while $V_{IN} = V_{TH+}$ remained constant. Further increase in V_{IN} results in V′ following a slightly different relationship defined by path DA. Steady state operation is at point A. If the V_{IN} voltage should return to zero, V′ would traverse path AC until $V′ = V'_{TH}$ again. However, V_{IN} is at a lower value than before, $V_{IN} = V_{TH-}$. Opposite to the other transition, V′ decreases discontinuously from point C to E, at constant V_{IN} due to a sudden increase in I_{IN}. As V_{IN} continues to decrease to zero, V′ follows the relationship of path EO. Hence, **hysteresis has been provided.**

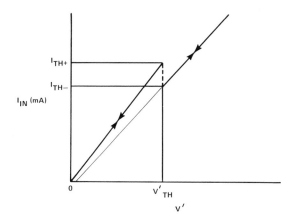

Figure 13.3.1-2. I_{IN} Versus V′.

Selection of the desired properties for this circuit can be made. The minimum forward current, $I_{F_{MIN}}$, acceptable for the LED emitter of the optocoupler prior to turn off, point C of Figure 13.3.1-3, must be specified. The upper and lower voltage threshold points for V_{IN}: V_{TH+}, V_{TH-} respectively are selected. Correspondingly, the V_{IN} hysteresis is defined as $V_H \equiv V_{TH+} - V_{TH-}$. The larger the **amount of hysteresis desired,** the closer the magnitudes must be for V'_{TH} and V_F, the forward voltage across the LED

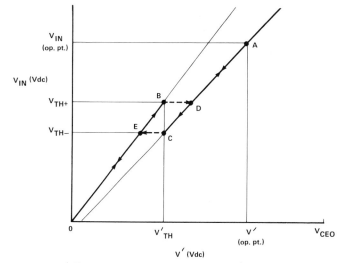

Figure 13.3.1-3. V_{IN} Versus V′.

emitter in the optocoupler. Also, the larger the ratio is of R5/R4, the **larger the hysteresis can be.** When determining the resistor values, care must be taken to insure that the loop gain of this detector circuit is always greater than unity to maintain the **hysteresis feature.**

Assume Q1 and Q2 are 2N3904 transistors with $V_{CEO} = 40V$ dc and $h_{FE_{SAT}} \geqslant 10$. With $V_{ac_{peak}} = 163V$, the steady state operating voltage for V_{IN} was chosen to be 150V, point A. The basic available parameters for selection were given the following values.

$$V_{TH+} \quad = 100V, \quad 67\% \Bigg\}$$
$$V_{TH-} \quad = 75V, \quad 50\% \quad \text{of } V_{IN} = 150V$$

$$V_H \quad = 25V$$

$$I_{4_{min}} \quad = 3.5\ mA \quad @ \quad \text{point C } (I_{F_{min}})$$

$$h_{FE_{SAT}} \quad \geqslant 10$$

$$\alpha \quad = 11$$

where $\alpha \equiv I_1/I_{B1}$ is the ratio of current I_1 to base current, I_{B1}, of transistor Q1.

Determination of R1 through R7 and C1 can be found by the following analysis and equations. Since IF_{min} is a fixed requirement at the lower input voltage threshold, $V_{IN} = V_{TH-}$, point C of Figure 13.3.1-3, design of this circuit should be centered around this V_{TH-} parameter. The amount of hysteresis voltage that was chosen determines the upper voltage threshold, V_{TH+}, point B.

For a first approximation, the currents I_1 and I_3 are neglected, then the sum of R4 + R5 is known.

$$R4 + R5 = \frac{V_{TH-} - V_F}{IF_{min}} \qquad (13.3.1\text{-}1)$$

$$= \frac{75V - 1.5V}{3.5 \text{ mA}} \text{ k}\Omega$$

$$R4 + R5 = 21\text{k}\Omega$$

At the V_{TH-} threshold point, the hysteresis equation can be written as:

$$V_H = R5 \left[\frac{V_F - V_{CE2}}{R4} - \frac{V_{BE2} - V_{CE1}}{R3} \right] \qquad (13.3.1\text{-}2)$$

With this equation, the ratio can be obtained for R5/R4 and knowledge of the sum of R4 + R5 will yield each value. The additional unknown, R3, must be expresed in terms of R4. This can be done as follows. When Q1 is switched off and Q2 switches on at $V_{IN} = V_{TH-}$, the voltage across R3 and R4 will be identical at this threshold point since

$V_{BE2} = V_{CE2}$. Equating these two voltages will give the R3 dependency on R4.

$$I_{B2} R3 = h_{FE} I_{B2} R4 \qquad (13.3.1\text{-}3)$$

$$R3 = h_{FE} R4 \qquad (13.3.1\text{-}4)$$

Substitution of R3 into the hysteresis equation yields R5/R4 ratio.

$$\frac{R5}{R4} = \frac{V_H}{\left[V_F - V_{CE2} - \frac{V_{BE2} - V_{CE1}}{h_{FE}} \right]} \qquad (13.3.1\text{-}5)$$

$$= \frac{25V}{1.5V - 0.2V - \left(\frac{0.7V - 0.2V}{10} \right)}$$

$$\frac{R5}{R4} = 20$$

Consequently,

$$R4 + R5 = 21\text{k}\Omega$$

$$\frac{R5}{R4} = 20$$

$$R4 = 1\text{k}\Omega, \ R5 = 20\text{k}\Omega$$

and

$$R3 = h_{FE} R4$$

$$= 10 \ (1\text{k}\Omega)$$

$$R3 = 10\text{k}\Omega$$

R1 can be determined as a function of R3 by using the relationship for α at the V_{TH-} threshold point $(V_{BE1} = V_{CE1})$.

$$I_1 = \frac{V' - V_{BE1}}{R1} = \alpha I_{B1} \qquad (13.3.1\text{-}6)$$

$$= \alpha \frac{I_3}{h_{FE}}$$

Substituting for I_3,

$$\frac{V' - V_{BE1}}{R1} = \frac{\alpha}{h_{FE}} \left(\frac{V' - V_{CE1}}{R3} \right)$$

$$R1 = \frac{h_{FE}}{\alpha} R3 \qquad (13.3.1\text{-}7)$$

$$= \frac{10}{11} (10\text{k}\Omega)$$

$$R1 = 9.1\text{k}\Omega$$

R2 can be calculated from the voltage, V_{BE1}, needed to turn on or off Q1. The proportionality between R1 and R2 is

$$\frac{R2}{V_{BE1}} = \frac{R1}{V' - V_{BE1}} \qquad (13.3.1\text{-}8)$$

13.10

$$R2 = R1 \frac{V_{BE1}}{V' - V_{BE1}} \qquad \text{(13.3.1-9)}$$

With $V_{BE1} = 0.65V$ and V' at V_{TH-} threshold, point C of Figure 13.3.1-3, is

$$V' = V_F + R4 \, (I_{F_{min}}) \qquad \text{(13.3.1-10)}$$

$$= 1.5V + 1k\Omega \, (3.5 \text{ mA}) \qquad \text{volts}$$

$$V' = 5V$$

then

$$R2 = 10k\Omega \frac{0.65V}{5V - 0.65V}$$

$$R2 = 1.49k\Omega \qquad \text{use } 1.5k\Omega \text{ resistor}$$

The actual value for V' above is slightly lower in value ($\approx 15\%$) because of the earlier approximation of neglecting I_1 and I_3 while determining $R4 + R5$. A nearly exact equation can be written for the relationship between V_{IN} and V' of Figure 13.3.1-3 along path DA which will enable the calculation for the power dissipation rating of R5 and the maximum I_F in the LED emitter at point A of Figure 13.3.1-3. The equation represents the voltage difference across R5.

$$\qquad \text{(13.3.1-11)}$$

$$V_{IN} - R5 \left(\frac{V' - V_{BE1}}{R1} + \frac{V' - V_{CE1}}{R3} + \frac{V' - V_F}{R4} \right) = V'$$

Solving for V' ,

$$V' = \frac{V_{IN} + R5 \left(\dfrac{V_{BE1}}{R1} + \dfrac{V_{CE1}}{R3} + \dfrac{V_F}{R4} \right)}{1 + R5 \left(\dfrac{1}{R1} + \dfrac{1}{R3} + \dfrac{1}{R4} \right)} \qquad \text{(13.3.1-12)}$$

Substituting the known values for R and using $V_{BE1} = 0.7V$, $V_{CE1} = 0.2V$,

$$V' = 0.04V_{IN} + 1.3V \qquad \text{(13.3.1-13)}$$

At $V_{IN} = 150V$,

$$V' = 0.04 \, (150V) + 1.3V$$

$$V' = 7.3V$$

The power dissipated in R5 at the steady state operating point A, is

$$P_{R5} = \frac{(V_{IN} - V')^2}{R5} \qquad \text{(13.3.1-14)}$$

$$= \frac{(150V - 7.3V)^2}{20,000\Omega} \qquad \text{watts}$$

$$P_{R5} = 1.02 \text{ watts} \qquad \text{use 2 watt resistor}$$

The maximum current through the LED emitter of the optocoupler is

$$I_{F_{max}} = \frac{V'_{max} - V_F}{R4} \qquad \text{(13.3.1-15)}$$

$$= \frac{7.3V - 1.5V}{1k\Omega} \qquad \text{mA}$$

$$I_{F_{max}} = 5.8 \text{ mA}$$

The value for R6 and C1 can be determined as follows. The input current, I_{IN}, to the threshold circuit at $V_{IN} = 150V$ is

$$I_{IN} = \frac{V_{IN} - V'}{R5} \qquad \text{(13.3.1-16)}$$

$$= \frac{150V - 7.3V}{20k\Omega} \qquad \text{mA}$$

$$I_{IN} = 7.14 \text{ mA}$$

With nominal line voltage of 115V ac,

$$V_{ac_{peak}} = \sqrt{2} \, (115V) = 163V \qquad \text{(13.3.1-17)}$$

At the chosen operating point of $V_{IN} = 150V$, $I_{IN} = 7.14$ mA, R6 is determined from

$$R6 = \frac{V_{ac_{peak}} - V_{D1} - V_{D2} - V_{IN}}{I_{IN}} \qquad \text{(13.3.1-18)}$$

$$= \frac{163V - 0.7V - 0.7V - 150V}{7.14 \text{ mA}} \qquad k\Omega$$

$$R6 = 1.62k\Omega \qquad \text{use } 1.6k\Omega \text{ resistor}$$

The value for R6C1 low pass filter break point at 12 Hertz with approximately 20db attenuation at 120 Hz.

$$C1 = \frac{\tau}{R6} \qquad (13.3.1\text{-}19)$$

$$= \frac{88.3ms}{1.6k\Omega} \qquad \mu f$$

$$C1 = 55.2\mu f \qquad \text{use } 60\mu f \ @ \ 250V \ dc$$

D6 and D7 are 1N6072A transient suppression diodes (Transzorbs®). D1, D2, D3, and D4 are a quad rectifying diode package or 1N4004 general purpose rectifiying diodes with sufficient peak reverse voltage capability.

With U1, a 6N138 optocoupler, and U2, a 7404 TTL unit load, the value for R7 can be evaluated as follows. To maintain the output voltage, V_o, in saturation, the condition of $I_o \geq I_7 - I_I$, as represented in Figure 13.3.1-1, must be maintained. I_o is related to I_F by the optocoupler current transfer ratio, CTR, via the equation

$$I_o = I_F \frac{CTR}{100} \qquad \text{(CTR in \%)} \qquad (13.3.1\text{-}20)$$

The range for R7 can be determined by accounting for the power supply variations and the minimum, maximum variations for V_o, I_o, I_I of the optocoupler and TTL gate load. This range of possible values for R7 is given by the following two formulas. See Section 3.4.6 for a more thorough consideration of the optocoupler load resistor.

$$R7 \leq \frac{V_{cc_{min}} - V_{IH}}{I_{OH_{max}} + I_{IH}} \qquad (13.3.1\text{-}21)$$

and

$$R7 \geq \frac{V_{cc_{max}} - V_{OL}}{I_{OL_{min}} + I_{IL}} \qquad (13.3.1\text{-}22)$$

where

$$I_{OL_{min}} = I_{F_{min}}\left(\frac{CTR_{min}}{100}\right) \qquad (13.3.1\text{-}23)$$

Given the following values for the above parameters, the R7 range is

V_{cc}	= 5.0V ± 5%	V_{IH}	= 2.4V
$I_{F_{min}}$	= 3.5 mA	V_{OL}	= 0.4V
CTR_{min}	= 300%	I_{IH}	= 40μA
$I_{OH_{max}}$	= 250μA	I_{IL}	= −1.6 mA
$R7 \leq 11.9 \, k\Omega$		$R7 \geq 545\Omega$	

Chose R7 to be 560Ω, 5% resistor.

13.3.2 Design Example 2

A 115V ac current sense circuit with threshold detection, LED transient protection, and a unit TTL gate load for monitoring heating element fail–open condition is illustrated in Figure 13.3.2-1.

Design of this current sense threshold monitor circuit will begin by specifying the desired minimum and maximum forward LED current, $I_{F_{min}}$, $I_{F_{max}}$ respectively. Also, V_{IN} is chosen for the value at the threshold voltage, V_{TH}, and at the normal operating voltage level. Initial values selected for these parameters in this example are:

$I_{F_{min}}$	= 1.6 mA	V_{TH}	= 9V (¾ of V_{IN})
$I_{F_{max}}$	= 5.0 mA	V_{IN}	= 12V

R4 is determined by requiring Q1 (2N3904) to conduct current in the collector path when $I_F > 5$ mA.

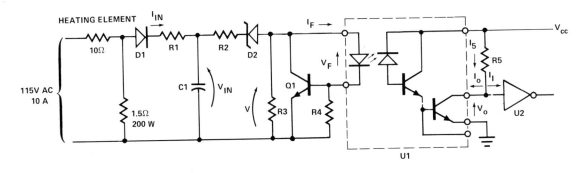

Figure 13.3.2-1. Current Sense Monitor.

$$R4 = \frac{V_{BE1}}{I_{Fmax}}$$ (13.3.2-1)

$$= \frac{0.65V}{5mA} \qquad k\Omega$$

$$R4 = 130\Omega$$

The R3 resistor was provided to allow the Zener diode, D2, leakage current to be shunted to the return path without passing through the LED emitter. Should this leakage current be less than $25\mu A$, R3 can be eliminated. Consider, the leakage current is large, say $50\mu A$. As V_{IN} increases or decreases depending upon the line current, the condition of a partially biased Zener diode can occur. Excessive leakage current magnitude could cause the LED to turn on prematurely as V_{IN} increases, or remain on longer than necessary as V_{IN} decreases. Allowing 1.25V to appear at the anode of the LED without R3 could possibly cause subtle difficulties. Consequently, R3 can be generally evaluated by:

$$R3 = \frac{V}{I_{leakage}} \qquad \text{(let V = 1V)} \qquad (13.3.2-2)$$

$$= \frac{1V}{.050 \ mA} \qquad k\Omega$$

$$R3 = 20k\Omega$$

R2 can be calculated from the threshold condition for V_{TH} and the I_{Fmin} condition. The voltage, V, across R4 and the LED at the I_{Fmin} condition is:

$$V = V_F + I_F R4 \qquad \text{(typically } V_F = 1.5V) \qquad (13.3.2-3)$$

$$= 1.5V + (1.6 \ mA)(.13k\Omega)$$

$$V = 1.71V$$

The corresponding current in R3 at this voltage V is:

$$I_3 = \frac{V}{R3} \qquad (13.3.2-4)$$

$$= \frac{1.71V}{20k\Omega} \qquad mA$$

$$I_3 = 0.086 \ mA$$

The input current, I_{IN}, at $V_{IN} = V_{TH}$, threshold condition is:

$$I_{TH} = I_3 + I_F \qquad (13.3.2-5)$$

$$= 0.086 \ mA + 1.6 \ mA$$

$$I_{TH} = 1.7 \ mA$$

R2 is then given by the following equation:

$$R2 = \frac{V_{TH} - V_{D2} - V}{I_{TH}} \qquad (13.3.2-6)$$

$$= \frac{9V - 6.2V - 1.71V}{1.7 \ mA} \qquad k\Omega$$

$$R2 = .641k\Omega \qquad \text{use } 680\Omega \text{ resistor}$$

Chose D2 — 1N821A (V_{D2} = 6.2V)

I_{IN} at the operating point can be found as:

$$I_{IN} = I_{Fmax} + \frac{V}{R3} \quad \text{, where } V = V_F + V_{BE1} \qquad (13.3.2-7)$$

$$= 5 \ mA + \frac{1.5V + 0.7V}{20k\Omega} \qquad mA$$

$$I_{IN} = 5.11 \ mA$$

The value for R1 is determined at the operating point. The peak voltage presented to the half-wave rectifying diode, D1 (1N645), is:

$$V_P = \sqrt{2} \ (10A)(1.5\Omega) = 21.2V \qquad (13.3.2-8)$$

At the operating point, $V_{IN} = 12V$, and

$$R1 = \frac{V_P - V_{D1} - V_{IN}}{I_{IN}} \qquad (13.3.2-9)$$

$$= \frac{21.2V - 0.7V - 12V}{5.11 \ mA} \qquad k\Omega$$

$$R1 = 1.66k\Omega \qquad \text{use } 1.65k\Omega \text{ resistor}$$

The value for $\tau = R1C1$ low pass filter for 60 Hz ripple frequency attenuation can be chosen at a break point approximately a decade lower, or 6 Hz, which corresponds to 167 ms.

$$C1 = \frac{\tau}{R1} \qquad\qquad (13.3.2\text{-}10)$$

$$= \frac{167\ ms}{1.65k\Omega} \qquad \mu f$$

$$C1 = 101.2\mu f \qquad \text{use } 100\mu f\ @\ 35V\ dc$$

The load resistance, R5, is calculated using the same relationships that were used in Design Example 1. Those relationships are:

$$R5 \leqslant \frac{V_{cc_{min}} - V_{IH}}{I_{OH_{max}} + I_{IH}} \qquad\qquad (13.3.2\text{-}11)$$

$$R5 \geqslant \frac{V_{cc_{max}} - V_{OL}}{I_{OL_{min}} + I_{IL}} \qquad\qquad (13.3.2\text{-}12)$$

$$\text{where } I_{OL_{min}} = I_{F_{min}}\left(\frac{CTR_{min}}{100}\right) \qquad (13.3.2\text{-}13)$$

Using the following values with U1 as a 6N138 optocoupler, and U2 as a 7404 TTL unit load, the R5 range is determined.

V_{cc}	= 5.0V ± 5%	V_{IH}	= 2.4V
$I_{F_{min}}$	= 1.6 mA	V_{OL}	= 0.4V
CTR_{min}	= 300%	I_{IH}	= 40μA
$I_{OH_{max}}$	= 250μA	I_{IL}	= −1.6 mA
R5	≤ 11.9kΩ	R5	≥ 1.52kΩ

Select R5 = 1.6kΩ

Other types of optocouplers can be employed with appropriate adjustment for $I_{F_{min}}$. For good threshold control level, a sharp "knee" Zener diode with low leakage current should be used. In general, Zener diodes below 5-6V do not exhibit sharp knee characteristics. This threshold circuit does not provide any hysteresis capability. However, LED emitter transient protection is provided by shunting excessive input current, I_{IN} > 5.11 mA, through Q1.

13.4   An Optocoupler with a Built-In Switching Threshold

The problem of establishing an input switching threshold is resolved in the design of the Hewlett-Packard HCPL-3700 optocoupler. This device combines an ac or dc voltage and/or current detection function with a high insulation voltage optocoupler in a single eight pin plastic dual in-line package.

As shown in the block diagram of Figure 13.4-1, this device consists of a full-wave bridge rectifier and threshold detection integrated circuit, an LED, and an optically coupled detector integrated circuit. The detector circuit is a combination of a photodiode and a high current gain, split Darlington, amplifier.

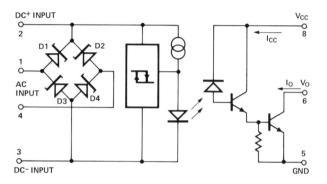

Figure 13.4-1.   Block Diagram of the HCPL-3700.

The input circuit will operate from an ac or dc source and provide a guaranteed, temperature compensated threshold level with hysteresis. The device may be programmed for higher switching thresholds through the use of a single external resistor.

With threshold level detection provided prior to the optical isolation path and subsequent gain stage, variations in the current transfer ratio of the device with time or from unit to unit are no longer important.

In addition to allowing ac or dc input signals, the Zener diodes of the bridge circuit also provide input voltage clamping to protect the threshold circuitry and LED from over voltage/current stress conditions. The LED current is provided by a switched current source.

The HCPL-3700 optocoupler output is an open collector, high gain, split Darlington configuration. The output is compatible with TTL and CMOS logic levels. High common mode rejection, or transient immunity of 600V/μs, allows excellent isolation. Insulation capability is 3000 volts dc. The recommended operating temperature range is 0°C to 70°C.

The HCPL-3700 meets the requirements of the industrial control environment for interfacing signals from ac or dc power equipment to logic control electronics. Isolated monitoring of relay contact closure or relay coil voltages, monitoring of limit or proximity switch operation or sensor signals for temperature or pressure, etc., can be accomplished by the HCPL-3700. The HCPL-3700 may also be used for sensing low power line voltage (Brown Out) or loss of line power (Black Out).

13.4.1   Device Characteristics

The function of the HCPL-3700 can best be understood through a review of the input V/I function and the input to output transfer function. Figure 13.4.1-1 shows the input characteristics, I_{IN} (mA) versus V_{IN} (volts), for both the ac and dc cases.

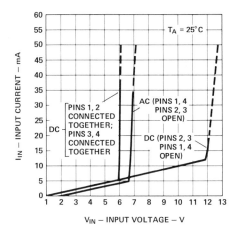

Figure 13.4.1-1.   Typical Input Characteristics, I_{IN} vs. V_{IN}.

The dc input of the HCPL-3700 appears as a 1000Ω resistor in series with a one volt offset. If the ac pins (1, 4) are left unconnected, the dc input voltage can increase to 12V (two Zener diode voltages) before the onset of input voltage clamping occurs. If the ac pins (1, 4) are connected to ground or to dc pins (2, 3) respectively, the dc input voltage will clamp at 6.0V (one Zener diode voltage). Under clamping conditions, it is important that the maximum input current limits not be exceeded. Also, to prevent excessive current flow in a substrate diode, the dc input can not be backbiased more than −0.5V. The choice of the input voltage clamp level is determined by the requirements of the system design. The advantages of clamping the input at a low voltage level is in limiting the magnitude of forward current to the LED as well as limiting the input power to the device during large voltage or current transients in the industrial control environment. The internal limiting will in some cases eliminate the need for additional protection components.

The ac input appears similar to the dc input except that the circuit has two additional diode forward voltages. The ac input voltage will clamp at 6.7V (one Zener diode voltage plus one forward biased diode voltage), and is symmetric for plus or minus polarity. The ac voltage clamp level can not be changed with different possible dc pin connections.

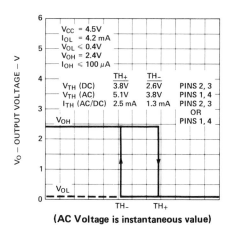

(AC Voltage is instantaneous value)

Figure 13.4.1-2.   Typical Transfer Characteristics of the HCPL-3700.

The transfer characteristic displayed in Figure 13.4.1-2 shows how the output voltage varies with input voltage, or current, levels. Hysteresis is provided to enhance noise immunity, as well as to maintain a fast transition response (t_r, t_f) for slowly changing input signals.

The hysteresis of the device is given in voltage terms as $V_{HYS} = V_{TH+} - V_{TH-}$, or in terms of current as $I_{HYS} = I_{TH+} - I_{TH-}$. The optocoupler output is in the high state until the input voltage (current) exceeds $V_{TH+}(I_{TH+})$. The output state will return high when the input voltage (current) becomes less than $V_{TH-}(I_{TH-})$.

As is shown in Figure 13.4.1-2, the HCPL-3700 has preprogrammed ac and dc switching threshold levels. Higher input switching thresholds may be programmed through the use of a single series input resistance as defined in Equation (13.4.2.1-1). In some cases, it may be desirable to split this resistance in half to achieve transient protection on each input lead and reduce the power dissipation requirement of each of the resistors.

13.4.2   Design Examples Using the HCPL-3700

Figure 13.4.2-1 illustrates three typical interface situations which a designer may encounter in utilizing a microprocessor as a controller in industrial environments.

Example 1.   A dc voltage applied to the motor is monitored as an indication of proper speed and/ or load condition.

Example 2.   A limit switch uses a 115V ac or 220V ac control loop to improve noise immunity and because it is a convenient high voltage for that purpose.

Example 3.   An HCPL-3700 is used to monitor a computer power line to sense a loss of line power condition. Use of a resistive shunt for im-

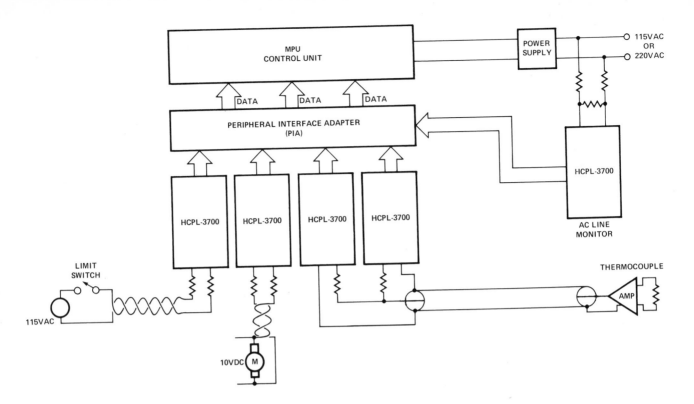

Figure 13.4.2-1.   Applications of the HCPL-3700 for Interfacing AC and DC Voltages to a Microprocessor.

provement of threshold accuracy is analyzed in this example.

Also illustrated is an application in which two HCPL-3700's are used to monitor a window of safe operating temperatures for some process parameters. This example also requires a rather precise control of the optocoupler switching threshold. An additional dedicated leased line system example is also shown (Example 4).

13.4.2.1   Example 1.   DC Voltage Sensing

The dc motor monitor function is established to provide an indication that the motor is operating at a minimum desired speed prior to the initiation of another process phase. If the applied voltage, V_M, is greater than 5V, it is assumed that the desired speed is obtained. The maximum applied

voltage in the system is 10V. The HCPL-3700 circuit configuration for this dc application is shown in Figure 13.4.2.1-1.

NOTE: See Appendix for a definition of terms and symbols for this and all other examples.

The following conditions are given for the external voltage threshold level and input requirements of the HCPL-3700:

External Voltage Levels — V_M

V_+	=	5V dc (50%)
V_{peak}	=	10V dc

HCPL-3700 Input Levels

V_{TH+}	=	3.8V
V_{TH-}	=	2.6V
V_{ICH3}	=	12V
I_{TH+}	=	2.5mA
I_{TH-}	=	1.3mA

Figure 13.4.2.1-1.   Interfacing a DC Voltage to an MPU Using the HCPL-3700.

For the 5V threshold, R_x is calculated via the expression:

$$R_x = \frac{V_+ - V_{TH+}}{I_{TH+}} \quad\quad (13.4.2.1\text{-}1)$$

$$= \frac{5V - 3.8V}{2.5mA}$$

$$R_x = 480\Omega \quad\quad (470\Omega \pm 5\%)$$

The resultant lower threshold level is formed by using the following expression:

$$V_- = I_{TH-} R_x + V_{TH-} \quad\quad (13.4.2.1\text{-}2)$$

$$= (1.3mA)\, 470\Omega + 2.60V$$

$$V_- = 3.21V$$

With the possible unit to unit variations in the input threshold levels as well as ±5% tolerance variations with R_x, the variation of V_+ is +12.4%, −15% and V_- varies +14%, −23.5%. (NOTE: With a low, external, voltage threshold level, V_+, which is comparable in magnitude to the V_{TH+} voltage threshold level of the optocoupler ($V_+ \leqslant 10V_{TH+}$) the tolerance variations are not significantly improved by the use of a 1% precision resistor for R_x. However, at a large external voltage threshold level compared to V_{TH+} ($V_+ > 10V_{TH+}$), the use of a precision 1% resistor for R_x does reduce the variation of V_+.)

For simultaneous selection of external upper, V_+, and lower, V_-, voltage threshold points a combination of a series and parallel input resistors can be used. Refer to the example on "ac operation with improved threshold control and accuracy" for detailed information.

Calculation of the maximum power dissipation in R_x is determined by knowing which of the following inequalities is true:

$$\frac{V_+}{V_{peak}} > \frac{V_{TH+}}{V_{IHC}} \quad (V_{IN} \text{ will not clamp}) \quad (13.4.2.1\text{-}3)$$

$$\frac{V_+}{V_{peak}} < \frac{V_{TH+}}{V_{IHC}} \quad (V_{IN} \text{ will clamp}) \quad (13.4.2.1\text{-}4)$$

where V_{IHC} is the particular input clamp voltage listed on the data sheet.

For this dc application with ac pins (1, 4) open, input voltage clamping will not occur, i.e.,

$$\frac{V_+}{V_{peak}} > \frac{V_{TH+}}{V_{IHC3}}$$

$$\frac{5V}{10V} > \frac{3.8V}{12.0V}$$

Consequently, a conservative value for the maximum power dissipation in R_x for the unclamped input voltage condition ignoring the input offset voltage is given by:

$$P_{R_x} = \frac{\left[V_{peak}\left(\frac{R_x}{R_x + 1\,k\Omega} \right) \right]^2}{R_x} \quad \text{(Unclamped Input)} \quad (13.4.2.1\text{-}5)$$

$$= \frac{\left[10V\left(\frac{470\Omega}{1470\Omega} \right) \right]^2}{470\Omega}$$

$$P_{R_x} = 21.8mW$$

If $V_+/V_{peak} < V_{TH+}/V_{IHC}$ was true (clamped input voltage condition), then the formula for the maximum power dissipation in R_x becomes:

$$P_{R_x} = \frac{\left(V_{peak} - V_{IHC} \right)^2}{R_x} \quad \text{(Clamped Input)} \quad (13.4.2.1\text{-}6)$$

The maximum input current or power must be determined to ensure that it is within the maximum input rating of the HCPL-3700. For the clamped input voltage condition,

$$I_{IN} = \frac{V_{peak} - V_{IHC}}{R_x} < I_{IN\,(max)} \quad (13.4.2.1\text{-}7)$$

or

$$P_{IN} = V_{IHC}\,(I_{IN}) < P_{IN\,(max)} \quad (13.4.2.1\text{-}8)$$

⎫
⎬ Clamped Condition
⎭

For the unclamped input voltage condition, the maximum input current, or power will not be exceeded, because maximum input current and power will occur only under clamp conditions.

An output load resistance is not needed in this application because the peripheral interface adapter, such as MC6821, has an internal pullup resistor connected to its input.

13.4.2.2   Example 2.   AC Operation

As shown in Figure 13.4.2.2-1, an ac application is that of a monitored 115V ac limit switch. Ac sensing is commonly used and the HCPL-3700 conveniently provides an internal rectification circuit. With the HCPL-3700 interfacing to the P.I.A., a choice can be made not to filter the ac signal or to filter the ac signal at the input or output of the device. All three conditions will be explored. Simplicity is obtained with no filtering at all, but software detection techniques must be used. Output filtering is a standard method, but may present problems with slow RC rise time of the output waveform when TTL logic is used. Input filtering avoids the RC rise time problem of output filtering, but introduces an extra time delay at the input.

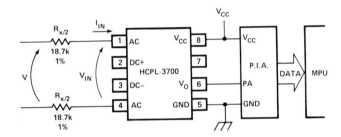

Figure 13.4.2.2-1.   Interfacing an AC Voltage to an MPU Using the HCPL-3700.

13.4.2.3   AC Operation With No Filtering

In this example, a V_+ value of 98V is selected based on a criteria of 60% of V_{peak}. Monitoring a limit switch for a 60% level of the signal will give sufficient noise immunity from an open 115V ac line while allowing the HCPL-3700 to turn on under low line voltage conditions of −15% from nominal values when the limit switch is closed.

The value of R_x for the upper threshold detection level without the filter capacitor, C, across the dc input, can be obtained from the following expression.

$$R_x = \frac{V_+ - V_{TH+}}{I_{TH+}} \qquad V_{TH+} = 5.1V \qquad (13.4.2.3\text{-}1)$$

$$\text{(ac instantaneous)}$$

$$I_{TH+} = 2.5mA$$

$$R_x = \frac{98V - 5.1V}{2.5mA}$$

$$R_x = 37.2k\Omega \qquad \text{(use } R_x/2 = 18.7k\Omega, 1\% \text{ resistor for each input lead)}$$

The resulting lower threshold point is

$$V_- = I_{TH-}R_x + V_{TH-} \qquad (13.4.2.3\text{-}2)$$

$$= (1.3mA)(37.4k\Omega) + 3.8V$$

$$V_- = 52.4V \qquad \text{(32\% of peak input voltage)}$$

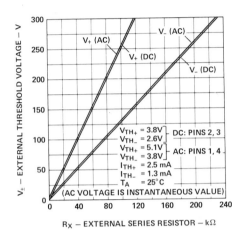

Figure 13.4.2.3-1.   Typical External Threshold Characteristic, $V_\pm$ − vs. R_x.

Figure 13.4.2.3-1 provides a convenient, graphical choice for the external series resistor, R_x, and a particular external threshold voltage $V_\pm$.

The corresponding R_x value and output waveform of the HCPL-3700 for a $V_+ = 98V$ (60% of peak) is shown in Figure 13.4.2.3-2.

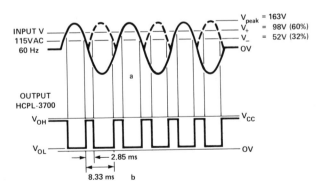

Figure 13.4.2.3-2.   Output Waveforms of the HCPL-3700 Design in Figure 13.4.2.3-1 with no Filtering Applied.

To determine the time in the high state, refer to Figure 13.4.2.3-3 and Equation (13.4.2.3-3).

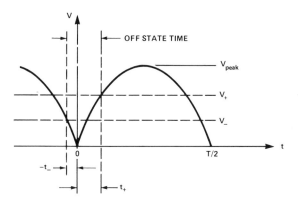

Figure 13.4.2.3-3.   Determination of Off/On State Time.

Due to symmetry of sinusoidal waveform, the high state time is $t_- + t_+$ where $t_{\pm}$ is given by:

$$t_{\pm} = \frac{T}{360°} \sin^{-1}\left(\frac{V_{\pm}}{V_{peak}}\right)$$ (13.4.2.3-3)

where arc sine is in degrees and T = period of sinusoidal waveform.

In the unfiltered condition, the output waveform of Figure 13.4.2.3-2 must be used as sensed information. Software can be created in which the microprocessor will examine the waveform from the optocoupler at specific intervals to determine if ac is present or absent at the input to the HCPL-3700. This technique eliminates the problem of filtering, and accompanying delays, but requires more sohpisticated software implementation in the micro-processor.

13.4.2.4   Input Filtering for AC Operation

A convenient method by which to achieve a continuous output low state in the presence of the applied ac signal is to filter the input dc terminals (pins 2–3) with a capaci-tance C while the ac signal is applied to the ac input (pins 1–4) of the full wave rectifier bridge. Input filtering allows flexibility in using the HCPL-3700 output for direct inter-facing with TTL or CMOS devices without the slow rise time which would be encountered with output filtering. In addition, the input filter capacitor provides extra transient and contact bounce filtering. Because filtering is done after R_x, the capacitor working voltage is limited by the V_{IHC2} clamp voltage rating which is 6.7V peak for ac operation. The disadvantage of input filtering is that this technique introduces time delays at turn on and turn off of the opto-coupler due to initial charge/discharge of the input filter capacitor.

The application of ac input filtering is illustrated in Figure

13.4.2.4-1 and is described in the following example. The ac input conditions are the same as in the previous example of the 115V ac limit switch.

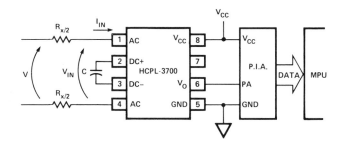

Figure 13.4.2.4-1.   Input Filtering with the HCPL-3700.

The minimum value of capacitance C to ensure proper ac filtering is determined by the parameters of the opto-coupler. At low ac input voltage, the capacitor must charge to at least V_{TH+} in order to turn on, but must not dis-charge to V_{TH-} during the discharge cycle. A conservative estimate for the minimum value of C is given by the fol-lowing equations.

(13.4.2.4-1)

$$V_{TH+} - V_{TH-} = V_{TH+}e^{-t/\tau}, \ \tau = R_{IN}C_{min}$$

where R_{IN} is the equivalent input resistance of the HCPL-3700.

$$C_{min} = \frac{t}{R_{IN} \ln\left(\frac{V_{TH+}}{V_{TH+} - V_{TH-}}\right)}$$ (13.4.2.4-2)

with $R_{IN} = 1k\Omega$, $V_{TH+} = 3.8V$, $V_{TH-} = 2.6V$ and t = 8.33ms for 60 Hz or t = 10ms for 50 Hz.

$$C_{min} = 7.23\mu F \text{ for 60 Hz}$$

$$C_{min} = 8.68\mu F \text{ for 50 Hz}$$

To ensure proper filtering, the recommended value of C should be large enough such that with the tolerance varia-tion, C will always be greater than C_{min} (C should other-wise be kept as small as possible to minimize the inherent delay times which are encountered with this technique). Since the filter capacitor affects the input impedance, a slightly different value of R_x is required for the input filtered condition. Figure 13.4.2.4-2 shows the R_x versus $V_{\pm}$ threshold voltage for C = 10μF, 22μF, and 47μF. For an application of monitoring a 115V RMS line for 65% of nominal voltage condition (75V RMS), an $R_x = 26.7k\Omega \pm 1\%$ with $C = 10\mu F$ will yield the desired threshold. The power dissipation for R_x is determined from the clamped condition $(V_+/V_{peak} < V_{TH+}/V_{ICH2})$ and is 455mW (see Figure 13.4.2.4-1) which suggests $R_x/2$ of 1/2 watt resistors for each input lead.

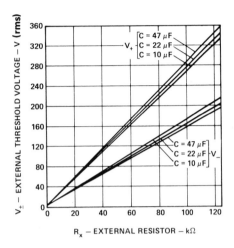

Figure 13.4.2.4-2. **External Threshold Voltage vs. R_x for Applications Using an Input Filter Capacitor C (Figure 13.4.2.4-1).**

13.4.2.5   Example 3.   AC Operation with Improved Threshold Control and Accuracy

Some applications may occur which require threshold level detection at specific upper and lower threshold points. The ability to independently set the upper and lower threshold levels will provide the designer with more flexibility to meet special design criteria. As illustrated in Figure 13.4.2.5-1, a computer power line is monitored for a power failure condition in order to prevent loss of memory information during power line failure.

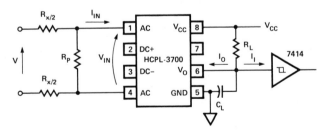

Figure 13.4.2.5-1. **An AC Power Line Monitor with Simultaneous Selection of Upper and Lower Threshold Levels and Output Filtering.**

In this design, the HCPL-3700 optocoupler monitors the computer power line and the output of the optocoupler is interfaced to a TTL Schmitt trigger gate (7414).

In the earlier ac application of the HCPL-3700 (limit switch example), a single external series resistor, R_x, was used to determine one of the threshold levels. The other threshold level was determined by the hysteresis of the device, and not the designer. A potential problem of single threshold selection with 115V line application would be to determine R_x for a lower threshold level of 50% of nominal peak input voltage, only to find that the upper threshold level is 90% of peak input voltage. With the possible ac line voltage variations (+10%, −15%), it would be possible that the optocoupler could never reach the upper threshold point with an ac line that is at −15% of nominal value. To give the designer more control over both threshold points, a combination of series resistance, R_x, and parallel resistance, R_P, may be used, as shown in Figure 13.4.2.5-1.

Two equations can be written for the two external threshold level conditions. At the upper threshold point,

$$V_+ = R_x \left(I_{TH+} + \frac{V_{TH+}}{R_P} \right) + V_{TH+} \qquad (13.4.2.5\text{-}1)$$

and at the lower threshold point,

$$V_- = R_x \left(I_{TH-} + \frac{V_{TH-}}{R_P} \right) + V_{TH-} \qquad (13.4.2.5\text{-}2)$$

Solving these equations for R_x and R_P yield the following expressions:

$$R_x = \frac{V_{TH-}(V_+) - V_{TH+}(V_-)}{I_{TH+}(V_{TH-}) - I_{TH-}(V_{TH+})} \qquad (13.4.2.5\text{-}3)$$

$$(13.4.2.5\text{-}4)$$

$$R_P = \frac{V_{TH-}(V_+) - V_{TH+}(V_-)}{I_{TH+}(V_- - V_{TH-}) + I_{TH-}(V_{TH+} - V_+)}$$

Equations (13.4.2.5-3) and (13.4.2.5-4) are valid only if the conditions of Equations (13.4.2.5-5) or (13.4.2.5-6) are met. The desired external voltage threshold levels, V_+ and V_-, are established and the values for $V_{TH\pm}$ and $I_{TH\pm}$ are found from the data sheet. With the $V_{TH\pm}$, $I_{TH\pm}$ values, the denominator of R_x, Equation (13.4.2.5-3) is checked to see of it is positive or negative. If it is positive, then the following ratios must be met:

$$\frac{V_+}{V_-} \geq \frac{V_{TH+}}{V_{TH-}} \quad \text{and} \quad \frac{V_+ - V_{TH+}}{V_- - V_{TH-}} < \frac{I_{TH+}}{I_{TH-}} \qquad (13.4.2.5\text{-}5)$$

Conversely, if the denominator of R_x Equation (13.4.2.5-3) is negative, then the following ratios must hold:

$$\frac{V_+}{V_-} \leqslant \frac{V_{TH+}}{V_{TH-}} \quad \text{and} \quad \frac{V_+ - V_{TH+}}{V_- - V_{TH-}} > \frac{I_{TH+}}{I_{TH-}} \qquad (13.4.2.5\text{-}6)$$

Consider that the computer power line is monitored for a 50% line drop condition and a 75% line presence condition. The 115V 60 Hz ac line (163V peak) can vary from 85% (139V) to 110% (179V) of nominal value.

Require:

V_- = 81.5V (50%) — **Turn off threshold**

V_+ = 122.5V (75%) — **Turn on threshold**

Given:

V_{TH+} = 5.1V I_{TH+} = 2.5mA V_{IHC2} = 6.7V

V_{TH-} = 3.8V I_{TH-} = 1.3mA

Using the Equations (13.4.2.5-3, 13.4.2.5-4) for R_x, R_p with the conditions of Equations (13.4.2.5-5, 13.4.2.5-6) being met yields

R_x = 17.4 kΩ   use 18 kΩ   5%

R_p = 1.2 kΩ   use 1.2 kΩ   5%

To complete the input calculations for maximum input current, I_{IN}, to the device and maximum power dissipation in R_x and R_p, a check must be made to determine if the input voltage will clamp at peak applied voltage. Using Equations (13.4.2.1-3) and (13.4.2.1-4) to determine if a clamp or no clamp exists, it is found that the ratios

$$0.75 = \frac{V_+}{V_{peak}} \approx \frac{V_{TH+}}{V_{IHC2}} = 0.76$$

indicate that V_{IN} slightly entered clamp condition. In this application, the operating input current, I_{IN}, is given approximately by

$$(13.4.2.5\text{-}7)$$

$$I_{IN} = \frac{V - \dfrac{V_{IHC2}}{\sqrt{2}}}{R_x} - \frac{\dfrac{V_{IHC2}}{\sqrt{2}}}{R_p} < I_{IN\,(max)}$$

$$= \frac{115V - \dfrac{6.7V}{\sqrt{2}}}{18\ k\Omega} - \frac{\dfrac{6.7V}{\sqrt{2}}}{1.2\ k\Omega}$$

I_{IN} = 2.18mA RMS $<$ 34.3mA

Power dissipation in R_x is determined from the following equation,

$$P_{R_x} = \frac{\left(V - \dfrac{V_{IHC2}}{\sqrt{2}} \right)^2}{R_x} \qquad (13.4.2.5\text{-}8)$$

which yields 0.675W. With the clamp condition existing, the maximum power dissipation for R_p is 18.7mW which is determined from

$$P_{R_p} = \frac{\left(\dfrac{V_{IHC2}}{\sqrt{2}} \right)^2}{R_p} \qquad (13.4.2.5\text{-}9)$$

13.4.2.6   Output Filtering

The advantages of filtering at the output of the HCPL-3700 are that it is a simple method to implement. The output waveform introduces only one additional delay time at turn off condition as opposed to the input filtering method which introduces additional delay times at both the turn on and turn off conditions due to initial charge or discharge of the input filter capacitor. The disadvantage of output filtering is that the long transition time, t_r, which is introduced by the output RC filter requires a Schmitt trigger logic gate to buffer the output filter circuit from the subsequent logic circuits to prevent logic chatter problems. The determination of load resistance and capacitance is illustrated in the following text.

The following given values specify the interface conditions.

HCPL-3700

V_{OL} = 0.4V

I_{OL} = 4.2mA

I_{OH} = 100µA max

V_{CC} = 5.0V ± 5%

7414

$V_{T+\,(min)}$ = 1.5V   } Schmitt trigger upper

$V_{T+\,(max)}$ = 2.0V   } threshold level

I_{IH} = 40µA max

I_{IL} = −1.2mA max

With the current convention shown in Figure 13.4.2.5-1, the minimum value of R_L which ensures that the output transistor remains in saturation is:

$$R_{L \text{ (min)}} \geq \frac{V_{CC \text{ (max)}} - V_{OL}}{I_{OL} + I_{IL}} \qquad (13.4.2.6\text{-}1)$$

$$= \frac{5.25V - 0.4V}{4.2mA - 1.2mA} = 1.62 \text{ k}\Omega$$

The maximum value for R_L is calculated allowing for a guardband of 0.4V in $V_{T+ \text{ (max)}}$ parameter, or $V_{IH} = V_{T+ \text{ (max)}} + 0.4V$.

$$R_{L \text{ (max)}} \leq \frac{V_{CC \text{ (min)}} - V_{IH}}{I_{OH} - I_{IH}} \qquad (13.4.2.6\text{-}2)$$

$$= \frac{4.75V - 2.4V}{0.1mA + 0.04mA} = 16.8 \text{ k}\Omega$$

R_L is chosen to be 1650Ω.

C_L can be determined in the following fashion. As illustrated in Figure 13.4.2.3-2, the output of the optocoupler will be in the high state for a specific amount of time dependent upon the selected V_+ levels. In this example, V_+ = 122.5V (75%) and V_- = 81.5V (50%) and allowing for a minimum peak line voltage of 138V (−15%), the high state time (without C_L) is from Equation (13.4.2.3-3) 4.58ms. With the appropriate C_L value, the output waveform (solid line) shown in Figure 13.4.2.6-1 is filtered.

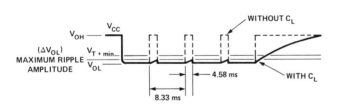

Figure 13.4.2.6-1. Output Waveforms of the HCPL-3700.

The maximum ripple amplitude above V_{OL} is chosen to be 0.6V; that is, $V_{OL} + \Delta V_{OL} = 1.0V$. This gives a 0.5V noise margin before $V_{T+ \text{ (min)}}$ = 1.5V is reached. The exponential ripple waveform is caused by the C_L being charged through R_L and input resistance, $R_{IN_{TTL}}$, of TTL gate. An expression for the allowable change in V_{OL} can be written:

$$\Delta V_{OL} = (V_{OH} - V_{OL})(1 - e^{-t/\tau}) \qquad (13.4.2.6\text{-}3)$$

where $\tau = R'_L C_L$ with R'_L equal to parallel combination of R_L and $R_{IN_{TTL}}$.

Below V_{T+} = 1.5V (min), $R_{IN_{TTL}}$ is constant and nominally 6 kΩ. Hence:

$$R'_L = \frac{R_L R_{IN_{TTL}}}{R_L + R_{IN}} \qquad (13.4.2.6\text{-}4)$$

$$= \frac{(1.65 \text{ k}\Omega)(6 \text{ k}\Omega)}{1.65 \text{ k}\Omega + 6 \text{ k}\Omega}$$

$$R'_L = 1.29 \text{ k}\Omega$$

Solving Equation (13.4.2.6-3) for τ yields

$$\tau = \frac{t}{\ln\left(\dfrac{V_{OH} - V_{OL}}{V_{OH} - V_{OL} - \Delta V_{OL}}\right)} \qquad (13.4.2.6\text{-}5)$$

and substituting previous parameter values and using $V_{OH} = V_{CC} - (I_{OH} + I_{IH})R_L$ results in

$$= \frac{4.58ms}{\ln\left(\dfrac{4.8V - 0.4V}{4.8V - 0.4V - 0.6V}\right)}$$

$$\tau = 31.24ms$$

C_L can be calculated directly,

$$C_L = \frac{\tau}{R'_L} \qquad (13.4.2.6\text{-}6)$$

$$= \frac{31.24ms}{1.29 \text{ k}\Omega}$$

$$C_L = 24.2\mu F \qquad\qquad \text{use } 27\mu F \pm 10\%$$

$$\text{or } 33\mu F \pm 20\%$$

With this value of C_L, the time the $R'_L C_L$ filter network takes to reach V_{T+} of the TTL gate is found as follows.

$$V_{OL} + (V_{OH} - V_{OL})(1 - e^{-t/\tau}) = V_{T+} \qquad (13.4.2.6\text{-}7)$$

Solving for t,

$$t = \tau \ln\left(\frac{V_{OH} - V_{OL}}{V_{OH} - V_{T+ (min)}}\right) \qquad \text{(13.4.2.6-8)}$$

and substituting V_{OH} = 4.8V, V_{OL} = 0.4V, $V_{T+ (min)}$ = 1.5V, and τ = 31.24ms yields

$$t = 9.0ms$$

This is the delay time that the system takes to respond to the ac line voltage going below the 50% ($V\_$) threshold level. In essence, the response time is slightly more than a half cycle (8.33ms) of 60 Hz ac line with worst case line variation taken into account. This delay time is acceptable for system power line protection. In this example, a complete worst case analysis was not performed. A worst case analysis should be done to ensure proper function of the circuit over variations in line voltage, unit to unit device parameter variations, component tolerances and temperature.

13.4.2.7 Threshold Accuracy Improvement

In the above example on output filtering, the two external threshold levels were selected for turn on conditions at V_+ = 122.5V (75%) and turn off at $V\_$ = 81.5V (50%). The calculated external resistor values were R_x = 17.4 kΩ and R_p = 1.2 kΩ. Using standard 5% resistors of 18 kΩ and 1.2 kΩ respectively, the upper threshold voltage was actually 126.6V nominal.

Examination of the worst possible combination of variations of the HCPL-3700 optocoupler V_{TH+}, I_{TH+}, levels from unit to unit, and the ± 5% variations of R_x and R_p can result in the V_+ level changing +23% to −25% from design nominal.

If higher threshold accuracy is desired, it can be accomplished by decreasing the value of R_p in order to allow R_p to dominate the input resistance variations of the optocoupler. Using a 1% resistor for R_p and resistance of sufficiently small magnitude, the V_+ tolerance variations can be significantly improved. The following analysis will allow the designer to obtain nearly optimum threshold accuracy from unit to unit. It should be noted that the HCPL-3700 demonstrates excellent threshold repeatability once the external resistors are adjusted for a particular level and unit. The compromise which is made for the added control on threshold accuracy is that more input power must be consumed within the R_p, R_x resistors.

In Figure 13.4.2.7-1, assume the circuit is at the upper threshold point. At constant V_{TH+}, it is desired to maintain I_+ to within ± 5% variation of nominal value while allowing ± 1% variation in I_{p+}. With this requirement,

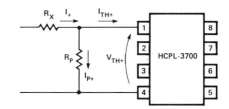

Figure 13.4.2.7-1. Threshold Accuracy Improvement Through the Use of External R_x and R_P Resistors.

Equations (13.4.2.7-1) and (13.4.2.7-2) can be written and solved for the magnitude of I_{p+} which is needed to maintain the desired condition on I_+. I_+ is the sum of I_{p+} and I_{TH+}.

$$1.05\, I_+ = 1.01\, I_{p+} + I_{TH+ (max)} \qquad \text{(13.4.2.7-1)}$$

at constant V_{TH+}

$$0.95\, I_+ = 0.99\, I_{p+} + I_{TH+ (min)} \qquad \text{(13.4.2.7-2)}$$

where

$$I_{TH+ (max)} = 3.11mA$$

$$I_{TH+ (min)} = 1.96mA$$

Solving for I_{p+} yields

$$I_{p+} = 11.2mA,$$

and

$$R_P = \frac{V_{TH+}}{I_{p+}} \qquad \text{(13.4.2.7-3)}$$

$$= \frac{5.1V}{11.2mA}$$

$$R_P = 433\Omega \qquad \text{(use } 453\Omega, \text{1% resistor)}$$

This new value of R_P replaces the earlier R_P = 1.2 kΩ, and the circuit requires a new R_x value to maintain the same V_+ threshold level.

$$\text{(13.4.2.7-4)}$$

$$R_x = \frac{V_+ - V_{TH+}}{I_+} \qquad \text{where} \quad I_+ = I_{p+} + I_{TH+}$$

$$= 11.2mA + 2.5mA$$

$$= \frac{122.5V - 5.1V}{13.7mA}$$

R_x = 8.57 kΩ   (use 8.66 kΩ, 1% resistor)

With the possible variation of ± 1% in R_p and R_x, as well as unit to unit variations in the optocoupler V_{TH+}, I_{TH+}, the upper threshold level V_+ will vary significantly less than in the 5% resistor design case. The variations in V_+, which is given by $V_+ = R_x I_+ + V_{TH+}$, where $I_+ = I_{P+} + I_{TH+}$, are compared in Table 13.4.2.7.

Table 13.4.2.7 illustrates the possible improvements in V_+ tolerance as R_x and R_p are adjusted to limit the variation of the external input threshold current, I_+, to the resistor network and optocoupler. This table is centered at a nominal external input threshold voltage of V_+ = 122.5V. It is the designer's compromise to keep power consumption low, but threshold accuracy high.

NOTE: The above method for selection of R_p and R_x can be adapted for applications where larger sense currents (wet sensing) may be appropriate.

13.4.2.8   Example 4.   Dedicated Lines for Remote Control

In situations involving a substantial separation between the signal source and the receiving station, it may be desirable to lease a dedicated private line metallic circuit (dc path) for supervisory control of remote equipment. The HCPL-3700 can provide the interface requirements of voltage threshold detection and optical isolation from the metallic line to the remote equipment. This greatly reduces the expense of using a sophisticated modem system over a convention telephone line.

Figure 13.4.2.8-1 represents the application of the HCPL-3700 for a line which is to control tank levels in a water district.

Some comments are needed about dedicated metallic lines. The use of a private metallic line places restrictions upon the designer's signal levels. The line in this example would be used in the interrupted dc mode (duration of each interruption greater than one second), the maximum allowed voltage between any conductor and ground is ≤ 135 volts. Maximum current should be limited to 150mA if the cable has compensating inductive coils in it. Balanced operation of the line is strongly recommended to reduce possible cross talk interference as well as to allow larger signal magnitudes to be used. Precaution also should be taken to protect the line and equipment. The line needs to be fused to ensure against equipment failure causing excessive current to flow through telephone company equipment. In addition, protection from damaging transients must be taken via spark gap arrestors and commercial transient suppressors. Details of private line metallic circuits can be founded in the American Telephone and Telegraph Company publication 43401.

In this application, a 48V dc floating power source supplies the signal for the metallic line. The HCPL-3700 upper voltage threshold level is set for V_+ = 36V (75%). Consequently, R_x is

$$R_x = \frac{V_+ - V_{TH+}}{I_{TH+}} \qquad (13.4.2.8\text{-}1)$$

$$= \frac{36V - 3.8V}{2.5mA}$$

= 12.9 kΩ   (use R_x/2 = 6.49 kΩ, 1% resistor in each input level)

The resulting lower voltage threshold level is

R_x	T O L.	R_p	T O L.	I_+ TOLERANCE	V_+ TOLERANCE		MAXIMUM TOTAL POWER IN $R_x + R_p$ (RMS)
18 kΩ	5%	1.2 kΩ	5%	+17.5% −21.2%	+ 23%	− 25%	0.69 W
8.66 kΩ	1%	453Ω	1%	±5%	+12.7%	−19.3%	1.45 W
4.32 kΩ	1%	205Ω	1%	±3%	+11.2%	−18.9%	2.92 W
2.15 kΩ	1%	97.5Ω	1%	±2%	+10.6%	−18.8%	5.89 W

Table 13.4.2.7.   Comparison of the V_+ Threshold Accuracy Improvement vs. R_x and R_p and   Power Dissipation for a Nominal V_+ = 122.5 V.

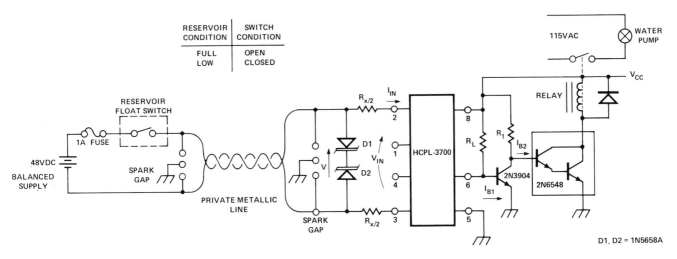

FLOAT SWITCH AND WATER PUMP ARE REMOTELY LOCATED WITH RESPECT TO EACH OTHER.

Figure 13.4.2.8-1. Application of the HCPL-3700 to Private Metallic Telephone Circuits for Remote Control.

$$V_- = R_x I_{TH-} + V_{TH-} \qquad (13.4.2.8-2)$$

$$= 13 \text{ k}\Omega \ (1.3\text{mA}) + 2.6\text{V}$$

$$V_- = 19.5\text{V}$$

yielding V_{HYS} = 16.5V. The average induced ac voltage from adjacent power lines is usually less than 10 volts (reference ATT publication 43401) which would not falsely turn on, or off, the HCPL-3700, but could affect conventional optocouplers.

Under normal operation (full reservoir), the optocoupler is off. When the float switch is closed (low reservoir), the optocoupler output (V_{OL}) needs inversion, via a transistor, to drive the power Darlington transistor which controls a motor starting relay. The relay applies ac power to the system water pump. With V_{CC} = 10V, I_{B_2} = 0.5mA, I_{B_1} = 0.5mA.

$$R_1 = \frac{V_{CC} - 2V_{BE}}{I_{B_2}} \qquad (13.4.2.8-3)$$

$$= \frac{10\text{V} - 1.4\text{V}}{0.5\text{mA}}$$

$$R_1 = 17.2 \text{ k}\Omega$$

$$(R_1 = 18 \text{ k}\Omega)$$

$$R_L = \frac{V_{CC} - V_{BE}}{I_{B_1}} \qquad (13.4.2.8-4)$$

$$= \frac{10\text{V} - 0.7\text{V}}{0.5\text{mA}}$$

$$R_L = 18.6 \text{ k}\Omega$$

$$(R_L = 18 \text{ k}\Omega)$$

For this application, the ac inputs could also be used, which would remove any concern about the polarity of the input signal.

13.4.3 General Protection Considerations for the HCPL-3700

The HCPL-3700 optocoupler combines a unique function of threshold level detection and optical isolation for interfacing sensed signals from electrically noisy, and potentially harmful, environments. Protection from transients which could damage the threshold detection circuit and LED is provided internally by the Zener diode bridge rectifier and an external series resistor. By examination of Figure 13.4-1, it is seen that an input ac voltage clamp condition will occur at a maximum of a Zener diode voltage plus a forward biased diode voltage.

At clamp condition, the bridge diodes limit the applied input voltage at the device and shunt excess input current which could damage the threshold detection circuit or cause excessive stress to the LED.

The HCPL-3700 optocoupler can tolerate significant input current transient conditions. The maximum dc input current into or out of any lead is 50mA. The maximum input surge current is 140mA for 3ms at 120 Hz pulse repetition rate, and the maximum input transient current is

500mA for 10µs at 120 Hz pulse repetition rate. The use of an external series resistor, R_x, provides current limiting to the device when a large voltage transient is present. The amplitude of the acceptable voltage transient is directly proportional to the value of R_x.

However, in order to protect the HCPL-3700 when the input voltage to the device is clamped, the maximum input current must not be exceeded. An external means by which to enhance transient protection can be seen in Figure 13.4.3-1.

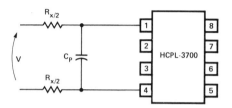

Figure 13.4.3-1. $R_x C_P$ Transient Filter for the HCPL-3700.

A transient $R_x C_P$ filter can be formed with C_P chosen by the designer to provide a sufficiently low break point for the low pass filter to reduce high frequency transients. However, the break point must not be so low as to attenuate the signal frequency. Consider the previous ac application where no filtering was used. In that application, $R_x = 37.4$ kΩ, and if the bandwidth of the transient filter needs to be 600 Hz, then C_P is:

$$C_P = 2\pi f R_x$$

$$C_P = \frac{1}{2\pi f R_x} \qquad\qquad (13.4.3-1)$$

$C_P = 0.0071\mu F$ (use 0.0068µF capacitor @ 50V dc)

Should additional protection be needed, a very effective external transient suppression technique is to use a commercial transient suppressor, such as a Transzorb ®, or metal oxide varistor, MOV ®, at the input to the resistor network prior to the optocoupler. The Transzorb ® will provide extremely fast transient response, clamp the input voltage to a definite level, and absorb the transient energy. Selection of a Transzorb ® is made by ensuring that the reverse stand off voltage is greater than the continuous peak operating voltage level. Transzorbs ® can be stacked in series or parallel for higher peak power ratings. Depending upon the designer's potential transient problems, a solution may warrent the expense of a commercial suppression device.

13.4.4   Thermal Considerations

Thermal considerations which should be observed with the HCPL-3700 are few. The plastic 8 pin DIP package is designed to be operated over a temperature range of −25°C to 85°C. The absolute maximum ratings are established for a 70°C ambient temperature requiring slight derating to 85°C. In general, if operation of the HCPL-3700 is at ambient temperature of 70°C or less, no heat sinking is required. However, for operation between 70°C and 85°C ambient temperature, the maximum ratings should be derated per the data sheet specifications.

13.4.5   Mechanical and Safety Considerations

13.4.5.1   Mechanical Mounting Considerations

The HCPL-3700 optocoupler is a standard 8 pin dual-inline plastic package designed to interface ac or dc power systems to logic systems. This optocoupler can be mounted directly onto a printed circuit board by wave soldering.

13.4.5.2   Electrical Safety Considerations

Special considerations must be given for printed circuit board lead spacing for different safety agency requirements. Various standards exist with safety agencies (U.L., V.D.E., I.E.C., etc.) and should be checked prior to PC board layout. The HCPL-3700 optocoupler component is recognized under the Component Program of Underwriters Laboratories, Inc. in file number E55361. This file qualifies the component to specific electrical tests to 220V ac operation.

The spacing required for the PC board leads depends upon the potential difference that would be observed on the board. Some standards that could pertain to equipment which would use the HCPL-3700 are UL1244, Electrical and Electronic Measuring and Testing Equipment, UL1092, Process Control Equipment, and IEC348, Electronic Measuring Apparatus. Spacing for the worst case in an uncontrolled environment with a 2000 volt-amperes maximum supplying source rating must be 3.2mm (0.125 inches) for 51 − 250 volts RMS potential difference over a surface (creepage distance), and 3mm (0.118 inches) through air (bare wire). These separations are between any uninsulated live part and uninsulated live part of opposite polarity, or uninsulated ground part other than the enclosure or an exposed metal part.

An uncontrolled environment is an environment which has contaminants, chemical vapors, particulates or any substances which would cause corrosion, decrease resistance between PC board traces or, in general, be an unhealthy environment to human beings.

For 0 – 50 volts RMS, the spacing is 1.6mm (0.063 inches) through air or over surfaces.

13.4.6 Electrical Connectors

The HCPL-3700 provides the needed isolation between a power signal environment and a control logic system. However, there exists a physical requirement to actually interconnect these two environments. This interconnection can be accomplished with barrier strips, edge card connectors, and PCB socket connectors which provide the electrical cable/field wire connection to the I/O logic system. These connectors provide for easy removal of the PC board for repair or substitution of boards in the I/O housing and are needed to satisfy the safety agency (U.L., V.D.E., I.E.C.) requirements for spacing and insulation. Connectors are readily available from many commercial manufacturers, such as Connectron Inc., Buchanan, etc. The style of connector to choose is dependent upon the application for which the PC board is used. If possible it is wise to choose a style which does not mount to the PC board. This would enable the PC card to be removed without having to disconnect field wires. The use of connectors which are called "gas tight connectors" provide for good electrical and mechanical reliability by reducing corrosion effects over time.

13.4.7 APPENDIX I. List of Parameters

V ≡ Externally Applied Voltage
V_+ ≡ External Upper Threshold Voltage Level
V_- ≡ External Lower Threshold Voltage Level
V_{IHC1} = Device* Input Voltage Clamp Level; Low Voltage DC Case
V_{IHC2} = Low Voltage AC Case
V_{IHC3} = High Voltage DC Case
I_{IN} = Device Input Current
V_{IN} = Device Input Voltage
V_{TH+} = Device Upper Voltage Threshold Level
V_{TH-} = Device Lower Voltage Threshold Level
I_{TH+} = Device Upper Input Current Threshold Level
I_{TH-} = Device Lower Input Current Threshold Level
R_x = External Series Resistor for Selection of External Threshold Level
R_P = External Parallel Resistor for Simultaneous Selection/Accuracy Improvement of External Threshold Voltage Levels
I_+ = Total Input Current at Upper Threshold Level to External Resistor Network (R_x, R_P) and Device
I_{P+} = Current in R_P at Upper Threshold Levels
V_{peak} = Peak Externally Applied Voltage
V_O = Output Voltage of Device
V_{OL} = Output Low Voltage of Device
V_{OH} = Output High Voltage of Device
I_{OH} = Output High Leakage Current of Device
I_{OL} = Output Low Sinking Current of Device
I_{IH} = Input High Current of Driven Gate
I_{IL} = Input Low current of Driven Gate
V_{CC} = Positive Supply Voltage
R_{IN} = Input Resistance of HCPL-3700
V_{T+} ≡ Schmitt Trigger Upper Threshold Voltage of TTL Gate (7414)
R_L = Output Pullup Resistance
C_L = Output Filter Capacitance
C = Input Filter Capacitor
TH_+ = Upper Threshold Level
TH_- = Lower Threshold Level
P_{R_x} = Power Dissipation in R_x
P_{IN} = Power Dissipation in HCPL-3700 Input IC
PA = Input Signal Port to P.I.A.
t_+ = Turn On Time
t_- = Turn Off Time
T = Period of Waveform
C_P = Similar to R_P

*Device = HCPL-3700

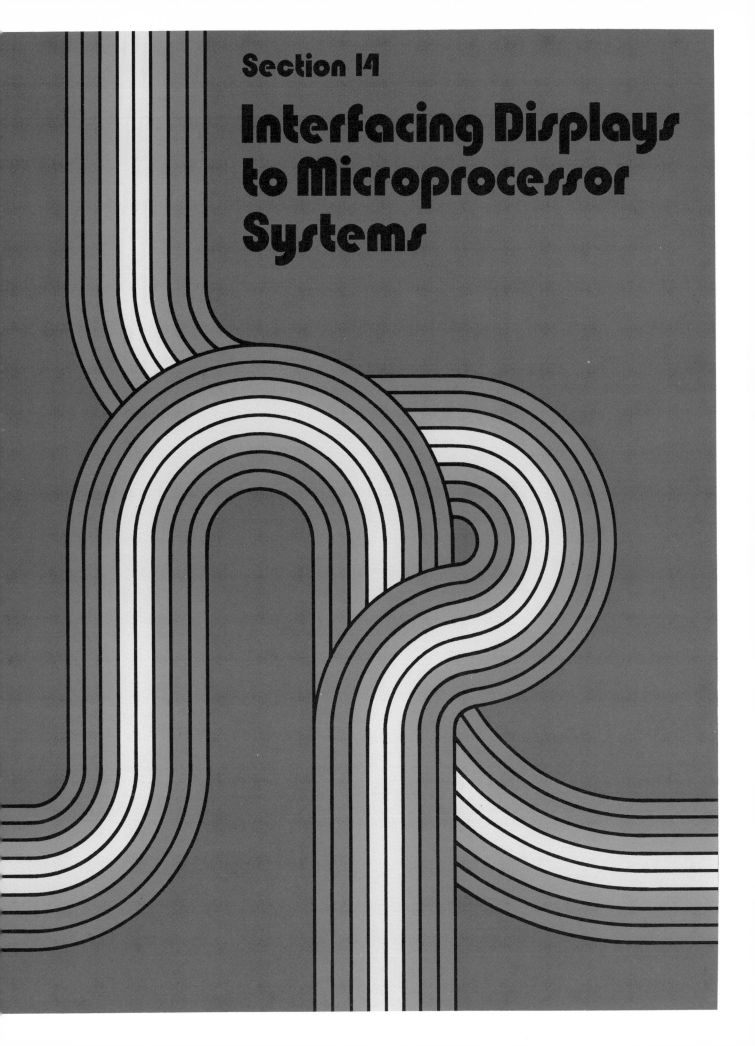

Section 14

Interfacing Displays to Microprocessor Systems

14.0 INTERFACING DISPLAYS TO MICROPROCESSOR SYSTEMS

The development of the microprocessor has revolutionized systems design. Complicated logic diagrams have been replaced by a specialized microprocessor program and input/output circuitry. While microprocessors have changed the usage of all types of LED displays, microprocessors have particularly affected the applications of alphanumeric displays. The use of microprocessors as system controllers has substantially increased the use of alphanumeric displays in new system designs for two reasons. First, the microprocessor has greatly simplified the traditionally difficult task of designing an alphanumeric display into a system. Second, the use of microprocessors has allowed the designer to improve the communication between the end user and his system by providing user prompting and system diagnostics. An alphanumeric display is capable of displaying the alphabet, numbers, punctuation, and many special symbols and provides a means for implementing this communication. The selection of the appropriate display interface is a tradeoff between microprocessor utilization and external circuit requirements. A system design for a high volume application would tend to stress maximum microprocessor utilization and minimum total circuitry, while a design for a low volume application would tend to emphasize low development costs. For these reasons, this chapter will investigate several different schemes for interfacing an LED display to a microprocessor based system.

Paralleling the acceptance of microprocessors has been the development of several new types of LED displays, as well as new display interface circuits. The inclusion of shift registers and constant current LED drivers within the 5 x 7 dot matrix alphanumeric display has greatly simplified the usage of this type of display. Advances in 16 segment displays have included the addition of a centered decimal point and colon to increase display readability. Advances in display interface circuitry include 5 x 7 character generators that require only a single 5 volt supply, an ASCII to 18 segment decoder/driver, complex display controllers within a single integrated circuit, and improved high current NPN Darlington digit drivers.

14.1 Microprocessor Operation

In order to effectively utilize the interface techniques outlined in the following sections, an understanding of microprocessor fundamentals is required. A microprocessor system usually consists of a microprocessor, ROM memory, RAM memory, and a specific I/O Interface as outlined in Figure 14.1-1. The microprocessor performs the desired system function by executing a program stored within the ROM. The RAM memory provides temporary storage for the microprocessor system. The I/O Interface consists of circuitry that is used as an input to the system or as an output from the system. The microprocessor interfaces to this system through an Address Bus, Data Bus, and Control Bus. The Address Bus consists of several outputs $(A_0, A_1 \ldots A_n)$ from the microprocessor which collectively specify a binary number. This number or 'address' uniquely specifies each word in the ROM memory, RAM memory, and I/O Interface. The Data Bus serves as an input to the microprocessor during a memory or input read and as an output from the microprocessor during a memory or output write. The Control Bus provides the required timing and signals to the microprocessor system to distinguish a memory read from a memory write, and in some systems an I/O read from an I/O write. These control lines and the timing between the Address Bus, Data Bus, and Control Bus vary for different microprocessors.

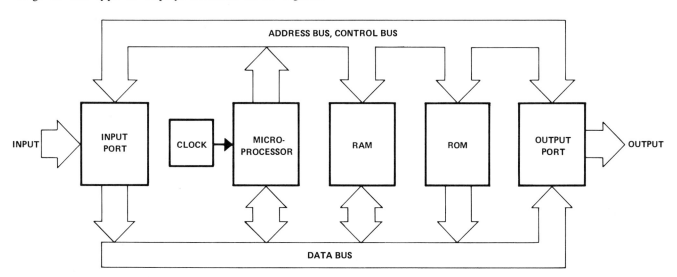

Figure 14.1-1. Block Diagram of a Typical Microprocessor System.

14.1.1 CPU Operation

The heart of a microprocessor system is the Central Processing Unit (CPU), which is often just called the microprocessor. A simplified block diagram of a typical microprocessor is shown in Figure 14.1.1-1. A microprocessor consists of several registers, an arithmetic logic unit (ALU), and a microprogram sequencer. The internal data paths or buses within the microprocessor are buffered and become the Address, Data, and Control Bus. The width of the Data Bus usually determines the word size of the machine. For example, an 8 bit microprocessor would have an 8 bit ALU and typically an 8 bit Data Bus and 8 bit registers. The width of the Address Bus determines the address space of the machine. Thus, a microprocessor with a 16 bit Address Bus could access 2^{16} or 65,536 words.

The registers within a microprocessor consist of an Accumulator (A), a Program Counter (PC), an Instruction Register (IR), a Flag Register (FLAG), a Memory Address Register (MAR), and usually one or more user accessible registers. In many microprocessors, the destination of any arithmetic or logic operation involving the ALU is the

Accumulator. The Program Counter contains the address of the next program instruction. As each instruction is executed, the PC is automatically incremented to the address of the next instruction. The Instruction Register contains the Instruction that is currently being executed by the Microprogram Sequencer. The Memory Address Register contains the address in memory where the microprocessor performs a read or write. The Flag Register consists of several user testable flags that indicate such results as Carry, Overflow, Zero, Negative, and Parity of the previous ALU operation.

The ALU provides the means for the arithmetic or logical manipulation of the microprocessor registers. Typical ALU operations include ADD, SUBTRACT, AND, OR, XOR, and COMPARE of two registers and CLEAR, COMPLEMENT, SHIFT, ROTATE, INCREMENT, and DECREMENT of a single register. The ALU also periodically updates the Flag Register after each operation.

The Microprogram Sequencer decodes the Instruction residing in the IR into a sequence of fundamental operations that can be understood by the ALU and

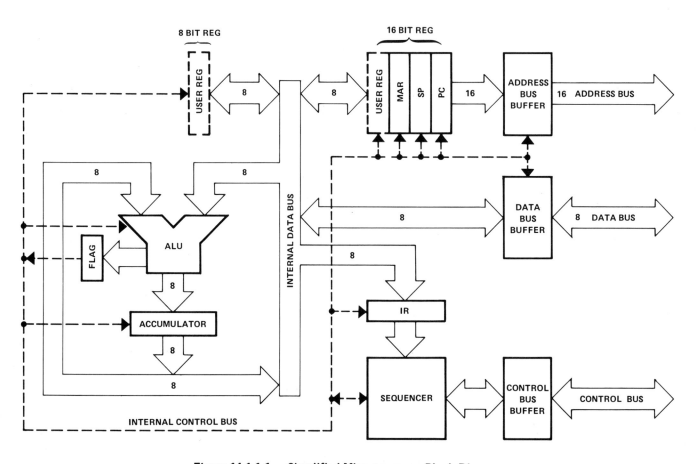

Figure 14.1.1-1. Simplified Microprocessor Block Diagram.

registers. For example, the operation of adding the user accessible register, R_1, to the Accumulator and storing the result in the Accumulator might consist of the following steps. First, the output of the Accumulator is connected to the left input of the ALU and disconnected from the internal bus. At the same time, the output of R_1 is connected to the internal bus and the internal bus is connected to the right input of the ALU. Next, the ALU is configured to perform an ADD operation. The ALU adds the left and right inputs together and outputs the Sum, Carry, and any other flags. Finally, the result is loaded into the Accumulator and the Flag Register is updated. The sequencer also controls the operation of the Address Bus, Data Bus, and Control Bus.

Operation of the microprocessor is controlled by a sequence of instructions stored within RAM or ROM memory. The microprocessor fetches the current instruction, executes the specified operation, then fetches the next instruction and so on. This operation is called the Fetch-Execute Cycle. The Fetch-Execute Cycle is accomplished in the following way:

1. The microprocessor outputs the PC to the Address Bus and performs a memory read.

2. The word on the Data Bus is stored in the IR.

3. The PC is incremented by 1.

4. The first word of the Instruction is decoded.

5. Complete specifications of the instruction may require additional memory reads. If so, the microprocessor outputs the PC to the Address Bus and performs another memory read. The word on the Data Bus is stored in the MAR and the PC is incremented by 1, and so on.

6. After the Instruction is completely decoded, the desired operation is performed.

7. After the Instruction has been executed, the process is repeated.

14.1.2   Bus Structure

The Address, Data, and Control Buses provide the flow of instructions and data into the microprocessor. Program Execution consists of a series of memory reads (instruction fetches) which are sometimes followed by a memory read or write (instruction execution). The microprocessor performs a memory read by outputting the memory address of the word to be read on the Address Bus. This address uniquely specifies a word within the memory system. The microprocessor also outputs a signal on the Control Bus,

which instructs the memory system to perform a memory read. The address selects one memory element, either RAM or ROM, within the memory system. Then, the desired word within the selected memory element is gated on the Data Bus by the Read signal. Meanwhile, the unselected memory elements tristate their output lines so that only the selected memory element is active on the Data Bus. After sufficient delay, the microprocessor reads the word that appears on the Data Bus. Similarly, for a memory write, the microprocessor outputs the memory address of the word to be written on the Address Bus. After sufficient delay, the microprocessor outputs a signal on the Control Bus, which instructs the memory system to perform a memory write. The microprocessor also outputs the desired memory word on the Data Bus. The address selects one RAM memory element within the memory system. The Write signal causes the memory element to read the word on the Data Bus and store it at the desired location. After the write cycle has been completed, the new word will have replaced the previous word within the RAM memory. During the memory write, outputs from the unselected memory elements remain tristated so that only the microprocessor is active on the Data Bus. These control lines and the timing for the Address Bus, Data Bus, and Control Bus vary for different microprocessors.

Some microprocessors, such as the Motorola 6800 microprocessor family, handle memory and I/O in exactly the same way. Memory and I/O occupy a common address space and are accessed by the same instructions. With this type of microprocessor, the hardware decoding of the Address Bus determines whether the read or write is to a memory or I/O element. Other microprocessors, such as the Intel 8080A, Intel 8085A, and the Zilog Z-80 have separate address spaces for memory and I/O. These microprocessors use different instructions for a memory access or an I/O access and provide signals on the Control Bus to distinguish between memory and I/O. One advantage of this approach is that the I/O address space can be made smaller to simplify device decoding. However, the I/O instructions that are available are usually not as powerful as the memory reference instructions. Of course, the user can always locate specific I/O devices within the memory address space through proper decoding of the Address and Control Buses. This would allow these I/O devices to be accessed with memory reference instructions.

The 6800 microprocessor family has a 16 line Address Bus, 8 line Data Bus, and a Control Bus that includes the signals VMA (Valid Memory Address), R/$\overline{\text{W}}$ (Read/Write), DBE (Data Bus Enable), and clock signals ϕ_1 and ϕ_2. R/$\overline{\text{W}}$ specifies either a memory read or write while VMA is used in conjunction with R/W to specify a valid memory address. DBE gates the internal data bus of the 6800 to the external Data Bus. In many applications, DBE is connected to ϕ_2.

Additional data hold time, t_H, can be achieved by delaying ϕ_2 to the microprocessor or by extending DBE beyond the falling edge of ϕ_2. The timing between the Address Bus, Data Bus, VMA, and R/W for a memory write is shown in Figure 14.1.2-1.

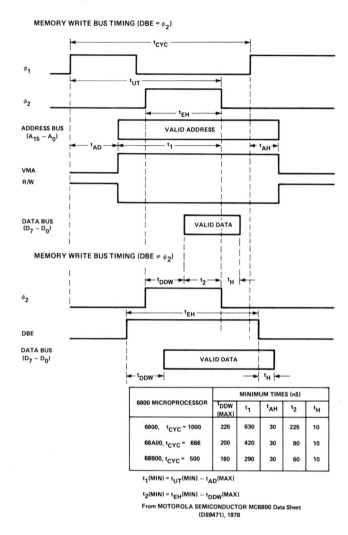

Figure 14.1.2-1. Memory Write Timing for the Motorola 6800 Microprocessor Family.

The 6802 microprocessor provides similar timing except that DBE and ϕ_1 are not available. However, the 6802 microprocessor includes another input, MR (Memory Ready), which allows the memory system to extend the time that the Address and Data Bus is valid by integral clock cycles. Bus timing for the 6802 microprocessor is shown in Figure 14.1.2-2.

For the 8080A microprocessor, the Address Bus consists of 16 lines, the Data Bus consists of 8 lines, and the Control Bus consists of several lines including DBIN (Data Bus In), $\overline{\text{WR}}$ (Write), SYNC (Synchronizing Signal), READY, and clock signals ϕ_1 and ϕ_2. DBIN and $\overline{\text{WR}}$ are used to specify a read or write operation. The 8080A microprocessor

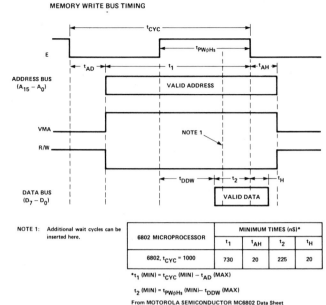

Figure 14.1.2-2. Memory Write Timing for the Motorola 6802 Microprocessor Family.

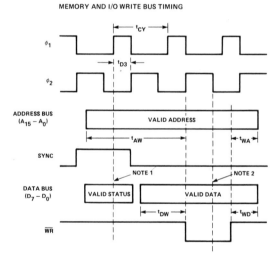

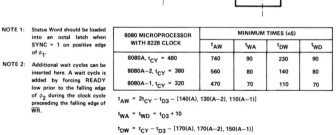

Figure 14.1.2-3. Memory and I/O Write Timing for the Intel 8080A Microprocessor Family.

distinguishes memory from I/O through the use of a status word that precedes every machine cycle. When SYNC is high, the status word should be loaded into an octal latch on the positive edge of ϕ_1. The outputs from the latch

14.4

can then be decoded to specify whether the machine cycle is a memory write, memory read, I/O write, or I/O read. The Intel 8228 or 8238 System Controller provides this status latch and additionally encodes the outputs of the status latch with DBIN and $\overline{WR}$ to generate four timing signals MEM R (Memory Read), $\overline{MEM\ W}$ (Memory Write), $\overline{I/O\ R}$ (I/O Read), and $\overline{I/O\ W}$ (I/O Write). However, the 8228 and 8238 do not provide the outputs of the status latch. The timing between the Address Bus, Data Bus, $\overline{WR}$, and SYNC for both a memory write and an I/O write is shown in Figure 14.1.2-3. The 8080A also provides an input, READY, which allows the memory system to extend the time the Address and Data Bus is valid by integral clock cycles.

The Intel 8085A microprocessor provides a 16 line multiplexed Address Bus, an 8 line Data Bus, and a Control Bus consisting of $\overline{RD}$ (Read), $\overline{WR}$ (Write), ALE (Address Latch Enable), CLK (Clock), and 3 Status Outputs – S_0, S_1, and IO/$\overline{M}$. During every memory or I/O access, the lowest 8 Address bits $(A_7 - A_0)$ are brought out on the Data Bus to be loaded into an octal latch on the negative edge of ALE. The 3 Status Outputs specify whether the access is a Memory Read, Memory Write, I/O Read, or I/O Write. The $\overline{RD}$ and $\overline{WR}$ outputs provide the necessary timing for the Read or Write. The 8 bit address for an I/O instruction is duplicated on both halves of the Address Bus so that $(A_{15} - A_8) = (A_7 - A_0)$. Thus, the address latch is not

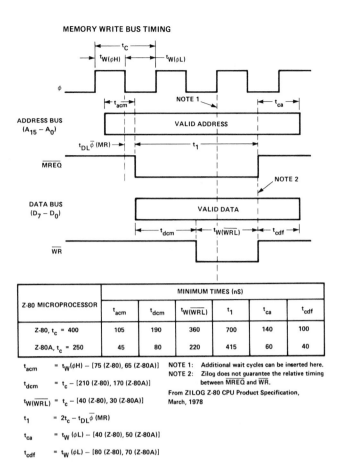

MEMORY WRITE BUS TIMING

Z-80 MICROPROCESSOR	MINIMUM TIMES (nS)					
	t_{acm}	t_{dcm}	$t_{W(\overline{WRL})}$	t_1	t_{ca}	t_{cdf}
Z-80, t_c = 400	105	190	360	700	140	100
Z-80A, t_c = 250	45	80	220	415	60	40

t_{acm} = $t_W(\phi H)$ – [75 (Z-80), 65 (Z-80A)]

t_{dcm} = t_c – [210 (Z-80), 170 (Z-80A)]

$t_{W(\overline{WRL})}$ = t_c – [40 (Z-80), 30 (Z-80A)]

t_1 = $2t_c - t_{DL}\overline{\phi}$ (MR)

t_{ca} = $t_W (\phi L)$ – [40 (Z-80), 50 (Z-80A)]

t_{cdf} = $t_W (\phi L)$ – [80 (Z-80), 70 (Z-80A)]

NOTE 1: Additional wait cycles can be inserted here.
NOTE 2: Zilog does not guarantee the relative timing between $\overline{MREQ}$ and $\overline{WR}$.
From ZILOG Z-80 CPU Product Specification, March, 1978

Figure 14.1.2-5a.   Memory Write Timing for the Zilog Z-80 Microprocessor Family.

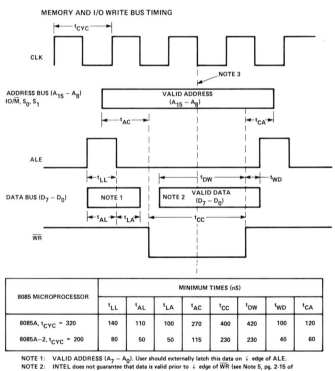

8085 MICROPROCESSOR	MINIMUM TIMES (nS)							
	t_{LL}	t_{AL}	t_{LA}	t_{AC}	t_{CC}	t_{DW}	t_{WD}	t_{CA}
8085A, t_{CYC} = 320	140	110	100	270	400	420	100	120
8085A–2, t_{CYC} = 200	80	50	50	115	230	230	40	60

NOTE 1: VALID ADDRESS $(A_7 - A_0)$. User should externally latch this data on ↓ edge of ALE.
NOTE 2: INTEL does not guarantee that data is valid prior to ↓ edge of $\overline{WR}$ (see Note 5, pg. 2-15 of 8085 USERS MANUAL, June 1977).
NOTE 3: Additional wait cycles can be inserted here.
From INTEL Component Data Catalog, 1978

Figure 14.1.2-4.   Memory and I/O Write Timing for the Intel 8085A Microprocessor Family.

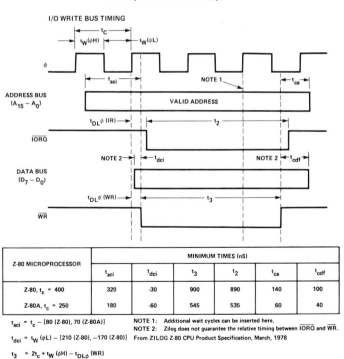

I/O WRITE BUS TIMING

Z-80 MICROPROCESSOR	MINIMUM TIMES (nS)					
	t_{aci}	t_{dci}	t_3	t_2	t_{ca}	t_{cdf}
Z-80, t_c = 400	320	-30	900	890	140	100
Z-80A, t_c = 250	180	-60	545	535	60	40

t_{aci} = t_c – [80 (Z-80), 70 (Z-80A)]

t_{dci} = $t_W (\phi L)$ – [210 (Z-80), –170 (Z-80)]

t_3 = $2t_c + t_W (\phi H) - t_{DL\phi}$ (WR)

t_2 = $2t_c + t_W (\phi H) - t_{DL\phi}$ (IR)

NOTE 1: Additional wait cycles can be inserted here.
NOTE 2: Zilog does not guarantee the relative timing between $\overline{IORQ}$ and $\overline{WR}$.
From ZILOG Z-80 CPU Product Specification, March, 1978

Figure 14.1.2-5b.   I/O Write Timing for the Zilog Z-80 Microprocessor Family.

PORT 2 STROBED WRITE TIMING

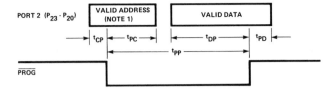

NOTE 1: ADDRESS FORMAT

P_{23} P_{22}	P_{21} P_{20}	X X			Y Y	
X X	Y Y	0 0	READ		0 0	PORT 4
		0 1	WRITE		0 1	PORT 5
		1 0	OR		1 0	PORT 6
		1 1	AND		1 1	PORT 7

8048 MICROPROCESSOR	MINIMUM TIMES (ns)				
	t_{PP}	t_{CP}	t_{PC}	t_{DP}	t_{PD}
8021, 3 MHz	1510	110	140	220	65
8022, 3 MHz	1510	110	140	220	65
8048, 6 MHz	1510	110	140	220	65
8049, 11 MHz	700	100	60	200	20

From INTEL MCS-48 User's Manual, July 1978.

Figure 14.1.2-6.   Port 2 Output Timing for the Intel 8048 Microprocessor Family.

BUS PORT WRITE TIMING (MOV X @ R, A INSTRUCTION)

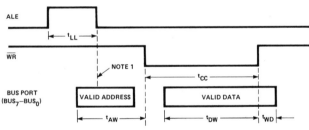

NOTE 1:   Address Word should be loaded into an octal latch (INTEL 8212 or 74LS363) on negative edge of ALE.

BUS PORT WRITE TIMING (OUT L BUS, A INSTRUCTION)

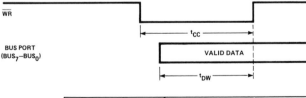

8048 MICROPROCESSOR	MINIMUM TIMES (ns)				
	t_{LL}	t_{AW}	t_{CC}	t_{DW}	t_{WD}
8048, 6 MHz	400	230	700	500	120
8049, 11 MHz	150	200	300	250	40

From INTEL MCS-48 User's Manual, July 1978.

Figure 14.1.2-7.   Bus Port Output Timing for the Intel 8048 Microprocessor Family.

STROBED WRITE TIMING

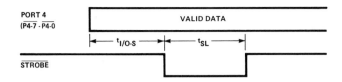

3870 MICROPROCESSOR	MINIMUM TIMES (ns)	
	$t_{I/O-S}$	t_{SL}
3870, 4 MHz	500	3750
3872, 4 MHz	500	3750

From MOSTEK MK3870 Data Sheet, February 1978
MOSTEK MK3872 Data Sheet, March 1978

Figure 14.1.2-8.   Port 4 Output Timing for the Mostek 3870 Microprocessor Family.

required for I/O operations. The timing for an I/O or Memory Write is shown in Figure 14.1.2-4. The 8085A also provides a Ready input to extend the setup time on the Address and Data Bus by integral clock cycles.

The Zilog Z-80 microprocessor uses a 16 line Address Bus, 8 line Data Bus, and provides a Control Bus that includes $\overline{MREQ}$ (Memory Request), $\overline{IORQ}$ (I/O Request), $\overline{RD}$ (Read), $\overline{WR}$ (Write), and ϕ (Clock). The Z-80 microprocessor provides a 16 bit I/O address space and several powerful output instructions. Memory write and I/O write timing is shown in Figure 14.1.2-5. The Z-80 provides an input, WAIT, which allows the memory access to be extended by integral clock cycles for slow memory or I/O devices.

The Intel 8021, 8022, 8041, 8048, and 8049 microprocessor family is designed for small microprocessor systems. These microprocessors provide program memory, scratch pad RAM memory, clock generator, timer, clock output, I/O ports, and CPU on the same integrated circuit. In many applications, the entire microprocessor system exclusive of I/O devices is located within this IC. The I/O ports can be used in several ways. Each line can be configured as an input or as an output. All of these microprocessors allow for strobed input or output using four port lines $(P_{23} - P_{20})$ and an output strobe, $\overline{PROG}$. This $\overline{PROG}$ output is especially designed to interface to the Intel 8243 I/O expander but can also be used with any positive edge triggered latch. $\overline{PROG}$ timing is shown in Figure 14.1.2-6. In addition, the 8048 and 8049 microprocessors provide an 8 bit port, $BUS_7 - BUS_0$, $\overline{RD}$ (Read), and $\overline{WR}$ (Write) for 8 bit data transfers. This timing is shown in Figure 14.1.2-7.

The Mostek 3870 and 3872 single chip mciroprocessors also provide program memory, scratch pad RAM memory, a clock generator, timer, I/O ports and CPU on the same integrated circuit. These microprocessors provide four 8 bit I/O ports and a STROBE output associated with one of these ports. Output strobe timing is shown in Figure 14.1.2-8.

14.1.3   Programming Considerations

Much of the power of the microprocessor lies in the flexibility of the microprocessor to implement a complicated operation by executing a series of instructions. These instructions are known as the program. Since execution speed is directly related to the number and types of instructions that are executed, the throughput of a microprocessor system depends largely on the way the programs are written. Just as in hardware design, the programmer has a number of tradeoffs to consider while writing a program. Some of these tradeoffs include the use of loops within a program and the use of subroutines.

The Conditional Branch instruction tests one or more of the microprocessor flags and, if the flag is properly set, changes the PC to a new address. Thus, the next instruction executed after the conditional branch is either the next instruction after the conditional branch or an instruction located somewhere else within the program. In many instances, a program segment consists of a series of repeated instructions. The conditional branch allows these instructions to be written once and executed several times. The number of times the instructions are to be executed can be determined by loading a register with the loop count, decrementing the register each time the loop is executed, and testing for zero. The advantage of the programming loop is that it shortens the length of the program segment. The disadvantage of the program loop is that it increases the execution time of the program segment because of the execution time of the decrement and conditional branch instructions. When a program is repeated 100's or 1000's or times a second, the use of a program loop substantially increases the program execution time as compared to the equivalent in-line code.

The Subroutine Branch instruction changes the PC to a new address in a way that allows the program to eventually return to the series of instructions following the subroutine branch. During program execution, the subroutine branch instruction temporarily stores the address of the next instruction following the subroutine branch in memory before it changes the PC to the address of the subroutine. Program execution continues with the desired subroutine. After the subroutine has been executed, a Return instruction loads the PC with the address that was temporarily stored in memory. Thus, the program continues executing with the next instruction following the subroutine branch. The subroutine branch allows a segment of program that is used by many different programs to be written once and accessed by each program. Subroutines can be nested

such that a subroutine would call another subroutine and so on. The level of subroutine nesting is determined by the way the return address is stored. Microprocessors, such as the 6800, 8080A, 8085A, and Z-80, provide a register within the microprocessor called the Stack Pointer (SP). The SP points to a location within RAM memory where the next return address is to be stored. The SP is automatically incremented and decremented as return addresses are stored and read from the 'Stack.' In this way, the level of subroutine nesting is limited only by the amount of RAM memory reserved for the Stack. The Stack can also be used to temporarily store data that is being used by a program. The subroutine can considerably reduce the amount of program code required for a system by eliminating redundant program segments. However, just as with program loops, when a subroutine is called 100's or 1000's of times a second from the same program, the program execution time can be improved by replacing the subroutine branch instruction in the calling program with the series of instructions that make up the body of the subroutine.

14.1.4   Interrupts

Most microprocessors have one or more inputs that are called Interrupt Inputs. These inputs allow an external device to request help from the microprocessor. During each instruction cycle, the microprocessor tests the state of these inputs. If the microprocessor detects an interrupt, the microprocessor performs a special type of subroutine branch. The program counter, accumulator, user accessible registers, and flag register are stored on the stack. The microprocessor then begins execution of a new program. This program services the external device. After the transaction is completed, the interrupt program restores the accumulator, user definable registers, and flag register to their previous values and performs a return. The original program then continues executing. The interrupt is virtually transparent to the main program except for the time required to service the external device. Usually one of the bits within the flag register is used to enable and disable interrupt requests. This interrupt flag is set and cleared by two different instructions. In this way, the main program can prevent itself from being interrupted during program segments that are time sensitive. The mechanism by which the interrupt is processed varies for different microprocessors.

The 6800 microprocessor has two interrupt inputs; $\overline{IRQ}$ (Interrupt Request), and $\overline{NMI}$ (Non-Maskable Interrupt). When the microprocessor detects an interrupt request on either of these inputs, the microprocessor reads two words from a special location in memory. These words specify the starting address for one of the two interrupt programs. The $\overline{IRQ}$ interrupt can be enabled and disabled from the main program. The $\overline{NMI}$ interrupt always forces an interrupt from the main program. The interrupt process automatically stores the program counter, accumulators, index register, and flag register on the stack.

The 8080A microprocessor has a single interrupt input, INT. When the microprocessor detects an interrupt request, it begins a special fetch cycle. The INTA (Interrupt Acknowledge) bit in the status word tells the interrupting device that the interrupt has been acknowledged. The interrupting device must then respond by loading a special instruction (RST) on the Data Bus. This instruction provides a subroutine branch to one of eight locations in memory. The INT input can be enabled and disabled from the main program. The interrupt process automatically stores the program counter on the stack. Registers that are used by the interrupt program must also be stored on the stack prior to their usage.

The 8085A microprocessor has five interrupt inputs: INTR (Interrupt Request), RST 5.5, RST 6.5, RST 7.5 (Restart Interrupts), and TRAP (Non-Maskable Interrupt). The INTR input functions in the same way as the INT input on the 8080A. RST 5.5, RST 6.5, and RST 7.5 interrupt inputs automatically generate an RST instruction that causes a subroutine branch to address $2C_{16}$, 34_{16}, or $3C_{16}$. The INTR and RST inputs are internally prioritized by the 8085A such that RST 7.5 has the highest priority and INTR has the lowest priority. The INTR input and each RST input can be individually enabled or disabled from the main program. The TRAP interrupt input generates a subroutine branch to address 24_{16}. This input has the highest priority of all five interrupts and cannot be disabled by the main program. The interrupt process automatically stores the program counter on the stack. All registers used by the interrupt program should also be stored in the stack prior to their use.

The Z-80 microprocessor has two interrupt inputs, $\overline{INT}$ (Interrupt Request) and $\overline{NMI}$ (Non-Maskable Interrupt). The $\overline{INT}$ input is sampled at the end of each instruction. If an interrupt is detected, a special fetch cycle is executed and $\overline{IORQ}$ goes low to indicate that the interrupt has been acknowledged. The Z-80 operates in any one of three interrupt modes. One mode automatically generates an RST instruction to address 38_H. Another mode requires the interrupting device to load an RST instruction on the Data Bus. The third mode requires the interrupting device to supply an 8 bit address on the Data Bus, which is then used to look up the starting address of one of 256 interrupt programs. The $\overline{INT}$ input can be enabled and disabled by the main program. The $\overline{NMI}$ input always forces an interrupt to address 66_{16}. The interrupt procedure automatically stores the program counter on the stack. The Z-80 microprocessor provides duplicate registers which can be used by the interrupt program or the user can store the register contents on the stack.

The 8048, 8049, 8022, and 3870 microprocessors have a single interrupt input that can be enabled or disabled from the main program. If the interrupt is accepted, the program counter is stored on an 8 level stack for the 8022, 8048, and 8049 microprocessors or in register PC1 for the 3870. Execution commences at address 3_{16} for the 8022, 8048, and 8049 microprocessors or AO_{16} for the 3870 microprocessors. Further interrupt requests are ignored until the previous interrupt has been serviced.

14.2 Display Interface Techniques

LED Displays can be interfaced to microprocessor systems with five different techniques as shown in Figure 14.2-1a-e:

1. The DC DRIVEN CONTROLLER statically drives an LED display from a microprocessor output port. Typically, each display digit would be assigned a different address so that a Memory or I/O Write to that address would change the contents of the corresponding display digit.

2. The REFRESH CONTROLLER interfaces the microprocessor system to a multiplexed LED display. The controller periodically interrupts the microprocessor and after each interrupt, the microprocessor supplies new display data for the next refresh cycle of the display.

3. The DECODED DATA CONTROLLER refreshes a multiplexed LED display independently from the microprocessor system. A local RAM stores decoded display data. This data is continuously read from the RAM and then used to refresh the display. Whenever the display message is changed, the microprocessor decodes each character in software and writes the decoded data into the local RAM.

4. The CODED DATA CONTROLLER also refreshes a multiplexed LED display independently from the microprocessor system. The local RAM stores ASCII data which is continuously read from the RAM, decoded, and used to refresh the display. The display message is changed by writing new ASCII characters within the local RAM.

5. The DISPLAY PROCESSOR CONTROLLER uses a separate microprocessor to drive the LED display. This microprocessor provides ASCII storage, ASCII decode, and display refresh independently from the main microprocessor system. Software within the dedicated microprocessor provides many powerful features not available in the other controllers. The main microprocessor updates the LED display by sending new ASCII data to the slave microprocessor.

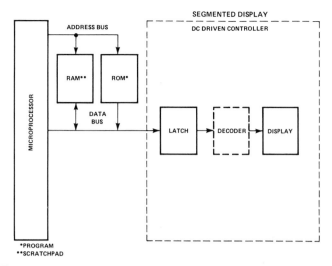

Figure 14.2-1a.   DC DRIVEN CONTROLLER Display Interface.

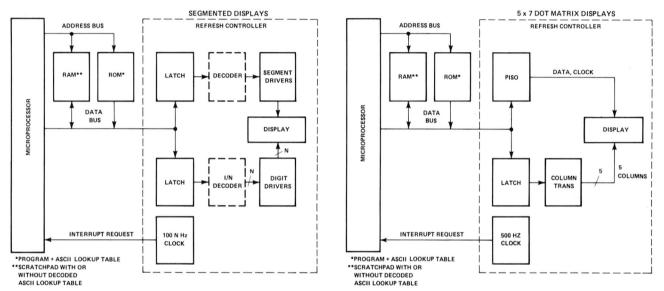

Figure 14.2-1b.   REFRESH CONTROLLER Display Interface.

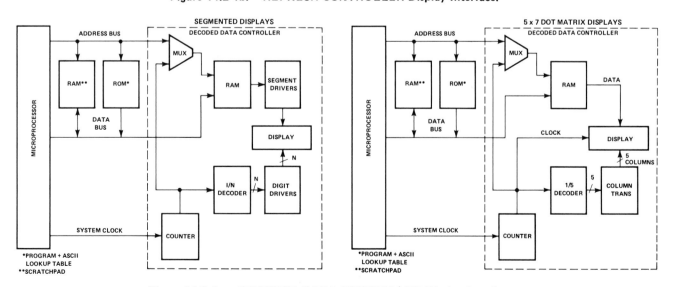

Figure 14.2-1c.   DECODED DATA CONTROLLER Display Interface.

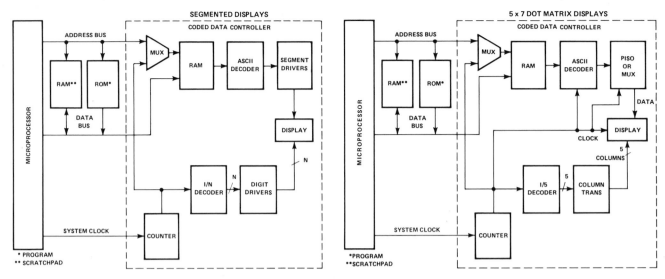

Figure 14.2-1d. CODED DATA CONTROLLER Display Interface.

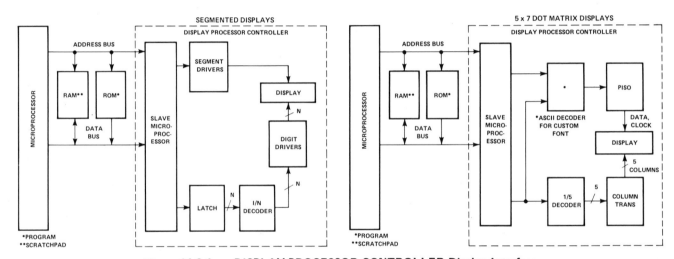

Figure 14.2-1e. DISPLAY PROCESSOR CONTROLLER Display Interface.

14.2.1 Comparison of Interface Techniques

The choice of a particular interface is an important consideration because it affects the design of the entire microprocessor system. Each interface requires one or more Memory or I/O addresses. These addresses are generated by decoding the microprocessor Address Bus. The display decoder can be located within the microprocessor program or as circuitry within the display interface. Location of the display decoder within the microprocessor program gives the designer total control of the display font within the program. This feature can be particularly important if the display will be used to display different languages and special graphics symbols. Naturally, the segments of the microprocessor program that change the LED display must be tailored to one particular interface. However, the interface may also limit some of the software techniques that can be used in the other segments of the microprocessor program. For example, interface techniques using an Interrupt require that the program status can be stored on the microprocessor stack at any time or that the Interrupt Input be enabled and disabled during sensitive portions of the microprocessor program. In addition, each interface technique requires a specific amount of microprocessor time.

The DC DRIVEN CONTROLLER requires interaction from the microprocessor only when the LED display is changed. The display decoder can be located either within the microprocessor program or as circuitry within the interface. Since the display interface circuitry must be duplicated for each digit of the display, this technique becomes expensive for long LED display lengths. The DC DRIVEN CONTROLLER can not be used with multiplexed LED displays. Thus, this technique is applicable primarily for LED Lamps, Large Seven Segment Displays, and On Board Integrated Circuit (OBIC) Displays.

The REFRESH CONTROLLER requires continuous interaction from the microprocessor system. Since the microprocessor actively strobes the LED display, the display interface circuitry is reduced. Generally, this technique provides the lowest hardware cost for any given display length. The display decoder can be located either within the microprocessor program or as circuitry within the interface. Display strobing is accomplished through use of the microprocessor Interrupt circuitry. Demands upon microprocessor time are directly proportional to display length. This technique applies to all multiplexed LED displays such as Monolithic Displays and Alphanumeric Displays.

The DECODED DATA CONTROLLER and CODED DATA CONTROLLER require microprocessor interaction only when the display message is changed. Both techniques employ a local RAM memory that is continuously scanned by the display interface electronics. For the DECODED DATA CONTROLLER, the display decoder is located within the microprocessor software and the local RAM stores decoded display data. The CODED DATA CONTROLLER includes the display decoder within the display interface circuitry and the local RAM stores BCD or ASCII data. Since BCD or ASCII data is more compact than decoded display data, the CODED DATA CONTROLLER uses a smaller RAM than the DECODED DATA CONTROLLER. Both techniques allow the microprocessor to individually change each display character by a Memory or I/O Write to a specific display address. These interface techniques can accept new data at a very high rate

The DISPLAY PROCESSOR CONTROLLER requires microprocessor interaction only when the LED display message is changed. By using a dedicated microprocessor, the DISPLAY PROCESSOR CONTROLLER provides many additional display features. The software within the DISPLAY PROCESSOR CONTROLLER further reduces the microprocessor interaction by providing more powerful Left and Right data entry modes compared to the RAM entry mode of the DECODED DATA and CODED DATA CONTROLLERS. The DISPLAY PROCESSOR CONTROLLER can also provide features such as a Blinking Cursor, Editing Commands, and a Data Out function. The display decoder can either be designed into the dedicated display microprocessor or can be located within a separate PROM. This allows the user to provide a custom display decoder for applications needing a special character font. The DISPLAY PROCESSOR CONTROLLER does not allow as high a data entry rate as either the DECODED DATA or CODED DATA CONTROLLERS.

14.3   DC Driven Controllers

The DC DRIVEN CONTROLLER is the simplest type of display interface. The DC DRIVEN CONTROLLER consists of one or more registers that are connected so that they can be loaded from the Data Bus by a Memory or I/O Write. The outputs of the registers are used to directly drive the LED Display or are connected to BCD to 7 segment decoder/drivers which, in turn, drive the LED Display. The display is changed by writing new data into the appropriate registers. An example of this type of interface is shown in Figure 5.2.6-1. In this interface, several Fairchild 9374 latched 7 segment decoder/drivers are connected directly to the microprocessor Data Bus. The enable input of each 9374 is assigned a different Memory or I/O Address so that a Memory or I/O Write to that address location will store a new BCD code within the 9374. The National DS 8859 hex latched LED Driver provides storage for the overflow display and the decimal points of each 7 segment display. Some displays include a latch and decoder driver within the display. Thus, these displays can be connected directly to the microprocessor Data Bus. Figures 5.1.3-1 and 5.1.3-2 show examples of this type of interface. The BCD inputs of the display are connected directly to the microprocessor Data Bus and the Enable input is assigned a Memory or I/O Address. Different versions of the OBIC LED Display provide for the display of numeric or hexidecimal data.

14.4   Refresh Controllers

Figure 14.4-1 shows a REFRESH CONTROLLER for a 16 character HDSP-6508 Alphanumeric Display. The circuit operates by interrupting the microprocessor at a 1600 Hz rate. Following each Interrupt, the microprocessor responds by outputting a new ASCII character to the Texas Instruments AC5947 ASCII to 18 Segment Decoder/Driver and a new digit word to the 74LS174. The character font for the AC5947 is shown in Figure 14.4-2. The outputs of the 74LS174 are decoded such that digit word 00_{16} turns the leftmost display character on, digit word $0F_{16}$ turns the rightmost display character on, and digit word $1F_{16}$ turns all digits off. The interface can be expanded to 24 characters with an additional Signetics NE590 driver.

A 6800 microprocessor program that interfaces to this REFRESH controller is shown in Figure 14.4-3. Following each Interrupt, the program "RFRSH" is executed. The program uses a scratch pad register "POINT" that points to the location within a 16 byte ASCII message of the next ASCII character to be stored in the display interface. The scratch pad register "DIGIT" contains the next digit word to be loaded into the display interface. The program interfaces to the circuit through two memory or I/O addresses. A Memory Write to address "SEG" writes a six bit word into the AC5947, and a Memory Write to address "DIG" writes a five bit word into the 74LS174. To prevent undesirable ghosting, the digit drivers are turned off prior to loading the next ASCII character into the AC5947. After sufficient delay, the next digit is turned on. Registers "POINT" and "DIGIT" are then updated by the program.

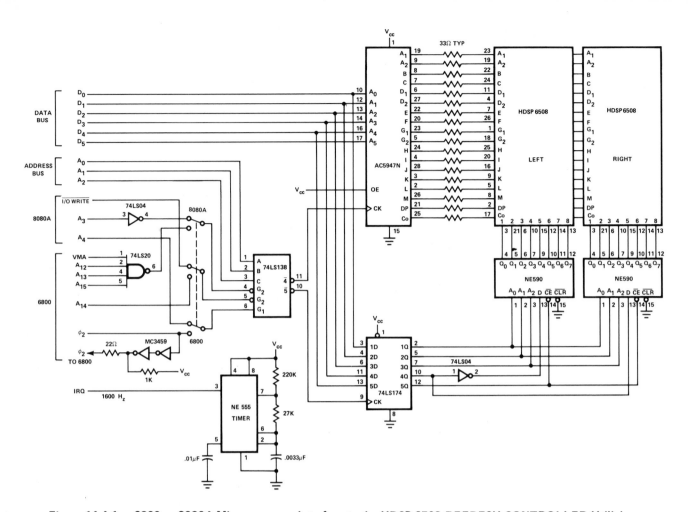

Figure 14.4-1. 6800 or 8080A Microprocessor Interface to the HDSP-6508 REFRESH CONTROLLER Utilizing the Texas Instruments AC5947 ASCII to 18 Segment Decoder/Driver.

BITS		D_3 D_2 D_1 D_0	0 0 0 0	0 0 0 1	0 0 1 0	0 0 1 1	0 1 0 0	0 1 0 1	0 1 1 0	0 1 1 1	1 0 0 0	1 0 0 1	1 0 1 0	1 0 1 1	1 1 0 0	1 1 0 1	1 1 1 0	1 1 1 1
D_6 D_5 D_4	HEX		0	1	2	3	4	5	6	7	8	9	A	B	C	D	E	F
0 1 0	2		(space)	!	"	‡	5	%	L	'	⟨	⟩	✳	+	╱	—	.	╱
0 1 1	3		0	1	2	3	4	5	6	7	8	9	:	;	∠	=	╲	?
1 0 0	4		⊡	A	B	C	D	E	F	G	H	I	J	K	L	M	N	O
1 0 1	5		P	Q	R	S	T	U	V	W	X	Y	Z	[	\	]	↑	←

Figure 14.4-2. 18 Segment Display Font for the Texas Instruments AC5947 ASCII to 18 Segment Decoder/Driver.

Following execution of the "RTI" instruction, execution of the main program is resumed. A similar program written for an 8080A microprocessor is shown in Figure 14.4-4.

The REFRESH CONTROLLER shown in Figure 14.4-5 uses a software ASCII decoder to provide total font control for the HDSP-6508 Alphanumeric Display. This circuit replaces the AC5947 in Figure 14.4-1 with 3 six bit latches

LOC	OBJECT CODE		SOURCE STATEMENTS		
	BF04		SEG	EQU	$BF04
	BF05		DIG	EQU	$BF05
0000	0003		POINT	FDB	DATA
0002	00		DIGIT	FCB	0
0003			DATA	RMB	16
0400				ORG	$0400
0400	DE	00	RFRSH	LDX	D , POINT
0402	E6	00		LDA B	X , 0
0404	86	1F		LDA A	I , $1F
0406	B7	BF05		STA A	E , DIG
0409	F7	BF04		STA B	E , SEG
040C	96	02		LDA A	D , DIGIT
040E	81	0F		CMP A	I , 15
0410	27	0A		BEQ	LOOP1
0412	7C	0002		INC	E , DIGIT
0415	08			INX	
0416	B7	BF05		STA A	E , DIG
0419	DF	00		STX	D , POINT
041B	3B			RTI	
041C	7F	0002	LOOP1	CLR	E , DIGIT
041F	F6	0001		LDA B	E , POINT+1
0422	B7	BF05		STA A	E , DIG
0425	C0	0F		SUB B	I , 15
0427	D7	01		STA B	D , POINT+1
0429	24	03		BCC	LOOP2
042B	7A	0000		DEC	E , POINT
042E	3B		LOOP2	RTI	

Figure 14.4-3. 6800 Microprocessor Program That Interfaces to the REFRESH CONTROLLER Shown in Figure 14.4-1.

and 3 segment drivers. This circuit also uses a 1600 Hz Interrupt generator. Following each Interrupt, the microprocessor responds by outputting 3 six bit words to the segment drivers and a new digit word. The outputs of the 74LS174 segment latches are decoded so that a low output turns the appropriate segment on. The outputs of the 74LS174 digit latch are decoded so that 0_{16} turns the leftmost digit driver on and F_{16} turns the rightmost digit driver on.

An 8080A microprocessor program that interfaces to the circuit in Figure 14.4-5 is shown in Figure 14.4-6. Following each Interrupt, the program "RFRSH" is executed. In order to minimize requirements for microprocessor time, this program utilizes a 48 byte scratch pad memory that contains decoded display data. This memory is organized such that the first three bytes contain segment information for the leftmost display character and the last three bytes contain segment information for the rightmost display character. The program uses a scratch pad register "POINT" that points to the location within the 48 byte scratch pad memory of the next decoded data character to be stored in the display interface. Scratch pad register "DIGIT" contains the next digit word to be loaded into the display interface. The program interfaces to the circuit through four memory or I/O addresses. A Memory Write to address "SEG1" updates segments a_1 to d_2, "SEG2" updates segments e to i, and "SEG3" updates segments j to colon.

LOC	OBJECT CODE			SOURCE STATEMENTS		
001C					SEG	EQU	001CH
001D					DIG	EQU	001DH
					ORG	0E000H
E000	03	E0		POINT	DW	DATA
E002	00			DIGIT	DB	00H
E003	00			DATA	DS	16
					ORG	0E400H
E400	F5			RFRSH	PUSH	PSW
E401	E5				PUSH	H
E402	2A	00	E0		LHLD	POINT
E405	3E	1F			MVI	A , 1FH
E407	D3	1D			OUT	DIG
E409	7E				MOV	A , M
E40A	D3	1C			OUT	SEG
E40C	3A	02	E0		LDA	DIGIT
E40F	D3	1D			OUT	DIG
E411	FE	0F			CPI	15
E413	CA	21	E4		JZ	LOOP1
E416	3C				INR	A
E417	32	02	E0		STA	DIGIT
E41A	23				INX	H
E41B	22	00	E0	LOOP2	SHLD	POINT
E41E	E1				POP	H
E41F	F1				POP	PSW
E420	C9				RET	
E421	3E	00		LOOP1	MVI	A , 0
E423	32	02	E0		STA	DIGIT
E426	7D				MOV	A , L
E427	D6	0F			SUI	15
E429	6F				MOV	L , A
E42A	D2	1B	E4		JNC	LOOP2
E42D	25				DCR	H
E42E	C3	1B	E4		JMP	LOOP2

Figure 14.4-4. 8080A Microprocessor Program That Interfaces to the REFRESH CONTROLLER Shown in Figure 14.4-1.

A Memory Write to address "DIG" writes a four bit word to the digit drivers. Interdigit blanking is achieved by turning off the segment drivers prior to updating the digit drivers. Then, new segment information is written into the segment drivers. Finally, registers "POINT" and "DIGIT" are updated by the program. Following execution of the "RET" instruction, the microprocessor resumes execution of the main program. The contents of the alphanumeric display are changed by replacing the contents of the 48 byte scratch pad memory with new display data. Subroutine "LOAD" decodes a string of 16 ASCII characters and stores the resulting decoded display data into the scratch pad. Scratch pad register "ASCII" points to the first ASCII character in the message. Subroutine "LOAD" executes in 1.0 milliseconds for a 2 MHz 8080A microprocessor. The ASCII to 18 Segment Decoder, "DECDR", is a 192 byte look up table located in the main program. The display font generated by "DECDR" is shown in Figure 14.4-7. This display font includes the mathematical symbols +, −, *, ÷, and √ and shows the use of full width numbers.

14.13

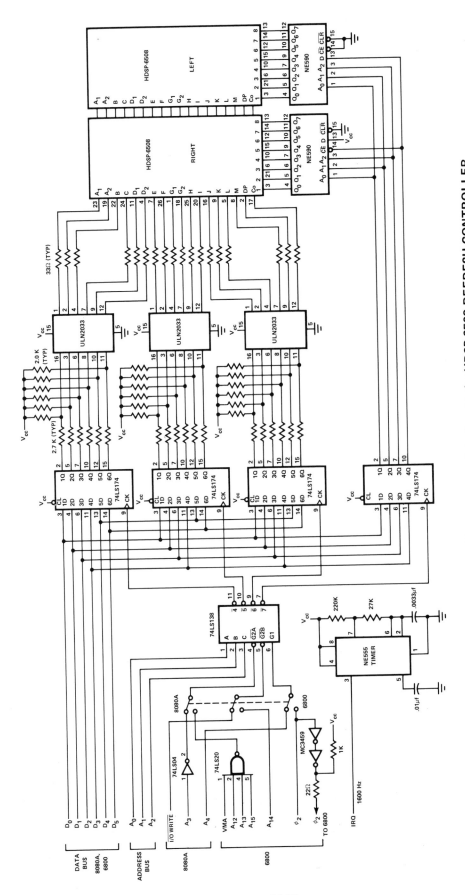

Figure 14.4-5. 6800 or 8080A Microprocessor Interface to the HDSP-6508 REFRESH CONTROLLER.

14.14

```
        LOC     OBJECT CODE      SOURCE STATEMENTS

        001C                    SEG1    EQU     1CH
        001D                    SEG2    EQU     1DH
        001E                    SEG3    EQU     1EH
        001F                    DIG     EQU     1FH
        E500                    DECDR   EQU     0E500H

                                        ORG     0E000H
        E000    03  E0          POINT   DW      DATA
        E002    00              DIGIT   DB      00H
        E003    00              DATA    DS      48

                                        ORG     0E033H
        E033    35  E0          ASCII   DW      MESSGE
        E035    00              MESSGE  DS      16

                                        ORG     0E400H
        E400    F5              RFRSH   PUSH    PSW
        E401    E5                      PUSH    H
        E402    2A  00  E0              LHLD    POINT
        E405    3E  FF                  MVI     A , 0FFH
        E407    D3  1C                  OUT     SEG1
        E409    D3  1D                  OUT     SEG2
        E40B    D3  1E                  OUT     SEG3
        E40D    3A  02  E0              LDA     DIGIT
        E410    D3  1F                  OUT     DIG
        E412    7E                      MOV     A , M
        E413    D3  1C                  OUT     SEG1
        E415    23                      INX     H
        E416    7E                      MOV     A , M
        E417    D3  1D                  OUT     SEG2
        E419    23                      INX     H
        E41A    7E                      MOV     A , M
        E41B    D3  1E                  OUT     SEG3
        E41D    3A  02  E0              LDA     DIGIT
        E420    FE  0F                  CPI     15
        E422    CA  30  E4              JZ      LOOP1
        E425    3C                      INR     A
        E426    32  02  E0              STA     DIGIT

        E429    23                      INX     H
        E42A    22  00  E0      LOOP2   SHLD    POINT
        E42D    E1                      POP     H
        E42E    F1                      POP     PSW
        E42F    C9                      RET
        E430    3E  00          LOOP1   MVI     A , 0
        E432    32  02  E0              STA     DIGIT
        E435    21  03  E0              LXI     H , DATA
        E438    C3  2A  E4              JMP     LOOP2

        E43B    01  03  E0      LOAD    LXI     B , DATA
        E43E    16  E5                  MVI     D , 0E5H
        E440    2A  33  E0              LHLD    ASCII
        E443    7E              LOOP3   MOV     A , M
        E444    07                      RLC
        E445    86                      ADD     M
        E446    5F                      MOV     E , A
        E447    23                      INX     H
        E448    1A                      LDAX    D
        E449    02                      STAX    B
        E44A    13                      INX     D
        E44B    03                      INX     B
        E44C    1A                      LDAX    D
        E44D    02                      STAX    B
        E44E    13                      INX     D
        E44F    03                      INX     B
        E450    1A                      LDAX    D
        E451    02                      STAX    B
        E452    03                      INX     B
        E453    3E  33                  MVI     A , 033H
        E455    B9                      CMP     C
        E456    C2  43  E4              JNZ     LOOP3
        E459    22  33  E0              SHLD    ASCII
        E45C    C9                      RET
```

Figure 14.4-6. 8080A Microprocessor Program That Interfaces to the REFRESH CONTROLLER Shown in Figure 14.4-5.

A 6800 microprocessor program that also interfaces to the circuit shown in Figure 14.4-5 is shown in Figure 14.4-8. Following each Interrupt, the program "RFRSH" is executed. This program eliminates the need for a scratch pad memory to hold decoded display data by decoding each ASCII character immediately before loading into the display interface. Scratch pad register "POINT" points to the next ASCII character to be stored in the display interface. Scratch pad register "DIGIT" contains the next digit word to be stored in the display. This program operates in a similar manner to the program shown in Figure 14.4-6 except that the ASCII character is decoded by "RFRSH."

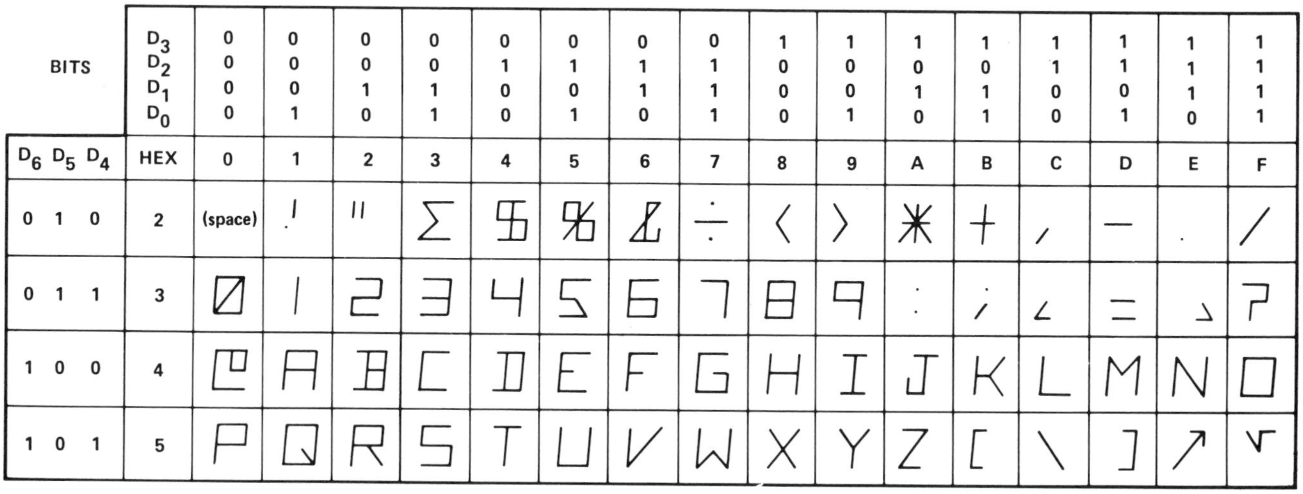

BITS	D_3 D_2 D_1 D_0	0 0 0 0	0 0 0 1	0 0 1 0	0 0 1 1	0 1 0 0	0 1 0 1	0 1 1 0	0 1 1 1	1 0 0 0	1 0 0 1	1 0 1 0	1 0 1 1	1 1 0 0	1 1 0 1	1 1 1 0	1 1 1 1
D_6 D_5 D_4	HEX	0	1	2	3	4	5	6	7	8	9	A	B	C	D	E	F
0 1 0	2	(space)	!	"	Σ	5	%	∠	÷	<	>	✳	+	,	−	.	/
0 1 1	3	0	1	2	3	4	5	6	7	8	9	:	;	∠	=	↘	?
1 0 0	4	@	A	B	C	D	E	F	G	H	I	J	K	L	M	N	O
1 0 1	5	P	Q	R	S	T	U	V	W	X	Y	Z	[	\	]	↗	∨

Figure 14.4-7.   One Possible 18 Segment Display Font for the REFRESH CONTROLLER Shown in Figure 14.4-5.

LOC	OBJECT CODE		SOURCE STATEMENTS	
	BF04		SEG1	EQU	$BF04
	BF05		SEG2	EQU	$BF05
	BF06		SEG3	EQU	$BF06
	BF07		DIG	EQU	$BF07
	0600		DECDR	EQU	$0600
0000	0007		POINT	FDB	MESSGE
0002	00		DIGIT	FCB	0
0003	0007		ASCII	FDB	MESSGE
0005			PAD1	RMB	2
0007			MESSGE	RMB	16
0400				ORG	$0400
0400	86	FF	RFRSH	LDA A	I , $FF
0402	B7	BF04		STA A	E , SEG1
0405	B7	BF05		STA A	E , SEG2
0408	B7	BF06		STA A	E , SEG3
040B	96	02		LDA A	D , DIGIT
040D	B7	BF07		STA A	E , DIG
0410	DE	00		LDX	D , POINT
0412	E6	00		LDA B	X , 0
0414	58			ASL B	
0415	EB	00		ADD B	X , 0
0417	08			INX	
0418	DF	00		STX	D , POINT
041A	CB	00		ADD B	I , >DECDR
041C	D7	06		STA B	D , PAD1+1
041E	C6	06		LDA B	I , <DECDR
0420	C9	00		ADC B	I , 0
0422	D7	05		STA B	D , PAD1
0424	DE	05		LDX	D , PAD1
0426	E6	00		LDA B	X , 0
0428	F7	BF04		STA B	E , SEG1
042B	E6	01		LDA B	X , 1
042D	F7	BF05		STA B	E , SEG2
0430	E6	02		LDA B	X , 2
0432	F7	BF06		STA B	E , SEG3
0435	81	0F		CMP A	I , 15
0437	27	04		BEQ	LOOP1
0439	4C			INC A	
043A	97	02		STA A	D , DIGIT
043C	3B			RTI	
043D	DE	03	LOOP1	LDX	D , ASCII
043F	DF	00		STX	D , POINT
0441	7F	0002		CLR	E , DIGIT
0444	3B			RTI	

Figure 14.4-8.   6800 Microprocessor Program That Interfaces to the REFRESH CONTROLLER Shown in Figure 14.4-5.

Because the microprocessor system is interrupted at a rate of 100 Hz times the display length, software design is especially important for the HDSP-6508 REFRESH CONTROLLERS. Figure 14.4-9 compares the microprocessor time required for the refresh programs shown in Figures 14.4-3, 14.4-4, 14.4-6, and 14.4-8 as a function of display length. The times shown are based upon a 100 Hz refresh rate, 1 MHz clock for the 6800 microprocessor, and 2 MHz clock for the 8080A microprocessor. The REFRESH CONTROLLERS utilizing the AC5947 require the smallest amount of microprocessor time because the ASCII decoding is done in hardware. The use of the 3n byte scratch pad memory, where n is the display length, for the REFRESH CONTROLLER shown in Figure 14.4-5 also saves microprocessor time. This is particularly important

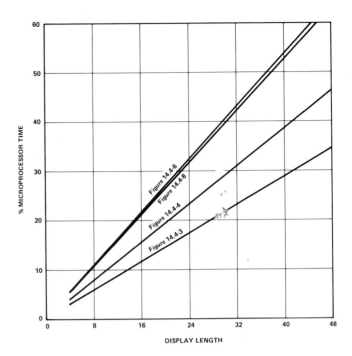

Figure 14.4-9.   Comparison of Required Microprocessor Time Versus Display Length for the Microprocessor Programs Shown in Figures 14.4-3, 14.4-4, 14.4-6, and 14.4-8.

for long display lengths. Faster versions of the 6800 and 8080A microprocessors can reduce the microprocessor times shown in Figure 14.4-9 by up to 50%.

The REFRESH CONTROLLER shown in Figure 14.4-10 is designed to interface the HDSP-2000 Alphanumeric Display to either a 6800 or 8080A microprocessor system. The circuit operates by interrupting the microprocessor at a 500 Hz rate. Following each interrupt, the microprocessor responds by outputting decoded display data for one column of the display and new column select data. Display data is loaded from the Data Bus into the serial input of the display via a 74165 parallel in, serial out shift register. The 74LS293 counter and associated gates provide seven clock pulses to the display for each word loaded into the 74165. The 74LS122 divides the ϕ_2 clock from the 8080A microprocessor system by two to provide the proper timing between ϕ_2 and the write signal to the 74165. Column select data is loaded into a 74174 latch, which drives the column transistors. The circuit can interface to an HDSP-2000 display of any length by using appropriate column transistors and properly buffering the display clock. The circuit timing is shown in Figure 14.4-11.

A 6800 microprocessor program that interfaces to the HDSP-2000 REFRESH CONTROLLER is shown in Figure 14.4-12. Following each Interrupt, the program "RFRSH" is executed. In order to minimize the microprocessor time required for the display interface, this program uses a

14.16

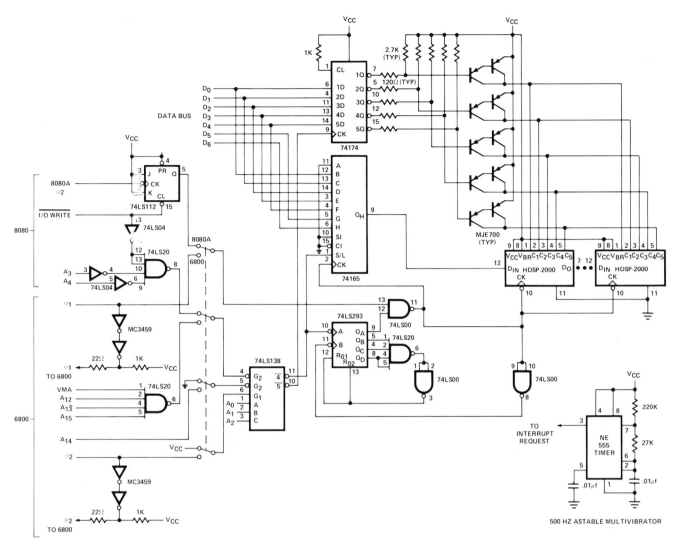

Figure 14-4.10.   6800 or 8080A Microprocessor Interface to the HDSP-2000 REFRESH CONTROLLER.

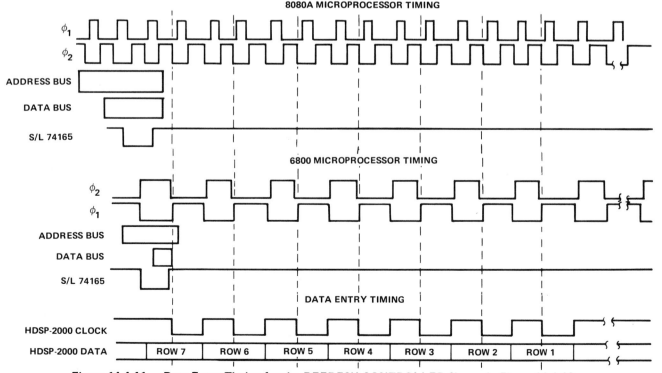

Figure 14.4-11.   Data Entry Timing for the REFRESH CONTROLLER Shown in Figure 14.4-10.

```
         OBJECT
   LOC   CODE          SOURCE STATEMENTS
                       *
                       *
         BF 05    CDVR    EQU   $BF05
         BF 04    RDVR    EQU   $BF04
         06 00    DECDR   EQU   $0600
   0000           POINT   RMB   2
   0002           COLMN   RMB   1
   0003           COUNT   RMB   2

   0005  00 AD    ASCII   FDB   DATA
   0007           DISPNT  RMB   2
   0009           DCRPNT  RMB   2
   000B           COLCNT  RMB   1
   000C           DIGCNT  RMB   1

   000D           BUFFR   RMB   160
   00AD           DATA    RMB   32

   0400                   ORG   $0400
   0400  86 FF    RFRSH   LDA A  I, $FF
   0402  B7 BF 05         STA A  E, CDVR
   0405  DE 00            LDX    D, POINT
   0407  A6 00    LOOPHH  LDA A  X, 0
   0409  B7 BF 04         STA A  E, RDVR
   040C  A6 01            LDA A  X, 1
   040E  B7 BF 04         STA A  E, RDVR
                           •
                           •
                           •
   04A2  A6 1F            LDA A  X, 31
   04A4  B7 BF 04         STA A  E, RDVR
   04A7  96 02            LDA A  D, COLMN
   04A9  B7 BF 05         STA A  E, CDVR
   04AC  81 EF            CMP A  I, $EF
   04AE  27 10            BEQ    LOOPB
   04B0  D6 06            LDA B  D, POINT +1
   04B2  CB 20            ADD B  I, 32
   04B4  D7 00            STA B  D, POINT +1
   04B6  24 03            BCC    LOOPA
   04B8  7C 00 00         INC    E, POINT
   04BB  OD       LOOPA   SEC
   04BC  79 00 02         ROL    E, COLMN
   04BF  3B               RTI
   04C0  CE 00 0D LOOPB   LDX    I, BUFFR
   04C3  DF 00            STX    D, POINT
   04C5  DE 03            LDX    D, COUNT
   04C7  09               DEX
   04C8  DF 03            STX    D, COUNT
   04CA  86 FE            LDA A  I, $FE
   04CC  97 02            STA A  D, COLMN
   04CE  3B               RTI

   04CF  5F       LOAD    CLR B
   04D0  CE 00 0D         LDX    I, BUFFR
   04D3  DF 07            STX    D, DISPNT
   04D5  86 06            LDA A  I, <DECDR
   04D7  97 09            STA A  D, DCRPNT
   04D9  86 05            LDA A  I, 5
   04DB  97 0B            STA A  D, COLCNT
   04DD  86 20    LOOP1   LDA A  I, 32
   04DF  97 0C            STA A  D, DIGCNT
   04E1  9B 06            ADD A  D, ASCII+1
   04E3  24 03            BCC    LOOP2
   04E5  7C 00 05         INC    E, ASCII
   04E8  97 06    LOOP2   STA A  D, ASCII+1
   04EA  DE 05    LOOP3   LDX    D, ASCII
   04EC  09               DEX
   04ED  A6 00            LDA A  X, 0
   04EF  DF 05            STX    D, ASCII
   04F1  1B               ABA
   04F2  97 0A            STA A  D, DCRPNT+1
   04F4  DE 09            LDX    D, DCRPNT
   04F6  A6 00            LDA A  X, 0
   04F8  DE 07            LDX    D, DISPNT
   04FA  A7 00            STA A  X, 0
   04FC  08               INX
   04FD  DF 07            STX    D, DISPNT
   04FF  7A 00 0C         DEC    E, DIGCNT
   0502  26 E6            BNE    LOOP3
   0504  CB 80            ADD B  I, $80
   0506  24 03            BCC    LOOP4
   0508  7C 00 09         INC    E, DCRPNT
   050B  7A 00 0B LOOP4   DEC    E, COLCNT
   050E  26 CD            BNE    LOOP1
   0510  39               RTS
```

Figure 14.4-12.   6800 Microprocessor Program That Interfaces to the REFRESH CONTROLLER Shown in Figure 14.4-10.

scratch pad memory to store decoded display data. This scratch pad memory is 5n bytes long, where n is the display length, and is organized such that the first n bytes contain column 1 data, second n bytes contain column 2 data, and so on. Scratch pad register "POINT" points to the location in the scratch pad memory of the next column of display data to be stored in the display interface. Scratch pad register "COLMN" contains the next column word to be stored in the display interface. The program interfaces to the circuit through two memory or I/O addresses. A Memory Write to "RDVR" loads a seven bit word into the 74165 shift register and a Memory Write to "CDVR" loads a five bit word into the 74174 latch. The program turns off all columns by writing column select word FF_{16} to "CDVR." The display could also be turned off by pulling V_B low. Next the microprocessor reads n bytes from the scratch pad memory such that the first word read is the rightmost display character and stores each word at "RDVR." Then, the next column select word is read from "COLMN" and stored at "CDVR." Finally, the microprocessor updates "POINT" and "COLMN." Following execution of the "RTI" instruction, execution of the main program is resumed. Subroutine "LOAD" is used to decode a string of 32 ASCII characters into the scratch pad memory. "LOAD" uses a scratch pad register "ASCII" that points to the first ASCII character to be decoded. Subroutine "LOAD" executes in 10.2 milliseconds for a 1 MHz 6800 microprocessor.

The 128 character ASCII to 5 x 7 Decoder, "DECDR." is a 640 byte look up table located in the main program. The table is organized so that the first 128 bytes contain column 1 data, the next 128 bytes contain column 2 data, and so on. ASCII character 00_{16} (NULL) would be the first byte in each column of data. Each byte within the table is formatted so that D_6 through D_0 contain ROW_7 through ROW_1 display data respectively such that a logical "ONE" would turn the corresponding display dot ON. This decoder table is shown in Figure 14.4-13. The display font generated by "DECDR" is the same as shown on the HDSP-2471 data sheet.

An equivalent program for an 8080A microprocessor is shown in Figure 14.4-14. This program is similar to the 6800 program shown in Figure 14.4-12 except that the program uses a loop to read the display data from the scratch pad memory instead of in-line code. Subroutine "LOAD" executes in 7.5 milliseconds for a 2 MHz 8080A microprocessor.

Another program that interfaces to the HDSP-2000 REFRESH CONTROLLER is shown in Figure 14.4-15. This program is written for an 8080A microprocessor. Following each Interrupt, the program "RFRSH" is executed. This program eliminates the 5n byte scratch pad memory by decoding each ASCII character immediately

DECODER ADDRESS FOR FIG. 7a, 7b, 10	DECODER ADDRESS FOR FIG. 6	HDSP-2471 ROM ADDRESS	HEXIDECIMAL DATA																	
E500	0600	080	08	30	45	7D	7D	38	7E	30	60	1E	3E	62	40	08	38	41	$COLUMN_1$
		090	10	18	5E	78	38	78	38	3C	38	3C	38	08	20	12	48	01	
		0A0	00	00	00	14	24	23	36	00	00	00	08	08	00	08	00	20	
		0B0	3E	00	62	22	18	27	3C	01	36	06	00	00	00	14	41	06	
		0C0	3E	7E	7F	3E	7F	7F	7F	3E	7F	00	20	7F	7F	7F	7F	3E	
		0D0	7F	3E	7F	26	01	3F	07	7F	63	03	61	00	02	41	04	40	
		0E0	00	38	7F	38	38	38	08	08	7F	00	20	00	00	78	7C	38	
		0F0	7C	18	00	48	04	3C	1C	3C	44	04	44	00	00	00	08	2A	
E580	0680	100	1C	48	29	09	09	44	01	4A	50	04	49	14	3C	7C	44	63	$COLUMN_2$
		110	08	24	61	14	44	15	45	43	45	41	42	08	7E	19	7E	12	
		120	00	5F	03	7F	2A	13	49	0B	00	41	2A	08	58	08	30	10	
		130	51	42	51	41	14	45	4A	71	49	49	36	5B	08	14	22	01	
		140	41	09	49	41	41	49	09	41	08	41	40	08	40	02	04	41	
		150	09	41	09	49	01	40	18	20	14	04	51	00	04	41	02	40	
		160	07	44	48	44	44	54	7E	14	08	44	40	7F	41	04	08	44	
		170	14	24	7C	54	3E	40	20	40	28	48	64	08	00	41	04	55	
E600	0700	180	3E	45	11	11	05	44	29	4D	48	04	49	08	20	04	44	55	$COLUMN_3$
		190	78	7E	01	15	45	14	44	42	44	40	40	2A	02	15	49	7C	
		1A0	00	00	00	14	7F	08	56	07	3E	3E	1C	3E	38	08	30	08	
		1B0	49	7F	49	49	12	45	49	09	49	49	36	3B	14	14	14	51	
		1C0	5D	09	49	41	41	49	09	41	08	7F	40	14	40	0C	08	41	
		1D0	09	51	19	49	7F	40	60	18	08	78	49	7F	08	7F	7F	40	
		1E0	0B	44	44	44	44	54	09	54	04	7D	44	10	7F	18	04	44	
		1F0	24	14	08	54	44	40	40	30	10	30	54	36	77	36	08	2A	
E680	0780	200	7F	40	29	21	05	38	2E	49	50	38	49	10	20	7C	3C	49	$COLUMN_4$
		210	08	24	61	14	3C	15	3D	43	45	41	42	1C	02	12	41	12	
		220	00	00	03	7F	2A	64	20	00	41	00	2A	08	00	08	00	04	
		230	45	40	49	49	7F	45	49	05	49	29	00	00	22	14	08	09	
		240	55	09	49	41	41	49	09	51	08	41	40	22	40	02	10	41	
		250	09	21	29	49	01	40	18	20	14	04	45	41	10	00	02	40	
		260	00	3C	44	44	48	54	02	54	04	40	3D	28	40	04	04	44	
		270	24	7C	04	54	20	20	20	40	28	08	4C	41	00	08	10	55	
E700	0800	280	00	30	45	7D	79	44	10	30	60	40	3E	60	1C	02	04	41	$COLUMN_5$
		290	04	18	5E	78	40	78	40	3C	38	3C	38	08	02	00	42	01	
		2A0	00	00	00	14	12	62	50	00	00	00	08	08	00	08	00	02	
		2B0	3E	00	46	36	10	39	30	03	36	1E	00	00	41	14	00	06	
		2C0	1E	7E	36	22	3E	41	01	72	7F	00	3F	41	40	7F	7F	3E	
		2D0	06	5E	46	32	01	3F	07	7F	63	03	43	41	20	00	04	40	
		2E0	00	40	38	20	7F	08	00	3C	78	00	00	44	00	78	78	38	
		2F0	18	40	04	20	00	7C	1C	3C	44	04	44	00	00	00	08	2A	

Figure 14.4-13. 128 Character ASCII Decoder Table for the Microprocessor Programs Shown in Figures 14.4-12, 14.4-14, and 14.4-15. 5x7 Display Font is Shown in the HDSP-247X Data Sheet.

LOC	OBJECT CODE			SOURCE STATEMENTS		
0004				RDVR	EQU	0004H
0005				CDVR	EQU	0005H
E500				DECDR	EQU	0E500H
					ORG	0E000H
E000	05	E0		POINT	DW	BUFFR
E002	FE			COLMN	DB	0FEH
E003	FF	FF		COUNT	DW	0FFFFH
E005	00			BUFFR	DS	160
					ORG	0E0A5H
E0A5	A7	E0		ASCII	DW	DATA
E0A7	00			DATA	DS	32
					ORG	0E400H
E400	F5			RFRSH	PUSH	PSW
E401	C5				PUSH	B
E402	E5				PUSH	H
E403	2A	00	E0		LHLD	POINT
E406	06	20			MVI	B, 32
E408	3E	FF			MVI	A, 0FFH
E40A	D3	05			OUT	CDVR
F40C	7E			LOOP	MOV	A, M
E40D	D3	04			OUT	RDVR
E40F	23				INX	H
E410	05				DCR	B
E411	C2	0C	E4		JNZ	LOOP
E414	3A	02	E0		LDA	COLMN
E417	D3	05			OUT	CDVR
E419	FE	EF			CPI	0EFH
E41B	CA	28	E4		JZ	FIRST
E41E	22	00	E0		SHLD	POINT
E421	07				RLC	
E422	32	02	E0		STA	COLMN
E425	C3	3A	E4		JMP	END
E428	21	05	E0	FIRST	LXI	H, BUFFR
E42B	22	00	E0		SHLD	POINT
E42E	3E	FE			MVI	A, 0FEH
E430	32	02	E0		STA	COLMN
E433	2A	03	E0		LHLD	COUNT
E436	2B				DCX	H
E437	22	03	E0		SHLD	COUNT
E43A	E1			END	POP	H
E43B	C1				POP	B
E43C	F1				POP	PSW
E43D	C9				RET	
E43E	11	24	E0	LOAD	LXI	D, BUFFR+31
E441	0E	20			MVI	C, 32
E443	2A	A5	E0	LOOP1	LHLD	ASCII
E446	7E				MOV	A, M
E447	23				INX	H
E448	22	A5	E0		SHLD	ASCII
E44B	26	E5			MVI	H, DECDR/256
E44D	6F				MOV	L, A
E44E	06	05			MVI	B, 5
E450	7E			LOOP2	MOV	A, M
E451	12				STAX	D
E452	7D				MOV	A, L
E453	C6	80			ADI	80H
E455	6F				MOV	L, A
E456	D2	5A	E4		JNC	LOOP3
E459	24				INR	H
E45A	7B			LOOP3	MOV	A, E
E45B	C6	20			ADI	32
E45D	5F				MOV	E, A
E45E	05				DCR	B
E45F	C2	50	E4		JNZ	LOOP2
E462	7B				MOV	A, E
E463	C6	5F			ADI	5FH
E465	5F				MOV	E, A
E466	0D				DCR	C
E467	C2	43	E4		JNZ	LOOP1
E46A	C9				RET	

Figure 14.4-14.   8080A Microprocessor Program That Interfaces to the REFRESH CONTROLLER Shown in Figure 14.4-10.

before loading into the display interface. Scratch pad register "ASCII" points to the leftmost ASCII character to be stored in the display interface. Scratch pad register

LOC	OBJECT CODE			SOURCE STATEMENTS		
0004				RDVR	EQU	0004H
0005				CDVR	EQU	0005H
E500				DECDR	EQU	0E500H
					ORG	0E000H
E000	07	E0		ASCII	DW	DATA
E002	FE			COLMN	DB	0FEH
E003	FF	FF		COUNT	DW	0FFFFH
E005	00	E5		BASE	DW	DECDR
E007	00			DATA	DS	32
					ORG	0E400H
E400	F5			RFRSH	PUSH	PSW
E401	C5				PUSH	B
E402	D5				PUSH	D
E403	E5				PUSH	H
E404	2A	05	E0		LHLD	BASE
E407	EB				XCHG	
E408	2A	00	E0		LHLD	ASCII
E40B	01	1F	00		LXI	B, 31
E40E	09				DAD	B
E40F	43				MOV	B, E
E410	0E	20			MVI	C, 32
E412	3E	FF			MVI	A, 0FFH
E414	D3	05			OUT	CDVR
E416	78			LOOP	MOV	A, B
E417	86				ADD	M
E418	5F				MOV	E, A
E419	1A				LDAX	D
E41A	D3	04			OUT	RDVR
E41C	2B				DCX	H
E41D	0D				DCR	C
E41E	C2	16	E4		JNZ	LOOP
E421	EB				XCHG	
E422	3A	02	E0		LDA	COLMN
E425	D3	05			OUT	CDVR
E427	FE	EF			CPI	0EFH
E429	CA	3B	E4		JZ	FIRST
E42C	07				RLC	
E42D	32	02	E0		STA	COLMN
E430	68				MOV	L, B
E431	01	80	00		LXI	B, 0080H
E434	09				DAD	B
E435	22	05	E0		SHLD	BASE
E438	C3	4D	E4		JMP	END
E43B	3E	FE		FIRST	MVI	A, 0FEH
E43D	32	02	E0		STA	COLMN
E440	21	00	E5		LXI	H, DECDR
E443	22	05	E0		SHLD	BASE
E446	2A	03	E0		LHLD	COUNT
E449	2B				DCX	H
E44A	22	03	E0		SHLD	COUNT
E44D	E1			END	POP	H
E44E	D1				POP	D
E44F	C1				POP	B
E450	F1				POP	PSW
E451	C9				RET	

Figure 14.4-15.   8080A Microprocessor Program That Interfaces to the REFRESH CONTROLLER Shown in Figure 14.4-10.

"COLMN" contains the next column word to be stored in the display interface. Scratch pad register "BASE" points to the first byte within the ASCII decoder table "DECDR" of the next column to be decoded by the program. The program turns off all columns by writing column select word FF_{16} to "CDVR." "ASCII" is incremented by 31 so it points to the rightmost ASCII character. The ASCII character is decoded and written at "RDVR." "ASCII" is decremented and the process is repeated 31 times so that 32 words are stored at "RDVR." Then, the microprocessor reads the next column select word from "COLMN" and stores it at "CDVR." Finally, the microprocessor updates "POINT," "COLMN," and "BASE." Following execution

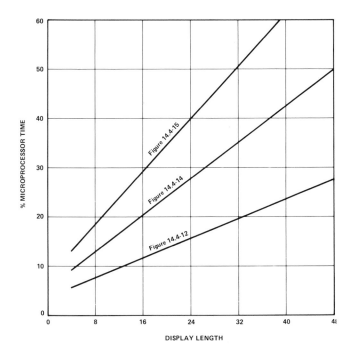

Figure 14.4-16. Comparison of Required Microprocessor Time for the Microprocessor Programs Shown in Figures 14.4-12, 14.4-14, and 14.4-15.

of the "RET" instruction, execution of the main program is resumed.

Because the microprocessor system is interrupted at a 500 Hz rate, proper software design is especially important for the HDSP-2000 REFRESH CONTROLLER. Figure 14.4-16 compares the microprocessor times required for the refresh programs shown in Figures 14.4-12, 14.4-14, and 14.4-15. The times shown are based on a 100 Hz refresh rate, for a 6800 microprocessor utilizing a 1 MHz clock, or an 8080A microprocessor utilizing a 2 MHz clock. The use of the scratch pad memory significantly reduces the time required to refresh the display. The fastest program, shown in Figure 14.4-12, uses in-line code to access data from the scratch pad and output it to the display. The program shown in Figure 14.4-14 takes longer to execute because it uses a program loop instead of the in-line code. These programs utilize a subroutine "LOAD" to change the display data stored in the scratch pad. The program shown in Figure 14.4-15 requires the most microprocessor time because it continuously decodes each ASCII character prior to loading into the display interface. The advantage of this program is that it only requires five bytes of RAM, regardless of display length. Reductions in the microprocessor times shown in Figure 14.4-16 can be achieved by using faster versions of the 6800 and 8080A microprocessors.

14.5 Decoded Data Controllers

Figure 14.5-1 shows a DECODED DATA CONTROLLER

designed for a 32 character HDSP-6508 Alphanumeric Display. To simplify the circuitry, the display is configured as a 14 segment display with decimal point and colon. This allows each display character to be specified by two 8 bit words. One possible display font is shown in Figure 14.5-2. The Motorola 6810 RAM stores 64 bytes of display data that are continually read and displayed. The display data is organized within the RAM such that Addresses A_5, A_4, A_3, A_2, and A_1 specify the desired character and Address A_0 differentiates between the two words of display data for each character. The display data is formatted such that word 0 ($D_7 - D_0$) is decoded as G_2, G_1, F, E, D, C, B, A and word 1 ($D_7 - D_0$) is decoded as COLON, DP, M, L, K, J, I, H. The display data is coded low true such that a low output turns the appropriate segment on. Strobing of the display is accomplished with the 74LS14 oscillator and 74LS393 counter. The counter continuously reads display data from the RAM and enables the appropriate digit driver. The time allotted to each digit is broken into four segments. During the first segment of time, the display is turned off and word 0 is read from the RAM and stored in the 74LS273 octal register. During the next three segments of time, word 1 is read from the RAM and the display is turned on. Thus, the display duty factor is (1/32)(3/4) or 1/42.6. For the values of R and C specified, the display is strobed at a 130 Hz refresh rate.

Data is entered into the RAM from the Address and Data Bus of the microprocessor via two control lines, Chip Select and $\overline{\text{Write}}$. When $\overline{\text{Chip Select}}$ goes low, the address generated by the counter is disabled and the microprocessor Address and Data Bus is gated to the RAM. Then, after sufficient delay, the $\overline{\text{Write}}$ input is pulsed, which stores the data within the RAM. The data entry timing for the HDSP-6508 DECODED DATA CONTROLLER is shown in Figure 14.5-3. Because of the requirement that the address inputs of the 6810 RAM must be stable prior to the falling edge of $\overline{\text{Write}}$, $\overline{\text{Chip Select}}$ should go low for time t_{CW} prior to the falling edge of $\overline{\text{Write}}$. To guarantee that the address and data inputs of the RAM remain stable until after $\overline{\text{Write}}$ goes high, $\overline{\text{Chip Select}}$ should remain low for time t_{CH} following the rising edge of $\overline{\text{Write}}$. This requirement for two separate timing signals is also required for the CODED DATA CONTROLLERS shown in Figures 14.6-1 and 14.6-3. Because this interface timing is somewhat more difficult than the previously described circuits, the following methods are presented for interfacing to commonly used microprocessors.

Interface to the 6800 microprocessor family is accomplished by NANDing together VMA and some specified combination of high order Address lines to generate $\overline{\text{Chip Select}}$ and using ϕ_2 to generate $\overline{\text{Write}}$.

For the 8080A and 8085A microprocessor families, the limited flexibility of the Output instruction requires that

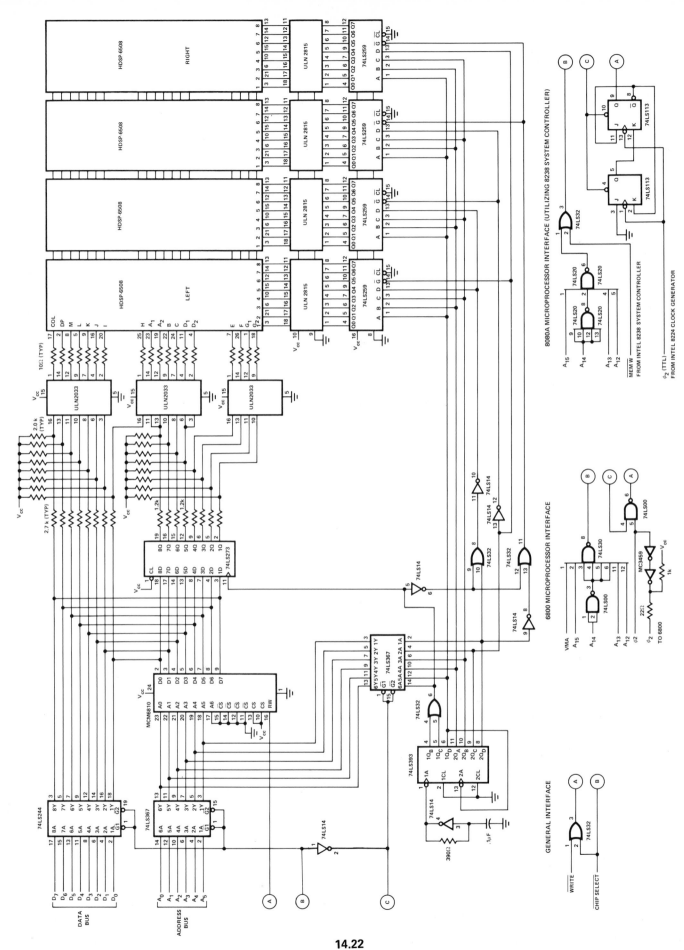

Figure 14.5-1. 6800, 8080A, and General Interface to the HDSP-6508 DECODED DATA CONTROLLER.

14.22

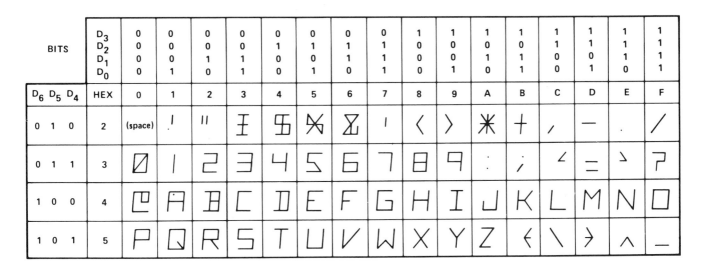

BITS	D_3 D_2 D_1 D_0	0 0 0 0	0 0 0 1	0 0 1 0	0 0 1 1	0 1 0 1	0 1 0 0	0 1 0 1	0 1 1 0	1 0 0 1	1 0 0 0	1 0 1 1	1 0 1 0	1 1 0 1	1 1 0 0	1 1 1 1	1 1 1 0	1 1 1 1
D_6 D_5 D_4	HEX	0	1	2	3	4	5	6	7	8	9	A	B	C	D	E	F
0 1 0	2	(space)	!	"	#	$	%	&	'	(	)	*	+	,	-	.	/
0 1 1	3	0	1	2	3	4	5	6	7	8	9	:	;	<	=	>	?
1 0 0	4	@	A	B	C	D	E	F	G	H	I	J	K	L	M	N	O
1 0 1	5	P	Q	R	S	T	U	V	W	X	Y	Z	[	\	]	^	_

Figure 14.5-2. One Possible 16 Segment Display Font (14 Segments Plus Decimal Point and Colon) for the DECODED DATA CONTROLLER Shown in Figure 14.5-1.

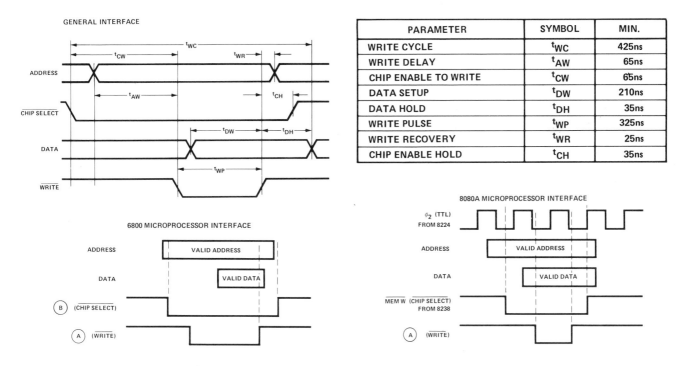

PARAMETER	SYMBOL	MIN.
WRITE CYCLE	t_{WC}	425ns
WRITE DELAY	t_{AW}	65ns
CHIP ENABLE TO WRITE	t_{CW}	65ns
DATA SETUP	t_{DW}	210ns
DATA HOLD	t_{DH}	35ns
WRITE PULSE	t_{WP}	325ns
WRITE RECOVERY	t_{WR}	25ns
CHIP ENABLE HOLD	t_{CH}	35ns

Figure 14.5-3. Data Entry Timing for the DECODED DATA CONTROLLER Shown in Figure 14.5-1.

the HDSP-6508 DECODED DATA CONTROLLER must be addressed as memory instead of I/O. The 8080A microprocessor requires an external status latch to hold status information provided during program execution. This status latch function can be implemented with an octal register such as the Intel 8212 or 74LS273. A Memory Write signal can be generated by NORing together all outputs of this status latch. This signal can then be NANDed with some specified combination of high order Address lines to generate Chip Select. The 8080A $\overline{WR}$ output can then be connected to $\overline{Write}$. The Intel 8238 System Controller, which is commonly used with the 8080A microprocessor, prevents direct access to the outputs of the status latch. An example of an interfacing to a system utilizing the 8238 is illustrated in Figure 14.5-1. $\overline{MEM\ W}$ from the 8238 is inverted and then NANDed with some specified combination of high order Address lines to generate $\overline{Chip\ Select}$. The 74LS113 generates $\overline{Write}$ from the microprocessor clock, ϕ_2 (TTL).

LOC	OBJECT CODE		SOURCE STATEMENTS		
BF00			DSPLY	EQU	$BF00
0600			DECDR	EQU	$0600
0000	0006		ASCII	FDB	MESSGE
0002	BF00		PAD1	FDB	DSPLY
0004	0600		PAD2	FDB	DECDR
0006			MESSGE	RMB	32
0400				ORG	$0400
0400	CE	BF00	LOAD	LDX	I , DSPLY
0403	DF	02		STX	D , PAD1
0405	CE	0600		LDX	I , DECDR
0408	DF	04		STX	D , PAD2
040A	DE	00	LOOP1	LDX	D , ASCII
040C	A6	00		LDA A	X , 0
040E	08			INX	
040F	DF	00		STX	D , ASCII
0411	48			ASL A	
0412	97	05		STA A	D , PAD2+1
0414	DE	04		LDX	D , PAD2
0416	A6	00		LDA A	X , 0
0418	E6	01		LDA B	X , 1
041A	DE	02		LDX	D , PAD1
041C	A7	00		STA A	X , 0
041E	08			INX	
041F	E7	00		STA B	X , 0
0421	08			INX	
0422	DF	02		STX	D , PAD1
0424	8C	BF40		CPX	I, DSPLY+64
0427	26	E1		BNE	LOOP1
0429	39			RTS	

Figure 14.5-4. 6800 Microprocessor Program That Interfaces to the DECODED DATA CONTROLLER Shown in Figure 14.5-1.

LOC	OBJECT CODE			SOURCE STATEMENTS		
BF00				DSPLY	EQU	0BF00H
					ORG	0E000H
E000	02	E0		ASCII	DW	DATA
E002	00			DATA	DS	32
					ORG	0E400H
E400	01	00	BF	LOAD	LXI	B , DSPLY
E403	11	00	E5		LXI	D , DECDR
E406	2A	00	E0		LHLD	ASCII
E409	7E			LOOP1	MOV	A , M
E40A	23				INX	H
E40B	07				RLC	
E40C	5F				MOV	E , A
E40D	1A				LDAX	D
E40E	02				STAX	B
E40F	13				INX	D
E410	03				INX	B
E411	1A				LDAX	D
E412	02				STAX	B
E413	03				INX	B
E414	79				MOV	A , C
E415	FE	40			CPI	64
E417	C2	09	E4		JNZ	LOOP1
E41A	22	00	E0		SHLD	ASCII
E41D	C9				RET	

Figure 14.5-5. 8080A Microprocessor Program That Interfaces to the DECODED DATA CONTROLLER Shown in Figure 14.5-1.

Interface to the 8085A microprocessor family can be accomplished by inverting the I/O/$\overline{M}$ Output and NANDing the resulting signal with the S_o output and some specified combination of high order Address lines to generate Chip Select. The $\overline{WR}$ output from the microprocessor is connected directly to $\overline{Write}$.

The simplest interface to the Z-80 microprocessor family is accomplished by addressing the HDSP-6508 DECODED DATA CONTROLLER as I/O instead of memory. An example of this interface is shown in Figure 14.6-3. The $\overline{IORQ}$ output is inverted and NANDed with some specified combination of Address lines to generate $\overline{Chip\ Select}$. The 74LS113 circuit generates $\overline{Write}$ from the inverted microprocessor clock $\overline{\phi}$.

A 6800 microprocessor program that interfaces to the HDSP-6508 DECODED DATA CONTROLLER in Figure 14.5-1 is shown in Figure 14.5-4. This program decodes 32 ASCII characters and stores the resulting decoded display data within the display. The scratch pad register "ASCII" points to the location of the next ASCII character to be decoded. The program reads the first ASCII character, increments the point, "ASCII," and then looks up two words of display data within the 64 character ASCII look up table "DECDR." These words of display data are then stored at the two addresses for the leftmost display location. Subsequent ASCII characters are decoded, and stored at the appropriate address within the display until all 32 characters have been decoded. After the program is finished, the pointer "ASCII" will have been incremented by 32. This program requires 2.4mS for a 1 MHz clock to decode and load 32 ASCII characters into the HDSP-6508 DECODED DATA CONTROLLER. The corresponding 8080A microprocessor program is shown in Figure 14.5-5. This program requires 1.4 mS for a 2 MHz clock to decode and load 32 ASCII characters into the HDSP-6508 DECODED DATA CONTROLLER.

Figure 14.5-6 shows a DECODED DATA CONTROLLER designed for a 32 character HDSP-2000 Alphanumeric Display. The circuit is specifically designed to interface to an 8080A microprocessor. This circuit stores 1120 serial bits of data within the Intel 2102 RAMs and continuously strobes the display with this data. The counters continuously read each column of data into the display (224 bits/column), turn the appropriate column on, wait, and then read the next column of data into the display. The counter string consists of two 74LS163 counters and two 74LS162 counters. The upper 74LS163 counter is connected as a divide by 2 and a divide by 7 counter. The divide by 2 portion divides the 2 MHz input clock, ϕ_2 (TTL), by two and provides the proper write timing for the 2102 RAMs as data is loaded into the system. The divide by 7 portion sequentially reads the seven bits of row data from the RAM into the Data Input of the display. The second

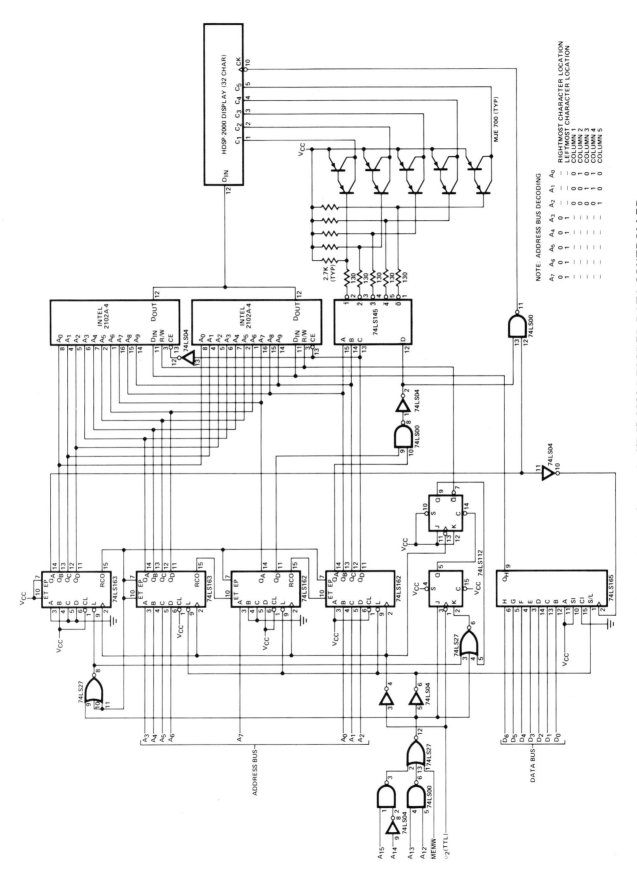

Figure 14.5-6. 8080A Interface to the HDSP-2000 DECODED DATA CONTROLLER.

14.25

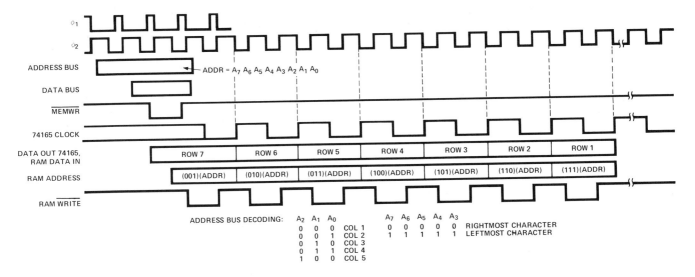

ADDRESS BUS DECODING:

A2	A1	A0		A7	A6	A5	A4	A3	
0	0	0	COL 1	0	0	0	0	0	RIGHTMOST CHARACTER
0	0	1	COL 2	1	1	1	1	1	LEFTMOST CHARACTER
0	1	0	COL 3						
0	1	1	COL 4						
1	0	0	COL 5						

Figure 14.5-7. Data Entry Timing for the DECODED DATA CONTROLLER Shown in Figure 14.5-6.

```
        OBJECT
LOC     CODE        SOURCE STATEMENTS

B000                DISPL   EQU     0B000H
E500                DECDR   EQU     0E500H

                    ORG     0E000H
E000    02 E0  ASCII    DW      DATA
E002    00     DATA     DS      32

                    ORG     0E400H
E400    11 F8 B0 LOAD   LXI     D, DISPL+00F8H
E403    0E 20           MVI     C, 32
E405    2A 00 E0 LOOP1  LHLD    ASCII
E408    7E             MOV     A, M
E409    23             INX     H
E40A    22 00 E0       SHLD    ASCII
E40D    26 E5          MVI     H, DECDR/256
E40F    6F             MOV     L, A
E410    06 05          MVI     B, 5
E412    7E      LOOP2  MOV     A, M
E413    12             STAX    D
E414    13             INX     D
E415    7D             MOV     A, L
E416    C6 80          ADI     80H
E418    6F             MOV     L, A
E419    D2 1D E4       JNC     LOOP3
E41C    24             INR     H
E41D    05      LOOP3  DCR     B
E41E    C2 12 E4       JNZ     LOOP2
E421    7B             MOV     A, E
E422    D6 0D          SUI     13
E424    5F             MOV     E, A
E425    0D             DCR     C
E426    C2 05 E4       JNZ     LOOP1
E429    C9             RET
```

Figure 14.5-8. 8080A Microprocessor Program That Interfaces to the DECODED DATA CONTROLLER Shown in Figure 14.5-6.

74LS163 counter and part of the first 74LS162 counter are connected as a divide by 32 counter which sequentially selects each of the 32 display locations. Part of the first 74LS162 counter and the first stage of the last 74LS162 counter are connected as a divide by 10 counter which determines the relationship between load time and display time. The last part of the 74LS162 counter is connected as a divide by 5 counter which scans the five columns of the display. Thus, when Q_D of the first 74LS162 and Q_A of

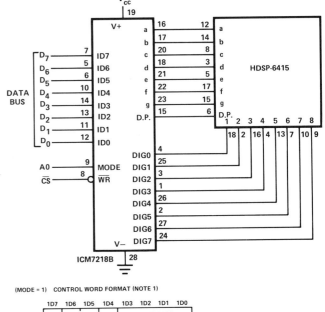

(MODE = 1)   CONTROL WORD FORMAT (NOTE 1)

1D7	1D6	1D5	1D4	1D3	1D2	1D1	1D0
A	B		C		DON'T CARE		

A = 0   DATA NOT COMING
  = 1   DATA COMING
        B = 0 0   CODE B DECODER
          = 1 0   HEXADECIMAL DECODER
          = X 1   UNDECODED SEGMENT INFORMATION
                C = 0   BLANK DISPLAY (RAM CONTENTS UNCHANGED)
                  = 1   NORMAL OPERATION

NOTE 1:   First Data Word following a Control Word with 1D7 = 1 is loaded into location DIG0. When connected as shown, subsequent Data Words are loaded from left to right across the display.

(MODE = 0)   DATA WORD FORMAT

1D7	1D6	1D5	1D4	1D3	1D2	1D1	1D0	
D.P.	a	b	c	e	g	f	d	UNDECODED SEGMENT INFORMATION (NOTE 2)
DON'T CARE				D	C	B	A	HEXADECIMAL/CODE B INFORMATION (NOTE 3)

NOTE 2:   Logic 1 represents an ON segment for all inputs except D.P. where logic 0 represents an ON D.P. segment.

NOTE 3:   Binary Coded Decimal word.

Figure 14.5-9. Seven Segment DECODED DATA CONTROLLER Utilizing the Intersil ICM7218B.

the second 74LS162 are logical 1, the display is turned off, the display clock is turned on, and 224 bits of serial data is read from the RAM into the display. The HDSP-2000 DECODED DATA CONTROLLER refreshes the display at a 90 Hz refresh rate for a 2 MHz clock. The column on time of the display is $(20\%)(9/10) = 18\%$. The circuit can be modified for different display lengths by modifying the display location counter and load/display counter.

The microprocessor loads 160 bytes of display data into the circuit to update all 32 display characters. Each byte is loaded into the 2102 RAMs by presetting the counters to the desired address and then sequentially loading each serial bit into the next seven addresses via the 74165 parallel in, serial out shift register. This data entry timing is summarized in Figure 14.5-7.

An 8080A program that interfaces to the DECODED DATA CONTROLLER shown in Figure 14.5-6 is shown in Figure 14.5-8. The HDSP-2000 DECODED DATA CONTROLLER is addressed as a 256 x 8 RAM with Addresses A_2, A_1, and A_0 specifying the desired column and Addresses A_7, A_6, A_5, A_4, and A_3 specifying the desired character. The program uses a scratch pad register "ASCII" that points to the location within a 32 byte ASCII message of the next ASCII character to be loaded into the display interface. As each ASCII character is decoded, "ASCII" is incremented by one so as to point to the next character to be decoded. The program uses a 128 character ASCII look up table "DECDR" that translates each ASCII character into five bytes of display data. The program uses the HL register pair to decode each ASCII character into display data by adding the appropriate offset to the beginning address of "DECDR." The register pair DE is used to point to the destination address within the DECODED DATA CONTROLLER. The first byte of display data accessed by HL is stored at the address accessed by DE for column 1 of the leftmost display location and the next four bytes of display data are stored at the addresses of column 2 through column 5. After these five bytes of display data have been stored within the DECODED DATA CONTROLLER, the next ASCII character is read, decoded, and stored at the addresses of column 1 through column 5 of the second leftmost display location. In a similar manner, the program decodes all 32 ASCII characters. After the program is finished, the pointer "ASCII" will have been incremented by 32. The program "LOAD" requires about 6.6mS for a 2 MHz clock to decode and load the 32 character message into the HDSP-2000 DECODED DATA CONTROLLER.

Figure 14.5-9 shows a DECODED DATA CONTROLLER for an 8 digit 7 segment display. The circuit uses the Intersil ICM7218 to provide segment storage and display multiplexing. The Intersil ICM7218A is designed to drive common anode LED displays at 40 mA I_{PEAK}/Segment on a 12% duty cycle. The Intersil ICM7218B is designed to drive common cathode LED displays at 20 mA I_{PEAK}/Segment on a 12% duty cycle. The circuit illustrated in Figure 14.5-9 uses the ICM7218B to drive the HDSP-6415 monolithic display. The microprocessor interfaces to the ICM7218A/B through eight Data Inputs, a Mode Input, and a $\overline{\text{Write}}$ Input. First a control word is loaded into the ICM7218A/B by pulsing $\overline{\text{Write}}$ with Mode high. The control word specifies one of three types of input data (uncoded segment information, hexadecimal information, or numeric information), whether or not to blank the display, and to enable data to be entered into the display. After the control word has been read, the next eight data words are loaded from left to right into the LED display. Subsequent data words are ignored.

14.6 Coded Data Controllers

Figure 14.6-1 shows a CODED DATA CONTROLLER designed for a 32 character HDSP-6508 Alphanumeric Display. Operation of this circuit is similar to the DECODED DATA CONTROLLER shown in Figure 14.5-1 except that the Motorola 6810 RAM stores 32 six bit ASCII words and the Texas Instruments AC5947 decodes this ASCII data into 18 segment display data. The resulting display font is shown in Figure 14.4-2. Strobing of the display is accomplished by the 74LS14 oscillator and 74LS393 counter. Because the long propagation delay through the AC5947 tends to cause display ghosting, the display is blanked momentarily after each new character is read from the RAM. This is accomplished by breaking the total time allotted for each digit into four segments. During the first segment, the display is turned off to allow data to ripple through the AC5947 and during the next three segments, the display is turned on. The resulting display duty factor is $(1/32)(3/4)$ or $1/42.6$. The display is strobed at a 130 Hz refresh rate.

Data is entered into the RAM from the Address and Data Bus of the microprocessor via two control lines $\overline{\text{Chip Select}}$ and $\overline{\text{Write}}$. When Chip Select goes low, the address from the counter is tristated and the microprocessor Address Bus and Data Bus is gated to the RAM. Data Entry Timing for the HDSP-6508 CODED DATA CONTROLLER is shown in Figure 14.6-2. Since this timing is very similar to the DECODED DATA CONTROLLER shown in Figure 14.5-1, interface to the various microprocessor families is the same as described in section 14.5.

Figure 14.6-3 shows a CODED DATA CONTROLLER designed for a 32 character HDSP-2000 Alphanumeric Display. The Motorola 6810 RAM stores 32 bytes of ASCII data which is continuously read, decoded and displayed. The ASCII data from the RAM is decoded by the Motorola 6674 128 character ASCII decoder. The 6674 decoder has five column outputs which are gated to the Data Input of the display via a 74151 multiplexer. Strobing

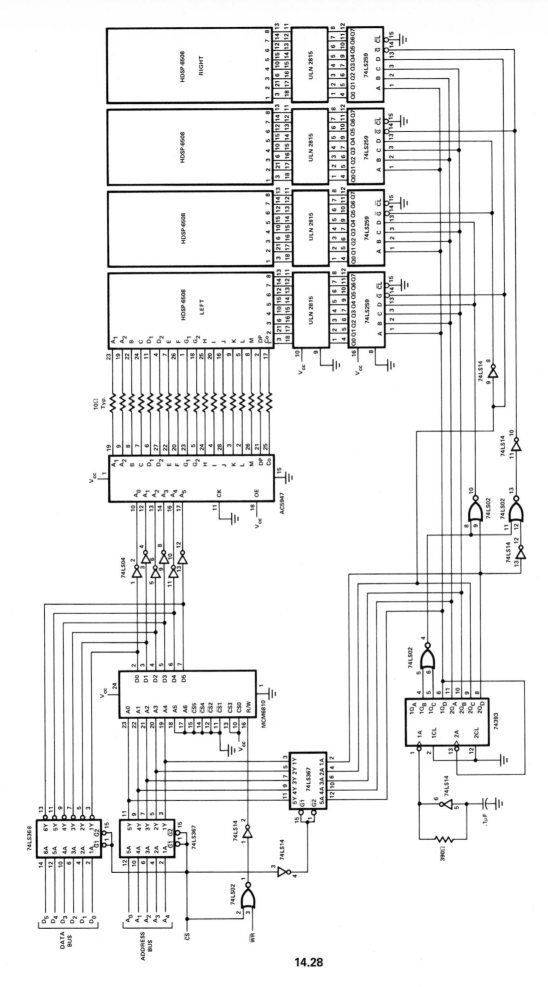

Figure 14.6-1.   General Interface to the HDSP-6508 CODED DATA CONTROLLER.

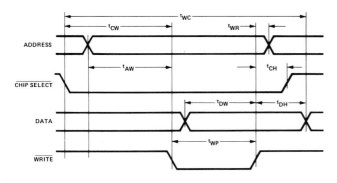

PARAMETER	SYMBOL	MIN.
WRITE CYCLE	t_{WC}	455ns
WRITE DELAY	t_{AW}	65ns
CHIP ENABLE TO WRITE	t_{CW}	65ns
DATA SETUP	t_{DW}	215ns
DATA HOLD	t_{DH}	50ns
WRITE PULSE	t_{WP}	340ns
WRITE RECOVERY	t_{WR}	40ns
CHIP ENABLE HOLD	t_{CH}	50ns

Figure 14.6-2. Data Entry Timing for the CODED DATA CONTROLLER Shown in Figure 14.6-1.

of the display is accomplished via the 74197, 74393, and 7490 counter string. The 74197 is connected as a divide by 8 counter that sequentially selects the seven rows within the 6674. The 74393 is a divide by 256 counter connected so that the five lowest order outputs select each of the 32 ASCII characters within the RAM and the three highest order outputs set up the relationship between load time and display time. The column on time of the display is (20%)(7/8) or 17.5%. The 7490 is connected as a divide by 5 counter to scan the five columns of the display. Thus, when $2Q_B = 2Q_C = 2Q_D = 1$ of the 74393, the circuit scans 32 characters from the RAM and serializes the column data by counting through each of the seven rows of the 6674 and gating the appropriate column to the display. For a 1 MHz input frequency, the CODED DATA CONTROLLER provides a 98 Hz display refresh rate. The circuit can easily be converted to different length displays by modifying the 74393 character counter/duty factor counter. Note that the display is connected so that pin 1 of each cluster is located in the upper right corner of the display string. While this is upside down from the normal convention of the HDSP-2000 display, it allows RAM address 00_{16} to refer to the leftmost display location.

Data is entered into the RAM in a similar manner as the circuits shown in Figures 14.5-1, and 14.6-1. Data Entry Timing for the HDSP-2000 CODED DATA CONTROLLER is shown in Figure 14.6-4. Since this timing is very similar to the DECODED DATA CONTROLLER shown in Figure 14.5-1, interface to the various microprocessor families is

the same as described in section 14.5. The input circuitry illustrated in Figure 14.6-3 illustrates the interface to a Z-80 type microprocessor.

A CODED DATA CONTROLLER for a 7 Segment Display is shown in Figure 14.6-5. This circuit uses the Intersil ICM 7218C/D to provide BCD storage and display multiplexing. The Intersil ICM 7218C is designed to drive common anode LED displays at 40mA I_{PEAK}/segment on a 12% duty cycle while the Intersil ICM 7218D is designed to drive common cathode LED displays at 20mA I_{PEAK}/segment on a 12% duty cycle. The circuit illustrated in Figure 14.6-5 shows the ICM 7218C to drive an 8 digit 5082-7610 LED display. The microprocessor interfaces to the ICM 7218C/D through five data inputs (BCD + $\overline{dp}$), three address inputs, $\overline{Write}$ input, and a tristate input that is used to select Hexidecimal decoding (HIGH), Numeric decoding (FLOAT), or display blanking (LOW). The Data Entry Timing for the ICM 7218C/D allows it to interface directly to the Address and Data Bus of most microprocessors.

14.7   Display Processor Controllers

The DISPLAY PROCESSOR CONTROLLER provides a powerful, smart interface which performs many of the functions normally found in a small terminal. The DISPLAY PROCESSOR CONTROLLER is designed around a slave microprocessor or custom LSI integrated circuit that provides display storage and multiplexing with a very minimum of circuit complexity. The simplest DISPLAY PROCESSOR CONTROLLER designed for a 16 digit HDSP-6508 Alphanumeric Display is shown in Figure 14.7-1. This circuit is designed around the Intel 8279 Progammable Keyboard/Display Interface. This LSI chip contains the circuitry necessary to interface directly to a microprocessor bus and provides a 16 x 8 RAM, programmable scan counter, and keyboard debounce and control logic. While the 8279 is specifically designed for 7 segment displays, inclusion of the Texas Instrument AC5947 ASCII to 18 segment decoder/driver allows the use of an 18 segment Alphanumeric Display. The 8279 Keyboard/Display Controller interfaces to a microprocessor via an eight line bidirectional Data Bus, Control lines $\overline{RD}$ (Read), $\overline{WR}$ (Write), $\overline{CS}$ (Chip Select), A_O (Command/Data), RESET, IRQ (Interrupt Output), and a clock input, CLK. The display is scanned by outputs A_{0-3} and B_{0-3} which are connected to the inputs of the AC5947, and outputs SL_{0-3} which are connected to the digit scanning circuitry. The 74LS122 is used to provide interdigit blanking to prevent display ghosting. In addition to display scanning, the 8279 also has the ability to scan many different types of encoded or decoded keyboards, X–Y matrix keyboards, or provide a strobed data input to the microprocessor. The 8279 provides for either Block data entry, where data enters from left to right across the

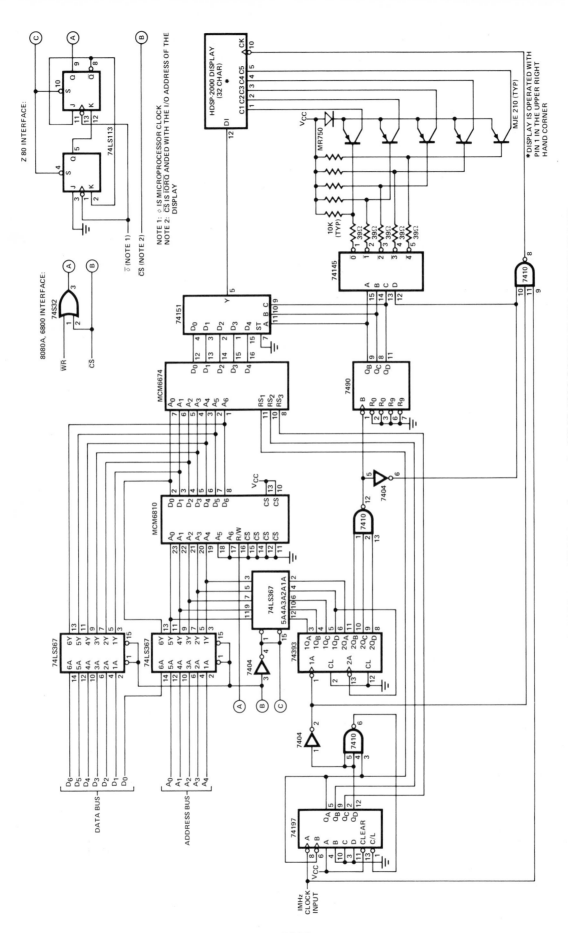

Figure 14.6-3. 6800, 8080A, and Z-80 Interface to the HDSP-2000 CODED DATA CONTROLLER.

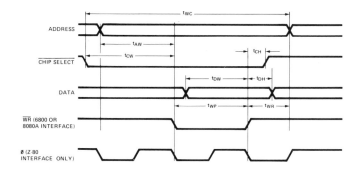

PARAMETER	SYMBOL	MIN.
WRITE CYCLE	t_{WC}	390ns
WRITE DELAY	t_{AW}	65ns
CHIP ENABLE TO WRITE	t_{CW}	65ns
DATA SETUP	t_{DW}	220ns
DATA HOLD	t_{DH}	20ns
WRITE PULSE	t_{WP}	310ns
WRITE RECOVERY	t_{WR}	10ns
CHIP ENABLE HOLD	t_{CH}	20ns

Figure 14.6-4. Data Entry Timing for the CODED DATA CONTROLLER Shown in Figure 14.6-3.

display overflowing to the leftmost display location; Right data entry, where data enters at the righthand side of the display and previous data shifts toward the left; and RAM data entry, where a four bit field in the Control Word specifies the address at which the next Data Word will be written. The 8279 allows data written into the display to be read by the microprocessor, and provides commands to either blank or clear the display.

The DISPLAY PROCESSOR CONTROLLER shown in

Figure 14.7-2 is designed to provide a flexible HDSP-6508 Display Interface for displays up to 40 characters in length. This circuit utilizes a dedicated 8048 single chip microprocessor to provide features such as a blinking cursor, display editing routines, multiple data entry modes, variable display string length, and data out. This controller is available as a series of printed circuit board subsystems of 16, 24, 32, and 40 characters in length. The user interfaces to the 8048 microprocessor through eight Data In Inputs, six Address Inputs, a Chip Select Input, Reset Input, Blank Input, six Data Out Outputs, Data Valid Output, Refresh Output, and Clock Output. The software within the 8048 microprocessor provides four data entry modes — Left Entry with a Blinking Cursor, Right Entry, Block Entry, and RAM Entry. The Data Out port allows the user to read the ASCII data stored within the display, determine the configured data entry mode and display length, and locate the position of the cursor within the display. Since the Data Out port is separate from the Data In port, the HDSP-6508 DISPLAY PROCESSOR CONTROLLER can be used for text editing independent of the main microprocessor system. In Left Entry mode, the controller provides the Clear, Carriage Return, Backspace, Forwardspace, Insert, and Delete editing functions; while in Right Entry mode, the controller provides Clear and Backspace editing functions. The controller can also be expanded into multiple line panels.

The 8048 microprocessor interfaces to the display via the Port 2 output. The output is configured to enable the microprocessor to send a six bit word to one of three destinations. The PROG output is then used to store this word at the specified destination. Destination$_0$ is the 74LS174 hex register. The outputs of this register are decoded by the 74LS259 addressable latches and

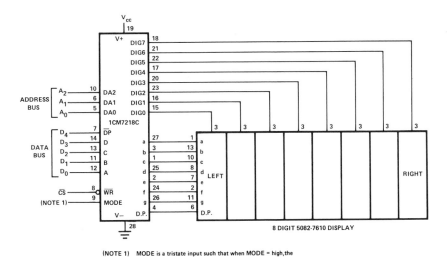

(NOTE 1) MODE is a tristate input such that when MODE = high, the the 7218 provides hexadecimal decoding; when MODE = floating, the 7218 provides Code B decoding; and when MODE = low, the 7218 blanks the display.

Figure 14.6-5. Seven Segment CODED DATA CONTROLLER Utilizing the Intersil ICM7218C.

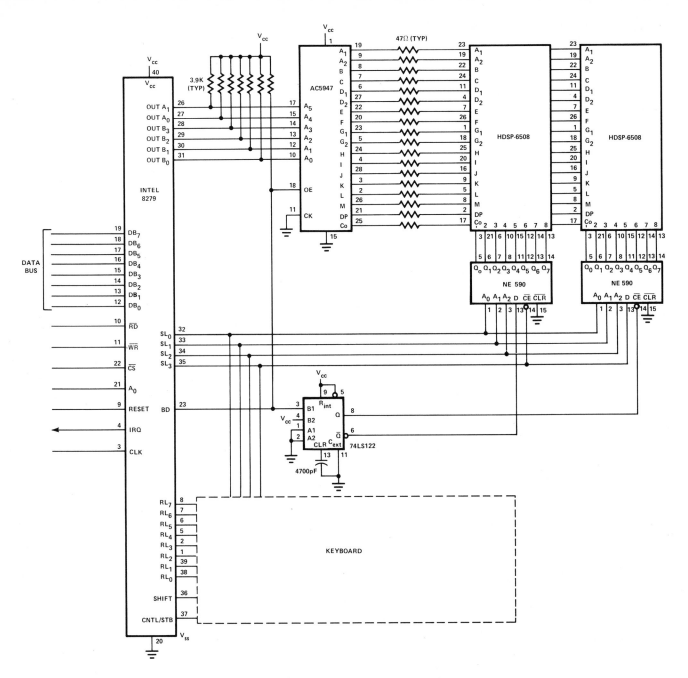

Figure 14.7-1. HDSP-6508 DISPLAY PROCESSOR CONTROLLER Utilizing the Intel 8279 Programmable Keyboard Display Interface.

Sprague ULN 2815 digit drivers. Output $3F_{16}$ is decoded to turn on the rightmost display digit while the address of the leftmost display digit varies from 18_{16} for a 40 character display to 30_{16} for a 16 character display. Destination$_1$ is the AC5947 18 segment decoder/driver. The positive edge of $\overline{\text{PROG}}$ stores a six bit ASCII code within the AC5947. Because Destination$_1$ is pulsed once **every time a digit is** refreshed, this output is also used as the Refresh Output. Destination$_2$ is the Data Valid Output of the Data Output port. Thus, Data Out actually consists of a series of six bit words that are sent to Destination$_2$. Display refresh is accomplished by first turning off the digit drivers

by outputting a 0_{16} to the 74LS174. Then a new ASCII character is stored within the AC5947. Finally, a new digit word is stored within the 74LS174. The actual time that each digit is on varies according to the configured display length so as to provide a fixed 100 Hz refresh rate.

The DISPLAY PROCESSOR CONTROLLER shown in Figure 14.7-3 is designed to provide a flexible HDSP-2000 Display Interface for displays up to 48 characters in length. This circuit also utilizes a dedicated 8048 microprocessor to provide most of the features of the DISPLAY PROCESSOR CONTROLLER shown in Figure 14.7-2 as well as any one

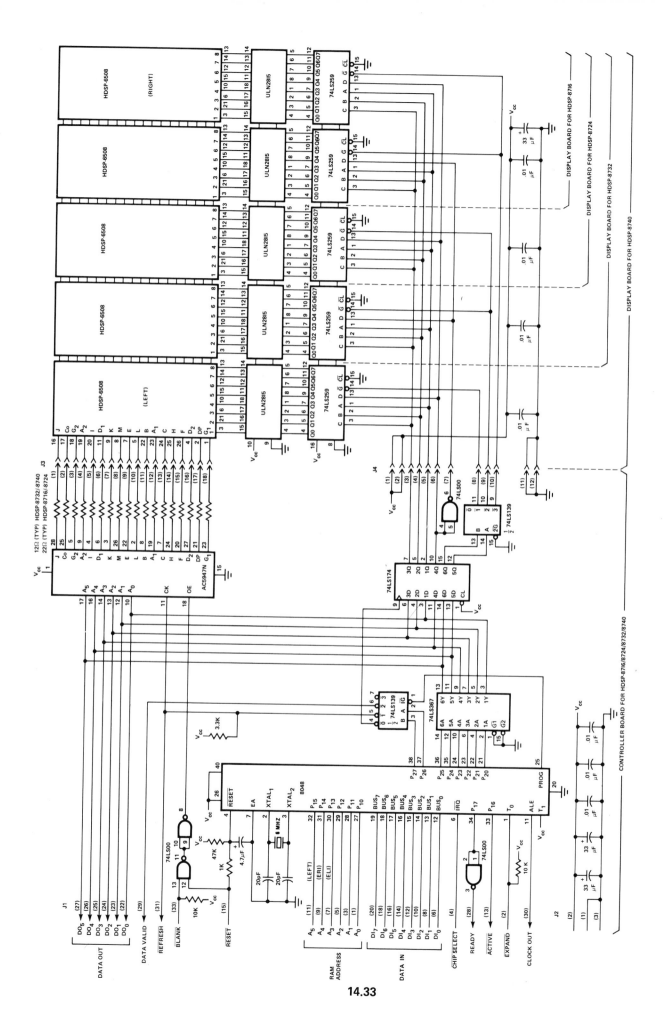

Figure 14.7-2. HDSP-8716/-8724/-8732/-8740 DISPLAY PROCESSOR CONTROLLER.

14.33

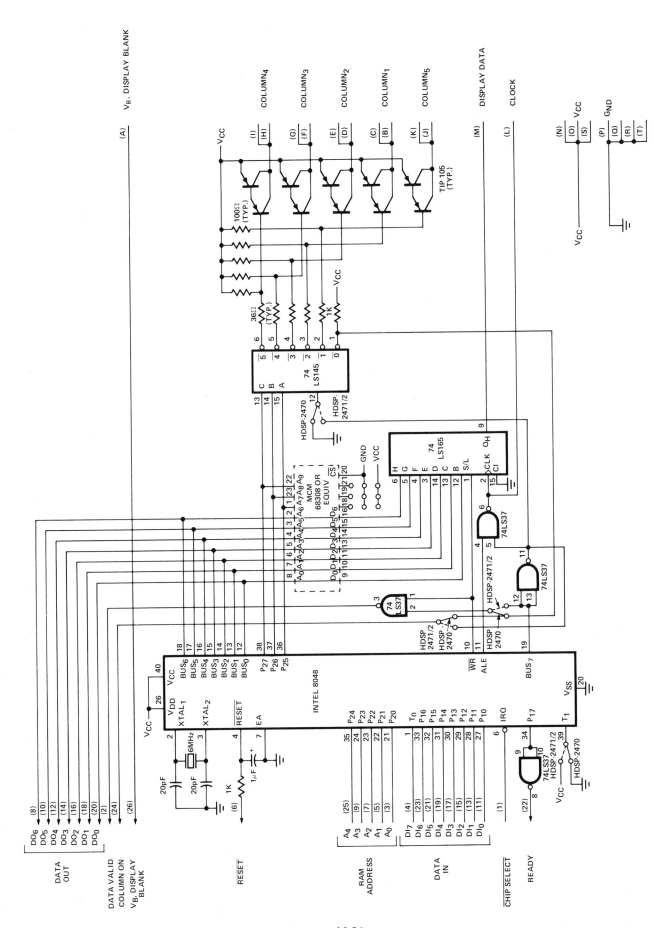

Figure 14.7-3 HDSP-2470/-2471 / -2472 DISPLAY PROCESSOR CONTROLLER.

14.34

of three types of ASCII decoders — a 64 character ASCII decoder, a 128 character ASCII decoder, and a 128 character user programmed decoder. This controller is available as a series of printed circuit board subsystems that include three DISPLAY PROCESSOR CONTROLLERS and HDSP-2000 display boards of 16, 24, 32, and 40 characters in length. The user interfaces to the 8048 microprocessor through eight Data In Inputs, five Address Inputs, a Chip Select Input, Reset Input, Blank Input, seven Data Out Outputs, Data Valid Output, and a Column On Output. The software within the 8048 microprocessor provides four data entry modes — Left Entry with a blinking cursor, Right Entry, Block Entry, and RAM Entry for displays up to 32 characters in length. For displays up to 32 characters in length, the Data Out port allows the user to read the ASCII data stored within the display, a Status Word that specifies configured data entry mode and display length, and a Cursor Address that specifies the location of the cursor within the display. For displays greater than 32 characters, the HDSP-2000 DISPLAY PROCESSOR CONTROLLER does not provide Data Out, but does output a Status Word between refresh cycles. The HDSP-2000 DISPLAY PROCESSOR CONTROLLER provides the Clear, Backspace, Forwardspace, Insert, and Delete editing commands in Left Entry Mode and the Clear and Backspace editing commands in Right Entry mode.

The 8048 microprocessor interfaces to the display via the Bus Port, outputs $P_{27} - P_{25}$, $\overline{WR}$, and ALE. Outputs $P_{27} - P_{25}$ are used to select one of the five columns of the display or to turn all five columns off. The BUS_7 output is used to gate the microprocessor clock, ALE, to the 74LS165 parallel in—serial out shift register and to the clock input of the HDSP-2000 display. In the 128 character ASCII decoder versions of the DISPLAY PROCESSOR CONTROLLER, the Bus_7 output is also used to gate Data Valid and to turn off the display during refresh. The $BUS_6 - BUS_0$ and $\overline{WR}$ outputs are used to strobe display data into the 74LS165 parallel in—serial out shift register. In the 64 character decoder version of the HDSP-2000 DISPLAY PROCESSOR CONTROLLER, the ASCII decoder is located within the 8048 microprocessor and the BUS outputs decoded display data directly to the 74LS165. In the 128 character decoder versions of the controller, the decoder is located within a 1K x 8 ROM. In this application, the desired display data word is generated from a three line column select code provided by $P_{27} - P_{25}$ and from the seven bit ASCII character provided by $BUS_6 - BUS_0$. This technique provides a 500 nS access time for the 1K x 8 ROM before the display data word is strobed into the 74LS165. Display refreshing is accomplished by first turning off the five column drivers and then outputting a string of decoded display data to the 74LS165. The 8048 microprocessor program provides exactly seven clock pulses to the display for each word stored within the 74LS165. Thus, the seven bit display data

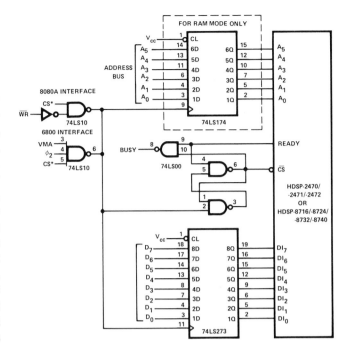

*CS IS A LOGICAL COMBINATION OF HIGH ORDER ADDRESS BITS THAT DISTINGUISH THE ADDRESS OF THE HDSP-2470/-2471/-2472/-8716/-8724/-8732/-8740 FROM THE REST OF THE MICROPROCESSOR SYSTEM.

Figure 14.7-4. Latched Interface to the DISPLAY PROCESSOR CONTROLLERS Shown in Figures 14.7-2 and 14.7-3.

word is converted into seven serial bits. After the display has been loaded with new information, the appropriate column is turned on and the 8048 microprocessor waits a specified time before refreshing the next column. Regardless of configured display length, the HDSP-2000 DISPLAY PROCESSOR CONTROLLER provides a fixed refresh rate of 100 Hz and a column on time that is optimized for each display length. For displays up to 32 characters in length, the 8048 microprocessor outputs the Status Word, Cursor Address, and 32 ASCII characters during the time between refresh cycles. Because the display loading time is proportional to display length, when the system is configured to refresh displays longer than 32 characters, insufficient time remains between refresh cycles to perform Data Out. However, the 8048 microprocessor still outputs the Status Word between refresh cycles.

Interfacing the DISPLAY PROCESSOR CONTROLLERS shown in Figures 14.7-2 and 14.7-3 to microprocessor systems depends on the needs of the particular application. Figure 14.7-4 shows a latched interface between the main microprocessor and the DISPLAY PROCESSOR CONTROLLER. The latch provides temporary storage to avoid making the main microprocessor wait for the DISPLAY PROCESSOR CONTROLLER to accept data. Data from the main microprocessor system is loaded into the 74LS273 octal register on the positive transition of the clock input (pin 11). At the same time, the $\overline{CS}$ Input to the

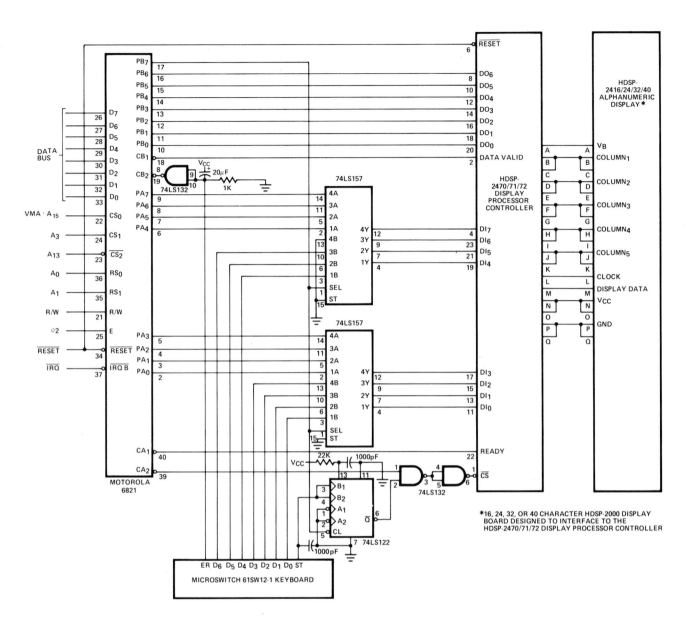

Figure 14.7-5. 6800 Microprocessor Interface to the DISPLAY PROCESSOR CONTROLLER Shown in Figure 14.7-3 Utilizing a Motorola 6821 PIA.

DISPLAY PROCESSOR CONTROLLER is forced low. The $\overline{CS}$ Input stays low until READY goes low. The main microprocessor should avoid loading new data into the 74LS273 as long as BUSY is high. The latched interface can be implemented with an octal register and $\overline{S}$ $\overline{R}$ flip flop if the DISPLAY PROCESSOR CONTROLLER is operated in Left, Right, or Block Entry Modes. RAM Entry Mode requires an additional register for the Address Inputs of the DISPLAY PROCESSOR CONTROLLER.

In sophisticated systems, it may be desirable to have the DISPLAY PROCESSOR CONTROLLER act as a buffer between a keyboard and the main microprocessor. In this configuration, the main processor could output a prompting message to the user via the DISPLAY PROCESSOR CONTROLLER. The user could then enter

data from the keyboard into the display utilizing the controller's editing capability. After the message has been entered and edited, the user would instruct the main microprocessor to read the final edited message from the Data Output Port. This function can be achieved by using a peripheral interface adapter (PIA) available for use with the main microprocessor. One port from the PIA can be used to control the Data Inputs of the DISPLAY PROCESSOR CONTROLLER and another port of the PIA can be used to read the Data Output Port. Figure 14.7-5 shows a 6800 microprocessor system using a Motorola 6821 PIA to control the HDSP-2000 DISPLAY PROCESSOR CONTROLLER shown in Figure 14.7-3. The PB_7 output of the PIA determines whether data is entered into the controller from the microprocessor system or from the keyboard. Control lines CA_1 and CA_2 are used to provide a

```
* PORT CONFIGURATION:
* 1.  PORT A:
          PA0-PA7 OUTPUTS TO DATA IN OF HDSP-247X
*         CA1 (INPUT) MODE 00 SET FLAG NEG EDGE OF READY
*         CA2 (OUTPUT) MODE 100 CLEARED MPU READ PRA, SET
*             NEG EDGE OF READY
* 1.  PORT B:
*         PB0-PB6 INPUTS DATA TO 6800 FROM DATA OUT OF HDSP-247X
*         CB1 (INPUT) MODE 00 SETS FLAG NEG EDGE OF DATA VALID
*         CB2 (INPUT) MODE 000 SETS FLAG NEG EDGE OF ER KEY
*         CB2 (INPUT) MODE 001 SETS FLAG NEG EDGE OF ER KEY
*             CAUSING IRQ
*
*         PB7 (OUTPUT) LOW ENABLES PA0-PA7 TO MUX
*                      HIGH ENABLES KEYBOARD TO MUX
*


LOC   OBJECT     CODE      SOURCE STATEMENT

      8008       PRA       EQU    $8008
      8008       DRA       EQU    $8008
      8009       CRA       EQU    $8009
      800A       PRB       EQU    $800A
      800A       DRB       EQU    $800A
      800B       CRB       EQU    $800B

                 ORG       $0000
0000             MESSAGE   RMB    2

                 ORG       $0100
0100             STATUS    RMB    1
0101             CURSOR    RMB    1
0102             DATA      RMB    32

                 ORG       $0400
0400  CE 0100    READ      LDX    I, STATUS
0403  B6 800A    LOOP1     LDA A  E, PRB        CLEAR CB1 AND CB2
0406  5F                   CLR B
0407  5C         LOOP2     INC B
0408  B6 800B              LDA A  E, CRB
040B  2A FA                BPL    LOOP2         WAIT FOR DATA VALID
040D  C1 0A                CMP B  I, 10
040F  23 F2                BLS    LOOP1
0411  C6 21                LDA B  I, 33
0413  B6 800A    LOOP3     LDA A  E, PRB        READ AND CLEAR CB1
0416  84 7F                AND A  I, $7F
0418  A7 00                STA A  X, 0          STORE IN RAM

041A  B6 800B    LOOP4     LDA A  E, CRB
041D  2A FB                BPL    LOOP4         WAIT FOR DATA VALID
041F  08                   INX
0420  5A                   DEC B
0421  26 F0                BNE    LOOP3         READ DATA
0423  B6 800A              LDA A  E, PRB
0426  84 7F                AND A  I, $7F
0428  A7 00                STA A  X, 0
042A  39                   RTS

042B  DE 00      LOAD      LDX    D, MESSGE
042D  A6 00      LOOP10    LDA A  X, 0
042F  08                   INX
0430  81 FF                CMP A  I, $FF        LAST WORD IN STRING
0432  27 0D                BEQ    ENDL          JUMP WHEN DONE
0434  B7 8008              STA A  E, PRA
0437  7D 8008              TST    E, PRA        CLEAR CA1 AND CA2
043A  B6 8009    LOOP11    LDA A  E, CRA
043D  2A FB                BPL    LOOP11        WAIT
043F  20 EC                BRA    LOOP10
0441  DF 00      ENDL      STX    D, MESSGE
0443  39                   RTS

                 ORG       $0500
0500  7F 8009    START     CLR    E, CRA
0503  7F 800B              CLR    E, CRB
0506  86 FF                LDA A  I, $FF
0508  B7 8008              STA A  E, DRA
050B  86 24                LDA A  I, $24
050D  B7 8009              STA A  E, CRA
0510  86 80                LDA A  I, $80
0512  B7 800A              STA A  E, DRB
0515  86 04                LDA A  I, $04
0517  B7 800B              STA A  E, CRB

* PROCEDURE TO LOAD HDSP-247X SYSTEM
051A  0E                   CLI
051B  7F 800A              CLR    E, PRB        DISABLE KEYBD FROM MUX
051E  BD 042B              JSR    E, LOAD

* PROCEDURE TO READ DATA OUT OF HDSP-247X SYSTEM
0521  7D 800A              TST    E, PRB        CLEAR CB1, CB2
0524  86 80                LDA A  I, $80
0526  B7 800A              STA A  E, PRB        ENABLE KEYBD TO MUX
0529  86 0C                LDA A  I, $0C
042B  B7 800B              STA A  E, CRB        ENABLE IRQ,
052E  0F                   SEI                  IRQ CAUSE JSR TO READ
```

Figure 14.7-6.   6800 Microprocessor Program That Interfaces to the Circuit Shown in Figure 14.7-5.

data entry handshake between the 6800 and the DISPLAY PROCESSOR CONTROLLER. This allows the 6800 to load data into the controller at the highest possible rate. Data is read into the main microprocessor system through Port B of the PIA using the CB_1 input as a data strobe.

The 6800 microprocessor program shown in Figure 14.7-6 is used to operate the PIA interface described in Figure 14.7-5. The microprocessor program following "START" is used to initialize the 6821 PIA. Once initialized, the PIA can be used either to load data into the controller via the main microprocessor, allow data to be loaded into the controller via the keyboard, or to read data from the Data Out port into the main microprocessor. The instruction CLR E, PRB at location $051B_{16}$ forces PB_7 low to connect the outputs of Port A to the Data Inputs of the controller. Subroutine "LOAD" then loads a series of eight bit words into the controller. "LOAD" continues to output words until it reads an FF_{16} to denote the end of the prompting message. The instruction sequence LDA A I, $80 and STA A E, PRB at location 0526_{16} forces PB_7 high to connect the output of the keyboard to the Data Inputs of the controller. In this mode, the user can enter or edit data into the DISPLAY PROCESSOR CONTROLLER. The 4B

input of the 74LS157 has been grounded to prevent the keyboard from loading a control word into the DISPLAY PROCESSOR CONTROLLER. The instructions LDA A I, $OC and STA A E, CRB at location 0529_{16} enables the "ER" key on the keyboard to interrupt the microprocessor when the edited message is complete. Subroutine "READ" would then be used to read data into the 6800 system. "LOOP 1" within the program continuously reads the Data Valid output and waits until the controller outputs the STATUS word. This STATUS word and the subsequent CURSOR ADDRESS word, and 32 ASCII characters are then stored in 32 consecutive words of scratch pad memory starting at address "STATUS."

A similar PIA interface designed for an 8080A microprocessor system that uses an Intel 8255A PIA is shown in Figure 14.7-7. This interface operates in much the same way as the 6821 PIA interface that was previously described. The PC_4 output of the PIA determines whether the Data Inputs of the HDSP-2000 DISPLAY PROCESSOR CONTROLLER shown in Figure 14.7-3 are connected to the PIA or to the keyboard. Control lines PC_6 and PC_7 are used to provide a data entry handshake between the 8080A microprocessor and the DISPLAY PROCESSOR

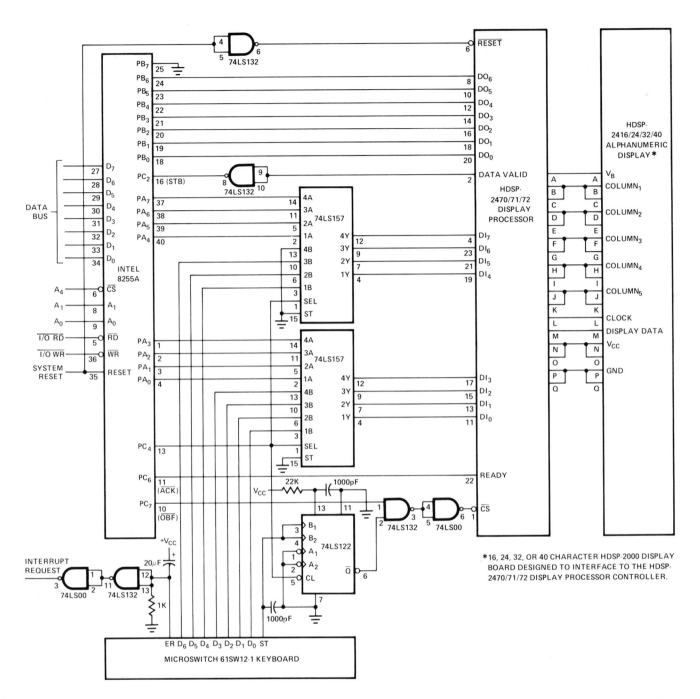

Figure 14.7-7. **8080A Microprocessor Interface to the DISPLAY PROCESSOR CONTROLLER Shown in Figure 14.7-3 Utilizing an Intel 8255 PIA.**

CONTROLLER. Data is read into the 8080A microprocessor system through Port B of the PIA using PC_2 as the data strobe.

The 8080A microprocessor program shown in Figure 14.7-8 is used to operate the PIA interface described in Figure 14.7-7. The microprocessor program following "START" is used to initialize the 8255A PIA. The instructions MVI A, 08H and OUT CNTRL at location $E456_{16}$ force PC_4 low to connect Port A of the PIA to the Data Inputs of

the DISPLAY PROCESSOR CONTROLLER. Subroutine "LOAD" would then be used to load a prompting message into the controller. The instructions MVI A, 09H and OUT CNTRL at location E45D connect the keyboard to the Data Inputs of the controller. In this mode, the user can enter data into the DISPLAY PROCESSOR CONTROLLER, or to edit an existing line. Subroutine "READ" would then be used to read the data from the Data Output port into the 8080A microprocessor system.

14.38

```
* PORT CONFIGURATION:
* 1.   PORT A (MODE 1 OUTPUT):
*         PA0-PA7 OUTPUTS TO DATA IN OF HDSP-247X
*         PC7 (OBF) OUTPUT; TO CHIP SELECT
*         PC6 (ACK) INPUT; TO READY
*         FLAG PC7 (OBF) CLEARED BY OUTPUT; SET BY READY
*
* 2.   PORT B (MODE 1 INPUT):
*         PB0-PB6 INPUTS DATA FROM DATA OUT OF HDSP-247X
*         PC2 (STB) INPUT; LOADS DATA ON NEG EDGE OF DATA VALID
*         FLAG PC0 (INTR) CLEARED BY INPUT; SET BY DATA VALID
*
* 3.   PORT C:
*         PC4 OUTPUT; LOW ENABLES PA0-PA7 TO HDSP-247X
*                     HIGH ENABLES KEYBOARD TO HDSP-247X


LOC    OBJECT    CODE     SOURCE STATEMENTS

000C             PA       EQU    0CH
000D             PB       EQU    0DH
000E             PC       EQU    0EH
000F             CNTRL    EQU    0FH

                          ORG    0E000H
E000   02 E0     ASCII    DW     TEXT
E002   00        TEXT     DS     32

                          ORG    0E100H
E100   00        STAT     DB     0
E101   00        ADDR     DB     0
E102   00        DATA     DS     32

                          ORG    0E400H
E400   F3        READ     DI
E401   F5                 PUSH   PSW
E402   E5                 PUSH   H
E403   C5                 PUSH   B
E404   0E 20              MVI    C, 32
E406   21 00 E1           LXI    H, STAT        FIRST WORD
E409   DB 0D              IN     PB             CLEAR INTR
E40B   06 00     LOOP1    MVI    B, 0
E40D   DB 0E     LOOP2    IN     PC
E40F   04                 INR    B
E410   1F                 RAR
E411   D2 0D E4           JNC    LOOP2          WAIT UNTIL INTR IS SET
E414   3E 0A              MVI    A, 10
E416   B8                 CMP    B

E417   DB 0D              IN     PB
E419   D2 0B E4           JNC    LOOP1          WAIT UNTIL STATUS WORD
E41C   77        LOOP3    MOV    M, A           STORE IN RAM
E41D   23                 INX    H
E41E   DB 0E     LOOP4    IN     PC
E420   1F                 RAR
E421   D2 1E E4           JNC    LOOP4          WAIT UNTIL INTR IS SET
E424   DB 0D              IN     PB
E426   0D                 DCR    C
E427   C2 1C E4           JNZ    LOOP3
E42A   77                 MOV    M, A           STORE LAST WORD
E42B   C1                 POP    B
E42C   E1                 POP    H
E42D   F1                 POP    PSW
E42E   FB                 EI
E42F   C9                 RET

E430   2A 00 E0  LOAD     LHLD   ASCII          FIRST WORD OF MESSAGE
E433   7E        LOOP5    MOV    A, M
E434   FE FF              CPI    0FFH           CHECK TO SEE IF DONE
E436   CA 45 E4           JZ     ENDL
E439   D3 0C              OUT    PA             OUTPUT TO DISPLAY
E43B   23                 INX    H
E43C   DB 0E     LOOP6    IN     PC
E43E   17                 RAL
E43F   D2 3C E4           JNC    LOOP6          WAIT
E442   C3 33 E4           JMP    LOOP5          NEXT WORD
E445   23        ENDL     INX    H
E446   22 00 E0           SHLD   ASCII
E449   C9                 RET

E44A   3E A7     START    MVI    A, 0A7H        PA OUTPUT, PB INPUT
E44D   D3 0F              OUT    CNTRL
E44E   3E 0C              MVI    A, 0CH         CLEAR INTE A
E450   D3 0F              OUT    CNTRL
E452   3E 05              MVI    A, 05H
E454   D3 0F              OUT    CNTRL          SET INTE B

          * PROCEDURE TO LOAD HDSP-247X SYSTEM
E456   3E 08              MVI    A, 08H
E458   D3 0F              OUT    CNTRL          ENABLE A SIDE OF MUX
E45A   CD 30 E4           CALL   LOAD

          * PROCEDURE TO READ DATA OUT OF HDSP-247X SYSTEM
E45D   3E 09              MVI    A, 09H
E45F   D3 0F              OUT    CNTRL          ENABLE B SIDE OF MUX
E461   FB                 EI                    INT MUST CALL READ
```

Figure 14.7-8.   8080A Microprocessor Program That Interfaces to the Circuit Shown in Figure 14.7-7.

With minor modifications, the PIA interfaces shown in Figure 14.7-5 and 14.7-7 can also be used with the HDSP-6508 DISPLAY PROCESSOR CONTROLLER shown in Figure 14.7-2. Subroutine "LOAD" will work for both controllers. Subroutine "READ" must be modified to replace the wait loop "LOOP1" with code that initiates the Data Out function. This is accomplished by connecting the main microprocessor to the Data Inputs of the controller, outputting the command FF_{16} to the controller, and then waiting until the controller begins to output data. The controller will output a STATUS WORD, CURSOR ADDRESS, and a string of ASCII characters that must be read into the main microprocessor system.

The HDSP-6508 DISPLAY PROCESSOR CONTROLLER shown in Figure 14.7-2 is programmed to default to Left Entry Mode for a 40 character display and the HDSP-2000 DISPLAY PROCESSOR CONTROLLER shown in Figure 14.7-3 is programmed to default to Left Entry for a 32 character display. If some other entry mode or string length is desired, it is necessary to either load the appropriate control word from the microprocessor or to provide a control word during POWER ON RESET. Both DISPLAY PROCESSOR CONTROLLERs are programmed to read the Data In lines during RESET and interpret the contents as the control word. The circuit depicted in Figure 14.7-9 can be utilized to load any desired preprogrammed word into the DISPLAY PROCESSOR CONTROLLER during power on.

Under certain operating conditions, it may be desirable to vary the brightness of displays controlled by the DISPLAY PROCESSOR CONTROLLERS shown in Figures 14.7-2 and 14.7-3. The circuit depicted in Figure 14.7-10 may be utilized to provide manual brightness control of the display through pulse width modulation. Automatic control may be achieved by substituting an appropriate value photo-conductor for potentiometer R_1.

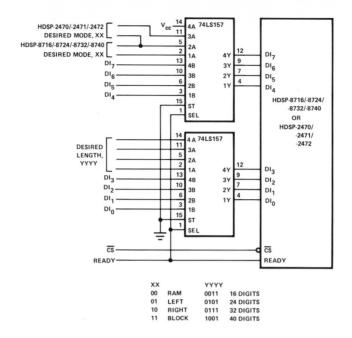

Figure 14.7-9.   External Circuitry to Load a Control Word
Into the DISPLAY PROCESSOR
CONTROLLERS Shown in Figures 14.7-2
and 14.7-3 Upon $\overline{\text{Reset}}$.

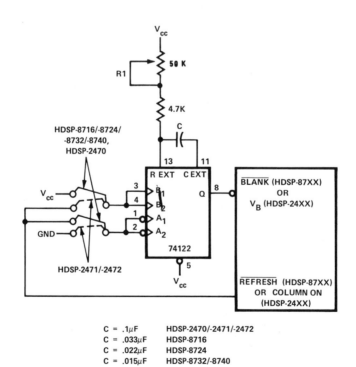

Figure 14.7-10.   External Circuitry to Vary Luminous
Intensity of the DISPLAY PROCESSOR
CONTROLLERS Shown in Figures 14.7-2
and 14.7-3.

Section 15
Optical Sensing Systems

15.0 OPTICAL SENSING

15.1 Introduction

The rapid growth of digital processing being used in commercial, industrial, and consumer products, has created the need for new sensors that convert physical parameters into electrical signals which may be directly interfaced to a digital system. Optical sensors have utility in each of these areas, in that they provide a quick non-contact response to the parameter sought as a data source. Commercial applications include bar code reading, paper edge sensing, end of tape sensing, position and magnetic tape loop stabilizing. Industrial uses may consist of optical tachometry, assembly line monitoring, and safety interlocks, while the consumer uses of optical sensors may be found in audio products, entertainment products, and video games.

Each of these applications include an optical emitter, a transmission path, and a detector to perform the sensing function. The sensing may occur by having the object obstruct the transmission path, or complete it by reflecting the emitter beam to the detector. In either the transmissive or reflective sensing configuration, there are optical, electrical, and mechanical considerations that must be addressed in order to insure optimum performance of the emitter-detector system.

15.2.1 System Elements

Every optical sensor system includes a source of optical flux, a transmission path, and a receiving detector. In most sensor system analysis, the available flux and the responsivity of the receiver are considered to be constant and consequently the dynamic incidance change found at the receiver results from modifications of the transmission path. Figure 15.2.1-1 shows a paper tape reader and densitometer examples of transmissive sensor systems. The presence or absence of the paper hole results in a binary ON-OFF transmission path attenuation. The densitometer functions by causing an analog modulation of the transmission path. Bar code scanning and paper edge sensing are applications of reflective sensor techniques shown in Figure 15.2.1-1. The bar code provides a reflective contrast between the background and the bars. It is the reflective differential that modulates the transmission path. In paper edge sensing applications, the paper edge creates a digital responding transmission system. When the paper edge is not present in the reflecting field of the sensor, the flux transmission will be zero, and the transmittance will increase when the edge is in the field.

15.2.2 Optical Transfer Function

The characteristics of the transmission path can be estimated through the use of an optical transfer function, OTF. The function is the ratio of the total optical flux available, ϕ_e, to the incident flux arriving at the receiver, $\phi_e(R)$. This function allows the designer to calculate the amount of photocurrent available for the detector amplifier.

$$\text{OTF} = \frac{\phi_e\text{(RECEIVED)}}{\phi_e\text{(AVAILABLE)}} \qquad \text{(15.2.2-1)}$$

The optical flux attenuation comes from a number of causes. One cause is the transmission loss as the flux is incident upon and passes through the transmission media. These losses come about because of reflection at the surface of the material, and scattering and absorption within the material. For the calculation of the optical transfer function, it is customary to describe these losses, τ, in terms of a transmittance, T, of the material for the wavelength of the source. The transmittance is equal to:

$$\text{T} = (1 - \tau) \qquad \text{(15.2.2-2)}$$

Another cause of attenuation is the incomplete coupling of the flux from the source to the receiver caused by the mismatch of the receiver relative aperture to the source relative aperture.

15.2.3 Coupling Fundamentals

The general case of source flux coupled to a receiver is dependent upon the source radiation pattern, the source-receiver spacing, and the receiver area. The following analyses shows how these parameters affect the OTF.

Figure 15.2.3-1 shows a source which is located at the center of a hemisphere with a receiver of an area A located at a distance, **d**, from the apex. The ratio of the receiver area to the distance squared defines the solid angle, ω, sub-

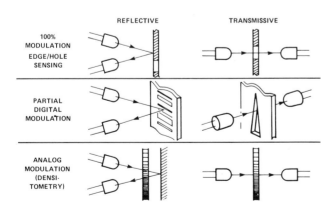

Figure 15.2.1-1. Different Techniques of Flux Path Modulation.

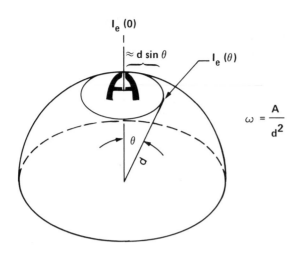

Figure 15.2.3-1. Definition of the Solid Angle, ω.

$$\omega = \frac{A}{d^2}$$

tended by the area, **A**. The total flux, ϕ_e, being radiated is the integral of the flux incident within the hemisphere. The radiation pattern of a Lambertian source is shown in Figure 15.2.3-2. The pattern describes the ratio of the radiant intensity at an off-axis angle, $I_e(\theta)$ to the axial radiant intensity, $I_e(0)$. The radiation pattern for a Lambertian source (or the reception pattern for a Lambertian receiver) is described by the cosine of the off-axis angle. The Lambertian radiation pattern function is:

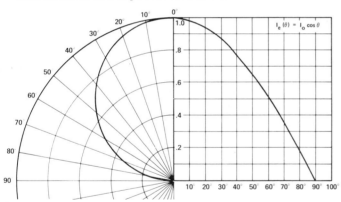

Figure 15.2.3-2. Radiation Pattern of a Lambertian Source.

$$I_e(\theta) = I_e(0) \cos \theta \qquad (15.2.3\text{-}1)$$

When the radiation pattern, $I_e(\theta)/I_e(0)$, and the axial radiant intensity, $I_e(0)$, are known, Equation (15.2.3-2) can be used to determine the total flux into the hemisphere, ($\theta = 90°$) or the flux into a cone created by the area, A.

$$\phi_e(\theta) = I_e(0) \int_0^\theta \frac{I_e(\theta)}{I_e(0)} 2\pi \sin \theta \, d\theta \qquad (15.2.3\text{-}2)$$

When a Lambertian radiator is used as a source, the flux into the cone specified by the off-axis angle, θ, is calculated from Equations (15.2.3-1) and (15.2.3-2).

$$\phi_e(\theta) = I_e(0) \int_0^\theta 2\pi \cos \theta \sin \theta \, d\theta$$

$$\phi_e(\theta) = I_e(0) \pi \sin^2 \theta \qquad (15.2.3\text{-}3)$$

The total flux, ϕ_e, of the Lambertian source is obtained from Equation (15.2.3-) when $\theta = 90°$. Thus, the total flux ϕ_e is π times the axial radiant intensity, $I_e(0)$.

The amount of flux coupled into the receiver area A relates to the relative aperture of the receiver. The relative aperture, also known as the numerical aperture, N.A., defines the ability of the receiver to accept flux arriving at off-axis angles. When a source with a specific radiation pattern is considered, a receiver with a large numerical aperture will capture more flux than one with a smaller N.A. The numerical aperture is defined as the sine of one-half the included angle of the receiver cone. Thus,

$$\textbf{N.A.} = \sin \theta \qquad (15.2.3\text{-}4)$$

where θ = ½ cone angle.

As seen in figure 15.2.3-3, when small angles of θ are considered, the numerical aperture can be calculated as:

$$\textbf{N.A.} = \sin \theta \approx \textbf{D/2d}$$

$$\textbf{N.A.} = \sin \theta \approx \textbf{(A/π)}^{\frac{1}{2}}\textbf{/d}$$

$$\textbf{N.A.}^2 = \sin^2 \theta \approx \textbf{A/πd}^2 \qquad (15.2.3\text{-}5)$$

In Figure 15.2.3-3 are shown a source and a receiver separated by a distance, d. At that distance, the cone (θ_1) of radiation from the source irradiates an area, A_S. Clearly, the receiver having an area, A_R, much smaller than A_S, describes a smaller cone (θ_2) and will receive only a fraction of the flux radiated by the source. This fraction describes the Optical Transfer Function (OTF) for this simple

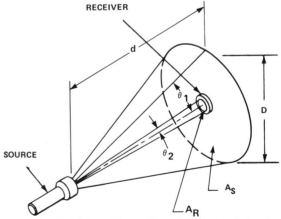

Figure 15.2.3-3. Illustration of Numerical Aperture Relationships.

situation, and a good estimate of the value of this fraction is the ratio of the areas A_R/A_S. Then, applying the relationship derived in Equation (15.2.3-5), the OTF can be defined in terms of the cones for source and receiver which are described by numerical apertures:

$$OTF = \frac{\phi_R}{\phi_S} = \frac{A_R}{A_S} = \frac{\pi d^2 N.A._R{}^2}{\pi d^2 N.A._S{}^2} \qquad (15.2.3-6)$$

$$= \left(\frac{N.A._R}{N.A._S}\right)^2 = \left(\frac{\sin\theta_2}{\sin\theta_1}\right)^2$$

In estimating OTF, it is necessary only to recognize and evaluate such cones of coupling — exit cone (or N.A.) and acceptance cones. As a general rule, the cone angle, θ, is defined as the angle at which coupling is 0.1 (10%) of the axial value ($\theta = 0$). A case of special interest is that of a Lambertian emitter having a radiation pattern varying as $\cos\theta$. The angle at which $\cos\theta = 0.1$ is $84.26°$, at which $\sin\theta = N.A. = 0.995$; thus, for practical purposes, a Lambertian source is regarded as having N.A. = 1. If, in Figure 15.2.3-3, a Lambertian source were used, the relationship in Equation (15.2.3-6) reduces:

$$OTF = \left(\frac{N.A._R}{N.A._S}\right)^2 = \left(\frac{N.A._R}{1}\right)^2 = (N.A._R)^2 \qquad (15.2.3-7)$$

This simplification is important as it is used extensively in the sections to follow. Notice, however, that OTF cannot exceed unity, so that if $N.A._R > N.A._S$, then OTF = 1. This occurs when the acceptance cone of the receiver is larger than the exit cone of the source.

15.2.4 Lens Fundamentals

In practical sensor applications, the source-receiver distance might be ten or more times the diameter of the receiver. If the receiver area is small relative to d, then by Equation (15.2.3-5), the receiver numerical aperture will be small. If the source is Lambertian, the total flux coupling will then relate to the square of this small numerical aperture.

Through the use of lens, more efficient coupling will result. In order to understand how this is accomplished, it is helpful to present a few rules of optics. The first is the basic lens equation. Referring to Figure 15.2.4-1, a double convex condensing lens is shown relaying an image from the source to the receiver. The source is located a distance, d_S, from the lens and the real image is located at a distance, d_R, which is related to the focal length of the lens. Thus,

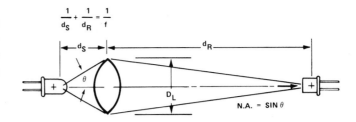

$$OTF = N.A._L{}^2 \cdot T \text{ , FOR A LAMBERTIAN SOURCE WITH IMAGE AREA SMALLER THAN RECEIVER AREA}$$

Figure 15.2.4-1. Focused Emitter Detector System Using a Double Convex Lens.

the relationship of d_S, d_R, f is called the basic lens equation.

$$\frac{1}{f} = \frac{1}{d_S} + \frac{1}{d_R} \qquad (15.2.4-1)$$

where f = Focal Length

d_S = Source Lens Distance

d_R = Received Image Distance

This relay lens system is focused when the receiver is placed at the distance, d_R, [from Equation (15.2.4-1) when f and d_S are known] from the lens.

Under the focused condition, the image of the source (at the plane of the receiver) is magnified. The degree of magnification, m, is equal to the ratio of d_R to d_S.

$$m = \frac{d_R}{d_S} \qquad (15.2.4-2)$$

$$m^2 = \frac{A_{S,M}}{A_S} \qquad (15.2.4-3)$$

where $A_{S,M}$ = Source Image (Source Magnified)

A_S = Source Area

Using Equations (15.2.4-1) and (15.2.4-2), the magnification can be represented in terms of the focal length and source to lens distance, d_S.

$$A_{S,M} = m^2 A_S = \left(\frac{f}{d_S - f}\right)^2 \cdot A_S \qquad (15.2.4-4)$$

When the source-lens distance, d_S, is twice the focal length, f, the image will appear at $d_R = 2f$ on the other side of the lens. Under a 2f focused condition, the area of the image

15.3

$A_{S,M}$, is equal to the area of the source A_S. This configuration is termed a 1:1 magnification or 2f system.

Lenses also have relative aperture qualities. The relative aperture may be specified as either a numerical aperture, N.A., or an an f-number, f/. The f-number is equal to the ratio of the focal length to the diameter of the lens. The f-number and the numerical aperture are inversely related, such that a smaller f-number will result in a proportionately larger N.A.

In Figure 15.2.4-1, the effective diameter of the lens is D_L, and relative to this, the numerical aperture is approximately $D_L/2d_S$ as in Equation (15.2.3-5). Comparing the definitions of f/no and N.A.:

$$f/no = \frac{f}{D_L} \qquad\qquad N.A. \approx \frac{D_L}{2d_S}$$

$$f/no = \frac{1}{2N.A.}\left(\frac{f}{d_S}\right) \qquad\qquad (15.2.4\text{-}5)$$

When $d_S = f$, the source is focused at infinity, making N.A. $= 1/2(f/no)$; when $d_S < f$, there is no real image, but a virtual image of the source lying on the same side of the lens, and the lens is less effective in improving the coupling.

15.2.5   Lens Coupling

The amount of flux coupled into the lens is dependent upon the numerical aperture of the lens and the exit aperture of the source. When a focused lens system is considered such as that shown in Figure 15.2.4-1, the flux that arrives at the received image point is equal to the ratio of the square of the numerical aperture of the lens divided by the square of the exit numerical aperture of the source, this total times the transmittance, T, of the lens. Thus, the lensed OTF is equal to:

$$OTF = \frac{\phi_R}{\phi_S} = \left[\frac{N.A._L}{N.A._S}\right]^2 \cdot T \qquad (15.2.5\text{-}1)$$

When the source is Lambertian, the equation simplifies to:

$$OTF = \frac{\phi_R}{\phi_S} = N.A._L^{\,2} \cdot T \qquad (15.2.5\text{-}2)$$

The following will illustrate a practical example:

Given: Detector Area A_D = .1 mm^2
Lambertian Source A_S = .1 mm^2, ϕ_S = 100μW
Lens Numerical Aperture N.A.$_L$ = .5
Lens Focal Length, f = 5 mm
Lens Transmittance, T = .95
Lens to Source Distance, d_S = 20 mm
Lens to Receiver Distance, d_R = 6.67 mm

Using Equation (15.2.5-1), the flux at the receiver will be:

$$\phi_R = \phi_S\, N.A._L^{\,2} \cdot T \qquad\qquad (15.2.5\text{-}3)$$

$$\phi_R = 100\mu W\,(.5)^2 \cdot .95$$

$$\phi_R = 23.75\mu W$$

The received area is determined by Equation (15.2.4-3).

$$A_{S,M} = \left(\frac{f}{d_S - f}\right)^2 \cdot A_S$$

$$(15.2.5\text{-}4)$$

$$A_{S,M} = \left(\frac{5\ mm}{20\ mm - 5\ mm}\right)^2 \cdot .1\ mm^2 = .011\ mm^2$$

Thus, into an area of .011 mm^2, a flux of 23.75μW is concentrated. Using the basic lens equation, (15.2.4-1), the received image lies 6.7 mm from the lens. If a photodiode with an area, A_D, were placed at this distance, d_R, the fraction of this flux that couples the detector is the ratio of A_D to $A_{S,M}$ if the detector area lies entirely within the source image area. In general, this fraction is the ratio of that portion of the source image that overlaps the receiver, divided by the total source image area, and therefore cannot exceed unity, even when the detector area A_D is much larger than the source image area, $A_{S,M}$.

$$\text{Detector Coupling} = K_D = \frac{A_D}{A_R} \leqslant 1 \qquad (15.2.5\text{-}5)$$

When a lens coupling is compared with non-lens coupling, the improvement is described by:

$$\frac{\phi_L}{\phi_{n-L}} = \frac{N.A._L^{\,2} \cdot T}{N.A._R^{\,2}} \quad ; \quad N.A._R^{\,2} = A_D/[\pi\,(d_S + d_R)^2]$$

where N.A.$_R$ is found from Equation (15.2.3-5) with $d = d_S + d_R$

$$\frac{\phi_L}{\phi_{n-L}} = \frac{(.5)^2 \cdot 0.95}{4.47 \times 10^{-5}} = 5.3 \times 10^3$$

Thus, this simple example using a lens improves the coupling gain by 37dB.

15.2.6   Reflector Fundamentals

Sensor applications, such as bar code scanning, paper edge detection, and optical tachometry, use the reflective properties of the object or element that they sense.

The reflection of the incident flux may either be specular or diffuse. A specular reflector is one where the angle of the reflected ray of flux is equal to the incident ray of flux. Thus, if a ray of flux was incident to the reflector at $20°$ from the normal, the reflected ray would also be $20°$ from the normal and $40°$ from the incident. A first surface mirror or a highly polished code wheel are examples of specular reflectors. They are characterized as having reflection coefficients, ρ, of almost unity. It has a numerical aperture, $N.A._{R-S}$, equal to the N.A. of the source that is incident upon it. The reflected flux would be equal to incident flux times the reflection coefficient.

$$\phi_{OUT} = \phi_{IN} \frac{N.A._L^2}{N.A._{R-S}^2} \cdot \rho; \quad \text{where } N.A._{RS} = N.A._L$$

$$\phi_{OUT} = \phi_{IN}\rho \tag{15.2.6-1}$$

The reflection coefficient is assumed to be constant over the wavelengths of interest. If this is not true, then a correction coefficient, $k\rho(\lambda)$ must be introduced to correct for this spectral property.

A perfectly diffuse reflector is characterized as having a Lambertian radiation pattern. The flux that is coupled from the reflector is equal to:

$$\frac{\phi_{OUT}}{\phi_{IN}} = \frac{N.A._{RECEIVER}^2}{N.A._{REFLECTOR}^2} \cdot \rho \tag{15.2.6-2}$$

In typical applications, the numerical aperture of the receiver will be that of the receiving lens. Also, the N.A. of the diffuse surface is unity; therefore, Equation (15.2.6-2) can be rewritten as:

$$\frac{\phi_{OUT}}{\phi_{IN}} = N.A._L^2 \cdot \rho \tag{15.2.6-3}$$

Most reflectors are neither perfectly specular or diffuse, but a combination of the two. A typical diffuse reflector may appear more specular at one wavelength and more diffuse at another. Also, the angle of incidence may modify the reflection properties.

It may be desirable in reflective sensor applications to compute the relative ratio of reflection between a specular and a diffuse reflector. Assuming both applications have the same ϕ_{IN} and the same receiving lens, and using Equations (15.2.6-1) and (15.2.6-2), this ratio becomes:

$$\frac{\phi_{OUT \, (SPECULAR)}}{\phi_{OUT \, (DIFFUSE)}} = \frac{\rho_S}{N.A._L^2 \rho_D} \tag{15.2.6-4}$$

Thus, if the receiving lens had a $N.A._L = .3$, the effective gain of using a specular over a diffuse reflector would be 10.45dB.

15.2.7   Confocal Coupling

It was shown in Section 15.2.5 that the area of the source could be reduced or magnified through the use of a lens. The technique of reducing the image of the receiver and source, such that they lie on a plane located a distance between the source and sensor, can be accomplished by using a confocally spaced pair of lenses.

Figure 15.2.7-1 shows two plano-convex lens positioned so that they are confocally focused. Using the coupling concepts developed in Section 15.2.3, the overall OTF can be developed in the following steps.

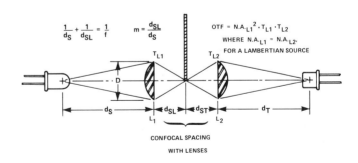

Figure 15.2.7-1.   Confocal Spaced Emitter Detector System Employing Two Plano Convex Lenses.

Step 1.   Flux into Lens 1.

$$\frac{\phi_{LI}}{\phi_S} = \frac{\text{entrance } N.A._{L1}^2}{N.A._S^2} \cdot T_{LI} \tag{15.2.7-1}$$

Step 2.   Flux into Lens 2 from Lens 1.

$$\frac{\phi_{L2}}{\phi_{L1}} = \frac{\text{entrance N.A.}_{L2}{}^2}{\text{exit N.A.}_{L1}{}^2} \qquad (15.2.7\text{-}2)$$

Step 3.   Total OTF.

$$\qquad (15.2.7\text{-}3)$$

$$OTF = \frac{\phi_{L2}}{\phi_S}$$

$$= \frac{\text{entrance N.A.}_{L1}{}^2}{\text{exit N.A.}_S{}^2} \cdot \frac{\text{entrance N.A.}_{L2}{}^2}{\text{exit N.A.}_{L1}{}^2} \cdot T_{L1} \cdot T_{L2}$$

The convenient fact about such a transfer function is that each element of the linear system can be evaluated and then the total function is the product of the individual terms.

In a practical application, the exit N.A. of lens 1 and the entrance N.A. of lens 2 will be equal. If a Lambertian source were used in Equation (15.2.7-1), it would reduce to $\text{N.A.}_{L1}{}^2 \, T_{L1}$. These two conditions allow the following.

$$\textbf{Confocal OTF} = \frac{\phi_{L2}}{\phi_S} = \text{N.A.}_{L1}{}^2 \cdot T_{L1} \cdot T_{L2} \quad (15.2.7\text{-}4)$$

The image size appearing equidistant between the two lens surfaces is determined by the magnification factor, $m = d_{SL}/d_S$. Thus, a large source and receiver can be focused down to a point in space, and when confocally coupled, the imaged source-receiver area can become the interruption point for a hole edge, or paper edge sensor.

15.2.8   Lens Reflective Coupling

If one of the lenses of a confocally coupled lens system were placed adjacent to the other lens element and skewed

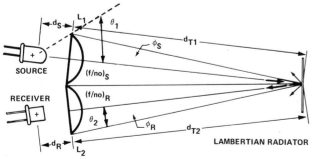

$$OTF = \text{N.A.}_{L1}{}^2 \cdot T_{L1} \cdot T_{L2} \cdot \text{N.A.}_{L2}{}^2 \cdot \rho_{REFLECTOR} \cdot O.F. \qquad \text{N.A.}_{L1} = \sin\theta_1$$
$$\text{N.A.}_{L2} = \sin\theta_2$$

T_{L1}, T_{L2} – TRANSMITTANCES OF LENS 1 AND LENS 2

Figure 15.2.8-1.   Reflective Coupling Employing Confocally Spaced Lenses.

so that the two images in front of the lenses intersected, a lensed reflective sensor would result. This is shown in Figure 15.2.8-1.

Every mechanical system has alignment tolerances which may allow the two images at the focused point to vary with respect to one another so that the area of the source and receiver images do not totally overlap. The ratio of the image overlap is termed the overlap fraction, OF.

$$O.F. = \frac{\left(\begin{array}{c}\textbf{AREA OF SOURCE IMAGE}\\\textbf{WHICH IS OVERLAPPED}\\\textbf{BY RECEIVER IMAGE}\end{array}\right)}{\left(\begin{array}{c}\textbf{TOTAL AREA OF}\\\textbf{SOURCE IMAGE}\end{array}\right)} \leqslant 1 \quad (15.2.8\text{-}1)$$

This fraction can vary from zero to unity. When it is zero, no coupling occurs and at unity, maximum coupling results. If the receiver image overlaps the entire source image, the O.F. will have its maximum value of unity. Having a larger receiver image provides a more consistent O.F. by reducing variability due to alignment difficulty.

The amount of flux coupling is also dependent upon the type and reflectance of the reflector placed at the image intersection, as well as on the N.A.'s of the lenses.

Using Equations (15.2.7-3) or (15.2.7-4), along with (15.2.6-3), and (15.2.8-1), the optical transfer function for a diffuse reflector can be determined.

$$\qquad (15.2.8\text{-}2)$$

$$OTF = \text{N.A.}_{L1}{}^2 \cdot T_{L1} \cdot T_{L2} \cdot \text{N.A.}_{L2}{}^2 \cdot \rho_D \cdot O.F.$$

where ρ_D = reflection coefficient of a diffuse reflector and other terms are as defined in Figure 15.2.8-1.

In a similar manner, the OTF for the specular reflector may be determined.

$$OTF = \text{N.A.}_{L1}{}^2 \cdot T_{L1} \cdot T_{L2} \cdot \rho_S \cdot O.F. \qquad (15.2.8\text{-}3)$$

where ρ_S = reflectance coefficient of a specular reflector

The conclusions to be drawn here are that a specular reflector will provide a much larger received flux. However, it suffers from a coupling problem where a movement of the normal of the reflecting plane, with respect to the axis of the confocally spaced lens system, causes the incident flux upon the reflector to be reflected outside of the aperture of the receiving lens. This is shown in Figure 15.2.8-2.

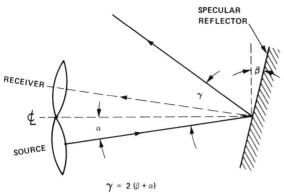

$$\gamma = 2(\beta + \alpha)$$

SOLID = REFLECTOR SKEWED TO SENSOR
DASHED = REFLECTOR NORMAL TO SENSOR

Figure 15.2.8-2. Positioning Sensitivity Caused by the Use of a Specular Reflecting Surface.

15.2.9 HEDS-1000 Reflective Coupling

Many of the alignment problems found in discrete confocally spaced reflective sensor systems can be eliminated with the use of the HEDS-1000 High Resolution Optical Reflective Sensor. This sensor includes a source and receiver focused with 2f optics. The optics system is a bifurcated aspheric lens with an effective numerical aperture of .3. These elements are housed in a TO—5 package with a glass window which is shown in Figure 15.2.9-1.

In the data sheet, radiant flux, ϕ_e, is specified as the flux coupled by the source lens to the image. This means that to develop the optical transfer function, only the receiving lens OTF need be described. The OTF for the HEDS-1000 and a diffuse reflector will appear very similar to that of a single lens system with the addition of the transmittance of the glass window, T_G, and the OTF of the reflector.

(15.2.9-1)

$$OTF = \frac{\phi_{RECEIVER}}{\phi_{REFLECTOR}} = T_L \cdot T_G \cdot O.F. \cdot OTF_{REFLECTOR}$$

The HEDS-1000 receiver area A_R is .160mm^2, and the source area A_S is .023mm^2. The ratio of A_R/A_S is greater than one which means that for focused operation, the overlap fraction O.F. is equal to one.

The following example shows the expected flux at the receiver photodiode from a diffuse reflector.

Step 1. Diffuse Reflector OTF.

$$OTF = N.A._L^2 \cdot \rho_D$$

Step 2. Total OTF for HEDS-1000 Using Step 1 and Equation (15.2.9-1).

$$OTF = T_L \cdot T_G \cdot O.F. \cdot N.A._L^2 \cdot \rho_D$$

Step 3. Using the Following:

$$N.A._L = .3 \quad , \quad \rho_D = 98\% \, , \quad \phi_e \text{ (DATA SHEET)} = 9\mu W$$

$$T_G = .9 \quad , \quad T_L = .8 \quad , \quad O.F. = 1$$

$$OTF = (.8)(.9)(1)(.3)^2(.98) = .064$$

Step 4. $\phi_{REFLECTOR} = \phi_e$ (DATA SHEET)

$$\phi_{RECEIVER} = \phi_{REFLECTOR} \text{ (OTF)}$$

$$\phi_{RECEIVER} = 576 \text{ nW}$$

If a specular reflector were used, the flux at the receiver would be the product of the transmittance of the glass and lens, the overlap fraction, O.F., and the reflectance of the specular reflector ρ_S. This is shown in Equation (15.2.9-3).

(15.2.9-3)

$$\phi_{RECEIVER} = \phi_{REFLECTOR}(T_L)(T_G)(O.F.)(\rho_S)$$

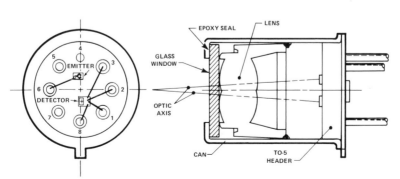

Figure 15.2.9-1. Elements of the HEDS-1000 High Resolution Optical Reflective Sensor.

$\rho_S = .95$

$\phi_{RECEIVER} = 9\mu W\ (.8)(.9)(1)(.95) = 6.16\mu W$

Through the use of the numerical aperture of the receiver lens, it is possible to determine the Optical Transfer Function, and using this OTF, the radiant flux that appears at the receiver surface. From the responsivity of the receiver diode, the photocurrent can be estimated.

15.3   Optical Sensor Parameters

15.3.1   Introduction

Bar code scanning, paper edge sensing, and optical tachometry applications place specific requirements on optical resolution and electrical performance of a reflective sensor system.

This section will describe the expected optical resolution and electrical performance of the HEDS-1000 as the object being sensed is placed at a location other than the optical focus point.

15.3.2   Modulation Transfer Function

The optical resolution of a reflective sensor system is determined by the overlap area of the images focused at the reflector surface. This may be limited by either source or receiver image size, whichever is the smaller. Optical resolution of a reflective sensor system is defined as the ability to discriminate the reflection of closely spaced lines with unequal reflectances. Figure 15.3.2-1 shows a series of reflectors and bars, and the response to this pattern as the imaged source and receiver, are moved laterally across the surface. The assumption is made that the reflector reflectance, $\rho_{REFLECTOR}$, is much greater than the reflectance of the bar between the reflectors. The conclusion that can be drawn from this illustration is that the ability of the sensor to discriminate between reflectors spaced a distance, s, apart increases as the space, s, increases.

$$Resolution\ \alpha\ \frac{s}{d} \qquad (15.3.2\text{-}1)$$

Figure 15.3.2-2a shows a reflector object with a reflecting surface composed of equal width reflecting (white), and non-reflecting (black) lines. When a receiver with a circular image with a diameter, d, is positioned over the reflector so that the image area totally intersects the white line, a maximum or 100% peak reflected response will result. In a similar manner, when the image is positioned over the black line, a minimum or 0% peak reflected response will

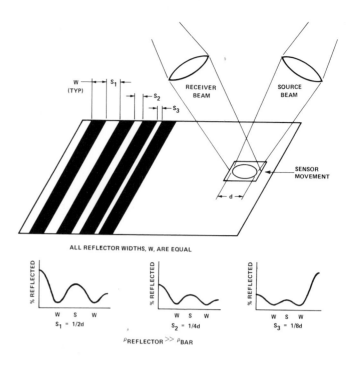

Figure 15.3.2-1.   Resolution of an Optical Reflective Sensor System.

occur. The difference between maximum and minimum response specifies the peak amplitude response under these line width-image diameter conditions.

If this scanning spot were scanned laterally across the black-white transition, the response would be a ramp with a slope (% RESPONSE/lateral distance) determined by the image diameter. This is shown in Figure 15.3.2-3a. The 50% response point shown in Figure 15.3.2-3b indicates that the image area is equally intersecting the black-white reflecting areas. If the lateral scanning were continued across the surface, the reflected response for Figure 15.3.2-2a would be a trapezoidal wave form. This is shown in Figure 15.3.2-2b.

If an image is used to scan a black white pattern where the line width is much smaller than the image diameter, the minimum to maximum response will be reduced from the response obtained when the line width is much larger than the image size.

An example of this is shown in Figure 15.3.2-4. A total $0-100\%$ response occurs for a $W_2 = 3$ condition, while only a 33% minimum to maximum response is obtained when $W_1 = 1$. Thus, as the line width gets smaller with respect to the image, the difference between the minimum and maximum is also reduced. When the performance at smaller line widths is compared to the performance for

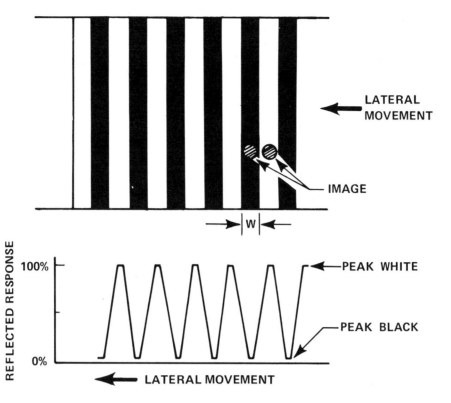

Figure 15.3.2-2a,b.   Image Response for Equally Spaced
White and Black Lines.

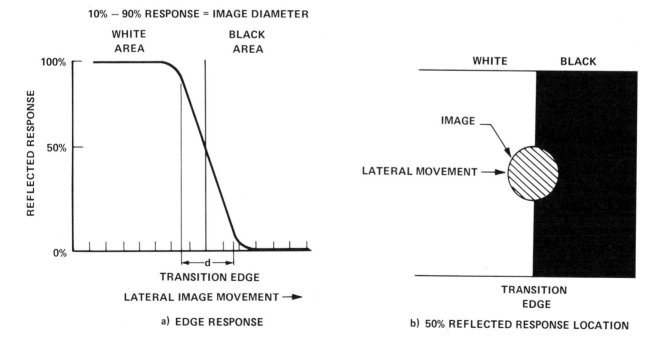

Figure 15.3.2-3.   Image Transition Response.
a) Edge Response
b) 50% Reflected Response Location

15.9

line widths resulting in 100% response, a modulation ratio is obtained. This ratio is shown in Equation (15.3.2-2) using data from Figure 15.3.2-4.

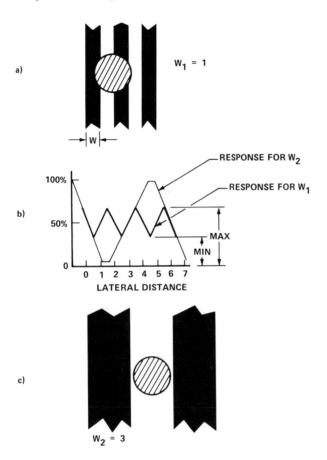

Figure 15.3.2-4.   Reflection Modulation for Different Line Widths.

$$\text{Modulation} = \frac{\text{MAX} - \text{MIN}}{\text{MAX} + \text{MIN}} \qquad (15.3.2\text{-}2)$$

The modulation performance for different line pair widths for a fixed image size is called the Modulation Transfer Function, MTF, of the optical image system. The MTF is specified as a percent response at a particular spatial frequency. The spatial frequency, F, is defined as the number of equal width white-black line pairs per lateral distance. The common units for spatial frequency is line pairs per millimeter, ln pr/mm. The spatial frequency is determined by Equation (15.3.2-3).

$$F = \text{ln pr/mm} = \frac{1}{2\ \text{line width (mm)}} \qquad (15.3.2\text{-}3)$$

Figure 15.3.2-5 shows the modulation transfer function, MTF, for the HEDS-1000. In applications such as bar code scanning and optical tachometry, the spatial frequency can be calculated with Equation 15.3.2-3. If a black-white line pattern with a line width of .254 mm (.01 in.) were to be scanned by the HEDS-1000, the performance can be determined through the use of Figure 15.3.2-5 and Equation (15.3.2-3). The spatial frequency is determined to be 1.97 ln pr/mm which results in an MTF response of 70%.

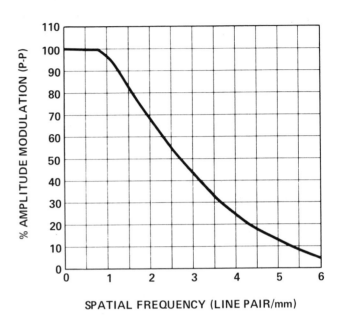

Figure 15.3.2-5.   Modulation Transfer Function of the HEDS-1000.

The MTF performance indicates that when a line pattern is scanned by the HEDS-1000, the reflected flux is degraded from the amount of flux obtained from a non-patterned reflector. Thus, the MTF response becomes another element to be added to the optical transfer function, OTF, as given in Equations (15.2.9-2) and (15.2.9-3).

15.3.3   Depth of Field

The optical transfer function, OTF, of a reflective sensing system was presented in Section 15.2.8 and 15.2.9. In each case, the assumption was that the source and received image were focused on the same plane. In most applications, the mechanical alignment of the sensor to the reflecting element will not be fixed at the focus point.

As the reflecting object moves away from the focus point, the image will become defocused resulting in a blurred image. In a reflective sensor system, the defocusing will occur for both the source and receiver images. The ratio of the intersection of the two image areas determines an overlap fraction, O.F., as discussed in Section 15.2.8. As the system is defocused, the overlap, O.F., decreases.

The defocused coupling response versus reflector distance is referred to as the depth of field, $\Delta\ell$. Figure 15.4.4-1 shows the relative response versus reflector distance. One will note an asymetrical response on either side of the maximum signal point. There is a much sharper $\% I_P (\Delta\ell)$ response roll-off for the distance between the reference plane to the maximum signal point, MSP, than from the MSP, to distances further away. This is due to the fact that the amount of blurring on the near side of the MSP is less than that on the far side. This is shown in Figure 15.3.3-1. The blurred image at $\Delta\ell$ inside is smaller than the blurred image at $\Delta\ell$ outside. When a reflective type of lensed system, as shown in Figure 15.2.9-1, is considered, the overlap fraction, O.F., is smaller at a $\Delta\ell$ inside than that for the same $\Delta\ell$ outside.

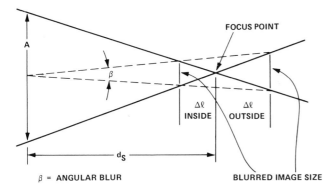

Figure 15.3.3-1.   Defocusing Versus Depth of Field.

The defocusing of the optical system also impacts the modulation transfer function. As the system is defocused, the image size will increase causing a reduction in the MTF for a specified line pair per millimeter. The MTF reduction is characterized as a linear translation of the MTF curve on the x-axis of the MTF plot, as shown in Figure 15.3.2-5. The assumption is made that the 50% MTF response point relates to the condition when the HEDS-1000 image size is equal to the line width. The image size is equated to be the line width at the 50% MTF point, and using Equation (15.3.2-3), the spatial frequency, F, can be determined. This calculation gives the spatial frequency F [50% MTF $(\Delta\ell)$].

The MTF performance at a specific depth of field and at the application's spatial frequency, F(APP), can be determined once the MTF response at F(APP) and $\Delta\ell = 0$ is obtained from Figure 15.3.2-5. This response at zero depth of field is termed MTF (APP). Using the quantities F [50% MTF $(\Delta\ell)$], F(APP), and MTF (APP), the MTF [$\Delta\ell$, F(APP)] can be determined from Equation (15.3.3-1). F [50% MTF $(\Delta\ell)$] = ½d with d obtained from Figure (15.3.3-2).

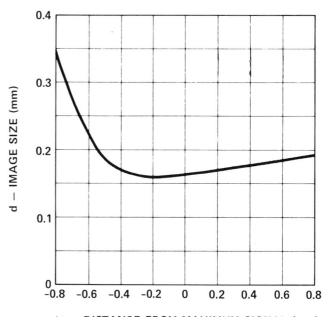

Figure 15.3.3-2.   Image Size vs. Depth of Field.

(15.3.3-1)

$$MTF\,[\Delta\ell, F(APP)] = \left[\frac{.5 - MTF(APP)}{2.5\ \ln\ pr/mm - F(APP)}\right] F(APP)$$

$$+ \left[.5 - F[50\%\ MTF\ (\Delta\ell)]\left(\frac{.5 - MTF(APP)}{2.5\ \ln\ pr/mm - F(APP)}\right)\right]$$

The two factors, O.F. $(\Delta\ell)$ and MTF [$\Delta\ell$, F(APP)] are considered to be linear factors that scale the optical transfer function, OTF, performance.

15.3.4   HEDS-1000 Total Transfer Function

The optical transfer function for the HEDS-1000 was developed in Section 15.2.9. This development specified the performance of the reflective sensor as a ratio of flux incident at the receiver ϕ_R to the flux ϕ_e incident at the reflector, OTF = ϕ_R/ϕ_e. The electrical designer needs to know the relationship of the current supplied to the emitter, I_F, which produces a photocurrent, I_{PR} in the detector. This relationship will be referred to as the sensor electrical transfer function or total transfer function, TTF.

The TTF is the product of the optical transfer function, the flux from the emitter, ϕ_e, and the flux responsivity, R_ϕ, of the photodiode. This is shown in Equation (15.3.4-1).

$$TTF = \frac{I_{PR}}{I_F} = R_\phi \cdot OTF \cdot \phi_e(I_F) \cdot K \qquad (15.3.4\text{-}1)$$

15.11

The flux responsivity, R_ϕ, of the photodiode at 700 nm is specified as .22 A/W. The flux responsivity, R_ϕ, will be considered a constant throughout the calculations. The radiant flux, ϕ_e, from the source is dependent upon the current through the LED emitter. The K-factor relationship of the output flux, ϕ_e, to LED forward current, I_F, is shown in Figure 15.3.4-1. This graph is normalized at 35 mA and 25°C. Thus, the data sheet typical of 9 μW occurs at 35 mA and 25°C.

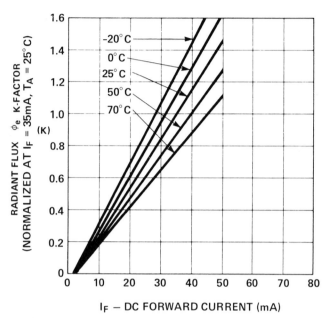

Figure 15.3.4-1. K-factor of ϕ_e Versus LED DC Forward Current.

The following example will illustrate how the TTF is used for a bar code scanner.

Given:

I_F = 45 mA ϕ_e (35 mA) = 9 μW
T_A = 25°C
Reflector = Lambertian, ρ_D = .85 @ 700 nm
Bar Width = .01″ , = 0.254 mm
Depth of Field = $\Delta\ell$ = .6 mm
N.A.$_L$ = .3 , T_G = .9 , T_L = .8
R_ϕ = .22 A/W

The total transfer function, TTF, is:

(15.3.4-2)

$$TTF = \frac{I_{PR}}{I_F} = R_\phi \cdot T_L \cdot T_G \cdot N.A._L^2 \cdot \rho_D \cdot O.F.(\Delta\ell)$$
$$\times \; MTF[\Delta\ell, F(APP)] \cdot \phi_e \cdot K$$

The first step is to evaluate the overlap fraction, O.F.$(\Delta\ell) = \% I_{PR}(\Delta\ell)$, a function of the depth of field, $\Delta\ell$. From Figure 15.4.4-1, the $\% I_{PR}(\Delta\ell) = 50\% = .5$. Thus, the O.F. is equal to .5.

The second step is to determine the MTF response for a bar width of .254 mm, at a depth of field of .6 mm. Equation (15.3.2-3) is used to arrive at an F(APP) = 1.97 ln pr/mm. Figure 15.3.2-5 is used to determine the MTF(APP) of 70%. Thus, MTF (1.97 ln pr/mm) is equal to 70%. This factor is then scaled by the translated MTF curve caused by an increase in spot size, d. Using Figure 15.3.3-2, the d($\Delta\ell$) is .22 mm and from Equation (15.3.2-3), the F[50% MTF($\Delta\ell$)] is found to be equal to 2.27 ln pr/mm. Using Equation (15.3.3-1), the MTF [$\Delta\ell$, F(APP)] for a bar width of 0.254 mm, and $\Delta\ell$ = .6 mm is found:

$$MTF[\Delta\ell, F(APP)] = \left[\frac{.5 - .7}{2.5 - 1.97}\right] + \left[.5 - 2.27\left(\frac{.5 - .7}{2.5 - 1.97}\right)\right]$$

$$= 0.62$$

Note that this is a reduction of only 8% from the 70% MTF response at $\Delta\ell$ = 0.

The third step is to determine the K-factor. This can be found from Figure 15.3.4-1. The K-factor for an I_F = 45 mA is equal to 1.3 at 25°C.

These values of O.F.$(\Delta\ell)$, MTF [$\Delta\ell$, F(APP)], and K are substituted into Equation (15.3.4-2) to determine the reflected photocurrent, I_{PR}.

$$I_{PR} = .22 \text{ A/W} \cdot .8 \cdot .9 \cdot (.3)^2 \cdot .85 \cdot .5 \cdot .62 \cdot 9\mu W \cdot 1.3$$

$$I_{PR} = 43 \text{ nA}$$

The conclusion that can be drawn from this example is that there are many factors contributing to the total transfer function.

15.4   HEDS-1000 Logic Interfacing

Optical sensing applications may be accomplished with the HEDS-1000 High Resolution Reflective Optical Sensor. This device includes a 700 nm emitter, a bifurcated aspheric lens, and a photodetector. The cathode of the LED emitter and the substrate of the photodetector are electrically connected to the mechanical package. The photodetector may be interconnected as a discrete photodiode or a photodiode-transistor amplifier.

15.4.1   Photodiode Interconnection

The photodiode, within the integrated photodetector, is isolated from the substrate-case by substrate diodes. These diodes appear from the common substrate-case to the transistor collector, and to the cathode of the photodiode.

Figure 15.4.1-1 shows recommended interconnection of the unused terminals of the sensor when discrete photodiode operation is desired. Care should be taken to ensure that these substrate diodes are always reverse biased so that they do not create a conductive path that may damage the substrate or other circuit elements.

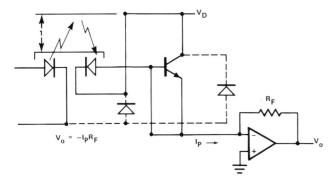

Figure 15.4.1-1. Photodiode Transresistance Amplifier.

The photodiode behaves like a current source, such that when optical flux falls on the device, it will generate a photocurrent in relationship to its responsivity, R_ϕ, of approximately .22μA/μW at 700 nm. The total photocurrent, I_P, generated by this photodiode is the summation of two currents, the reflected photocurrent, I_{PR}, and a stray photocurrent, I_{PS}. Thus, $I_P = I_{PR} + I_{PS}$.

15.4.2   Stray Photocurrent — I_{PS}

The stray photocurrent, I_{PS}, results from flux falling on the detector from sources other than the reflector surface. The principle source of stray photocurrent is from the scattered flux of the LED emitter that is reflected within the mechanical package. Ambient light can also be a source of the stray photocurrent, but this source has been greatly minimized by the use of an optical long wave filter. This filter action is provided by the red coloration found in the bifurcated lens.

15.4.3 I_{PR}/I_{PS} Ratio

The magnitude of the stray photocurrent resulting from internal scattering is directly proportional to the forward current, I_F, through the LED and the emitter relative efficiency. DC operation of the emitter will result in a steady state stray photocurrent that will range in direct relationship to the specifications and the worst case or typical value of $I_{P(MIN)}$ and $I_{P(MAX)}$ related to the stray photocurrent ratio, I_{PR}/I_{PS}. The photocurrent specified for the HEDS-1000 is the total photocurrent, I_P, which is equal to the sum of I_{PR} and I_{PS}. The ratio of I_{PR} to I_{PS} is designated a quality or Q-factor of the sensor. Thus, as Q increases for a given I_P, the value of stray photo-

current, I_{PS}, decreases. A worst case analysis for I_{PS} under the condition of minimum Q = 4, and an LED current of 35 mA results in an $I_{PS(MIN)} = 20$ nA for $I_{P(MIN)} = 100$ nA, and $I_{PS(MAX)} = 50$ nA for $I_{P(MAX)}$ 250 nA. A typical value of Q = 6.5 would cause I_{PS} to range from 13 nA to 33 nA.

The quality factor, Q, relationship to I_P, I_{PS}, and I_{PR} is shown in Equation (15.4.3-1).

$$Q = \frac{I_{PR}}{I_{PS}} \qquad I_P = I_{PS} + I_{PR} \qquad (15.4.3\text{-}1)$$

$$I_{PR} = I_P [Q/(Q+1)] \quad , \quad I_{PS} = I_P [1/(Q+1)]$$

where
    Q   = Quality Factor
    I_P = Total Photocurrent
    I_{PR} = Reflected Photocurrent
    I_{PS} = Stray Photocurrent

15.4.4   Depth of Field With Respect to Maximum Signal Point

Figure 15.4.4-1 shows that the 100% maximum reflected photocurrent, I_{PR}, response occurs at the location from the reference plane defined as the Maximum Signal Point, MSP. It also shows that the value of the reflected photocurrent I_{PR} is reduced as the reflector is moved in either direction away from the maximum signal point. The HEDS-1000 has a relatively symetrical response of I_{PR} versus the distance, ℓ, from the MSP. The depth of field of this optical system is

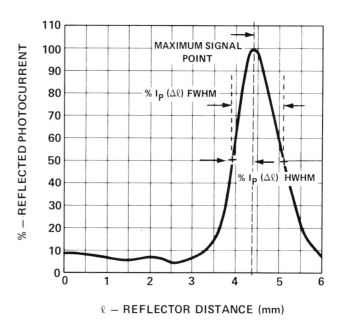

ℓ — REFLECTOR DISTANCE (mm)

Figure 15.4.4-1.   Depth of Field vs. Maximum Signal Point.

defined as the distance, $\Delta\ell$, between two equal percentage response points on either side of the MSP. The 50% I_{PR} response is referred to as the depth of field full width half maximum, FWHM. The depth of field, $\Delta\ell$, FWHM shown in Figure 15.4.4-1 is found to be 1.2 mm. Thus, if the reflector were moved one half of the total FWHM distance, $\Delta\ell$, from the MSP, the 50% I_{PR} point would occur approximately .6 mm on either side of the MSP location. The reflected photocurrent, I_{PR}, response at a specific depth of field is referred to as the %I_{PR} $(\Delta\ell)$. This value is always less than or equal to 100%.

The specific value of the I_{PR} is dependent upon the flux from the emitter and the type and reflection coefficient, ρ, of the reflector. Section 15.2.6 presented the characteristics of reflectors and demonstrated that a specular reflecting surface offers an order of magnitude improvement in reflected photocurrent when compared to a diffuse surface.

When a diffuse reflector is used, the expected value of the reflected photocurrent, at a specific depth of field, I_{PR} $(\Delta\ell)$, is the product of the percent response of I_{PR} at the depth of field %I_{PR} $(\Delta\ell)$, the reflection coefficient, ρ, of the reflector, the total photocurrent measured at the MSP from a diffuse reflector at a specific LED current I_P (I_F), and the quality ratio $Q/Q + 1$. This relationship is shown in Equation (15.4.4-1).

(15.4.4-1)

$$I_{PR}(\Delta\ell) = \%I_{PR}(\Delta\ell) \cdot \rho \cdot I_P(I_F) \cdot Q/(Q+1)$$

When a specular reflector is used, an additional coefficient is introduced. The HEDS-1000 I_{PR} performance is specified for a diffuse reflector; thus, when a specular reflector is used, an improvement factor dictated by Equation (15.2.6-4) is obtained. This factor indicates the I_{PR} improvement, for $\rho_S = \rho_D$, is the inverse of the lens numerical aperture squared. The $I_{PR}(\Delta\ell)$ response for a specular reflector is shown in Equation (15.4.4-2).

(15.4.4-2)

$$I_{PR}(\Delta\ell) = \%I_{PR}(\Delta\ell) \cdot \rho \cdot I_P(I_F) \cdot Q/(Q+1) \cdot 1/N.A._L{}^2$$

The expected value of I_{PR} $(\Delta\ell)$ using a diffuse reflector with a reflector having a reflectance of 75%, a total depth of field of 1.2 mm (.6 mm each way), an LED emitter current of 35 mA, and quality factor Q = 6.5, can be determined from Equation (15.4.4-1). The depth of field of 1.2 mm is equal to a %I_{PR} $(\Delta\ell)$ of 50%, and typical I_P (35 mA) = 140 nA. The I_{PR} $(\Delta\ell)$ under these conditions is equal to 45.5 nA. If a specular reflector were used, an $I_{PR}(\Delta\ell) = 506$ nA would be found from Equation (15.4.4-2), for N.A._L = 0.3.

These two equations are useful in determining the expected range of I_{PR} for a given reflector and a depth of field. These two system elements are very important in bar code scanning and paper edge sensing where the type of reflector and specific depth of field are variables.

15.4.5 Amplifier Considerations

Each sensor application will generally specify the electrical interface required and the range of the types of reflectors which will be utilized. The magnitude of the photocurrent generated by the photodiode is normally too small to interface directly to a logic gate. This condition indicates that an amplifier is needed. The amplifier electrical performance parameters, such as current and voltage gain, and the type of coupling are determined by the logic family to be interfaced, the application, and the magnitude of the reflected photocurrent.

Applications using specular reflectors include tachometry and optical limit sensing, while diffuse reflectors are more generally found in paper edge sensing and bar code reading. An ac coupled amplifier is acceptable in tachometry and bar code reading, while a dc coupled amplifier is required for steady state applications such as paper edge sensing and optical limit sensing.

The relationship of the reflected photocurrent, I_{PR}, to stray photocurrent, I_{PS}, has a large effect on the type of dc amplifier design selected. The initial step in amplifier design is to determine the worst case magnitude of the stray photocurrent, I_{PS}. It is this worst case value of I_{PS} that becomes the input quiescent bias current, which sets the threshold for the dc amplifier output voltage.

15.4.6 Transresistance — TTL Interface

A very common dc amplifier used with photodiodes is the transresistance type of amplifier. The simplest form is shown in Figure 15.4.1-1. The circuit configuration described by the electrical transfer function, $V_o = -I_P R_F$, is often called a current to voltage converter. A single power supply transresistance amplifier is shown in Figure 15.4.6-1. Here the photodiode is connected to the inverting input, and an offset voltage derived from V_{cc} is determined by a resistive voltage divider, $1 + R_2/R_1$ and is applied to the non-inverting input. The electrical transfer function is:

$$V_o = \frac{V_{cc}}{1 + R_2/R_1} - I_P R_F \qquad (15.4.6-1)$$

where $I_P = I_{PR} + I_{PS}$

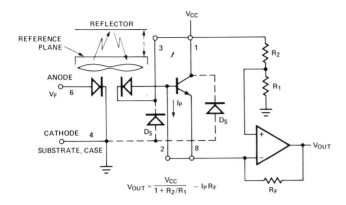

Figure 15.4.6-1. Photodiode Transresistance Amplifier with with Offset Voltage.

$$V_{OUT} = \frac{V_{CC}}{1 + R_2/R_1} - I_P R_F$$

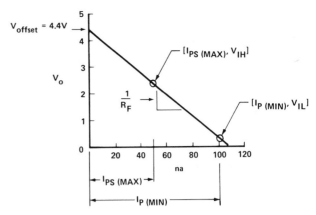

Figure 15.4.6-2. Graphical Solution of Transresistance Amplifier Design.

Equation (15.4.6-1) indicates that under the condition of zero reflected photocurrent, $I_{PR} = 0$, the output voltage, V_o, will be the offset voltage less the voltage developed by the stray photocurrent, I_{PS}, times the transresistance, R_F. Thus, the relationship for $I_{PR} = 0$ is:

$$V_o = \frac{V_{cc}}{1 + R_2/R_1} - I_{PS} R_F \qquad (15.4.6\text{-}2)$$

When the transresistance amplifier shown in Figure 15.4.6-1 is used to interface the photodiode to a TTL logic device, the output voltage, V_o, of the amplifier must change from a logic high, V_{IH}, of 2.0V to a logic low, V_{IL}, of .8V. To improve the noise immunity of the interface, it is desirable to broaden the range of V_{IH} to V_{IL} to 2.4V and .4V. The offset voltage and the value of the transresistance resistor, R_F, are selected to ensure that the maximum value of stray photocurrent, $I_{PS(MAX)} = 50\,nA$, does not cause the output voltage, V_o, to fall below V_{IH} of 2.4V, and the minimum total photocurrent, $I_{P(MIN)} = 100\,nA$, will cause the V_o to be equal to a $V_{IL} = .4V$.

It is very unlikely that an $I_{PS(MAX)}$ 50 nA and $I_{P(MIN)} = 100\,nA$ will occur simultaneously for a single device. A device that has an $I_{PS(MAX)}$ of 50 nA would also have an $I_{P(MIN)}$ of 250 nA. In a similar manner, a device that has an $I_{P(MIN)}$ of 100 nA would most likely have an $I_{PS(MAX)}$ of 20 nA.

Figure 15.4.6-2 shows the graphic construction of the electrical transfer function for Equation (15.4.6-1). The interface conditions of $[I_{PS(MAX)}, V_{IH}]$ and $[I_{P(MIN)}, V_{IL}]$ describe a line whose y intercept dictates the offset voltage, V_{offset}, and whose slope determines the transresistance, R_F. Using Equations (15.4.6-3) and (15.4.6-4), the offset voltage, V_{offset}, and the transresistance can be determined.

$$V_{offset} = \frac{V_{IL}\,I_{PS(MAX)} - V_{IH}\,I_{P(MIN)}}{I_{PS(MAX)} - I_{P(MIN)}} \qquad (15.4.6\text{-}3)$$

$$R_F = -\frac{V_{IH} - V_{IL}}{I_{PS(MAX)} - I_{P(MIN)}} \qquad (15.4.6\text{-}4)$$

The value of V_{offset} and R_F which satisfy the TTL interface conditions can be determined from Equations (15.4.6-3) and (15.4.6-4). For the example, $V_{offset} = 4.4V$, and $R_F = 40\,M\Omega$.

This example of photodiode-logic interfaces places parametric demands on the instrumentation operational amplifier selected. This amplifier should have a very low input offset current, thus allowing sensing of I_P at very low levels. It should have a gain greater than that required for the interface. For example, the required current gain for the photodiode-TTL amplifier is approximately 85dB; thus, an amplifier with an open loop gain of 100dB would be desirable. The slew rate of a transresistance amplifier may be slower than that required for TTL logic interconnection; thus, it may be necessary to specify a Schmitt trigger gate as the interconnecting logic element.

15.4.7   CMOS Interface

The internal transistor of the HEDS-1000 may also be used as a gain element in a single or multiple stage amplifier. Figure 15.4.7-1 shows an example of interconnecting the photodiode to a CMOS buffer gate, CD4049, using the internal transistor as a gain element.

It was shown previously that the value of I_{PS} and I_P can vary from unit to unit under similar conditions of I_F, reflector type and distance, ℓ. There will also be variations of the h_{FE} of the transistor from unit to unit. The design of the photodiode-transistor amplifier to CMOS logic gate must take into consideration the variations of I_{PS}, I_P, and h_{FE} when a direct coupled interconnection is desired. Figure 15.4.7-1 presents a design for an HEDS-1000 CMOS interface. The first step is to calculate the worst case stray photocurrent, $I_{PS(MAX)}$. The $I_{PS(MAX)}$ becomes the transistor base current which, when multiplied by the $h_{FE(MAX)}$, will determine the maximum collector current resulting from stray photocurrent. It is a requirement of this circuit that the collector current resulting from $I_{P(MAX)}$ does not cause the collector output voltage to fall below the input logic high level, V_{IH}, of 4 volts. Thus, the maximum value of the load resistor is selected based on the $I_{PS(MAX)}$ and $h_{FE(MAX)}$ of the sensor.

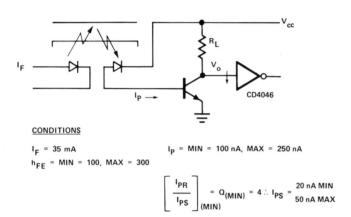

CONDITIONS

I_F = 35 mA I_P = MIN = 100 nA, MAX = 250 nA

h_{FE} = MIN = 100, MAX = 300

$$\left[\frac{I_{PR}}{I_{PS}}\right]_{(MIN)} = Q_{(MIN)} = 4 \therefore I_{PS} = \begin{array}{l} \text{20 nA MIN} \\ \text{50 nA MAX} \end{array}$$

Figure 15.4.7-1.   HEDS-1000 Interface to a CMOS Gate.

It is desirable in this type of interface to have at least a two to one difference between the $I_{PS(MAX)}$ and the $I_{P(MIN)}$. Such a stipulation will place constraints on the type of reflector, the depth of field, and the allowable spread between $h_{FE(MIN)}$ to $h_{FE(MAX)}$. Equations (15.4.1) and (15.4.2) are used to calculate the $I_P(\Delta\ell)$ for worst case conditions of $I_{P(MIN)}$ when a depth of field of 1.2 mm is desired. Thus, a diffuse reflector would cause 37.5 nA $I_{P(MIN)}$ and a specular reflector would result in an $I_{P(MIN)}$ of 416 nA. Using the criteria of $I_P(\Delta\ell)/I_{PS(MAX)} \geq 2$ [where $I_P(\Delta\ell)$ is evaluated for $I_{P(MIN)}$] indicates that a reflector with specular properties should be used.

The next step is to determine the minimum value of R_L for the minimum value of $I_P(\Delta\ell)$ and $h_{FE(MIN)}$ which causes the transistor collector voltage to fall below the V_{IL} of the CMOS gate.

This worst case analysis of Table 15.4.7-1 indicates that there is a very narrow range between the $R_{L(MAX)}$ and $R_{L(MIN)}$. If a smaller depth of field $\Delta\ell$ were selected, the range would be larger, thus giving a greater design margin.

Table 15.4.7-1.   HEDS-1000 to CMOS Interface Design Procedure.

Step 1.   Maximum Stray Photocurrent

$$I_{PS(MAX)} = \frac{I_{P(MAX)}}{Q_{MIN} + 1} = \frac{250\ nA}{4 + 1} = 50\ nA$$

$$Q_{MIN} = \frac{I_{PR(MIN)}}{I_{PS(MAX)}}$$

Step 2. $R_{L(MAX)}$ for V_{IH}

$$R_{L(MAX)} = \frac{V_{cc} - V_{IH}}{h_{FE(MAX)} \cdot I_{PS(MAX)}} = \frac{5.0 - 4.0}{300 \cdot 50 \times 10^{-9}}$$

$$= 66.7k$$

Step 3.   Minimum Photocurrent from Specular Reflector at $\Delta\ell$ = 1.2 mm

From Equation (15.4.4-2):

$$I_P(\Delta\ell) = \%I_P(\Delta\ell) \cdot \rho \cdot I_P \left[\frac{N.A._{(SURFACE)}}{N.A._{LENS)}}\right]^2$$

$$I_P(\Delta\ell) = .5 \cdot .75 \cdot 100\ nA \left[\frac{1}{.3}\right]^2 = 416\ nA$$

ρ = 75%

$N.A._{(SURFACE)}$ = 1 $I_{P(MIN)}$ @ I_F = 35 mA = 100 nA

$N.A._{(LENS)}$ = .3 $\%I_P(\Delta\ell)$, $\Delta\ell$ 1.2 mm = .5

Step 4. $R_{L(MIN)}$ for V_{IL} at $\Delta\ell$

$$R_L = \frac{V_{cc} - V_{IL}}{h_{FE(MIN)} \cdot I_P(\Delta\ell)} = \frac{5.0 - 2.25}{100 \times 416\ nA} = 66.1k$$

Step 5.   Select R_L = 66.2k 1%

15.4.8   Current Feedback Amplifier

Another common design problem is to interface the photodiode-transistor amplifier to a differential comparator such as the LM311 family. Here the design goals are similar to those specified in Figure 15.4.7-1 but an even greater degree of output voltage, V_O, stability is desired. By using a simple current feedback amplifier such as the one shown in Figure 15.4.8-1, the variations of V_O caused by Δh_{FE} and ΔI_{PS} will be minimized. In this amplifier design, the trade-offs are between voltage/current gain, stability, and amplifier speed. As the ratio of R_F to R_L approaches unity, the V_O stability improves, but there is a loss in signal gain. The difference between the I_{PS} and I_P $(\Delta\ell)$ will specify a current that will cause a change in the output voltage, V_O. When a larger swing in V_O is desired, the value of R_L is increased, such that $\Delta V_O \alpha (I_P - I_{PS}) R_L$. However, as R_L increases, the speed of the circuit decreases. Tables 15.4.8-1 and 15.4.8-2 show a design example using an $R_L = 100k$, $R_F = 10M\Omega$. In Figure 15.4.8-1, the differential comparator threshold is set by the resistor ratio R_1, R_2 and should be equal to 1.25V. This is below the minimum quiescent, V_O, of 1.3V caused by the variations of h_{FE} and I_{PS}. The change of the quiescent voltage caused by the variation of h_{FE} can be calculated through the use of the stability factor defined as s''. This factor is the incremental change of I_C caused by an incremental change in h_{FE}. Specifically,

$$s'' = \left(\frac{\Delta I_C}{\Delta h_{FE}}\right)\Bigg|_{I_P = 0}$$

The V_O stability is improved as the value of s'' is reduced The incremental V_O change of the circuit shown in Figure 15.4.8-1 can be calculated from the following relationship:

$$\Delta V_O = -\Delta h_{FE}\left[s''R_L + \frac{\partial^2 V_O}{\partial h_{FE}\,\partial I_P}\right]$$

$$\text{STABILITY FACTOR, } s'' = \frac{\Delta I_C}{\Delta h_{FE}}$$

$$s'' \approx \frac{(R_L + R_F)\,(V_{cc} - V_{BE})}{(R_F + R_L + R_L\,h_{FE})^2}$$

Table 15.4.8-2.   Practical Example of a Current Feedback Amplifier Design.

h_{FE}	MIN = 100	MAX = 300	$I_{PS(MAX)}$ = 41 nA
R_L = 100kΩ		R_F = 10MΩ	V_{cc} = 5.0V V_{BE} = .6V

	h_{FE} = 100	h_{FE} = 300
V_O	2.60V	1.39V
s''	1.1×10^{-7}	

The s'' parameter is important in dc coupled amplifier circuits because the change in V_O caused by the Δh_{FE} may result in the succeeding gain stages being driven into saturation.

15.4.9   Current–Voltage Feedback Amplifier

When even greater output voltage stability is desired, a modified current-voltage feedback amplifier may be necessary. This bias approach uses R_F and R_N as shown in Figure 15.4.9-1, to force a V_B which sets I_C to a level determined by V_B/R_E. This circuit can offer an s'' ten times better than the current feedback amplifier. The design is such that an h_{FE} variation of $100 - 300$ will cause a 0.3V change in V_O in the current-voltage feedback configuration, while this same Δh_{FE} for Figure 15.4.8-1 will cause a 1.2V change.

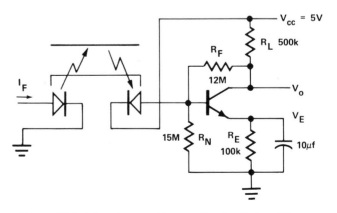

Figure 15.4.8-1.   Current Feedback Amplifier Interface to an Analog Comparator.

Table 15.4.8-1.   Current Feedback Amplifier Design Procedure.

OUTPUT VOLTAGE, V_O

$$V_O = \frac{V_{CC}\left(\dfrac{R_F}{h_{FE}R_L}\right) + V_{BE}\left(1 + \dfrac{1}{h_{FE}}\right) - R_F I_P}{\left(\dfrac{R_F}{h_{FE}R_L}\right) + \left(1 + \dfrac{1}{h_{FE}}\right)}$$

Figure 15.4.9-1.   Current-Voltage Feedback Amplifier.

15.17

The design example given in Table 15.4.9-1 sets the V_o at 2.5V and offers an s'' of 7.5×10^{-9}. The output voltage V_o will be reduced from a design center of 2.5V to a worst case $(V_o - s'' \Delta h_{FE} R_L) = 2.25V$. This implies that the threshold of the comparator should be set to a level of 1.55V. This amplifier offers a transresistance of $8 M\Omega$ which means for a 100 nA $I_P(\Delta \ell)$, the output voltage V_o will fall to 1.5V which is a sufficient differential to cause the LM311 output to change logic states.

Table 15.4.9-1. Design Equations for the Current-Voltage Feedback Amplifier.

Step 1.   Select R_L   Given: V_o and I_C

$$R_L = \frac{(V_{CC} - V_o)}{I_C} = \frac{2.5V}{5\mu A} = 500k$$

Step 2.   Select R_E   Given: V_E and I_E

$$R_E = \frac{V_E}{I_E} = \frac{.5}{4.91 \times 10^{-6}} = 101k \approx 100k$$

Step 3.   Select R_N   Given: I_N

$$R_N = \frac{V_E + V_{BE}}{I_N} = \frac{.5 + .6}{75 \text{ nA}} = 14.6M\Omega \approx 15M\Omega$$

Step 4.   Select R_F   Given: I_N and I_B

$$R_F = \frac{V_o - V_E - V_{BE}}{I_N + I_B} = \frac{2.5 - .5 - .6}{75 \text{ nA} + 50 \text{ nA}}$$

$$= 11.2M\Omega \approx 12M\Omega$$

Step 5.   Stability Factor, s''

$$s'' = \frac{\left[V_{CC} - V_{BE}\left(1 + \frac{R_F + R_L}{R_N}\right)\right](g+1)}{R_F\left[1 + (1 + h_{FE})g\right]^2}$$

$$\text{where}\quad g = \frac{R_E}{R_N}\left(1 + \frac{R_L}{R_F}\right) + \frac{R_E + R_L}{R_F}$$

$$s'' = 7.5 \times 10^{-9}$$

15.4.10   LSTTL Interface

The previous circuits dealt with CMOS and comparator type interfaces. Figure 15.3.10-1 shows a two transistor amplifier to LSTTL interface. This circuit can either be ac or dc coupled; only the direct coupled configuration will be presented. The design approach is similar to that of Figure 15.4.7-1 with the additional analysis of the second stage.

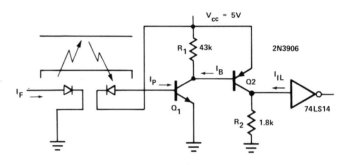

Figure 15.4.10-1.   DC Coupled HEDS-1000 to LSTTL Interface.

The first transistor, Q_1, is biased by the photodiode in a common emitter configuration. Under the conditions of $I_{PS(MAX)}$, the collector of Q_1 is pulled up to within .5V of V_{cc}, thus insuring that Q_2 is not conducting. This condition sets the maximum value of R_1. When a reflected photocurrent is present, the resulting I_C of Q_1 is the combination of current through R_1 and the I_B of Q_2. Thus, R_1 must be large enough that the current sinking capability of Q_1 (dictated by h_{FE} and I_P) will result in sufficient I_B in Q_2 to cause Q_2 to saturate.

In the absence of reflected photocurrent, both Q_1 and Q_2 are normally off. Under this condition, the load resistor, R_2, must be able to sink the I_{IL} of the LSTTL gate at the desired V_{IL}. To satisfy the desired logic condition, R_2 must be less than V_{IL}/I_{IL}. The minimum value of R_2 is determined by the current sourcing capability of Q_2 produced by I_B. The collector current of Q_2 must generate a voltage drop across R_2 greater than the V_{IH} of the gate. It is recommended that a high gain, low leakage PNP transistor, such as a 2N3906, be selected for Q_2.

The rate at which the output voltage of Q_2 changes is directly related to the speed at which the reflecting surface is moving into the reflection plane of the sensor. In many applications, the rate of change of V_o through the switching region of the LS gate is so slow that it may cause logic level chatter at the output of the gate. If this chatter is observed, it is recommended that a Schmitt trigger gate, such as the 74LS14, be used.

15.5   Reflective Sensor Applications

15.5.1   Rotary Tachometry

A reflective sensor can be used as the transducer to determine the rotary speed of a motor shaft. This can be accomplished by utilizing a disc with equally spaced reflective and non-reflective lines placed around the circumference of the disc. The number of line pairs per revolution will then give a specific pulse count per revolution. In many applications, it is desired to have a very high density of line pairs around the perimeter of a small diameter disc. The performance of the reflective sensor is determined by the MTF for the spatial frequency, F, of line pairs on the disc. The spatial frequency for a disc is determined by Equation (15.5.1-1).

$$F = \frac{\text{lines/rev.}}{4\pi r} \qquad (15.5.1\text{-}1)$$

Using Figure 15.5.1-1, the spatial frequency can be determined assuming the radius is to the center point of the line pattern. Given the radius, $r = 10\,\text{mm}$, and 220 lines/revolution, a spatial frequency of 1.75 ln pr/mm is calculated. When an HEDS-1000 is used as the sensor for this code wheel, an MTF response of 75% is obtained from Figure 15.3.2-5.

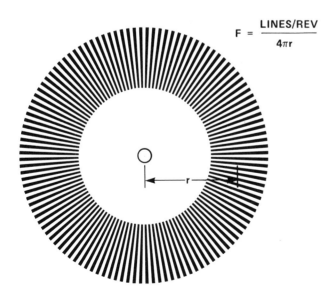

$$F = \frac{\text{LINES/REV}}{4\pi r}$$

SPATIAL FREQUENCY OF CODE WHEEL

Figure 15.5.1-1.   Spatial Frequency of a Code Wheel.

The code wheel is affixed to a hub which is placed on the rotating shaft. The reflective sensor is positioned perpendicular to the disc and at a distance such that the maximum signal point, MSP, is at the plane of the code pattern. The highest reflected photocurrent, I_{PR}, is obtained from a specular reflecting code pattern. This can be implemented by photolithographing a pattern of opaque bars on a shiny metallic wheel-hub assembly. A diffuse code wheel assembly should be used when the mechanical tolerance of the axial alignment of the HEDS-1000 to the normal of the code wheel exceeds $10°$.

Tachometry applications allow an ac coupled amplifier to be used, such as the current feedback type in Figure 15.4.8-1. AC coupling the output of the HEDS-1000 eliminates the dc output offset voltage variations caused by the stray photocurrent.

15.5.2   HEDS-1000 Analog Tachometer

The HEDS-1000 can be used as the transducer in high speed rotary tachometry applications. Figure 15.5.2-1 shows a circuit diagram that uses the reflective sensor as a pulse source input to a frequency to voltage converter.

The HEDS-1000 is configured as a current feedback amplifier and ac coupled to an LM2907 frequency to voltage converter. The transistor Q_1 is used as a current source to supply the I_F to the LED emitter.

The reflective sensor generates n pulses per revolution, where n is the number of line pairs per revolution. The magnitude of the frequency, f, applied to the F-V converter is n times the number of revolutions per minute. This is the relationship shown in Equation (15.5.2-1).

$$n = \frac{\pi r}{\text{line width}} = \frac{2\pi r}{\text{ln pr width}} \qquad (15.5.2\text{-}1)$$

$$f = n \times \text{rev/min} \times \frac{1}{60} \bullet \frac{\text{min}}{\text{sec}} \qquad \text{where } f = \text{Hz}$$

The capacitor, C, is the dominant factor in determining the maximum output voltage for a required full scale frequency indication. The capacitor required to determine a full scale output voltage for a specific full scale frequency is shown in Equation (15.5.2-2).

$$C = \frac{V_{\text{OUT FULL SCALE}}}{R \bullet V_{cc} \bullet f_{\text{FULL SCALE}}} \qquad (15.5.2\text{-}2)$$

A $V_{\text{OUT FULL SCALE}} = 1\text{V}$, 0–25,000 rev/min tachometer can be designed using a code wheel with a radius of

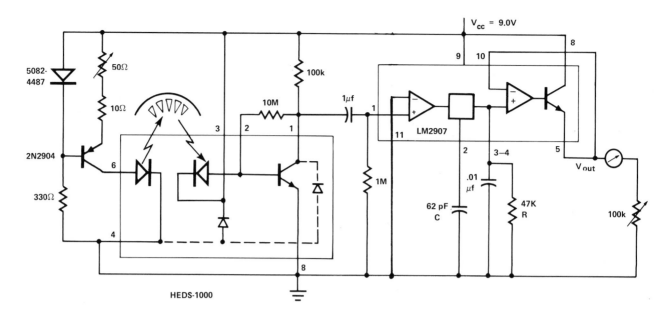

Figure 15.5.2-1. Analog Tachometer Circuit.

20 mm and a line width of .63 mm. The approach is to determine maximum full scale frequency from Equation (15.5.2-1), and then using Equation (15.5.2-2), the full scale frequency is 41.5 kHz and C is calculated to be 57 pF. A 62 pF capacitor is used for this example.

The F-V converter will respond to a minimum input signal swing of 250 mV. This input level can be insured through the use of a specular reflector. Section 15.3.4 showed that the I_{PR} increases by 10.45 dB when a specular reflector is used over a diffuse reflector. The limitation of a specular reflecting code wheel is that the HEDS-1000 alignment to the code wheel must not be greater than $10°$ from the normal. If the deviation is greater than $10°$, the image of the source will not be reflected to the detector.

15.5.3 Digital Tachometer

The HEDS-1000 can be used as the interface transducer to a digital tachometer system. The circuit diagram for a four digit high speed tachometer is shown in Figure 15.5.3-1.

The HEDS-1000 and LM311 differential comparator provide the reflective sensor to TTL interface. The pulses are counted by a Ferranti ZN1040E universal count/display circuit. The 74LS123 functions as the time base, which sets the count interval and also generates the timing for the count inhibit, data transfer, and reset. The 74LS00 and 74LS04 are configured as differentiators that generate the transfer and reset pulses. The function provided by the Ferranti circuit can also be implemented through the use of the Intersil ICM7216.

The count time output is provided at the Q output pin 13

of the 74LS123. The duration of the count time is specified by the totalized pulse count displayed at a specific number of revolutions per minute.

15.5.4 Paper Edge Sensor

The accurate detection of the edge of a piece of paper can be accomplished with an HEDS-1000 reflective sensor. If the range of reflectivity of the paper is known, either a paper reflective or an obscuration system can be selected. When a paper type which is highly reflective is considered, it is desirable to utilize a reflective system of the type that positions the sensor so that the maximum signal point lies at the surface of the paper platten. This approach is shown in Figure 15.5.4-1. When a low reflectance paper type is being sensed, the obscuration type system may be more suitable. Such a system is shown in Figure 15.5.4-2.

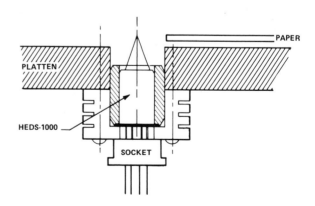

Figure 15.5.4-1. Reflective Type Paper Edge Sensor.

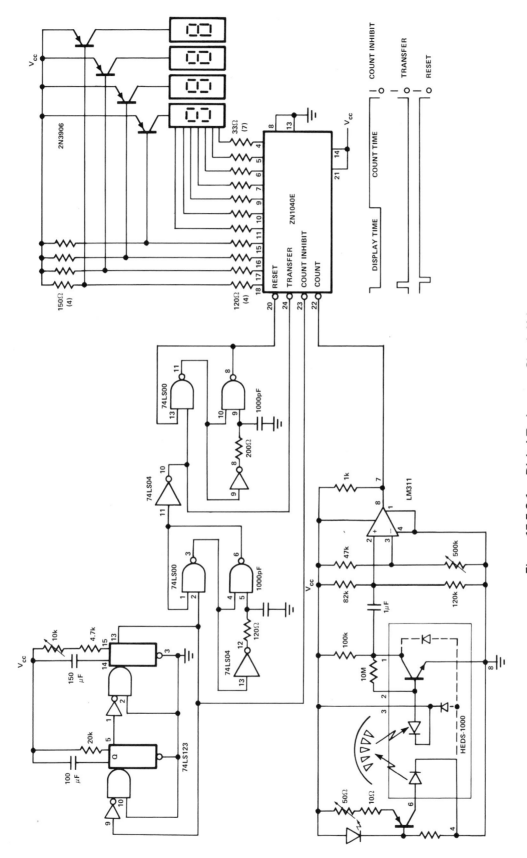

Figure 15.5.3-1. Digital Tachometer Circuit Using a Universal Counter/Display Circuit.

15.21

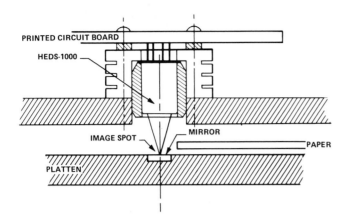

PRINTED CIRCUIT BOARD

HEDS-1000

IMAGE SPOT

MIRROR

PAPER

PLATTEN

Figure 15.5.4-2. Obscuration Type Paper Edge Sensor.

The edge position sensing accuracy is dependent on the spot location as referenced to the mechanical system. The HEDS-1000 offers a reflective sensing spot location of ±.51 mm with respect to the package center line.

When an obscuration sensor system is used, the two transistor amplifiers shown in Figure 15.4.10-1 provide a convenient dc coupling to a 74LS logic family. When the reflective system is applied, a transresistance amplifier of the type shown in Figure 15.4.6-1 should be considered.

15.5.5 Bar Code Scanner

A reflective optical sensor can be used as the transducer in a bar code scanner application. The bar code is an encoded form of binary data storage. The relative width difference bar to bar, and space to space, describe the typical encoding scheme of Differential Width encoding. This is the data format used in the Universal Product Code, UPC.

The sensor provides an electrical output signal with a pulse width determined by the bar and space widths, and signal amplitude dependent upon the bar and space reflection coefficients.

The Differential Width encoding scheme requires that output pulse width, bar to bar, or space to space, be an accurate representation of the distance per unit time. The accuracy of the scanning output improves when the reflecting spot size is smaller than the minimum bar or space width. The smaller the scanning image, the more abrupt the transition from bar to space.

The output signal amplitude is determined by the difference between the bar reflectance and space reflectance. The minimum output signal to maximum output signal ratio is directly proportional to the bar to space reflectance.

The signal amplifier that is interconnected to the sensor must have a large dynamic operating range to accomodate the variations of reflector types, and also provide adequate signal differential for the bar to space reflectivity difference.

Figure 15.3.2-4 shows the trapezoidal pulse train that is obtained from scanning a bar code of equal width bars and spaces. As the image size increases due to defocusing, the pulse train amplitude is reduced and the waveform becomes triangular. It is desirable that the amplifier provides the signal amplitude change at the same scanning location as the bar to space transition.

Figure 15.5.5-1 shows a schematic for an amplifier system that will convert the bar and space widths into TTL

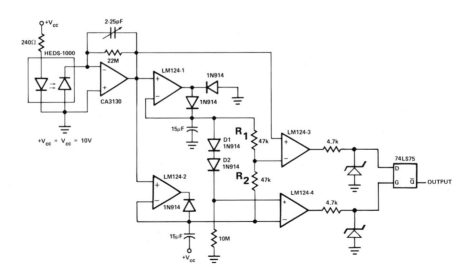

Figure 15.5.5-1. Bar Code Scanner Circuit.

compatible logic signals. The circuit uses the CA3130 as a transresistance amplifier for the HEDS-1000 photodiode. The output of the amplifier is applied to positive peak (LM124-1) and negative peak (LM124-2) detectors. The resistors R1 and R2 set the reference (negative-going) input to the code comparator (LM124-3) at a voltage which is halfway between the positive peak and the negative peak, so the switching threshold is therefore at 50% of the peak-to-peak modulation. The noise gate (LM124-4) compares the negative peak to a voltage which is two diode voltage drops (D_1 and D_2) below the positive peak, so unless the peak-to-peak amplitude exceeds two diode drops, the G input of the 74LS75 remains low and $\overline{Q}$ cannot change. This ensures that the $\overline{Q}$ output of the 74LS75 will remain fixed unless the excursions at the output of the CA3130 are of adequate amplitude (two diode drops) that noise will not interfere.

INDEX